Student Solutions Manual
for
Multivariable Calculus
by Laura Taalman and Peter Kohn

Roger Lipsett

W. H. Freeman and Company
New York

©2014 by W. H. Freeman and Company

ISBN-13: 978-1-4641-5019-7
ISBN-10: 1-4641-5019-2

Printed in the United States of America

First printing

W. H. Freeman and Company
41 Madison Avenue
New York, NY 10010
Houndmills, Basingstoke RG21 6XS, England

www.whfreeman.com

Contents

Part IV

Vector Calculus

Chapter 9

Parametric Equations and Polar Functions

9.1   Parametric Equations

Thinking Back

Eliminating a variable

- ▷ Since $x = t^2$, we get $y = t^6 = (t^2)^3 = x^3$.

- ▷ $x = \sin t = y$, so the curve is $x = y$.

- ▷ Since $\sin^2 t + \cos^2 t = 1$, the relationship between x and y is $x^2 + y^2 = 1$.

- ▷ Since $\cosh^2 t - \sinh^2 t = 1$, the relationship between x and y is $y^2 - x^2 = 1$.

Arc length

Using the arc length formula from Theorem 6.7, we have

- ▷ The arc length is $\int_0^5 \sqrt{1^2 + ((3x-4)')^2}\,dx = \int_0^5 \sqrt{1^2 + 3^2}\,dx = 5\sqrt{10}$.

- ▷ With $f(x) = \sqrt{4 - x^2}$, we get $f'(x) = \frac{1}{2}(4 - x^2)^{-1/2} \cdot (-2x) = -x(4 - x^2)^{-1/2}$. Then the arc length is

$$
\int_0^2 \sqrt{1 + (f'(x))^2}\,dx = \int_0^2 \sqrt{1 + \left(-x(4-x^2)^{-1/2}\right)^2}\,dx
$$
$$
= \int_0^2 \sqrt{1 + \frac{x^2}{4 - x^2}}\,dx
$$
$$
= \int_0^2 \sqrt{\frac{4}{4 - x^2}}\,dx
$$
$$
= \int_0^2 \frac{2}{\sqrt{4 - x^2}}\,dx
$$
$$
= \left[2\sin^{-1}\frac{x}{2}\right]_0^2
$$
$$
= 2\sin^{-1} 1 - 2\sin^{-1} 0 = \pi.
$$

▷ With $f(x) = x^{3/2}$ we get $f'(x) = \frac{3}{2}\sqrt{x}$, so that the arc length is

$$\int_0^4 \sqrt{1 + \frac{9}{4}x}\, dx = \left[\frac{8}{27}\left(1 + \frac{9}{4}x\right)^{3/2}\right]_0^4 = \frac{8}{27}\left(10\sqrt{10} - 1\right).$$

▷ With $f(x) = x^2$ we get $f'(x) = 2x$, so that the arc length is

$$\int_0^1 \sqrt{1 + 4x^2}\, dx.$$

To integrate, use the trigonometric substitution $x = \frac{1}{2}\tan u$, so that $dx = \frac{1}{2}\sec^2 u$. Then (using the integral of $\sec^3 u$ from previous chapters)

$$\begin{aligned}
\int_0^1 \sqrt{1 + 4x^2}\, dx &= \frac{1}{2}\int_{x=0}^{x=1} \sqrt{1 + \tan^2 u}\,\sec^2 u\, du \\
&= \frac{1}{2}\int_{x=0}^{x=1} \sec^3 u\, du \\
&= \frac{1}{4}\left[\left(\sec u \tan u + \ln|\sec u + \tan u|\right)\right]_{x=0}^{x=1} \\
&= \frac{1}{4}\left[\left(2x\sqrt{1+4x^2} + \ln\left|\sqrt{1+4x^2} + 2x\right|\right)\right]_0^1 \\
&= \frac{1}{4}(2\sqrt{5} + \ln(2 + \sqrt{5})).
\end{aligned}$$

Concepts

1. (a) True. One way of writing it is $x = t$, $y = f(t)$.

   (b) False. It may not be possible to solve either $x(t)$ or $y(t)$ for t.

   (c) False. For example, $x = \cos t$ and $y = \sin t$ for $t \in [0, 2\pi)$ is the unit circle, which does not pass the vertical line test.

   (d) False. Almost any curve has an infinite number of parameterizations. For example, the unit circle can be parametrized as $x = \cos(at)$, $y = \sin(at)$ for $t \in \left[0, \frac{2\pi}{a}\right]$ for any positive real number a.

   (e) False. If $f'(t) = 0$ for some value of t, then the slope of the parametric curve is infinite, so the curve is not differentiable there.

   (f) False. It could be that $x'(t) = 0$ as well, in which case the slope of the tangent line is undefined. For example, let $x(t) = t^2$ and $y(t) = t^3$; at $t = 0$, we have $x'(t) = y'(t) = 0$, and in fact the curve, which is $y = x^{3/2}$, is not differentiable at $x = 0$.

   (g) True. By Definition 9.2, the slope of the tangent line is $m = \frac{y'(t_0)}{x'(t_0)}$, so in the case given, $m = 0$.

   (h) False. If each loop were a semicircle, then clearly the center of the first circle would be at $(\pi r, 0)$. But the points $(0, 0)$ and $(\pi r, 2r)$, which both lie on the curve, are not equidistant from $(\pi r, 0)$. Thus this is not a semicircle.

3. (a) Since $x'(t) > 0$, the direction of motion is to the right. Since $y'(t) > 0$, the direction of motion is also up. Thus the direction of motion is up and to the right.

   (b) Since $x'(t) > 0$, the direction of motion is to the right. Since $y'(t) < 0$, the direction of motion is also down. Thus the direction of motion is down and to the right.

   (c) Since $x'(t) < 0$, the direction of motion is to the left. Since $y'(t) > 0$, the direction of motion is also up. Thus the direction of motion is up and to the left.

(d) Since $x'(t) < 0$, the direction of motion is to the left. Since $y'(t) < 0$, the direction of motion is also down. Thus the direction of motion is down and to the left.

5. Since $x'(t) = 2t$ and $y'(t) = 3t^2$, we see that for $t < 0$, $x'(t) < 0$ while $y'(t) > 0$, so that the particle is moving up and to the left. For $t > 0$, both derivatives are positive, so the particle is moving up and to the right. Finally, at $t = 0$, both derivatives are zero, so the particle is stopped.

7. Since $x'(t) = e^t$ and $y'(t) = \frac{1}{t}$, both derivatives are positive everywhere, so that the particle is always moving up and to the right.

9. (i) For (a), substitute x for t in $y = t^2 - 1$ to get $y = x^2 - 1$. For (b), substitute $-x$ for t in $y = t^2 - 1$ to again get $y = x^2 - 1$.

   (ii) In part (a), since $t \geq 0$, we only get the portion of $y = x^2 - 1$ corresponding to $x = t \geq 0$, so the right half of the parabola. For part (b), since $t \geq 0$, we get the portion of $y = x^2 - 1$ corresponding to $x = -t \leq 0$, so the left half of the parabola.

   (iii) For both parts, since $y'(t) = 2t$ and $t \geq 0$, the particle is always rising. In part (a), since $x'(t) = 1$, the particle is moving to the right, and in part (b), since $x'(t) = -1$, the particle is moving to the left.

11. (i) For part (a), substitute x for t in $y = \sin t$ to get $y = \sin x$ for $x \geq 0$. For part (b), substitute $t = x + 1$ for t in $y = \sin(t - 1)$ to get $y = \sin x$. In this substitution, since $t \geq 1$ and $x = t - 1$, we again have $x \geq 0$.

   (ii) Since the parametrizations and ranges of the independent variables are the same, these two systems parametrize the exact same curve over the same domain.

   (iii) For both parts, $x'(t) = 1$, so that both curves are always moving to the right. For part (a), $y'(t) = \cos t$, so that the particle oscillates between moving up and moving down as $\cos t$ changes sign. For part (b), $y'(t) = \cos(t - 1)$, so that again the particle oscillates between moving up and moving down (as it must, since the two parametrizations represent the same function over the same domain).

13. Horizontal tangent lines occur when $\frac{y'(t_0)}{x'(t_0)} = 0$, so when $y'(t_0) = 0$ but $x'(t_0) \neq 0$. Vertical tangent lines occur when $\frac{y'(t_0)}{x'(t_0)} = \pm\infty$, so when $y'(t_0) \neq 0$ but $x'(t_0) = 0$. Note that if both derivatives vanish at some point, then the tangent to the parametric curve at that point is not defined.

15. Since

$$x(t)^2 - 2x(t) - 3 = (\sin t + 1)^2 - 2(\sin t + 1) - 3 = \sin^2 t + 2\sin t + 1 - 2\sin t - 2 - 3$$
$$= \sin^2 t - 4 = y(t),$$

the given parametrization indeed traces out points on the graph of the function. But since no matter what t is, $\sin t$ lies between -1 and 1, it will trace out only portions of the graph for which $x = \sin t + 1$ is between 0 and 2. Since sin is periodic, it will trace out the same region over and over.

Skills

17. We have

t	-2	-1	0	1	2	3	4	5
$x = 3t + 1$	-5	-2	1	4	7	10	13	16
$y = t$	-2	-1	0	1	2	3	4	5

so a graph connecting these points is

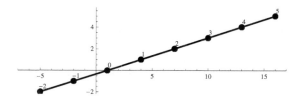

19. We have

t	-2	-1	0	1	2	3
$x = 1 - 2t$	5	3	1	-1	-3	-5
$y = 5 - 3t$	11	8	5	2	-1	-4

so a graph connecting these points is

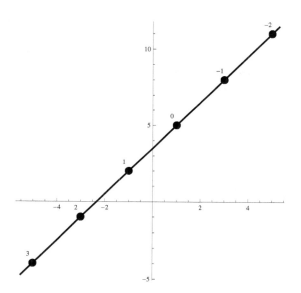

21. We have

t	-2.5	-2	-1	0	1	2	3
$x = t^3 - t$	-13.125	-6	0	0	0	6	24
$y = t^3 + t$	-18.125	-10	-2	0	2	10	30

so a graph connecting these points is

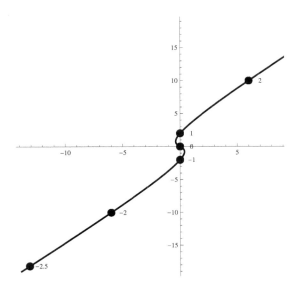

23. We have

t	0	$\frac{\pi}{6}$	$\frac{\pi}{2}$	$\frac{2\pi}{3}$	π	$\frac{5\pi}{4}$	$\frac{3\pi}{2}$	$\frac{7\pi}{4}$
$x = \cos^5 t$	1	$\frac{9\sqrt{3}}{32}$	0	$-\frac{1}{32}$	-1	$-\frac{\sqrt{2}}{8}$	0	$\frac{\sqrt{2}}{8}$
$y = \sin^5 t$	0	$\frac{1}{32}$	1	$\frac{9\sqrt{3}}{32}$	0	$-\frac{\sqrt{2}}{8}$	-1	$-\frac{\sqrt{2}}{8}$

so a graph connecting these points is

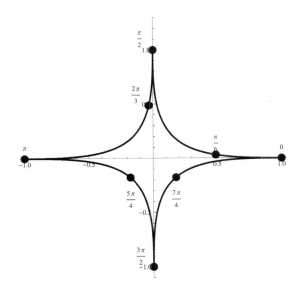

25. Solve $x = 2t - 1$ for t to get $t = \frac{x+1}{2}$; substitute into

$$y = 3t^2 + 5 = 3\left(\frac{x+1}{2}\right)^2 + 5 = 3\left(\frac{1}{4}x^2 + \frac{1}{2}x + \frac{1}{4}\right) + 5 = \frac{3}{4}x^2 + \frac{3}{2}x + \frac{23}{4}.$$

A plot is

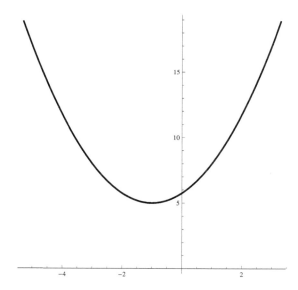

27. Note that $x = y$. The range of $\tan t$ for $t \in \left(-\frac{\pi}{2}, \frac{\pi}{2}\right)$ is $(-\infty, \infty)$, so this is $y = x$ on all of $\mathbb{R}$. A plot is

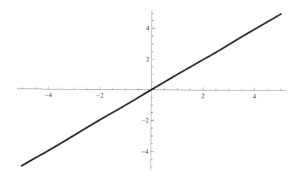

29. Use the identity $\cos^2 t + \sin^2 t = 1$. We have $\frac{x}{3} = \cos t$ and $\frac{y}{4} = \sin t$, so that the equation is

$$\frac{x^2}{9} + \frac{y^2}{16} = 1.$$

This is an ellipse with semimajor axis 4 and semiminor axis 3, oriented vertically:

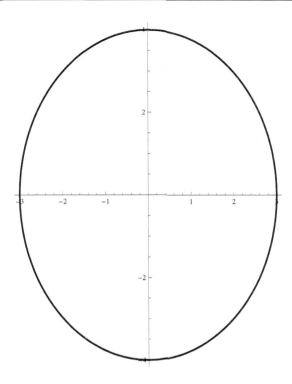

31. Since $\cosh^2 t - \sinh^2 t = 1$, we get for the equation $x^2 - y^2 = 1$. This is a hyperbola whose graph is

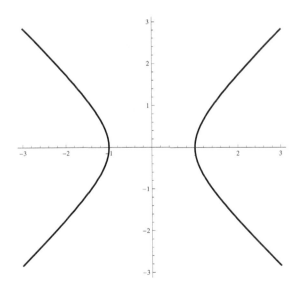

33. Since $1 + \cot^2 t = \csc^2 t$, we get $1 + y^2 = x^2$, so that $x^2 - y^2 = 1$. As $t \in (0, \pi)$, $x = \csc t$ ranges over $[1, \infty)$. This is the same function as in the previous exercise, the right branch of the hyperbola from Exercise 31:

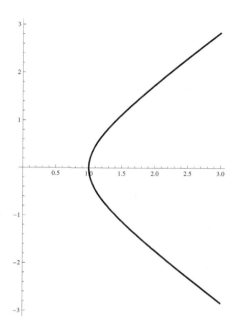

35. In Exercise 1(a), eliminating t gave the equation $x^2 + y^2 = 1$. Now, for $t \in [\pi, 2\pi]$, we see that $y = \sin t \leq 0$, while $x = \cos t$ ranges from -1 to 1. Thus this must be the lower half-circle. Since $x = \cos t$ increases monotonically from -1 to 1 as t goes from π to 2π, the semicircle is traversed counterclockwise.

37. From Exercise 36, we see that $x = 3\sin t$, $y = 3\cos t$ is a circle of radius 3 starting at $(0, 3)$ and traversed *clockwise* for $t \in [0, 2\pi]$. To turn this into a counterclockwise parametrization, we must replace x by its negative everywhere (reflecting across the y axis). We must also use $2\pi t$ instead of t to change the range of the parameter to $[0, 1]$. Thus the required parametrization is $x = -3\sin 2\pi t$, $y = 3\cos 2\pi t$.

39. A line through $(0, 1)$ with slope t has equation $y = tx + 1$. Each such line, for $t \in \mathbb{R}$, intersects the unit circle in exactly one point, but $(0, -1)$ is not covered by these lines since the line through $(0, 1)$ and $(0, -1)$ has infinite slope. We can use this information to derive an explicit parametrization. To find the point of intersection of $y = tx + 1$ and the unit circle $x^2 + y^2 = 1$, substitute $tx + 1$ for y to get
$$x^2 + (tx + 1)^2 = (1 + t^2)x^2 + 2tx + 1 = 1.$$
Thus $x = -\frac{2t}{1+t^2}$. Since $y = tx + 1$, we get
$$y = tx + 1 = t \cdot \frac{-2t}{t^2 + 1} + 1 = \frac{1 - t^2}{1 + t^2}.$$

41. The slope of the tangent line is
$$m = \frac{y'(-1)}{x'(-1)} = \frac{3}{2},$$
so that the equation of the tangent line at $t = -1$ is $y - y(-1) = \frac{3}{2}(x - x(-1))$, or $y - 2 = \frac{3}{2}(x - (-3))$. This simplifies to $y = \frac{3}{2}x + \frac{13}{2}$.

43. The slope of the tangent line is
$$m = \frac{y'\left(\frac{1}{2}\right)}{x'\left(\frac{1}{2}\right)} = \frac{-2(2 - t)|_{t=1/2}}{2t|_{t=1/2}} = -3,$$

so that the equation of the tangent line at $t = \frac{1}{2}$ is $y - y\left(\frac{1}{2}\right) = -3\left(x - x\left(\frac{1}{2}\right)\right)$, or

$$y - \left(2 - \frac{1}{2}\right)^2 = -3\left(x - \left(\frac{1}{2}\right)^2\right), \text{ which simplifies to } y = -3x + 3.$$

45. Since $\frac{d^2y}{dt^2} = \frac{d^2x}{dt^2} = 0$ and $\frac{dx}{dt} = 2 \neq 0$, we have

$$\frac{d^2y}{dx^2} = \frac{\frac{dx}{dt}\frac{d^2y}{dt^2} - \frac{d^2x}{dt^2}\frac{dy}{dt}}{\left(\frac{dx}{dt}\right)^3} = 0.$$

47. Since $\frac{d^2x}{dt^2} = 2 = \frac{d^2y}{dt^2}$, we have

$$\frac{d^2y}{dx^2} = \frac{\frac{dx}{dt}\frac{d^2y}{dt^2} - \frac{d^2x}{dt^2}\frac{dy}{dt}}{\left(\frac{dx}{dt}\right)^3} = \frac{(2t)(2) - 2(-2(2-t))}{8t^3} = \frac{1}{t^3}.$$

Thus at $t = \frac{1}{2}$ we get $\frac{d^2y}{dx^2} = \frac{1}{(1/2)^3} = 8$.

49. On $t \in [0, 1]$, the curve is

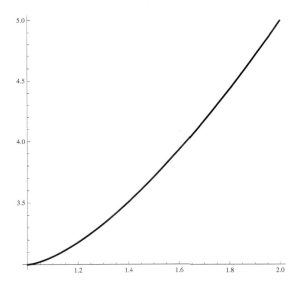

By Theorem 9.4, the arc length is given by

$$\int_0^1 \sqrt{x'(t)^2 + y'(t)^2}\, dt = \int_0^1 \sqrt{(2t)^2 + (6t^2)^2}\, dt$$

$$= \int_0^1 \sqrt{4t^2 + 36t^4}\, dt$$

$$= 2\int_0^1 t\sqrt{1 + 9t^2}\, dt$$

$$= \frac{2}{27}\left[(1 + 9t^2)^{3/2}\right]_0^1$$

$$= \frac{2}{27}(10^{3/2} - 1) = \frac{2}{27}(10\sqrt{10} - 1).$$

51. For $\theta \in [0, 2\pi]$, the curve is

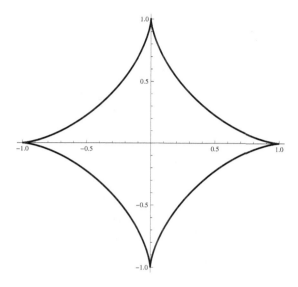

By Theorem 9.4, the arc length is given by

$$\int_0^{2\pi} \sqrt{x'(t)^2 + y'(t)^2}\,dt = \int_0^{2\pi} \sqrt{(3\sin^2 t \cos t)^2 + (-3\cos^2 t \sin t)^2}\,dt$$

$$= \int_0^{2\pi} \sqrt{9\sin^4 t \cos^2 t + 9\cos^4 t \sin^2 t}\,dt$$

$$= \int_0^{2\pi} \sqrt{9\sin^2 t \cos^2 t(\sin^2 t + \cos^2 t)}\,dt$$

$$= 3\int_0^{2\pi} \sqrt{\sin^2 t \cos^2 t}\,dt.$$

Now we must be careful, since $\sin t$ and $\cos t$ are both positive and negative on this range, so compute the integral from 0 to $\frac{\pi}{2}$ using the positive square roots and multiply by 4:

$$3\int_0^{2\pi} \sqrt{\sin^2 t \cos^2 t}\,dt = 12\int_0^{\pi/2} \sin t \cos t\,dt = 6\int_0^{\pi/2} \sin 2t\,dt = 6\left[-\frac{1}{2}\cos 2t\right]_0^{\pi/2} = 6.$$

53. For $t \in [0, 1]$, the curve is

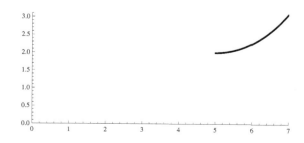

By Theorem 9.4, the arc length is given by

$$\int_0^1 \sqrt{x'(t)^2 + y'(t)^2}\,dt = \int_0^1 \sqrt{2^2 + (e^t - e^{-t})^2}\,dt$$

$$= \int_0^1 \sqrt{e^{2t} + 2 + e^{-t}}\,dt$$

$$= \int_0^1 \sqrt{(e^t + e^{-t})^2}\,dt$$

$$= \int_0^1 (e^t + e^{-t})\,dt$$

$$= \left[e^t - e^{-t}\right]_0^1 = e - \frac{1}{e}.$$

55. Using Exercise 54, parametric equations are

$$x = 1 + (6-1)t,\; y = -3 + (7-(-3))t, \quad \text{or } x = 5t+1, y = 10t - 3,\; t \in [0,1].$$

57. Using Exercise 54, parametric equations are

$$x = 1 + (-3-1)t,\; y = 4 + (5-4)t, \quad \text{or } x = -4t + 1, y = t + 4,\; t \in [0,1].$$

59. Using Exercise 54, parametric equations are

$$x = \pi + (\pi - \pi)t,\; y = 3 + (8-3)t, \quad \text{or } x = \pi, y = 5t + 3,\; t \in [0,1].$$

61. Using the parametrization from Exercise 55, we get for the arc length

$$\int_0^1 \sqrt{x'(t)^2 + y'(t)^2}\,dt = \int_0^1 \sqrt{5^2 + 10^2}\,dt = \int_0^1 \sqrt{125}\,dt = 5\sqrt{5}\,[t]_0^1 = 5\sqrt{5}.$$

Using the distance formula we get

$$\sqrt{(7-(-3))^2 + (6-1)^2} = \sqrt{10^2 + 5^2} = \sqrt{125} = 5\sqrt{5}.$$

The two answers are the same.

63. Consider the following diagram of the situation. At the start, the rolling circle lies along the x axis (the lower of the two small circles), and it rolls counterclockwise.

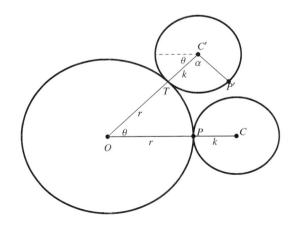

The coordinates of P' relative to C' are $(-k\cos(\theta+\alpha), -k\sin(\theta+\alpha))$, so that the coordinates of P' relative to O are

$$P' = ((r+k)\cos\theta, (r+k)\sin\theta) + (-k\cos(\theta+\alpha), -k\sin(\theta+\alpha))$$
$$= ((r+k)\cos\theta - k\cos(\theta+\alpha), (r+k)\sin\theta - k\sin(\theta+\alpha)).$$

We now need to eliminate α, in order to get a formula involving only r, k, and θ. However, note that when the point of tangency is T, the arcs TP' and TP are equal. Since the radius of the larger circle is r, we know that $TP = r\theta$, so that $TP' = r\theta$. But also $TP' = k\alpha$, giving $\alpha = \frac{r}{k}\theta$. Substituting gives

$$P' = \left((r+k)\cos\theta - k\cos\left(\theta + \frac{r}{k}\theta\right), (r+k)\sin\theta - k\sin\left(\theta + \frac{r}{k}\theta\right)\right)$$
$$= \left((r+k)\cos\theta - k\cos\left(\frac{r+k}{k}\theta\right), (r+k)\sin\theta - k\sin\left(\frac{r+k}{k}\theta\right)\right).$$

Applications

65. (a) Using Theorem 9.4, we have $f(\theta) = \sin\theta$ and $g(\theta) = 3\cos\theta$, so that $f'(\theta)^2 = \cos^2\theta$ and $g'(\theta)^2 = 9\sin^2\theta$. Thus the length of the track is

$$\int_0^{2\pi} \sqrt{f'(\theta)^2 + g'(\theta)^2}\, d\theta = \int_0^{2\pi} \sqrt{\cos^2\theta + 9\sin^2\theta}\, d\theta = \int_0^{2\pi} \sqrt{1 + 8\sin^2\theta}\, d\theta.$$

(b) Using 20 subintervals, we have $\Delta\theta = \frac{2\pi}{20} = \frac{\pi}{10}$ and $\theta_k = \frac{k\pi}{10}$ for $k = 0, 1, 2, \ldots, 20$. Then the midpoints are $\theta_k^* = \frac{(2k-1)\pi}{20}$ for $k = 1, 2, \ldots, 20$, and the midpoint sum is

$$\sum_{k=1}^{20} \sqrt{1 + 8\sin^2\frac{(2k-1)\pi}{20}}\,\frac{\pi}{10} = \frac{\pi}{10}\sum_{k=1}^{20}\sqrt{1 + 8\sin^2\frac{(2k-1)\pi}{20}}.$$

Evaluating numerically gives an approximate value of 13.3651.

67. (a) Integrating gives $x(t) = 2t\cos\frac{\pi}{12} + C$; applying the initial condition $x(0) = 0$ gives $C = 0$ so that $x(t) = 2t\cos\frac{\pi}{12}$. Then

$$y'(t) = 0.444x(t)(x(t) - 3) + 2\sin\frac{\pi}{12}$$
$$= 0.444 \cdot 2t\cos\frac{\pi}{12}\left(2t\cos\frac{\pi}{12} - 3\right) + 2\sin\frac{\pi}{12}$$
$$= 1.776t^2\cos^2\frac{\pi}{12} - 2.664t\cos\frac{\pi}{12} + 2\sin\frac{\pi}{12}.$$

Integrating gives (again with a zero constant of integration due to the initial condition $y(0) = 0$)

$$y(t) = 0.592t^3\cos^2\frac{\pi}{12} - 1.332t^2\cos\frac{\pi}{12} + 2t\sin\frac{\pi}{12}.$$

(b) Since the channel is 3 miles wide, she makes landfall when $2t\cos\frac{\pi}{12} = 3$, or $t = \frac{3}{2\cos\frac{\pi}{12}}$. At that time, her north-south position is

$$y\left(\frac{3}{2\cos\frac{\pi}{12}}\right) = 0.592\left(\frac{3}{2\cos\frac{\pi}{12}}\right)^3\cos^2\frac{\pi}{12} - 1.332\left(\frac{3}{2\cos\frac{\pi}{12}}\right)^2\cos\frac{\pi}{12} + 2\left(\frac{3}{2\cos\frac{\pi}{12}}\right)\sin\frac{\pi}{12}$$
$$\approx -0.2304 \text{ miles},$$

so about 0.23 miles south of her starting point.

(c) She paddles for $\frac{3}{2\cos\frac{\pi}{12}} \approx 1.553$ hours. The distance she travels is approximately

$$\int_0^{1.553} \sqrt{x'(t)^2 + y'(t)^2}\, dt \approx 3.044 \text{ miles}.$$

Proofs

69. Using the parametrization $x = t, y = f(t)$, the arc length from $x = a$ to $x = b$ is the arc length from $t = a$ to $t = b$, which is

$$\int_a^b \sqrt{x'(t)^2 + y'(t)^2}\, dt = \int_a^b \sqrt{(t')^2 + (f'(t))^2}\, dt = \int_a^b \sqrt{1 + f'(t)^2}\, dt = \int_a^b \sqrt{1 + f'(x)^2}\, dx.$$

71. The arc length of the curve parametrized by $x = kf(t)$ and $y = kg(t)$ from $t = a$ to $t = b$ is

$$\int_a^b \sqrt{((kf(t))')^2 + ((kg(t))')^2}\, dt = \int_a^b \sqrt{k^2 f'(t) + k^2 g'(t)}\, dt = k \int_a^b \sqrt{f'(t)^2 + g'(t)^2}\, dt,$$

since $k > 0$, and this is just k times the arc length of the curve parametrized by $x = f(t)$ and $y = g(t)$ from $t = a$ to $t = b$.

The arc length of the curve parametrized by $x = f(kt)$ and $y = g(kt)$ from $t = \frac{a}{k}$ to $t = \frac{b}{k}$ is

$$\int_{a/k}^{b/k} \sqrt{((f(kt))')^2 + ((g(kt))')^2}\, dt = \int_{a/k}^{b/k} \sqrt{k^2 f'(kt) + k^2 g'(kt)}\, dt = \int_{a/k}^{b/k} k\sqrt{f'(kt) + g'(kt)}\, dt.$$

Now use the change of variables $u = kt$, so that $du = k\, dt$. The new bounds of integration are from $t = a/k$, or $u = a$, to $t = b/k$ or $u = b$. Thus the arc length is

$$\int_a^b \sqrt{f'(u)^2 + g'(u)^2}\, du,$$

which is the arc length of the curve parametrized by $x = f(t)$ and $y = g(t)$ from $t = a$ to $t = b$.

73. We have

$$\frac{dy}{dx} = \frac{\frac{dy}{d\theta}}{\frac{dx}{d\theta}} = \frac{r \sin \theta}{-r \cos \theta} = -\cot \theta.$$

Then the slope of the tangent at any even multiple of π is $-\lim_{\theta \to 2k\pi} \cot \theta = -\infty$, so this limit does not exist.

Thinking Forward

Parametric equations in three dimensions

▷ Since for every value of t we have $z = 2x$ and $y = -3x$, this is a line through the origin pointing in the direction $(1, -3, 2)$.

▷ As t increases, x and y go around the unit circle, but the curve is rising in the z coordinate as well, so this looks like a coiled spring with radius 1.

▷ As t increases, x and y go around the origin, but form an expanding helix. In addition, the curve is rising. So this curve looks something like what you would get if you wound a piece of string around an inverted cone.

9.2   Polar Coordinates

Thinking Back

Plotting in a rectangular coordinate system

▷ The plot looks like

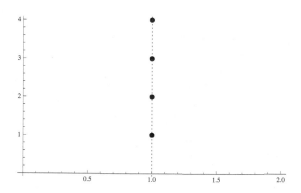

These points all lie on the same vertical line.

▷ The plot looks like

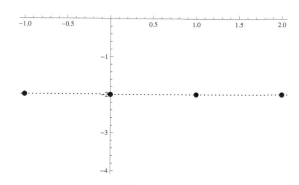

These points all lie on the same horizontal line.

▷ The plot looks like

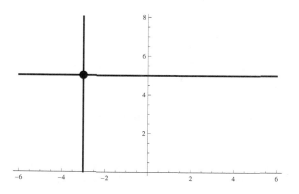

The two lines intersect at the point $(-3, 5)$.

Completing the square

▷ Subtract $4x$ from both sides, and add 4 to both sides, to get $x^2 - 4x + 4 + y^2 = 4$. Now $x^2 - 4x + 4 = (x-2)^2$, so the equation becomes $(x-2)^2 + y^2 = 4$.

▷ Regroup to get $(x^2 + 6x) + (y^2 - 8y) = 44$. Then add 9 to the part involving x and 16 to the part involving y. To keep the equation balanced, add $9 + 16 = 25$ to the right-hand side. This gives $(x^2 + 6x + 9) + (y^2 - 8y + 16) = 44 + 25 = 69$, so $(x+3)^2 + (y-4)^2 = 69$. This is a circle with center $(-3, 4)$ and radius $\sqrt{69}$.

Concepts

1. (a) True. The x and y coordinates are determined by its position in the plane.

   (b) False. For example, $(r, \theta) = (r, 2\pi + \theta)$.

   (c) False. If $b = -c$, they are the same.

   (d) False. For example, if $\beta = \alpha + 2\pi$, the two graphs are identical — they are both lines through the origin at angle α.

   (e) True. Since $r = \csc\theta$, we have $r\sin\theta = 1$. But $r\sin\theta = y$, so this is just the graph of $y = 1$, and since $\theta \in \left(-\frac{\pi}{2}, \frac{\pi}{2}\right)$ means that $\csc\theta \in (-\infty, \infty)$, we get the entire line $y = 1$.

   (f) True. By Theorem 9.5(b), (r, θ) and $(-r, \theta + \pi)$ represent the same point, so if this is also the point $(r, \theta + \pi)$, we must have $r = -r$ so that $r = 0$.

   (g) True. Multiply through by r, giving $Ar\sin\theta + Br\cos\theta = r^2$, so that $Ay + Bx = x^2 + y^2$. Collect terms and complete the square to get

   $$\left(x - \frac{B}{2}\right)^2 + \left(y - \frac{A}{2}\right)^2 = \frac{A^2 + B^2}{4},$$

   which is the equation of a circle.

   (h) True. Using Theorems 9.6 and 9.7, we can always translate polar to rectangular coordinates. However, the form of the expression in rectangular coordinates may not be particularly simple.

3. A point in the polar plane has a radius and an angle. However, the angle is not unique, since you can add any multiple of 2π to it and get the same angle with respect to the polar axis. Similarly, the radius is not unique, since $(r, \theta) = (-r, \theta + \pi)$.

5. Consider the diagram

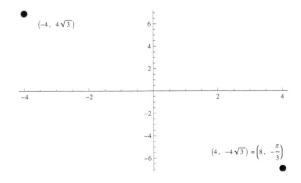

Even though with $(r, \theta) = \left(8, -\frac{\pi}{3}\right)$ and $(x, y) = (-4, 4\sqrt{3})$ we have

$$x^2 + y^2 = 16 + 48 = 64 = r^2 \qquad \text{and} \qquad \tan\left(-\frac{\pi}{3}\right) = -\sqrt{3} = \frac{y}{x},$$

the two points are not the same because they are reflections of each other in the origin. The reason for this is that $\frac{y}{x} = \tan\theta$ has two solutions for θ, which differ by π. In order to decide which one to choose, you must look at the quadrant of the point to determine the proper signs for x and y. In this case, since the point is in the fourth quadrant, we want $\frac{y}{x} = -\sqrt{3}$ with $x > 0$ and $y < 0$, so the correct point is $(4, -4\sqrt{3})$.

7. These two graphs are the same for any value of c. If the point (c, θ) is on the graph of $r = c$, then since $(c, \theta) = (-c, \pi + \theta)$, we see that (c, θ) is also on the graph of $r = -c$. Similarly, if the point $(-c, \theta)$ is on the graph of $r = -c$, then since $(-c, \theta) = (c, \pi + \theta)$, we see that $(-c, \theta)$ is also on the graph of $r = c$. Thus the two graphs are identical. This argument holds equally well if $c = 0$.

9. We know from the previous exercise that $\theta = \alpha$ and $\theta = -\alpha$ are the same for $\alpha = \frac{\pi}{2}$. So assume that $\alpha \neq \frac{\pi}{2}$, and assume $\alpha \in [0, 2\pi]$. Then $\tan\alpha = \frac{y}{x}$, and $\tan(-\alpha) = -\tan\alpha = \frac{y}{x}$. Thus $\tan\alpha = 0$, so that $\alpha = 0$ or $\alpha = \pi$. So the only possible values for α are $0 + 2n\pi$, $\pi + 2n\pi$, and $\frac{\pi}{2} + 2n\pi$ (or, more simply, $\alpha = n\frac{\pi}{2}$), where n is an integer.

11. A point (r, θ) with $r > 0$ and $0 < \theta < \frac{\pi}{2}$ is a positive distance from the origin along a first-quadrant angle, so it lies in the first quadrant. One possible description of third-quadrant points is $r > 0$ and $\pi < \theta < \frac{3\pi}{2}$; another is $r < 0$ and $0 < \theta < \frac{\pi}{2}$.

13. We must have $r = 1$ or $r = -1$. If $r = 1$, then the angle is 0 plus a multiple of 2π; if $r = -1$, then the angle is π plus a multiple of 2π. Thus the possible representations are $(1, 2n\pi)$ and $(-1, \pi + 2n\pi)$ where n is an integer.

15. The point $(-1, 0)$ in polar coordinates is the same as $(1, \pi)$ in polar coordinates. Other representations for this point add multiples of 2π to either of these, so we get $(-1, 2n\pi)$ and $(1, \pi + 2n\pi)$.

Skills

17. We have

$$\left(3, \frac{\pi}{6}\right): \qquad \left(3\cos\frac{\pi}{6}, 3\sin\frac{\pi}{6}\right) = \left(\frac{3\sqrt{3}}{2}, \frac{3}{2}\right)$$

$$\left(-3, \frac{\pi}{6}\right): \qquad \left(-3\cos\frac{\pi}{6}, -3\sin\frac{\pi}{6}\right) = \left(-\frac{3\sqrt{3}}{2}, -\frac{3}{2}\right)$$

$$\left(3, -\frac{\pi}{6}\right): \qquad \left(3\cos\left(-\frac{\pi}{6}\right), 3\sin\left(-\frac{\pi}{6}\right)\right) = \left(\frac{3\sqrt{3}}{2}, -\frac{3}{2}\right)$$

$$\left(-3, -\frac{\pi}{6}\right): \qquad \left(-3\cos\left(-\frac{\pi}{6}\right), -3\sin\left(-\frac{\pi}{6}\right)\right) = \left(-\frac{3\sqrt{3}}{2}, \frac{3}{2}\right).$$

A plot of all four points is

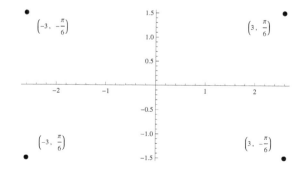

19. We have

$$\left(5, \frac{\pi}{2}\right): \qquad \left(5\cos\frac{\pi}{2}, 5\sin\frac{\pi}{2}\right) = (0, 5)$$

$$\left(-5, \frac{-3\pi}{2}\right): \qquad \left(-5\cos\frac{-3\pi}{2}, -5\sin\frac{-3\pi}{2}\right) = \left(-5\cos\frac{\pi}{2}, -5\sin\frac{\pi}{2}\right) = (0, -5)$$

$$\left(5, \frac{5\pi}{2}\right): \qquad \left(5\cos\frac{5\pi}{2}, 5\sin\frac{5\pi}{2}\right) = \left(5\cos\frac{\pi}{2}, 5\sin\frac{\pi}{2}\right) = (0, 5)$$

$$\left(-5, \frac{-\pi}{2}\right): \qquad \left(-5\cos\frac{-\pi}{2}, -5\sin\frac{-\pi}{2}\right) = (0, 5).$$

A plot of all four points is

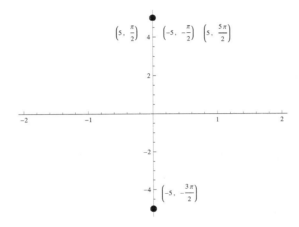

21. The point $\left(r, \frac{\pi}{4}\right)$ has rectangular coordinates $\left(r\cos\frac{\pi}{4}, r\sin\frac{\pi}{4}\right) = \left(r\frac{\sqrt{2}}{2}, r\frac{\sqrt{2}}{2}\right)$. Thus the rectangular coordinates of the four given points are

$$\left(\frac{\sqrt{2}}{2}, \frac{\sqrt{2}}{2}\right), \quad \left(\sqrt{2}, \sqrt{2}\right), \quad \left(\frac{3\sqrt{2}}{2}, \frac{3\sqrt{2}}{2}\right), \quad \left(2\sqrt{2}, 2\sqrt{2}\right).$$

A plot of all four points is

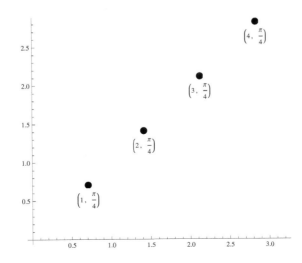

23. The point $\left(r, -\frac{\pi}{2}\right)$ has rectangular coordinates $\left(r\cos\left(-\frac{\pi}{2}\right), r\sin\left(-\frac{\pi}{2}\right)\right) = (0, -r)$, so the rectangular coordinates of the four given points are $(0, -1)$, $(0, -2)$, $(0, -3)$, and $(0, -4)$. A plot of all four points is

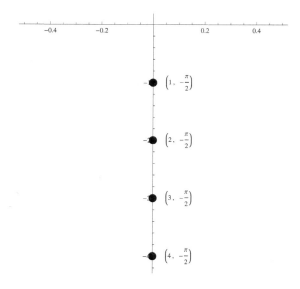

25. The angle representing this point must be either 0 plus a multiple of 2π or (by Theorem 9.5(b)) π plus a multiple of 2π; in the latter case, r will be -1. Thus the representations are $(1, 2n\pi)$ and $(-1, \pi + 2n\pi)$ where n is an integer.

27. The angle representing this point must be either $\frac{\pi}{2}$ plus a multiple of 2π or (by Theorem 9.5(b)) $-\frac{\pi}{2}$ plus a multiple of 2π; in the latter case, r will be -2. Thus the representations are $\left(2, \frac{\pi}{2} + 2n\pi\right)$ and $\left(-2, -\frac{\pi}{2} + 2n\pi\right)$ where n is an integer.

29. The angle θ representing this point is such that $\tan\theta = \frac{2\sqrt{3}}{6} = \frac{1}{\sqrt{3}}$; since x and y are both positive, we know that θ is a first quadrant angle and thus $\theta = \tan^{-1}\frac{1}{\sqrt{3}} = \frac{\pi}{6}$. We also have $r = \sqrt{6^2 + (2\sqrt{3})^2} = \sqrt{48} = 4\sqrt{3}$, so that one set of representations is $\left(4\sqrt{3}, \frac{\pi}{6} + 2n\pi\right)$ where n is an integer. By Theorem 9.5(b), the other set of representations is then $\left(-4\sqrt{3}, \frac{\pi}{6} + 2n\pi + \pi\right) = \left(-4\sqrt{3}, \frac{7\pi}{6} + 2n\pi\right)$ where n is an integer.

31. The angle θ representing this point is such that $\tan\theta = \frac{-3}{3} = -1$; since $x > 0$ and $y < 0$, we know that θ is a fourth quadrant angle, so that $\theta = \tan^{-1}(-1) = -\frac{\pi}{4}$. We also have $r = \sqrt{9+9} = 3\sqrt{2}$, so that one set of representations is $\left(3\sqrt{2}, -\frac{\pi}{4} + 2n\pi\right)$ where n is an integer. By Theorem 9.5(b), the other set of representations is then $\left(-3\sqrt{2}, -\frac{\pi}{4} + 2n\pi + \pi\right) = \left(-3\sqrt{2}, \frac{3\pi}{4} + 2n\pi\right)$ where n is an integer.

33. $\theta = \frac{\pi}{4}$ is the line through the origin at an angle of $\frac{\pi}{4}$, so for any point (x, y) on that line we have $\frac{y}{x} = \tan\frac{\pi}{4} = 1$ so that $y = x$. This is the line $y = x$.

35. $\theta = \frac{7\pi}{6}$ is the line through the origin at an angle of $\frac{7\pi}{6}$, so for any point (x, y) on that line we have $\frac{y}{x} = \tan\left(\frac{7\pi}{6}\right) = \frac{1}{\sqrt{3}}$. Thus the equation of the line is $y = \frac{1}{\sqrt{3}}x$.

37. By Theorem 9.8 with $a = \frac{5}{2}$, this is the circle $x^2 + \left(y - \frac{5}{2}\right)^2 = \frac{25}{4}$.

39. $r = 6 \csc \theta$ means that $r \sin \theta = 6$, so that $y = 6$ since $y = r \sin \theta$ in polar coordinates.

41. Since $\sin 2\theta = 2 \sin \theta \cos \theta = 2 \frac{y}{r} \frac{x}{r}$, we get $r^3 = 2xy$. But $r = \pm \sqrt{x^2 + y^2}$, so substituting gives $\pm (x^2 + y^2)^{3/2} = 2xy$.

43. Multiply both sides by r to get $r^3 = r \cos \theta$. Now use the fact that $r = \pm \sqrt{x^2 + y^2}$ and $x = r \cos \theta$ to get $\pm (x^2 + y^2)^{3/2} = x$.

45. There are several ways to do this. For example, since $r = \theta$, take the sine of both sides and multiply through by r to get $r \sin r = r \sin \theta = y$. Then $y = \sqrt{x^2 + y^2} \sin \sqrt{x^2 + y^2}$. There is no satisfying formula for this spiral in cartesian coordinates, but there is a reasonable parametrized solution given by $x = \pm \theta \cos \theta$ and $y = \pm \theta \sin \theta$.

47. Multiply through by r^3 to get $r^4 = r^3 \sin^3 \theta = (r \sin \theta)^3$, so that $(x^2 + y^2)^2 = y^3$.

49. $y = 0$ is $r \sin \theta = 0$, so we get $\sin \theta = 0$, so this is the line $\theta = 0$ (or $\theta = \pi$). (Note that we divided through by r, so we assumed $r \neq 0$. However, the origin is on the graph of the resulting equation, so that case is included in the answer).

51. $y = x$ means $r \sin \theta = r \cos \theta$, so that $\sin \theta = \cos \theta$, or $\tan \theta = 1$. Thus the equation is $\theta = \frac{\pi}{4}$ or $\theta = -\frac{\pi}{4}$. (Note that we divided through by r, so we assumed $r \neq 0$. However, the origin is on the graph of the resulting equation, so that case is included in the answer).

53. $y = -3$ means $r \sin \theta = -3$, so that $r = -3 \csc \theta$.

55. $y = mx$ gives upon substitution $r \sin \theta = rm \cos \theta$, so that $\tan \theta = m$ and then $\theta = \tan^{-1} m$.

57. In each case, substitute $r = \pm \sqrt{x^2 + y^2}$ and square the result:

$$
\begin{array}{lll}
r = 1: & \pm \sqrt{x^2 + y^2} = 1 & \Rightarrow \quad x^2 + y^2 = 1, \\
r = 2: & \pm \sqrt{x^2 + y^2} = 2 & \Rightarrow \quad x^2 + y^2 = 4, \\
r = 3: & \pm \sqrt{x^2 + y^2} = 3 & \Rightarrow \quad x^2 + y^2 = 9, \\
r = 4: & \pm \sqrt{x^2 + y^2} = 4 & \Rightarrow \quad x^2 + y^2 = 16, \\
r = 5: & \pm \sqrt{x^2 + y^2} = 5 & \Rightarrow \quad x^2 + y^2 = 25, \\
r = 6: & \pm \sqrt{x^2 + y^2} = 6 & \Rightarrow \quad x^2 + y^2 = 36, \\
r = 7: & \pm \sqrt{x^2 + y^2} = 7 & \Rightarrow \quad x^2 + y^2 = 49, \\
r = 8: & \pm \sqrt{x^2 + y^2} = 8 & \Rightarrow \quad x^2 + y^2 = 64, \\
r = 9: & \pm \sqrt{x^2 + y^2} = 9 & \Rightarrow \quad x^2 + y^2 = 81.
\end{array}
$$

59. For most values of k, take the tangent of both sides of the equation and simplify:

$$\theta = 0: \qquad \tan\theta = 0 = \frac{y}{x} \quad \Rightarrow \quad y = 0,$$

$$\theta = \frac{\pi}{6}, \frac{7\pi}{6}: \qquad \tan\theta = \frac{1}{\sqrt{3}} = \frac{y}{x} \quad \Rightarrow \quad y = \frac{1}{\sqrt{3}}x,$$

$$\theta = \frac{2\pi}{6} = \frac{\pi}{3}, \frac{8\pi}{6} = \frac{4\pi}{3}: \qquad \tan\theta = \sqrt{3} = \frac{y}{x} \quad \Rightarrow \quad y = x\sqrt{3},$$

$$\theta = \frac{3\pi}{6} = \frac{\pi}{2}, \frac{9\pi}{6} = \frac{3\pi}{2}: \qquad \tan\theta = \infty, \text{ so this is the vertical line } x = 0,$$

$$\theta = \frac{4\pi}{6} = \frac{2\pi}{3}, \frac{10\pi}{6} = \frac{5\pi}{3}: \qquad \tan\theta = -\sqrt{3} = \frac{y}{x} \quad \Rightarrow \quad y = -x\sqrt{3},$$

$$\theta = \frac{5\pi}{6}, \frac{11\pi}{6}: \qquad \tan\theta = -\frac{1}{\sqrt{3}} = \frac{y}{x} \quad \Rightarrow \quad y = -\frac{1}{\sqrt{3}}x.$$

Applications

61. (a) From the diagram, assuming the arms of the cam remain symmetrically placed, each of the $\frac{3}{4}$ length arms will cover half the crack. So for each arm, we have a right triangle whose hypotenuse is $\frac{3}{4}$ and one of whose angles is θ. The leg opposite that angle is half the width of the crack, so is $\frac{1}{2}$. Then $\theta = \arcsin\frac{1/2}{3/4} = \arcsin\frac{2}{3} \approx 41.8°$.

    (b) As in the previous problem, the downward force on the rope results in a rightward force inside the crack, which again causes the cams to wedge more tightly into the crack.

Proofs

63. Multiply both sides by $\sin\theta$ to get $r\sin\theta = k$. Since $r\cos\theta = y$, this graph has equation $y = k$ in rectangular coordinates, so it is a horizontal line.

65. If $a \neq 0$ and $b > 0$, multiply through by $1 - b\cos\theta$ to get $r - br\cos\theta = a$, so that $r = a + br\cos\theta$. Substituting gives $\pm\sqrt{x^2 + y^2} = a + bx$. Now square both sides to get $x^2 + y^2 = a^2 + 2abx + b^2x^2$. Rearrange to get

$$(1 - b^2)x^2 - 2abx + y^2 = a^2, \quad \text{so that} \quad x^2 - \frac{2ab}{1 - b^2}x + \frac{y^2}{1 - b^2} = \frac{a^2}{1 - b^2}.$$

Complete the square for x on the left to get

$$\left(x - \frac{ab}{1 - b^2}\right)^2 + \frac{y^2}{1 - b^2} = \frac{a^2}{1 - b^2} + \frac{a^2b^2}{(1 - b^2)^2} = \frac{a^2}{(1 - b^2)^2}.$$

The right-hand side of this equation is always positive, since $a \neq 0$ and $b \neq 1$. On the left, if $0 < b < 1$ then $1 - b^2 > 0$, so we have an equation of the form $x^2 + \frac{y^2}{c^2} = d^2$, which is an ellipse. If $b > 1$ then $1 - b^2 < 0$, so we have an equation of the form $x^2 - \frac{y^2}{c^2} = d^2$, which is a hyperbola.

67. Multiply this equation through by r to get $r^2 = kr\cos\theta$. Substituting gives $x^2 + y^2 = kx$, so that $x^2 - kx + y^2 = 0$. Complete the square to get $\left(x - \frac{k}{2}\right)^2 + y^2 = \frac{k^2}{4}$, so this is a circle with radius $\frac{k}{2}$ and center $\left(\frac{k}{2}, 0\right)$. Comparing the radius and the center, we see that the circle is tangent to the y axis at the origin.

69. Multiply $r = k\sin\theta + l\cos\theta$ through by r to get $r^2 = kr\sin\theta + lr\cos\theta$. Now substitute to get $x^2 + y^2 = ky + lx$, so that $x^2 - lx + y^2 - ky = 0$. Complete both squares to get

$$\left(x - \frac{l}{2}\right)^2 + \left(y - \frac{k}{2}\right)^2 = \frac{k^2 + l^2}{4}.$$

This is a circle of radius $\frac{\sqrt{k^2+l^2}}{2}$ and center $\left(\frac{l}{2}, \frac{k}{2}\right)$.

Thinking Forward

Understanding symmetry

$r = f(\theta) = f(-\theta)$ means that the distance of the graph from the origin is the same for θ as it is for $-\theta$. Since $-\theta$ is θ reflected through the x axis, this equality means that the graph is symmetric around the x axis.

Understanding symmetry

If $f(\theta) = r$, then $f(-\theta) = -f(\theta) = -r$. Thus the points (r, θ) and $(-r, -\theta)$ are both on the graph. But $(-r, -\theta) = (r, \pi - \theta)$ by Theorem 9.5, so that (r, θ) and $(r, \pi - \theta)$ are both on the graph. But these two points are mirror images across the y axis, since in rectangular coordinates $(r, \pi - \theta)$ is $(r\cos(\pi - \theta), r\sin(\pi - \theta)) = (-r\cos\theta, r\sin\theta)$. Thus the graph is symmetric with respect to the y axis.

9.3   Graphing Polar Equations

Thinking Back

Finding points of intersection

Graphs of the two functions ($1 + \sin x$ is the solid graph) are:

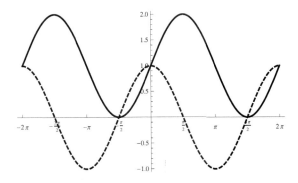

It seems that the two intersect at $2n\pi$ and at $-\frac{\pi}{2} + 2n\pi$, and in fact, equating the two gives

$$1 + \sin x = \cos x = \sqrt{1 - \sin^2 x}, \quad \text{so} \quad (1 + \sin x)^2 = 1 - \sin^2 x.$$

Expanding the left-hand side gives $1 + 2\sin x + \sin^2 x = 1 - \sin^2 x$. Now simplify to get $\sin^2 x + \sin x = 0$, so that $\sin x = 0$ (which happens for $x = \pm 2n\pi$ and $x = \pi \pm 2n\pi$) or $\sin x = -1$ (which happens for $x = -\frac{\pi}{2} + 2n\pi$). The solution $x = \pi \pm 2n\pi$ is an extraneous solution that was introduced when we squared the equation, so we reject it.

Finding points of intersection

Graphs of the two functions ($\frac{1}{2} + \cos x$ is the solid graph) are:

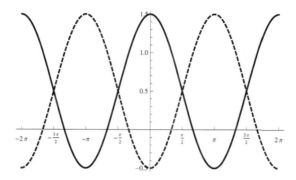

It seems that the two intersect at $\frac{\pi}{2} + n\pi$, and in fact, equating the two gives

$$\frac{1}{2} + \cos x = \frac{1}{2} - \cos x, \quad \text{so} \quad 2\cos x = 0.$$

So the solutions correspond to $\cos x = 0$, which is for $x = \frac{\pi}{2} + n\pi$.

Translations of graphs

The graph of $y = f(x - k)$ is the graph of $y = f(x)$ shifted right by k units, since when (say) $x = k$, the first graph contains the point $(k, f(0))$ while the second contains the point $(0, f(0))$.

Translations of graphs

The graph of $y = f(x) + k$ is the graph of $y = f(x)$ shifted up by k units, since for any value of x, the point $(x, f(x))$ is on the graph of $y = f(x)$, while the point $(x, f(x) + k)$ is on the graph of $y = f(x) + k$.

Concepts

1. (a) False. It is true if $r > 0$, but if $r < 0$, then this point is in the fourth quadrant.

    (b) True. Its graph is

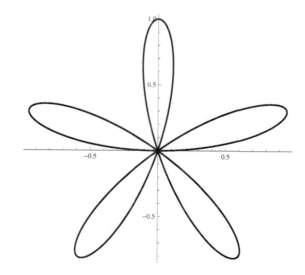

(c) False; it has twelve petals. Its graph is

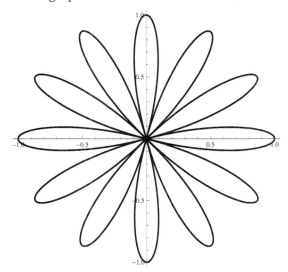

(d) True. If a graph is symmetric with respect to the origin, then the points (r, θ) is on the graph if and only if the point $(-r, \theta)$ is on the graph, by Theorem 9.11(c). But $(-r, \theta) = (-r, \theta + 2\pi)$ by Theorem 9.5(a).

(e) False. By Theorem 9.11, this condition expresses symmetry around the x axis. Symmetry around the y axis is given by the condition that (r, θ) is on the graph if and only if $(-r, -\theta)$ is on the graph.

(f) True. Let G be the graph of $r = \sin k\theta$. This graph is always symmetric around the y axis, since if $(r, \theta) \in G$, then $\sin(-k\theta) = -\sin k\theta = -r$, so that $(-r, -\theta) \in G$. For x axis symmetry, note that

$$(r, -\theta) \in G \iff (-r, \pi - \theta) \in G \iff -r = \sin(k(\pi - \theta)) \iff r = -\sin(k\pi - k\theta)$$
$$\iff r = -\sin k\pi \cos k\theta + \cos k\pi \sin k\theta = \cos k\pi \sin k\theta.$$

Now, if $(r, \theta) \in G$, then $r = \sin \theta$, so that $(r, -\theta) \in G$ if and only if $\cos k\pi = 1$, i.e., if k is even. The graph G of $r = \cos k\theta$ is always symmetric around the x axis, since if $(r, \theta) \in G$, then $\cos(-k\theta) = \cos k\theta = r$, so that $(r, -\theta) \in G$. For y axis symmetry, note that

$$(-r, -\theta) \in G \iff (r, \pi - \theta) \in G \iff r = \cos(k(\pi - \theta))$$
$$\iff r = \cos k\pi \cos k\theta + \sin k\pi \sin k\theta = \cos k\pi \cos k\theta.$$

Now, if $(r, \theta) \in G$, then $r = \cos \theta$, so that $(-r, -\theta) \in G$ if and only if $\cos k\pi = 1$, i.e., if k is even.

(g) True. Converting to polar coordinates gives $r^4 = k(r^2 \cos^2 \theta - r^2 \sin^2 \theta)$, or $r^2 = k(\cos^2 \theta - \sin^2 \theta) = k \cos 2\theta$. This is a lemniscate.

(h) True. If $y = f(x)$ is symmetric with respect to the y axis, then $f(x) = f(-x)$ for all x. If it is symmetric with respect to the origin, then $f(x) = -f(-x)$ for all x, so that $f(-x) = -f(x)$. It follows that for all x, $f(x) = f(-x)$, so that $f(x) = 0$ for all x and $y = 0$.

3. See the discussion in the text for an extended explanation. Basically, the areas above and below the θ axis for each multiple of $\frac{\pi}{2}$ correspond to quadrants in which the graph will lie for those values of θ.

5. (a) The pole in the polar plane corresponds to $r = 0$, and θ is arbitrary. This is the θ axis in the $r\theta$ plane.

(b) The horizontal axis in the polar plane is the line where θ is a multiple of π. This corresponds in the $r\theta$ plane to the vertical lines $\theta = k\pi$ for k an integer.

(c) The vertical axis in the polar plane is the line where θ is an odd multiple of $\frac{\pi}{2}$, so this corresponds in the $r\theta$ plane to the vertical lines $\theta = \frac{k\pi}{2}$ for k an odd integer.

7. (a) By Theorem 9.11(a), the point $\left(2, -\frac{\pi}{3}\right)$ is also on the graph.

(b) By Theorem 9.11(b), the point $\left(-2, -\frac{\pi}{3}\right)$ — or $\left(2, \frac{2\pi}{3}\right)$ — is also on the graph.

(c) Symmetry with respect to the y axis means that the function is even, so that $(-3, 5)$ is also on the graph.

(d) By Theorem 9.11(c), the point $\left(-2, \frac{\pi}{3}\right)$ — or $\left(2, \frac{4\pi}{3}\right)$ — is also on the graph.

(e) Symmetry with respect to the origin means that the function is odd, so that $(-3, -5)$ is also on the graph.

9. (a) For example, $r = \cos\theta$, since $\cos(-\theta) = \cos\theta$, so that (r, θ) is on the graph if and only if $(r, -\theta)$ is on the graph.

(b) For example, $r = \sin\theta$, since $\sin(-\theta) = -\sin\theta$, so that (r, θ) is on the graph if and only if $(-r, -\theta)$ is on the graph.

(c) For example, $r = 1$, since for any point (r, θ), the point $(r, \theta + \pi) = (-r, \theta)$ is also on the graph, since the graph contains all points whose radial coordinate is r.

(d) For example, the lemniscate $r^2 = \sin 2\theta$. Since if (r, θ) is on the curve, then $r^2 = \sin 2\theta$, so that also $(-r)^2 = \sin 2\theta$ and thus $(-r, \theta)$ is on the graph. Thus the graph is symmetric around the origin. However, since $\sin(-2\theta) = -2\sin 2\theta$, the curve is not symmetric around the x axis.

(e) For example, $r = 2$; this is a circle centered at the origin, so it has symmetry around the x axis, the y axis, and the origin.

(f) Not possible. By Theorem 9.10, and Exercise 56, any curve that has two of these three symmetries also has the third.

11. The cardioid is named for the heart; the limaçon for a snail, and the lemniscate for a ribbon.

13. The graph G of $r = \cos 5\theta$ is always symmetric around the x axis, since if $(r, \theta) \in G$, then $\cos(-5\theta) = \cos 5\theta = r$, so that $(r, -\theta) \in G$. However, let $\theta = 0$. Then $(\cos 5 \cdot 0, 0) = (1, 0) \in G$, but $(-1, 0) \notin G$. Thus G is not symmetric with respect to the y axis. By the table in the text, this rose has 5 petals. (It is also rotationally symmetric through an angle of $\frac{2\pi}{5}$).

Next, let G be the graph of $r = \sin 8\theta$. This graph is always symmetric around the y axis, since if $(r, \theta) \in G$, then $\sin(-8\theta) = -\sin 8\theta = -r$, so that $(-r, -\theta) \in G$. For x axis symmetry, note that

$$(r, -\theta) \in G \iff (-r, \pi - \theta) \in G \iff -r = \sin(8(\pi - \theta)) \iff r = -\sin(8\pi - 8\theta)$$
$$\iff r = -\sin 8\pi \cos 8\theta + \cos 8\pi \sin 8\theta = \cos 8\pi \sin\theta = \sin 8\theta.$$

But the last equation simply expresses the fact that $(r, \theta) \in G$. Thus $(r, -\theta) \in G \iff (r, \theta) \in G$, so that the graph is symmetric with respect to the x axis as well. Since it is symmetric around both axes, it is symmetric around the origin as well. By the table in the text, this rose has 16 petals. (It is also rotationally symmetric through an angle of $\frac{2\pi}{16} = \frac{\pi}{8}$.)

15. If (r, θ) is on the graph of $r = \cos 2\theta$, then since $\cos(-2\theta) = \cos 2\theta = r$, we see that $(r, -\theta)$ is also on the graph, so that the graph is symmetric around the x (horizontal) axis.

Skills

17. A graph of the curve is:

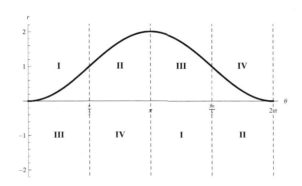

19. A graph of the curve is:

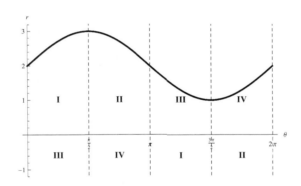

21. A graph of the curve is:

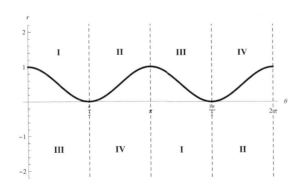

23. A graph of the curve is:

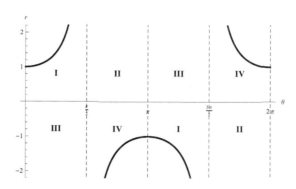

25. A graph of the curve is:

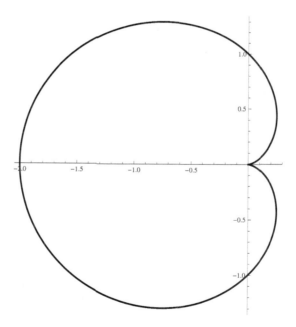

The curve moves through the quadrants in increasing order, as the $r\theta$ plot implies, and the radii are larger in the second and third quadrants, again in agreement with the $r\theta$ plot.

27. A graph of the curve is:

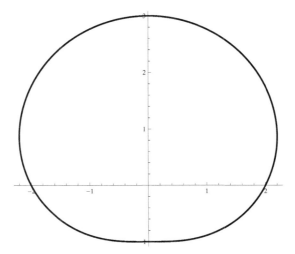

As implied by the $r\theta$ plot, the curve moves through all four quadrants in order.

29. A graph of the curve is:

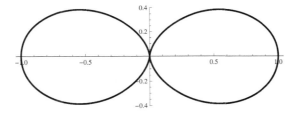

The quadrant I loop corresponds to the first arc of the $r\theta$ plot; then for the next arc, the radii are still positive, so we move into the second quadrant, and so forth. This again corresponds to the $r\theta$ plot.

31. A graph of the curve is:

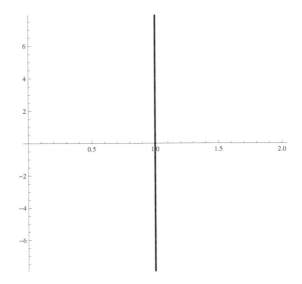

The polar graph is wholly in quadrants I and IV, which is what the $r\theta$ plot tells us. The upper half of the line is traced for $0 \le \theta < \frac{\pi}{2}$ and the lower half is traced, starting at $-\infty$ for $\frac{\pi}{2} < \theta \le \pi$.

33. The function first repeats for θ a multiple of 2π when $\theta = 10\pi$, so this is the period. The polar graph is

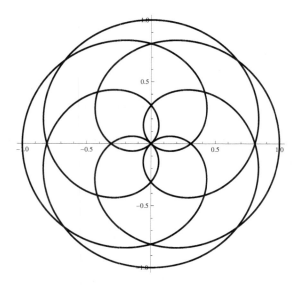

35. From the $r\theta$ plot

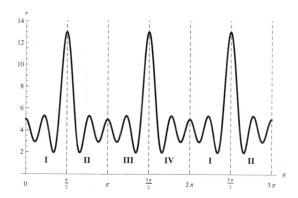

we see that the graph first repeats in quadrant I after 2π, so this is the period. A polar graph is

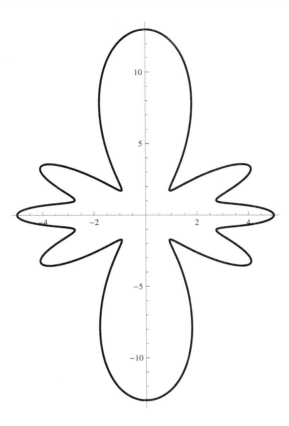

37. All three functions repeat first for θ a multiple of 2π when $\theta = 24\pi$, so this is the period. The polar graph is

39. The period is 2π. The graph is symmetric around the y axis, if $r = \sin\theta$, then $-r = -\sin\theta = \sin(-\theta)$, so that $(-r, -\theta)$ is also on the graph. Plotting, for example, $-\frac{\pi}{2} < \theta < \frac{\pi}{2}$ and reflecting it gives

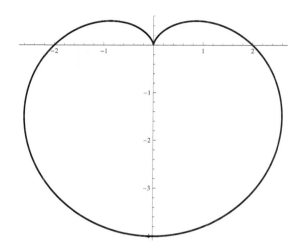

41. (a) $f_1(\theta) = f_2(\theta)$ if and only if $\sin\theta = \cos\theta$, i.e., when $\theta = \frac{\pi}{4} + n\pi$.

(b) Plots of the two curves, with $\sin\theta$ the solid line, are

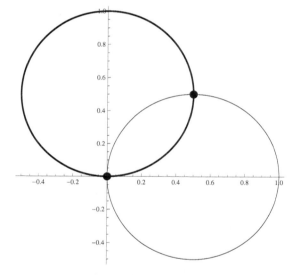

(c) The equations in rectangular coordinates are, by Theorem 9.8, $\left(x - \frac{1}{2}\right)^2 + y^2 = \frac{1}{4}$ and $x^2 + \left(y - \frac{1}{2}\right)^2 = \frac{1}{4}$. These two curves intersect when the two are equal, so that

$$\left(x - \frac{1}{2}\right)^2 + y^2 = x^2 + \left(y - \frac{1}{2}\right)^2 \Rightarrow -x = -y \Rightarrow x = y;$$

then setting y to x in the first equation and simplifying gives $2x^2 - x = 0$, so that $x = 0$ and $x = \frac{1}{2}$. Since $y = x$, the two points of intersection are $(0, 0)$ and $\left(\frac{1}{2}, \frac{1}{2}\right)$, marked on the graph. In polar coordinates, these are the points $(0, 0)$ and $\left(\frac{\sqrt{2}}{2}, \frac{\pi}{4}\right)$.

43. (a) $f_1(\theta) = f_2(\theta)$ if and only if $1 + \sin\theta = 1 - \cos\theta$, so if and only if $\sin\theta = -\cos\theta$. this happens for $\theta = -\frac{\pi}{4} + n\pi$ for n an integer.

(b) Plots of the two curves, with $1 + \sin\theta$ the solid line, are

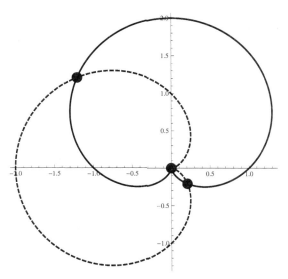

(c) Multiply the equations through by r to get $r^2 = r + r\sin\theta$ and $r^2 = r - r\cos\theta$; converting to rectangular coordinates gives

$$x^2 + y^2 = \sqrt{x^2 + y^2} + y, \quad x^2 + y^2 = \sqrt{x^2 + y^2} - x,$$

so that the two are equal for $y = -x$. Substitute $-x$ for y in the first equation to get $2y^2 = \sqrt{2y^2} + y$. Move the y to the left, square, and simplify to get $4y^4 - 4y^3 - y^2 = y^2(4y^2 - 4y - 1) = 0$. This has solutions $y = 0$ and $y = \frac{1}{2}(1 \pm \sqrt{2})$. Thus the points of intersection are $(0,0)$ and $\left(-\frac{1}{2}(1 \pm \sqrt{2}), \frac{1}{2}(1 \pm \sqrt{2})\right)$, which are marked on the graph. In polar coordinates, these are the points $(0,0)$, $\left(1 + \frac{\sqrt{2}}{2}, \frac{3\pi}{4}\right)$, and $\left(1 - \frac{\sqrt{2}}{2}, \frac{7\pi}{4}\right)$.

45. (a) $f_1(\theta) = f_2(\theta)$ if and only if $\sin^2\theta = \cos^2\theta$, so if and only if $\sin\theta = \pm\cos\theta$. This happens exactly when $\theta = \frac{\pi}{4}$ plus a multiple of $\frac{\pi}{2}$, i.e. when θ is an odd multiple of $\frac{\pi}{4}$.

(b) Plots of the two curves, with $\sin\theta$ the solid line, are

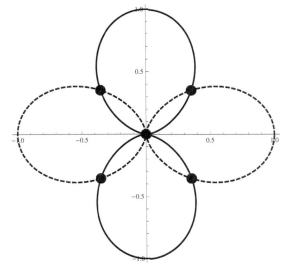

(c) To convert these equations to rectangular coordinates, multiply each through by r^2 to get $r^3 = (r\sin\theta)^2$ and $r^3 = (r\cos\theta)^2$, so that $\pm(x^2 + y^2)^{3/2} = y^2$ and $\pm(x^2 + y^2)^{3/2} = x^2$.

The points of intersection satisfy $x^2 = y^2$ (so that $y = \pm x$) Substituting x^2 for y^2 in the first equation gives $\pm(2x^2)^{3/2} = x^2$, or $\pm 2\sqrt{2}\,x^3 = x^2$. Thus $x = 0$ or $x = \pm\frac{1}{2\sqrt{2}}$. Hence the intersection points are $(0,0)$ together with the four points $\left(\pm\frac{1}{2\sqrt{2}}, \pm\frac{1}{2\sqrt{2}}\right)$, marked on the graph. In polar coordinates, these are the points $(0,0)$ and $\left(\frac{1}{2}, \frac{\pi}{4} + k\frac{\pi}{2}\right)$ for $k = 0, 1, 2, 3$.

47. The graphs are below, with $r = 1 + \sin\theta$ in black and $r = 1 - \sin\theta$ in gray.

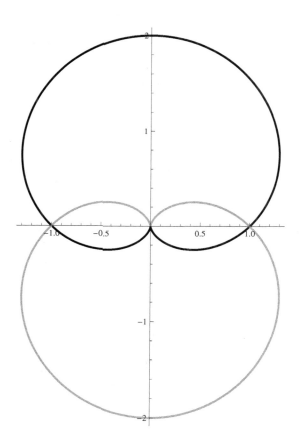

$1 - \sin\theta$ appears to be the reflection of $1 + \sin\theta$ through the x axis. To see this, note that $r = 1 + \sin\theta$ if and only if $r = 1 - \sin(-\theta)$, so that (r, θ) is on the graph of $1 + \sin\theta$ if and only if $(r, -\theta)$ is on the graph of $1 - \sin\theta$ — that is, the two are reflections of each other through the x axis (which is $\theta \leftrightarrow -\theta$).

49. Graphs of $r = 3 + b \sin \theta$ for various values of b are below:

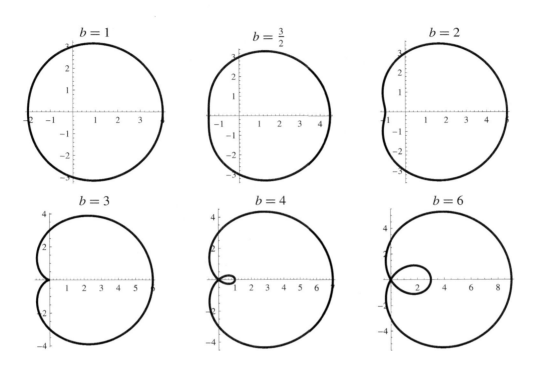

From these, it appears that for $b > 3$, the limaçon has an inner loop, for $\frac{3}{2} < b < 3$ it has a dimple, for $b = \frac{3}{2}$ the left side is flat, and for $b < \frac{3}{2}$ it is convex. For $b = 3$, of course, we get a cardioid, not a limaçon.

51. (a) Parametrize the circle by $(\cos \theta, \sin \theta)$. Then the midpoint of the chord extending from $(1,0)$ to $(\cos \theta, \sin \theta)$ is

$$\left(\frac{1 + \cos \theta}{2}, \frac{\sin \theta}{2} \right).$$

   (b) In rectangular coordinates, if the coordinates of a point on the unit circle are (x, y), the midpoint of the chord between such a point and $(1, 0)$ is

$$\left(\frac{x + 1}{2}, \frac{y}{2} \right).$$

   (c) Consider a point $B = (r, \theta)$ on the unit circle:

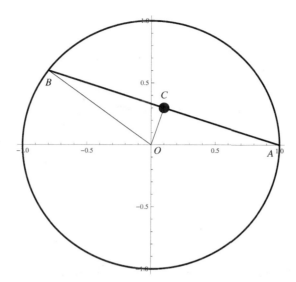

With the parametrization in part (a), the coordinates of the midpoint C are $\left(\frac{1+\cos\theta}{2}, \frac{\sin\theta}{2}\right)$, where $\theta = \angle AOB$. Then the length of OC, which is r in polar coordinates, is

$$r = \sqrt{\left(\frac{1+\cos\theta}{2}\right)^2 + \left(\frac{\sin\theta}{2}\right)^2} = \frac{1}{2}\sqrt{1 + 2\cos\theta + \cos^2\theta + \sin^2\theta} = \frac{1}{2}\sqrt{2 + 2\cos\theta}.$$

Also, since $\triangle AOB$ is an isosceles triangle and $AC = CB$ since C is the midpoint, we know that $\triangle OCB$ and $\triangle OCA$ are congruent, so that $\angle BOC = \angle AOC = \frac{1}{2}\angle AOB = \frac{\theta}{2}$. Thus the polar coordinates of point C are

$$\left(\frac{1}{2}\sqrt{2 + 2\cos\theta}, \frac{\theta}{2}\right).$$

Applications

53. When $A = 1$, this is the plot of $r = 1 - 0.5\sin\theta$:

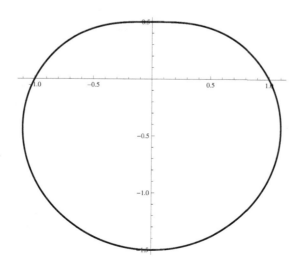

Proofs

55. Consider $r = \cos n\theta$. Note that $\cos(n(\theta + \pi)) = \cos(n\pi + n\theta)$; since n is odd, this is the same as $\cos(\pi + n\theta) = -\cos n\theta$. Thus the point on the graph associated with $\theta + \pi$ is $(-r, \theta + \pi) = (r, \theta)$. Hence every point is traced at least twice on $[0, 2\pi]$. But if $\alpha, \beta \in [0, \pi)$, with corresponding points on the graph (r_1, α) and (r_2, β), these points can be equal only if $\alpha = \beta$ plus a multiple of 2π or of π. But since α and β differ by less than π, they are equal. Thus $(r_1, \alpha) = (r_2, \beta)$ if and only if $\alpha = \beta$, so the curve is traced exactly once on $[0, \pi)$. Finally, to see that the curve has n petals, note that the tip of any petal corresponds to a point where $\cos n\theta = \pm 1$, so that θ must be a multiple of π. For $\theta \in [0, \pi)$, this happens at the points $\frac{k}{n}\pi$ for $k = 0, 1, \ldots, n-1$. Since there are n of these points, and no two are the same since the curve is traced only once, there are exactly n petals.

If instead $r = \sin n\theta$, note that $r = \sin n\left(\theta + \frac{\pi}{2}\right)$ is the graph of $\sin n\theta$ rotated clockwise by $\frac{\pi}{2}$ (see Exercise 6), so it has the same number of petals as $\sin n\theta$. But since n is odd, we have

$$\sin n\left(\theta + \frac{\pi}{2}\right) = \sin n\theta \cos \frac{n\pi}{2} + \cos n\theta \sin \frac{n\pi}{2} = \pm \cos n\theta.$$

Thus $r = \sin n\theta$ has the same number of petals as either $r = \cos n\theta$ or $r = -\cos n\theta$, both of which have n petals by the first part of this exercise.

57. Let G be the graph of $r = \sin n\theta$. Then since n is even, we have

$$(r, -\theta) \in G \Longleftrightarrow (-r, \pi - \theta) \in G \Longleftrightarrow -r = \sin(n(\pi - \theta)) \Longleftrightarrow r = -\sin(n\pi - n\theta)$$
$$\Longleftrightarrow r = -\sin(-n\theta) = \sin n\theta \Longleftrightarrow (r, \theta) \in G.$$

So by Theorem 9.11(a), the graph is symmetric about the x axis.

59. Let G be the graph of $r = \cos n\theta$. Then since n is even, we have

$$(-r, -\theta) \in G \Longleftrightarrow (r, \pi - \theta) \in G \Longleftrightarrow r = \cos(n(\pi - \theta)) = \cos(n\pi - n\theta)$$
$$\Longleftrightarrow r = \cos n\pi \cos n\theta + \sin n\pi \sin n\theta = \cos n\theta \Longleftrightarrow (r, \theta) \in G.$$

So by Theorem 9.11(b), G is symmetric about the y axis.

Thinking Forward

Volume of a cylinder

By Theorem 9.8, the cross-sectional circle is a circle of radius $r = 2$. The volume of a right circular cylinder is $\pi r^2 h$, so the volume of this cylinder is $\pi \cdot 2^2 \cdot 2 = 8\pi$.

Volume of a sphere

By Theorem 9.8, the cross-sectional circle is a circle of radius $r = 2$. The volume of a sphere is $\frac{4}{3}\pi r^3$, so the volume of this sphere is $\frac{4}{3}\pi \cdot 2^3 = \frac{32}{3}\pi$.

9.4   Arc Length and Area Using Polar Functions

Thinking Back

▷ Use the fact that $\cos^2 2\theta = \frac{1}{2}(1 + \cos 4\theta)$; then

$$\frac{1}{2}\int_0^{\pi/4} \cos^2 2\theta \, d\theta = \frac{1}{4}\int_0^{\pi/4}(1 + \cos 4\theta)\, d\theta = \frac{1}{4}\left[\theta + \frac{1}{4}\sin 4\theta\right]_0^{\pi/4} = \frac{\pi}{16}.$$

▷ Use the fact that $\cos^2 3\theta = \frac{1}{2}(1 + \cos 6\theta)$; then

$$\frac{1}{2}\int_0^\pi \cos^2 3\theta \, d\theta = \frac{1}{4}\int_0^\pi (1 + \cos 6\theta) \, d\theta = \frac{1}{4}\left[\theta + \frac{1}{6}\sin 6\theta\right]_0^\pi = \frac{\pi}{4}.$$

▷ Use the fact that $\cos^2 \theta = \frac{1}{2}(1 + \cos 2\theta)$; then

$$\begin{aligned}
\frac{1}{2}\int_0^{2\pi/3}\left(\frac{1}{2} + \cos\theta\right)^2 d\theta &= \frac{1}{2}\int_0^{2\pi/3}\left(\frac{1}{4} + \cos\theta + \cos^2\theta\right) d\theta \\
&= \frac{1}{2}\int_0^{2\pi/3}\left(\frac{1}{4} + \cos\theta + \frac{1}{2}(1 + \cos 2\theta)\right) d\theta \\
&= \frac{1}{2}\int_0^{2\pi/3}\left(\frac{3}{4} + \cos\theta + \frac{1}{2}\cos 2\theta\right) d\theta \\
&= \frac{1}{2}\left[\frac{3}{4}\theta + \sin\theta + \frac{1}{4}\sin 2\theta\right]_0^{2\pi/3} \\
&= \frac{1}{2}\left(\frac{3}{4}\cdot\frac{2\pi}{3} + \sin\frac{2\pi}{3} + \frac{1}{4}\sin\frac{4\pi}{3}\right) \\
&= \frac{1}{2}\left(\frac{\pi}{2} + \frac{\sqrt{3}}{2} - \frac{\sqrt{3}}{8}\right) \\
&= \frac{\pi}{4} + \frac{3\sqrt{3}}{16}.
\end{aligned}$$

▷ From the previous problem, the indefinite integral of the integrand is $\frac{3}{4}\theta + \sin\theta + \frac{1}{4}\sin 2\theta$, so we get

$$\begin{aligned}
\frac{1}{2}\int_\pi^{4\pi/3}\left(\frac{1}{2} + \cos\theta\right)^2 d\theta &= \frac{1}{2}\left[\frac{3}{4}\theta + \sin\theta + \frac{1}{4}\sin 2\theta\right]_\pi^{4\pi/3} \\
&= \frac{1}{2}\left(\frac{3}{4}\cdot\frac{4\pi}{3} + \sin\frac{4\pi}{3} + \frac{1}{4}\sin\frac{8\pi}{3} - \frac{3}{4}\pi\right) \\
&= \frac{1}{2}\left(\frac{1}{4}\pi - \frac{\sqrt{3}}{2} + \frac{\sqrt{3}}{8}\right) \\
&= \frac{\pi}{8} - \frac{3\sqrt{3}}{16}.
\end{aligned}$$

▷ Use the fact that $\cos^2 \theta = \frac{1}{2}(1 + \cos 2\theta)$; then

$$\begin{aligned}
\frac{1}{2}\int_0^{\pi/3}\left((3\cos\theta)^2 - (1 + \cos\theta)^2\right) d\theta &= \frac{1}{2}\int_0^{\pi/3}\left(8\cos^2\theta - 2\cos\theta - 1\right) d\theta \\
&= \frac{1}{2}\int_0^{\pi/3}(4 + 4\cos 2\theta - 2\cos\theta - 1) \, d\theta \\
&= \frac{1}{2}\left[3\theta + 2\sin 2\theta - 2\sin\theta\right]_0^{\pi/3} \\
&= \frac{1}{2}\left(\pi + 2\sin\frac{2\pi}{3} - 2\sin\frac{\pi}{3}\right) = \frac{\pi}{2}.
\end{aligned}$$

Concepts

1. (a) True. See the discussion following Theorem 9.12.

    (b) True. $\Delta\theta = \frac{\frac{\pi}{2}-\frac{\pi}{4}}{4} = \frac{\frac{\pi}{4}}{4} = \frac{\pi}{16}$.

    (c) False. By Theorem 9.13, it is given by $\frac{1}{2}\int_a^b (f(\theta))^2 \, d\theta$.

    (d) True. Here we are graphing $r = f(\theta)$ in a rectangular coordinate plane, so formulas for that situation apply, and the area between the curve and the θ axis is $\int_a^b |f(\theta)| \, d\theta$.

    (e) False. $r = 2\cos\theta$ traces this circle twice from 0 to 2π, since $2\cos(\pi+\theta) = -2\cos\theta$, so that the point $(-r, \pi+\theta) = (r, \theta)$ is traced for θ and for $\theta + \pi$. Thus the integral is twice the area of the circle, so its value is 2π.

    (f) False. Since $\cos(4(\theta+\pi)) = \cos(4\theta + 4\pi) = \cos(4\theta)$, the points (r, θ) and $(r, \pi+\theta)$ are both on the curve, and are different. So the curve is only traced once.

    (g) True. If $r = f(\theta)$, then $x = r\cos\theta = f(\theta)\cos\theta$ and $y = r\sin\theta = f(\theta)\sin\theta$.

    (h) False. This is only true if $f(\theta)$ is one-to-one on $[0, 2\pi]$; otherwise parts of the curve are traced more than once.

3. Since we are approximating area, which is summing up areas from the origin to the curve, the area between a small segment of the curve and the origin is a small triangle, which approximates a small wedge. This is the analog of the situation in rectangular coordinates where the area between a small segment of the curve and the x axis is a small rectangle.

5. A picture is:

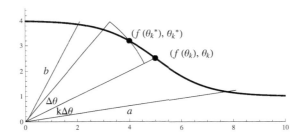

    The circular sector is bounded by the rays at angles $a + k\Delta\theta = \theta_k$ and $a + (k+1)\Delta\theta = \theta_{k+1}$, and we have chosen an angle θ_k^* between the two rays to use as the value of the function throughout that sector.

7. If $\beta - \alpha > 2\pi$, then the region bounded by the rays $\theta = \alpha$ and $\theta = \beta$ is the same region as that bounded by $\theta = \alpha$ and $\theta = \beta - 2\pi$. If we use α and β to compute the integral, we will be tracing the region multiple times.

9. If a graph is symmetric about the x axis, then the area need only be calculated for the region above the x axis, and then doubled. Similarly, if a graph is symmetric about the y axis, then we need only double the calculation of the area of the region to the right of the y axis. Finally, if a graph is symmetric about both axes, it suffices to calculate the area in the first quadrant and multiply by 4. Depending on the specifics of the graph, there may be further simplifications (for example, Example 1 discussed a situation in which we only needed to integrate from 0 to $\frac{\pi}{4}$ and then multiply by 8).

11. Using t as the parameter, let $x(t) = t$; then $y(t) = f(t)$.

13. First, this approach fails since $r = \sin \theta$ is not one-to-one on $[0, 2\pi]$ — the circle is traced twice on this interval. So his integral should have gone from 0 to π. So he gets a result that is twice the correct result. His second mistake, though, is that $r = \sin \theta$ is not a unit circle — it is a circle of radius $\frac{1}{2}$. Thus its true circumference is π. So he got an answer that was twice the correct circumference of a smaller circle. The two errors cancelled to give apparently a correct result for a unit circle.

15. A plot of the region is shown below:

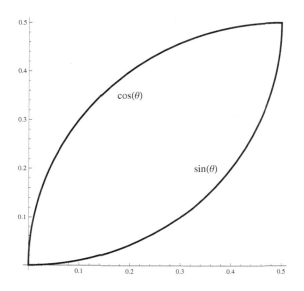

This region is clearly symmetric around the line $\theta = \frac{\pi}{4}$, since $\cos\left(\frac{\pi}{2} - \theta\right) = \sin\theta$. So it suffices to find the area under the polar curve $r = \sin\theta$ from 0 to $\frac{\pi}{4}$ and double it. Thus the required area is

$$2 \cdot \frac{1}{2} \int_0^{\pi/4} \sin^2\theta \, d\theta = \int_0^{\pi/4} \frac{1}{2}(1 - \cos 2\theta) \, d\theta = \frac{1}{2}\left[\theta - \frac{1}{2}\sin 2\theta\right]_0^{\pi/4} = \frac{\pi}{8} - \frac{1}{4}.$$

Skills

17. For $0 \le \theta \le \pi$, the graph of $r = \theta$ lies above the x axis, so the area between the curve and the x axis is the area of the polar region bounded by $\theta = 0$ and $r = \theta$, which is

$$\frac{1}{2}\int_0^\pi \theta^2 \, d\theta = \frac{1}{2}\left[\frac{1}{3}\theta^3\right]_0^\pi = \frac{\pi^3}{6}.$$

19. A graph of the limaçon is:

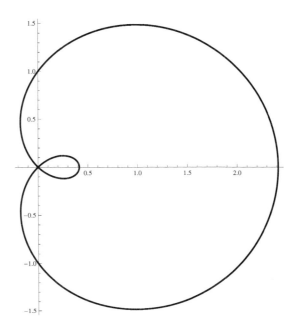

We need only determine the region between the two loops above the x axis and then double it. The rightmost point on the curve corresponds to $\theta = 0$; the curve then loops up and returns to the origin where $\cos\theta = -\frac{1}{\sqrt{2}}$, which is at $\theta = \frac{3\pi}{4}$. So the area under the curve between those two points is the area of the upper half of the outer loop. The upper half of the inner loop corresponds to $\pi \leq \theta \leq \frac{5\pi}{4}$. So we want to compute

$$2\left(\frac{1}{2}\int_0^{3\pi/4}(1+\sqrt{2}\cos\theta)^2\,d\theta - \frac{1}{2}\int_\pi^{5\pi/4}(1+\sqrt{2}\cos\theta)^2\,d\theta\right)$$

$$= \int_0^{3\pi/4}(1+\sqrt{2}\cos\theta)^2\,d\theta - \int_\pi^{5\pi/4}(1+\sqrt{2}\cos\theta)^2\,d\theta$$

$$= \int_0^{3\pi/4}(1+2\sqrt{2}\cos\theta+2\cos^2\theta)\,d\theta - \int_\pi^{5\pi/4}(1+2\sqrt{2}\cos\theta+2\cos^2\theta)\,d\theta$$

$$= \left[\theta+2\sqrt{2}\sin\theta+\left(\theta+\frac{1}{2}\sin 2\theta\right)\right]_0^{3\pi/4} - \left[\theta+2\sqrt{2}\sin\theta+\left(\theta+\frac{1}{2}\sin 2\theta\right)\right]_\pi^{5\pi/4}$$

$$= \left[2\theta+2\sqrt{2}\sin\theta+\frac{1}{2}\sin 2\theta\right]_0^{3\pi/4} - \left[2\theta+2\sqrt{2}\sin\theta+\frac{1}{2}\sin 2\theta\right]_\pi^{5\pi/4}$$

$$= \left(\frac{3\pi}{2}+2-\frac{1}{2}\right) - \left(\frac{5\pi}{2}-2+\frac{1}{2}-(2\pi+0+0)\right)$$

$$= 3+\pi.$$

21. Graphs of the two cardioids (with $3 - 3\sin\theta$ in black) are

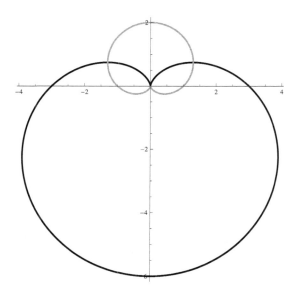

The curves intersect at the origin. The other two intersection points may be found by solving $3 - 3\sin\theta = 1 + \sin\theta$, which gives $4\sin\theta = 2$, so that $\theta = \frac{\pi}{6}$ and $\frac{5\pi}{6}$. It suffices to compute the area to the right of the vertical axis that is inside the larger cardioid and outside the smaller, and then double it. This area is found by integrating from $-\frac{\pi}{2}$ to $\frac{\pi}{6}$, and doubling it.

$$
\begin{aligned}
2 \cdot \frac{1}{2} \int_{-\pi/2}^{\pi/6} \left((3 - 3\sin\theta)^2 - (1 + \sin\theta)^2 \right) d\theta &= \int_{-\pi/2}^{\pi/6} (8 - 20\sin\theta + 8\sin^2\theta) \, d\theta \\
&= \left[8\theta + 20\cos\theta + (4\theta - 2\sin 2\theta) \right]_{-\pi/2}^{\pi/6} \\
&= \left[12\theta + 20\cos\theta - 2\sin 2\theta \right]_{-\pi/2}^{\pi/6} \\
&= \left(2\pi + 10\sqrt{3} - \sqrt{3} \right) - (-6\pi + 0 - 0) \\
&= 8\pi + 9\sqrt{3}.
\end{aligned}
$$

23. The region is shown below:

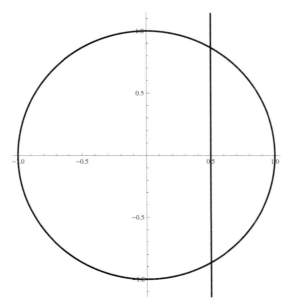

In polar coordinates, the two functions are $r = 1$ and $r = \frac{1}{2}\sec\theta$, so the two intersect where $\sec\theta = 2$, which is at $\theta = \pm\frac{\pi}{3}$. By symmetry, it suffices to find the area of the region between the two above the horizontal axis and double it, so for the range $0 \le \theta \le \frac{\pi}{3}$. Thus the area we want is

$$2 \cdot \frac{1}{2} \int_0^{\pi/3} \left(1^2 - \frac{1}{4}\sec^2\theta\right) d\theta = \int_0^{\pi/3} \left(1 - \frac{1}{4}\sec^2\theta\right) d\theta = \left[\theta - \frac{1}{4}\tan\theta\right]_0^{\pi/3}$$

$$= \frac{\pi}{3} - \frac{1}{4}\tan\frac{\pi}{3} = \frac{\pi}{3} - \frac{\sqrt{3}}{4}.$$

25. The curve looks like

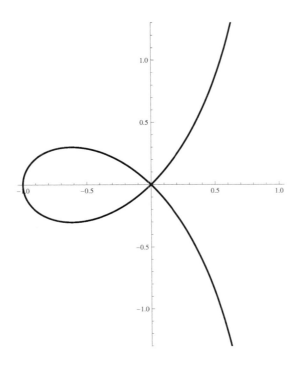

The vertical axis range has been restricted in order to show clearly the loop. With a larger vertical range, it would be clear that the equation has a vertical asymptote at $x = 1$ (this is not relevant to the problem, but is an interesting fact). The curve passes through the origin when $\sec\theta = 2\cos\theta$. Multiplying through by $\sec\theta$ gives $1 = 2\cos^2\theta$, so that in the range $-\frac{\pi}{2} < \theta < \frac{\pi}{2}$ we get $\theta = \pm\frac{\pi}{4}$. So the area enclosed by the loop is

$$
\begin{aligned}
\frac{1}{2}\int_{-\pi/4}^{\pi/4}(\sec\theta - 2\cos\theta)^2\, d\theta &= \frac{1}{2}\int_{-\pi/4}^{\pi/4}(\sec^2\theta + 4\cos^2\theta - 4)\, d\theta \\
&= \frac{1}{2}\left[\tan\theta + (2\theta + \sin 2\theta) - 4\theta\right]_{-\pi/4}^{\pi/4} \\
&= \frac{1}{2}\left[\tan\theta + \sin 2\theta - 2\theta\right]_{-\pi/4}^{\pi/4} \\
&= \frac{1}{2}\left(\left(1 + 1 - \frac{\pi}{2}\right) - \left(-1 - 1 + \frac{\pi}{2}\right)\right) \\
&= 2 - \frac{\pi}{2}.
\end{aligned}
$$

27. Since the function is one-to-one on its domain, the arc length is

$$
\int_0^{2\pi}\sqrt{((e^\theta)')^2 + (e^\theta)^2}\, d\theta = \int_0^{2\pi}\sqrt{2e^{2\theta}}\, d\theta = \sqrt{2}\int_0^{2\pi}e^\theta\, d\theta = \sqrt{2}\left[e^\theta\right]_0^{2\pi} = \sqrt{2}(e^{2\pi} - 1).
$$

29. Since the function is one-to-one on its domain, the arc length is

$$
\begin{aligned}
\int_0^{2\pi}\sqrt{((e^{\alpha\theta})')^2 + (e^{\alpha\theta})^2}\, d\theta &= \int_0^{2\pi}\sqrt{(\alpha^2 + 1)e^{2\alpha\theta}}\, d\theta = \sqrt{\alpha^2 + 1}\int_0^{2\pi}e^{\alpha\theta}\, d\theta \\
&= \frac{\sqrt{\alpha^2 + 1}}{\alpha}\left[e^{\alpha\theta}\right]_0^{2\pi} = \frac{\sqrt{\alpha^2 + 1}}{\alpha}(e^{2\alpha\pi} - 1).
\end{aligned}
$$

31. One loop of the rose is traced for $-\frac{\pi}{4} \le \theta \le \frac{\pi}{4}$, since $r = 0$ at each of these points. With $r = \cos 2\theta$ we have $r' = -2\sin 2\theta$, so that the arc length is

$$
\int_{-\pi/4}^{\pi/4}\sqrt{(-2\sin 2\theta)^2 + \cos^2 2\theta}\, d\theta = \int_{-\pi/4}^{\pi/4}\sqrt{4\sin^2 2\theta + \cos^2 2\theta}\, d\theta \approx 2.42211.
$$

33. One loop of the rose is traced for $-\frac{\pi}{8} \le \theta \le \frac{\pi}{8}$, since $r = 0$ at each of these points. With $r = \cos 4\theta$ we have $r' = -4\sin 4\theta$, so that the arc length is

$$
\int_{-\pi/8}^{\pi/8}\sqrt{(-4\sin 4\theta)^2 + \cos^2 4\theta}\, d\theta = \int_{-\pi/8}^{\pi/8}\sqrt{16\sin^2 4\theta + \cos^2 4\theta}\, d\theta \approx 2.14461.
$$

35. The limaçon passes through the origin when $r = 1 + 2\sin\theta = 0$, which is when $\sin\theta = -\frac{1}{2}$, so for $\theta = \frac{7\pi}{6}$ and $\frac{11\pi}{6}$. Thus the arc length of the inner loop is

$$
\int_{7\pi/6}^{11\pi/6}\sqrt{(2\cos\theta)^2 + (1 + 2\sin\theta)^2}\, d\theta = \int_{7\pi/6}^{11\pi/6}\sqrt{5 + 4\sin\theta}\, d\theta \approx 2.68245.
$$

37. We have

$$\int_0^{2\pi/3} \left(\frac{1}{2} + \cos\theta\right)^2 d\theta - \int_\pi^{4\pi/3} \left(\frac{1}{2} + \cos\theta\right)^2 d\theta$$

$$= \int_0^{2\pi/3} \left(\frac{1}{4} + \cos\theta + \cos^2\theta\right) d\theta - \int_\pi^{4\pi/3} \left(\frac{1}{4} + \cos\theta + \cos^2\theta\right) d\theta$$

$$= \left[\frac{\theta}{4} + \sin\theta + \frac{\theta}{2} + \frac{1}{4}\sin 2\theta\right]_0^{2\pi/3} - \left[\frac{\theta}{4} + \sin\theta + \frac{\theta}{2} + \frac{1}{4}\sin 2\theta\right]_\pi^{4\pi/3}$$

$$= \left[\frac{3\theta}{4} + \sin\theta + \frac{1}{4}\sin 2\theta\right]_0^{2\pi/3} - \left[\frac{3\theta}{4} + \sin\theta + \frac{1}{4}\sin 2\theta\right]_\pi^{4\pi/3}$$

$$= \left(\frac{3\sqrt{3}}{8} + \frac{\pi}{2}\right) - 0 - \left(\left(-\frac{3\sqrt{3}}{8} + \pi\right) - \frac{3\pi}{4}\right)$$

$$= \frac{\pi}{4} + \frac{3\sqrt{3}}{4}.$$

39. Here the function $f(\theta)$ is $1 + \cos\theta$, so this is twice the area enclosed by $r = 1 + \cos\theta$ for $0 \le \theta \le \pi$, which is the area enclosed by $r = 1 + \cos\theta$ for $0 \le \theta \le 2\pi$ (this is a cardioid):

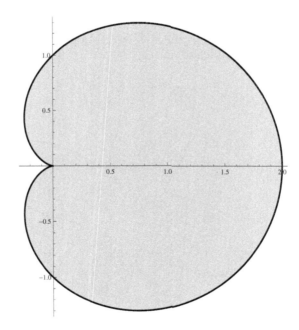

The area of the cardioid is the value of the integral:

$$\int_0^\pi (1 + \cos\theta)^2 d\theta = \int_0^\pi \left(1 + 2\cos\theta + \cos^2\theta\right) d\theta = \left[\frac{3\theta}{2} + 2\sin\theta + \frac{1}{4}\sin 2\theta\right]_0^\pi = \frac{3\pi}{2}.$$

41. The function $f(\theta)$ is $\sqrt{\sin 2\theta}$, so that this is twice the area enclosed by $r = \sqrt{\sin 2\theta}$ for $0 \le \theta \le \frac{\pi}{2}$, which is the area of the curve enclosed by $r = \sqrt{\sin 2\theta}$ for $0 \le \theta \le 2\pi$ (note that actually the valid range for θ is $0 \le \theta \le \frac{\pi}{2}$ and $\pi \le \theta \le \frac{3\pi}{2}$):

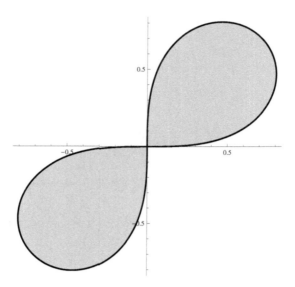

The area is the value of the integral:

$$\int_0^{\pi/2} \sin 2\theta \, d\theta = \left[-\frac{1}{2} \cos 2\theta \right]_0^{\pi/2} = 1.$$

43. The function $f(\theta)$ is $2 + \sin 4\theta$. This curve is traced once for $0 \le \theta \le 2\pi$, so that this integral represents the area enclosed by the curve:

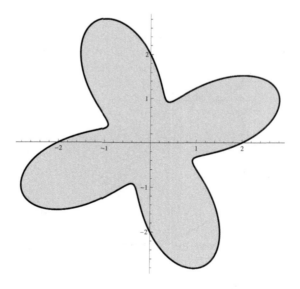

The area is

$$\frac{1}{2} \int_0^{2\pi} (2 + \sin 4\theta)^2 \, d\theta = \frac{1}{2} \int_0^{2\pi} (4 + 4\sin 4\theta + \sin^2 4\theta) \, d\theta$$

$$= \frac{1}{2} \left[4\theta - \cos 4\theta + \frac{\theta}{2} - \frac{1}{16} \sin 8\theta \right]_0^{2\pi}$$

$$= \frac{9\pi}{2}.$$

45. Since $r = f(\theta) = a$, the area of the circle is

$$\frac{1}{2} \int_0^{2\pi} a^2\, d\theta = \frac{1}{2} \left[a^2\theta\right]_0^{2\pi} = \pi a^2.$$

47. Since $r = f(\theta) = a$, $r'(\theta) = 0$, so that the circumference of the circle is just the arc length from 0 to 2π, which is

$$\int_0^{2\pi} \sqrt{r'(\theta)^2 + r(\theta)^2}\, d\theta = \int_0^{2\pi} a\, d\theta = [a\theta]_0^{2\pi} = 2\pi a.$$

49. The circle $r = a$ is the circle with radius a and center at the origin; the circle $r = 2a \cos\theta$ is the circle with radius a and center at $(a, 0)$, so that each circle passes through the center of the other. A diagram of this situation (with $a = 1$) is:

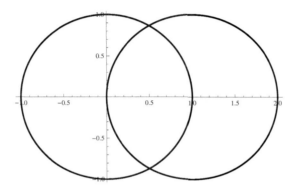

The upper point of intersection of the circles is at $\theta = \frac{\pi}{3}$, since $2a \cos\frac{\pi}{3} = a$, which is on the circle $r = a$ as well. We would like to find the area by integrating the squared difference between the two from 0 to $\frac{\pi}{3}$, but that does not work, as it fails to include the portion of the right-hand circle from $\frac{\pi}{3}$ to $\frac{\pi}{2}$. Instead, use the following approach: the area that is inside both circles is the entire area of the right-hand circle minus the area that is inside $r = 2a \cos\theta$ but outside $r = a$. This we *can* compute by finding the area between the curves from 0 to $\frac{\pi}{3}$ (and doubling it to get the region below the x axis as well), so the area is

$$\pi a^2 - 2 \cdot \frac{1}{2} \int_0^{\pi/3} \left((2a\cos\theta)^2 - a^2\right) d\theta = \pi a^2 - a^2 \int_0^{\pi/3} (4\cos^2\theta - 1)\, d\theta$$

$$= \pi a^2 - a^2\left[\theta + \sin 2\theta\right]_0^{\pi/3} = a^2\left(\frac{2\pi}{3} - \frac{\sqrt{3}}{2}\right).$$

Applications

51. As in the previous exercise, we must compute the arc length of this curve from 0 to 2π. With $f(\theta) = 0.83(1 - 0.5\sin\theta)(1 + 0.2\cos^2\theta)$ we get

$$f'(\theta) = \cos\theta(0.166\sin^2\theta - 0.332\sin\theta - 0.083\cos\theta^2 - 0.0415),$$

so that

$$\int_0^{2\pi} \sqrt{f(\theta)^2 + f'(\theta)^2}\, d\theta \approx 6.149 \text{ feet.}$$

Since

$$2 \text{ meters} \approx 2 \cdot \frac{39.37 \text{ in/m}}{12 \text{ in/ft}} \approx 6.562 \text{ feet,}$$

she now can cover the kayak with two widths of the fabric.

Proofs

53. Let the circle be $r = a$. Then the area of the sector from $\theta = 0$ to $\theta = \phi$ is given, by Theorem 9.13, by

$$\frac{1}{2} \int_0^\phi a^2 \, d\theta = \frac{1}{2} \left[\theta a^2 \right]_0^\phi = \frac{1}{2} \phi a^2.$$

Unfortunately, we used the area of a sector in deriving that integral formula, so that proof may not be too satisfying. A more geometric proof is the following: Clearly the fraction of the area of the entire circle that is in a sector with central angle ϕ is $\frac{\phi}{2\pi}$. Since the area of the entire circle is πr^2, the area of a sector with central angle ϕ is $\frac{\phi}{2\pi} \cdot \pi r^2 = \frac{1}{2} \phi r^2$.

55. Note that $r = \cos 2n\theta$ has $4n$ petals whose tips are separated by an angle of $\frac{2\pi}{4n}$, and that

$$\cos \left(2n \left(\theta - \frac{2\pi}{4n} \right) \right) = \cos \left(2n\theta - \pi \right) = -\cos 2n\theta.$$

By Exercise 6 in Section 9.3, the graph of $r = \cos \left(2n \left(\theta - \frac{2\pi}{4n} \right) \right)$ is the graph of $r = -\cos 2n\theta$ rotated counterclockwise through $\frac{2\pi}{4n}$. But by Exercises 59 and 60 in Section 9.3, $\cos 2n\theta$ is symmetric about both the x and y axes, so it is symmetric around the origin. Thus the graph of $r = -\cos 2n\theta$ is the same as the graph of $r = \cos 2n\theta$, so that the rose $r = \cos 2n\theta$ is moved on top of itself by a rotation through $\frac{2\pi}{4n}$. But since the graph has $4n$ petals, this rotation moves each petal on top of another petal. Since rotations do not change area, the areas of all the petals must be the same.

57. With $f(\theta) = \frac{1}{\theta}$ we get $f'(\theta) = -\frac{1}{\theta^2}$. The portion of the curve inside $r = 1$ is the portion for $\theta \geq 1$, so that the length is

$$\int_1^\infty \sqrt{f(\theta)^2 + f'(\theta)^2} \, d\theta = \lim_{t \to \infty} \int_1^t \sqrt{\frac{1}{\theta^2} + \frac{1}{\theta^4}} \, d\theta = \lim_{t \to \infty} \int_1^t \frac{1}{\theta^2} \sqrt{\theta^2 + 1} \, d\theta.$$

To integrate, use integration by parts with $u = \sqrt{\theta^2 + 1}$ and $dv = \frac{1}{\theta^2} \, d\theta$; then $du = \frac{\theta}{\sqrt{\theta^2+1}}$ and $v = -\frac{1}{\theta}$, so that we get

$$\lim_{t \to \infty} \int_1^t \frac{1}{\theta^2} \sqrt{\theta^2 + 1} \, d\theta = \lim_{t \to \infty} \left(\left[-\frac{\sqrt{\theta^2+1}}{\theta} \right]_1^t + \int_1^t \frac{1}{\sqrt{\theta^2+1}} \, d\theta \right)$$

$$= \lim_{t \to \infty} \left(\sqrt{2} - \frac{\sqrt{t^2+1}}{t} + \left[\sinh^{-1}\theta \right]_1^t \right)$$

$$= \lim_{t \to \infty} \left(\sqrt{2} - \frac{\sqrt{t^2+1}}{t} + \sinh^{-1}t - \sinh^{-1}1 \right)$$

$$= \infty$$

since the first and fourth terms are constants, the second term goes to 1, and the third term grows without bound. Since the integral diverges to ∞, the curve has infinite length.

Thinking Forward

▷ The area of a right cylinder is the area of its base times its height. Here the base is the cardioid $r = 1 + \cos\theta$, which has area

$$\frac{1}{2} \int_0^{2\pi} (1 + \cos\theta)^2 \, d\theta = \frac{1}{2} \int_0^{2\pi} (1 + 2\cos\theta + \cos^2\theta) \, d\theta$$

$$= \frac{1}{2} \left[\frac{3}{2}\theta + 2\sin\theta + \frac{1}{4}\sin 2\theta \right]_0^{2\pi} = \frac{3\pi}{2}.$$

Since the height is 1 unit, its volume is $\frac{3\pi}{2}$.

▷ The surface area is twice the area of the base (for the two ends) plus the height times the circumference of the base. To compute the circumference of the base, we have $r(\theta) = 1 + \cos\theta$, so that $r'(\theta) = -\sin\theta$, and the circumference is

$$\int_0^{2\pi} \sqrt{(1 + \cos\theta)^2 + (-\sin\theta)^2}\, d\theta = \int_0^{2\pi} \sqrt{2 + 2\cos\theta}\, d\theta.$$

This integral arose in Example 4, and was evaluated in Exercise 16 to equal 8. So the circumference of the base is 8; since the height is 1, the lateral surface area is 8, so that the total surface area is

$$2 \cdot \frac{3\pi}{2} + 8 = 8 + 3\pi.$$

9.5   Conic Sections

Thinking Back

Finding the center and radius for a circle by completing the square

▷ We have $x^2 + y^2 - y - \frac{3}{4} = x^2 + \left(y^2 - y + \frac{1}{4}\right) - \frac{1}{4} - \frac{3}{4} = 0$, so that $x^2 + \left(y - \frac{1}{2}\right)^2 = 1$. This is a circle of radius 1 centered at $\left(0, \frac{1}{2}\right)$.

▷ We have $x^2 + 4x + y^2 - 21 = (x^2 + 4x + 4) + y^2 - 4 - 21 = 0$, so that $(x + 2)^2 + y^2 = 25$. This is a circle of radius 5 centered at $(-2, 0)$.

▷ We have $x^2 + 3x + y^2 + y + \frac{3}{2} = (x^2 + 3x + \frac{9}{4}) + \left(y^2 + y + \frac{1}{4}\right) + \frac{3}{2} - \frac{9}{4} - \frac{1}{4} = 0$, so that $\left(x + \frac{3}{2}\right)^2 + \left(y + \frac{1}{2}\right)^2 = 1$. This is a circle of radius 1 centered at $\left(-\frac{3}{2}, -\frac{1}{2}\right)$.

Completing the square

▷ Complete the square to get $(x - 3)^2 + (y + 1)^2 = -10 + 9 + 1 = 0$. The only point satisfying the equation is the point $(3, -1)$.

▷ Complete the square to get $(x - 4)^2 + (y - 5)^2 = -40 + 25 + 16 = 1$. This is the circle of radius 1 centered at $(4, 5)$, so those are the points that satisfy the equation.

▷ Rearrange terms to get $y = -x^2 + 6x - 40$. This is an upward-opening parabola; the points on that parabola satisfy the equation.

▷ Rearrange terms to get $x = -y^2 + 4y + 5$. This is a leftward-opening parabola; the points on that parabola satisfy the equation.

Concepts

1. (a) True. See the discussion at the beginning of the chapter.

   (b) True. A point is a degenerate conic section.

   (c) False. An ellipse is the set of points in the plane the *sum* of whose distances from two distinct points is a constant. See Definition 9.15.

   (d) True. See Definition 9.17.

(e) False. Parabolas and hyperbolas have different definitions and different shapes; see Definitions 9.17 and 9.19.

(f) True. The set of all points equidistant from a line $\mathcal{L}$ and a point P on that line is the set of points on a line perpendicular to $\mathcal{L}$ through P.

(g) False. A hyperbola is the set of points in the plane the *difference* of whose distances from two distinct points is a constant. See Definition 9.19.

(h) True. The eccentricity of $\frac{1}{1+\sin\theta} = \frac{eu}{1+e\sin\theta}$ is clearly 1, so this is a parabola.

3. Given two points in the plane, called foci, an ellipse is the set of points for which the sum of the distances from the foci is a constant.

5. Given two points in a plane, called foci, a hyperbola is the set of points in the plane for which the difference between the distances to the foci is a constant.

7. Since a hyperbola is the set of points the difference of whose distances from two foci is a constant, by varying the constant we can get different hyperbolas. For example, with the foci $(1,0)$ and $(-1,0)$, if the constant difference is 1, then $\left(\frac{1}{2},0\right)$ is on the hyperbola, while if the constant difference is $\frac{2}{3}$, then $\left(\frac{1}{3},0\right)$ is on the hyperbola.

9. (a) By Definition 9.21 (b), the eccentricity is $\frac{\sqrt{A^2-B^2}}{A}$.

    (b) By Definition 9.21 (b), the eccentricity is $\frac{\sqrt{B^2-A^2}}{B}$.

    (c) In either of the above cases, the numerator is the square root of some number that is smaller than the denominator. Thus the fraction is always less than 1. If $A \neq B$, then the numerator is nonzero, so the eccentricity is greater than 0. Finally, both numerator and denominator are positive. Thus $0 < e < 1$.

    (d) $\lim\limits_{A \to B} \frac{\sqrt{A^2-B^2}}{B} = \frac{\sqrt{B^2-B^2}}{B} = 0$. The limit of the eccentricity is zero. As A approaches B, the length of the two axes approach each other, so the conic approaches a circle.

    (e) $\lim\limits_{A \to \infty} \frac{\sqrt{A^2-B^2}}{B} = \infty$. As A gets larger, the ellipse gets longer and longer; it gets arbitrarily stretched out, and when viewed in a large enough scale, becomes indistinguishable from a line.

11. Rewrite the equation as $\frac{y^2}{B^2} + \frac{x^2}{A^2} = 1$ with $B > A$; by Definition 9.22 (with x and y reversed) tells us that the directrices of the ellipse are $y = \pm\frac{B}{e}$.

13. Graphs of the two equations are

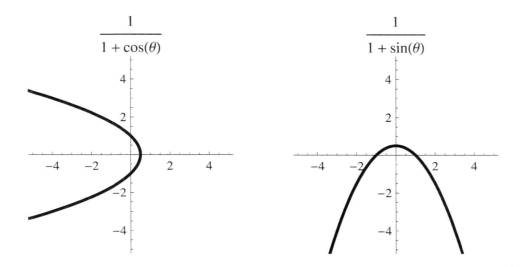

$$\frac{1}{1+\cos(\theta)} \qquad\qquad \frac{1}{1+\sin(\theta)}$$

Since the eccentricity (the coefficient of $\sin\theta$ or $\cos\theta$) is 1 in both cases, these are both parabolas, with focus at the origin. From the graphs, one is vertically oriented and the other is horizontally oriented. They are rotated versions of one another.

15. Graphs of the two equations are

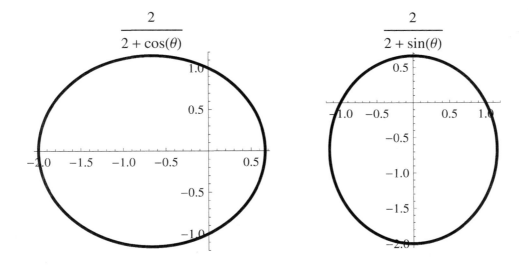

$$\frac{2}{2+\cos(\theta)} \qquad\qquad \frac{2}{2+\sin(\theta)}$$

Rewriting these equations in standard form gives $r = \frac{1}{1+\frac{1}{2}\cos\theta}$ and $r = \frac{1}{1+\frac{1}{2}\sin\theta}$. Since the eccentricity (the coefficient of $\sin\theta$ or $\cos\theta$) is $\frac{1}{2}$ in both cases, these are both ellipses. From the graphs, one is vertically oriented and the other is horizontally oriented. They are rotated versions of one another.

17. Divide top and bottom of the right hand side by β to get $r = \frac{\frac{\alpha}{\beta}}{1+\frac{\gamma}{\beta}\sin\theta}$. Thus the eccentricity is $\left|\frac{\gamma}{\beta}\right|$.

The directrix (or directrices) are $y = \frac{\alpha}{\beta} \cdot \frac{\beta}{\gamma} = \frac{\alpha}{\gamma}$.

Skills

19. We get

$$2x^2 + 4x - y^2 - 6y - 23 = (2x^2 + 4x + 2) - (y^2 + 6y + 9) - 23 - 2 + 9$$
$$= 2(x+1)^2 - (y+3)^2 - 16 = 0,$$

so that

$$2(x+1)^2 - (y+3)^2 = 16, \qquad \text{or} \qquad \frac{(x+1)^2}{8} - \frac{(y+3)^2}{16} = 1.$$

This is a hyperbola with foci $\left(-1 \pm \sqrt{8+16}, -3\right) = \left(-1 \pm 2\sqrt{6}, -3\right)$ and asymptotes $y + 3 = \pm\sqrt{2}(x+1)$, or $y = \pm\sqrt{2}(x+1) - 3$.

21. We get

$$y^2 - 8y - 4x^2 - 8x - 13 = (y^2 - 8y + 16) - 4(x^2 + 2x + 1) - 13 - 16 + 4 = 0,$$

so that

$$(y-4)^2 - 4(x+1)^2 = 25, \qquad \text{or} \qquad \frac{(y-4)^2}{25} - \frac{(x+1)^2}{25/4} = 1.$$

This is a hyperbola with foci $\left(-1, 4 \pm \sqrt{25 + \frac{25}{4}}\right) = \left(-1, 4 \pm \frac{5}{2}\sqrt{5}\right)$ and asymptotes $y - 4 = \pm 2(x+1)$, or $y = 2x + 6$ and $y = -2x + 2$.

23. Using Theorem 9.18, since the directrix is $y = 0 = y_0 - \frac{1}{4a}$ and the focus is $(0,1) = \left(x_0, y_0 + \frac{1}{4a}\right)$, we have $x_0 = 0$ and $y_0 = \frac{1}{4a}$, so that $2y_0 = 1$ and $y_0 = \frac{1}{4a} = \frac{1}{2}$. Thus $a = \frac{1}{2}$, and the equation is $y = \frac{1}{2}x^2 + \frac{1}{2}$.

25. Use Theorem 9.18. Since the directrix is $y = -6 = y_0 - \frac{1}{4a}$ and the focus is $(2, -8) = \left(x_0, y_0 + \frac{1}{4a}\right)$, we have $x_0 = 2$, $y_0 - \frac{1}{4a} = -6$, and $y_0 + \frac{1}{4a} = -8$. Thus $y_0 = -7$ and $a = -\frac{1}{4}$, so the equation is $y = -\frac{1}{4}(x-2)^2 - 7$.

27. Using Theorem 9.18, we construct an equation $y = a(x - x')^2 + y'$. Since the directrix is $y = y_0 = y' - \frac{1}{4a}$ and the focus is $(x_1, y_1) = \left(x', y' + \frac{1}{4a}\right)$, we get $x' = x_1$, $y' - \frac{1}{4a} = y_0$, and $y' + \frac{1}{4a} = y_1$, so that $y' = \frac{y_0 + y_1}{2}$ and $a = \frac{1}{2(y_1 - y_0)}$. Thus the equation is

$$y = \frac{1}{2(y_1 - y_0)}(x - x_1)^2 + \frac{y_0 + y_1}{2}.$$

29. By Theorem 9.24, the focus is at the pole; since the numerator is $4 = eu$ and the denominator uses a sin rather than a cos, the directrix is $y = 4$. Then by Exercise 27 we get for the equation

$$y = \frac{1}{2(0-4)}(x-0)^2 + 2 = -\frac{1}{8}x^2 + 2.$$

31. By Theorem 9.24, the focus is at the pole; since the numerator is $\alpha = eu$ and the denominator uses sine rather than cosine, the directrix is $y = \alpha$. Then by Exercise 27 we get for the equation

$$y = \frac{1}{2(0-\alpha)}(x-0)^2 + \frac{\alpha}{2} = -\frac{1}{2\alpha}x^2 + \frac{\alpha}{2}.$$

33. Since the foci are on the line $y = 1$, the ellipse is horizontally oriented, and is the ellipse with foci $\left(\pm\sqrt{5}, 0\right)$ and major axis 6 shifted up by 1 unit. To determine the equation of such an ellipse, Theorem 9.16 tells us that $\sqrt{A^2 - B^2} = \sqrt{5}$ and the major axis is $2A = 6$. Thus $A^2 = 9$ and thus $B^2 = 4$, so the equation of the ellipse is (shifting it up by one unit in the y direction)

$$\frac{x^2}{A^2} + \frac{(y-1)^2}{B^2} = \frac{x^2}{9} + \frac{(y-1)^2}{4} = 1.$$

35. Since the foci are on the y axis, the ellipse is vertically oriented. By Theorem 9.16, we have $\sqrt{B^2 - A^2} = \frac{3\sqrt{3}}{2}$, so that $B^2 - A^2 = \frac{27}{4}$; also, the minor axis is $2A = 3$, so that $A^2 = \frac{9}{4}$. Thus $B^2 = \frac{9}{4} + \frac{27}{4} = \frac{36}{4} = 9$. The equation of the ellipse is

$$\frac{x^2}{A^2} + \frac{y^2}{B^2} = \frac{4x^2}{9} + \frac{y^2}{9} = 1.$$

37. Since the foci are on the y axis, the ellipse is vertically oriented. By Theorem 9.16, we have $\sqrt{B^2 - A^2} = \alpha$, so that $B^2 - A^2 = \alpha^2$; also, the minor axis is $2A = 2\alpha$, so that $A = \alpha$. Then $B^2 = A^2 + \alpha^2 = 2\alpha^2$, so that the equation of the ellipse is

$$\frac{x^2}{A^2} + \frac{y^2}{B^2} = \frac{x^2}{\alpha^2} + \frac{y^2}{2\alpha^2} = 1.$$

39. The directrices are

$$x = \pm\frac{A}{e} = \pm\frac{A}{\frac{\sqrt{A^2+B^2}}{A}} = \pm\frac{A^2}{\sqrt{A^2+B^2}}.$$

For this exercise, $\sqrt{A^2 + B^2} = 6$, and the directrices are $x = \pm1$, so that $1 = \frac{A^2}{6}$. Thus $A^2 = 6$ and $B^2 = 30$, so the equation of the hyperbola is

$$\frac{x^2}{A^2} - \frac{y^2}{B^2} = \frac{x^2}{6} - \frac{y^2}{30} = 1.$$

41. The directrices are

$$y = \pm\frac{B}{e} = \pm\frac{B}{\frac{\sqrt{A^2+B^2}}{B}} = \pm\frac{B^2}{\sqrt{A^2+B^2}}.$$

For this exercise, $\sqrt{A^2 + B^2} = 4$, and the directrices are $y = \pm2$, so that $2 = \frac{B^2}{4}$. Thus $B^2 = 8$ and $A^2 = 8$, so the equation of the hyperbola is

$$\frac{y^2}{B^2} - \frac{x^2}{A^2} = \frac{y^2}{8} - \frac{x^2}{8} = 1.$$

43. The foci are at $(3, 5 \pm 4)$, so this is the hyperbola with foci $(0, \pm4)$ and directrix $y = 4 - 5 = -1$ shifted right by 3 units and up by 5 units. The directrices are

$$y = \pm\frac{B}{e} = \pm\frac{B}{\frac{\sqrt{A^2+B^2}}{B}} = \pm\frac{B^2}{\sqrt{A^2+B^2}}.$$

For this exercise, $\sqrt{A^2 + B^2} = 4$, and the directrices are $y = \pm1$, so that $1 = \frac{B^2}{4}$. Thus $B^2 = 4$ and $A^2 = 12$, so the equation of the hyperbola is (remembering to shift back)

$$\frac{(y-5)^2}{B^2} - \frac{(x-3)^2}{A^2} = \frac{(y-5)^2}{4} - \frac{(x-3)^2}{12} = 1.$$

45. Since $eu = 2$ and $e = 3$, this is a hyperbola with one focus at the origin and directrix $x = -\frac{2}{3}$ (due to the minus sign in the denominator). Its graph is

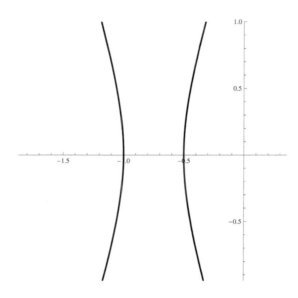

47. Since $e = 1$, this is a parabola; the numerator is $eu = 5$, so it is a parabola with directrix $x = 5$ and focus at the origin. Thus it is a leftward-opening parabola; its graph is

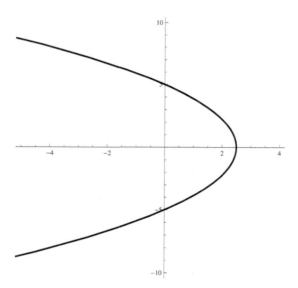

49. Dividing numerator and denominator by 4 gives $r = \frac{1/2}{1 + \frac{1}{4}\sin\theta}$. Thus $e = \frac{1}{4}$ and this is an ellipse. Since the denominator includes $\sin\theta$, it is vertically oriented. Its graph is

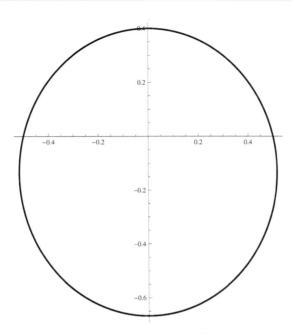

51. Dividing numerator and denominator by 2 gives $r = \frac{1}{1 + \frac{1}{2}\sin\theta}$. Thus $e = \frac{1}{2}$ and this is an ellipse. Since the denominator includes $\sin\theta$, it is vertically oriented. Its graph is

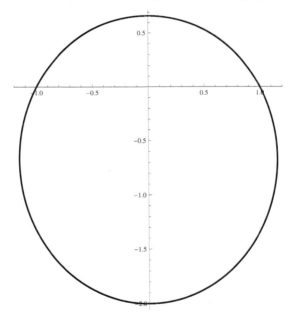

Applications

53. (a) The semiminor axis is given, by Exercise 52, as

$$B^2 = A^2(1 - e^2) = 1.00000011^2(1 - 0.0167^2) \approx 0.999721.$$

Thus the equation of the earth's orbit is approximately

$$\frac{x^2}{A^2} + \frac{y^2}{B^2} = \frac{x^2}{1.00000022} + \frac{y^2}{0.999721} = 1.$$

(b) Placing the center of the ellipse at the origin and the sun at the focus where $\theta = 0$ (i.e., in Cartesian coordinates, where $x > 0$), the associated directrix is (by Definition 9.22) the line $x = u = \frac{A}{e}$. Then by Theorem 9.24, the equation of the earth's orbit in polar coordinates is

$$r = \frac{eu}{1 + e\cos\theta} = \frac{A}{1 + e\cos\theta} = \frac{1.00000011}{1 + 0.0167\cos\theta}.$$

Proofs

55. Let $C = \sqrt{A^2 - B^2}$ as in the text. If (x, y) lies on the curve, then solving for y^2 gives $y^2 = B^2\left(1 - \frac{x^2}{A^2}\right) = B^2 - \frac{B^2 x^2}{A^2}$. Then the square of the distance D_2 from (x, y) to $(-C, 0)$ is

$$
\begin{aligned}
D_2^2 &= (x - (-C))^2 + (y - 0)^2 \\
&= (x + C)^2 + y^2 \\
&= x^2 + 2Cx + C^2 + B^2 - \frac{B^2 x^2}{A^2} \\
&= x^2 + 2Cx + A^2 - \frac{B^2 x^2}{A^2} \\
&= \frac{A^4 + 2A^2 Cx + (A^2 - B^2)x^2}{A^2} \\
&= \frac{A^4 + 2A^2 Cx + C^2 x^2}{A^2} \\
&= \frac{(A^2 + Cx)^2}{A^2}.
\end{aligned}
$$

Thus $D_2 = \frac{A^2 + Cx}{A}$.

57. Let G be the graph of the curve with equation $\frac{x^2}{A^2} - \frac{y^2}{B^2} = 1$ with $A, B > 0$. We show that if $(x, y) \in G$, then

$$d((x, y), (-\sqrt{A^2 + B^2}, 0)) - d((x, y), (\sqrt{A^2 + B^2}, 0)),$$

where d is the Euclidean distance function, is a constant independent of x and y. By Definition 9.19, this will show that G is the graph of a hyperbola with foci $(\pm\sqrt{A^2 + B^2}, 0)$.

Let $C = \sqrt{A^2 + B^2}$, so that $C^2 = A^2 + B^2$, and let (x, y) be a point on G. Solving the equation of the curve for y^2 gives

$$y^2 = \frac{B^2 x^2}{A^2} - B^2.$$

Let D_1 represent the distance from (x, y) to $(C, 0)$ and D_2 the distance from (x, y) to $(-C, 0)$. We want to show that $D_2 - D_1$ is constant. Then

$$D_1^2 = (x - C)^2 + (y - 0)^2 \qquad\qquad D_2^2 = (x - (-C))^2 + (y - 0)^2$$
$$= (x - C)^2 + y^2 \qquad\qquad\qquad = (x + C)^2 + y^2$$
$$= x^2 - 2Cx + C^2 + \frac{B^2 x^2}{A^2} - B^2 \qquad = x^2 + 2Cx + C^2 + \frac{B^2 x^2}{A^2} - B^2$$
$$= x^2 - 2Cx + A^2 + \frac{B^2 x^2}{A^2} \qquad\quad = x^2 + 2Cx + A^2 + \frac{B^2 x^2}{A^2}$$
$$= \frac{(A^2 + B^2)x^2 - 2A^2 Cx + A^4}{A^2} \qquad = \frac{(A^2 + B^2)x^2 + 2A^2 Cx + A^4}{A^2}$$
$$= \frac{C^2 x^2 - 2A^2 Cx + A^4}{A^2} \qquad\quad = \frac{C^2 x^2 + 2A^2 Cx + A^4}{A^2}$$
$$= \left(\frac{Cx - A^2}{A}\right)^2 \qquad\qquad\quad = \left(\frac{Cx + A^2}{A}\right)^2.$$

Note that $C > A$ and that $|x| \geq A$ (since if $|x| < A$, the equation has no solutions). Thus $Cx \pm A^2$ are both positive, so taking square roots gives

$$D_1 = \frac{Cx - A^2}{A}, \qquad D_2 \frac{Cx + A^2}{A},$$

so that $D_2 - D_1 = 2A$, which is a constant independent of x and y. So G is a hyperbola with $(\pm C, 0)$ as the foci.

59. For an ellipse of the form $\frac{x^2}{A^2} + \frac{y^2}{B^2} = 1$ where $A > B > 0$, the foci are $(\pm\sqrt{A^2 - B^2}, 0)$ (by Definition 9.15) and the vertices are the places where the curve intersects the x axis, which are $x = \pm A$. Thus

$$\frac{\text{distance between the foci}}{\text{distance between the vertices}} = \frac{2\sqrt{A^2 - B^2}}{2A} = \frac{\sqrt{A^2 - B^2}}{A} = e$$

by Definition 9.21.

For an ellipse of the form $\frac{x^2}{A^2} + \frac{y^2}{B^2} = 1$ where $B > A > 0$, the foci are $(\pm\sqrt{B^2 - A^2}, 0)$ (by Definition 9.15) and the vertices are the places where the curve intersects the y axis, which are $y = \pm B$. Thus

$$\frac{\text{distance between the foci}}{\text{distance between the vertices}} = \frac{2\sqrt{B^2 - A^2}}{2B} = \frac{\sqrt{B^2 - A^2}}{B} = e$$

by Definition 9.21.

For a hyperbola of the form $\frac{x^2}{A^2} - \frac{y^2}{B^2} = 1$, the foci are $(\pm\sqrt{A^2 + B^2}, 0)$ (by Definition 9.19) and the vertices are the places where the curve intersects the x axis, which are $x = \pm A$. Thus

$$\frac{\text{distance between the foci}}{\text{distance between the vertices}} = \frac{2\sqrt{A^2 + B^2}}{2A} = \frac{\sqrt{A^2 + B^2}}{A} = e$$

by Definition 9.21.

For a hyperbola of the form $\frac{y^2}{B^2} - \frac{x^2}{A^2} = 1$, the foci are $(0, \pm\sqrt{A^2 + B^2})$ (by Definition 9.19) and the vertices are the places where the curve intersects the y axis, which are $y = \pm B$. Thus

$$\frac{\text{distance between the foci}}{\text{distance between the vertices}} = \frac{2\sqrt{A^2 + B^2}}{2B} = \frac{\sqrt{A^2 + B^2}}{B} = e$$

by Definition 9.21.

61. Suppose $\frac{x^2}{A^2} - \frac{y^2}{B^2} = 1$ is a hyperbola, and let F, e, and l be as in the problem statement. If $P = (x, y)$ is a point on the curve, then the point D on l closest to (x, y) to l is clearly $\left(\frac{A}{e}, y\right)$, so that $DP = \frac{A}{e} - x$, so that $DP \cdot e = A - ex = A - x\frac{\sqrt{A^2 + B^2}}{A}$. Now, we have

$$FP = \sqrt{(\sqrt{A^2 + B^2} - x)^2 + y^2}$$

$$= \sqrt{A^2 + B^2 - 2x\sqrt{A^2 + B^2} + x^2 + \left(\frac{B^2 x^2}{A^2} - B^2\right)}$$

$$= \sqrt{A^2 - 2x\sqrt{A^2 + B^2} + \frac{(A^2 + B^2)x^2}{A^2}}$$

$$= \sqrt{\frac{A^4 - 2xA^2\sqrt{A^2 + B^2} + (A^2 + B^2)x^2}{A^2}}$$

$$= \sqrt{\frac{(A^2 - x\sqrt{A^2 + B^2})^2}{A^2}} = \frac{A^2 - x\sqrt{A^2 + B^2}}{A} = A - x\frac{\sqrt{A^2 + B^2}}{A}.$$

So $FP = e \cdot DP$ and thus $\frac{FP}{DP} = e$.

Thinking Forward

Analogs of conic sections in three dimensions

▷ The figure below is a plot of the ellipsoid consisting of points the sum of whose distances from $(-5, 0, 0)$ and $(5, 0, 0)$ is 20.

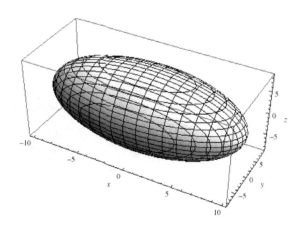

▷ The figure below is a plot of the paraboloid consisting of points whose distances from $(1, 0, 0)$ and the plane $z = -1$ are equal.

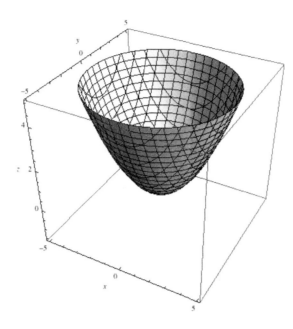

▷ The figure below is a plot of the hyperboloid consisting of points the difference of whose distances from $(5, 0, 0)$ and $(-5, 0, 0)$ is 4.

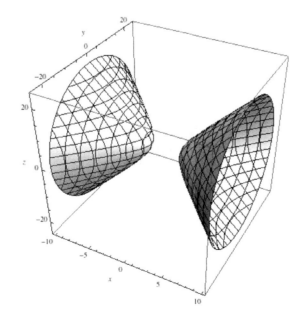

Chapter Review and Self-Test

1.

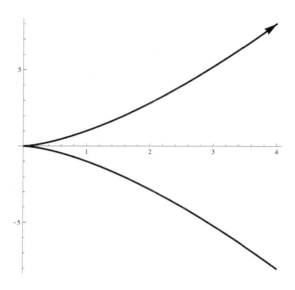

3. Note that the curve is traced twice as t ranges from 0 to 4π.

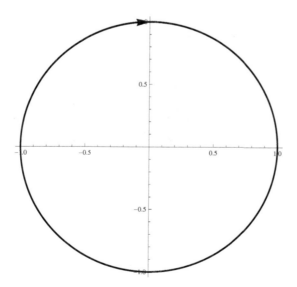

5.

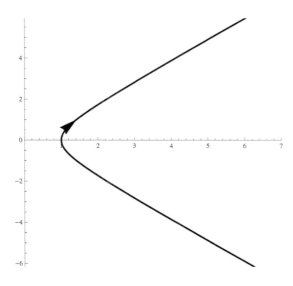

7.

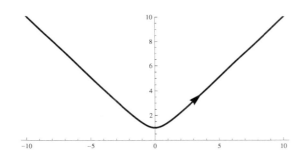

9. From the graph (see Exercise 1), or by symmetry, it is clear that the arc length on $[-2, 2]$ is twice the arc length on $[0, 2]$, so we get (with the substitution $u = 4 + 9t^2$)

$$
\begin{aligned}
2 \int_0^2 \sqrt{x'(t)^2 + y'(t)^2}\, dt &= 2 \int_0^2 \sqrt{(2t)^2 + (3t^2)^2}\, dt \\
&= 2 \int_0^2 \sqrt{4t^2 + 9t^4}\, dt \\
&= 2 \int_0^2 t \sqrt{4 + 9t^2}\, dt \\
&= \frac{1}{9} \int_{t=0}^{t=2} \sqrt{u}\, du \\
&= \frac{1}{9} \left[\frac{2}{3} u^{3/2} \right]_{t=0}^{t=2} \\
&= \frac{2}{27} \left[(4 + 9t^2)^{3/2} \right]_0^2 = \frac{2}{27} \left(80\sqrt{10} - 8 \right).
\end{aligned}
$$

11. This is a circle of radius k centered at the origin, so its arc length is its circumference, which is $2\pi k$. To evaluate using an integral, note that we can compute the arc length (using symmetry) as 8 times

the arc length from 0 to $\frac{\pi}{4}$, which is

$$8 \int_0^{\pi/4} \sqrt{x'(t)^2 + y'(t)^2}\, dt = 8 \int_0^{\pi/4} \sqrt{(k \cos kt)^2 + (-k \sin kt)^2}\, dt$$

$$= 8 \int_0^{\pi/4} \sqrt{k^2 (\cos^2 kt + \sin^2 kt)}\, dt = 8 \int_0^{\pi/4} k\, dt = 2k\pi.$$

13. This is a circle of radius $\frac{1}{2}$ with center at $\left(0, \frac{1}{2}\right)$. It is traced once for $\theta \in [0, \pi)$. Also, it is symmetric around the y axis, since $\sin(-\theta) = -\sin\theta$, so if (r, θ) is on the curve, so is $(-r, -\theta)$. Thus it suffices to plot the curve from 0 to $\frac{\pi}{2}$ and reflect it.

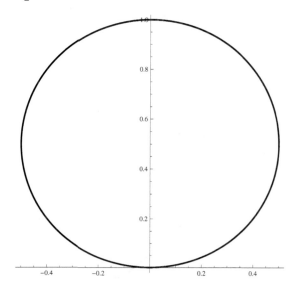

15. This is a three-petaled rose. It is traced once for $\theta \in [0, \pi)$. Also, it is symmetric around the y axis, since $\sin(-3\theta) = -\sin 3\theta$, so if (r, θ) is on the curve, so is $(-r, -\theta)$. Thus it suffices to plot the curve from 0 to $\frac{\pi}{2}$ and reflect it.

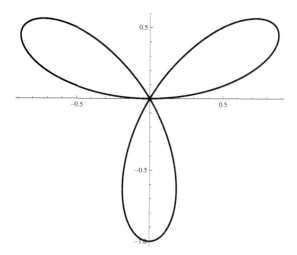

17. This is a cardioid. It is traced once for $\theta \in [0, 2\pi)$. It is symmetric around the y axis, since if (r, θ) is on the curve, then $2 - 2\sin(\pi - \theta) = 2 - 2\sin\theta$, so that $(r, \pi - \theta) = (-r, -\theta)$ is on the curve. So we need only plot the curve from 0 to π and reflect it across the y axis.

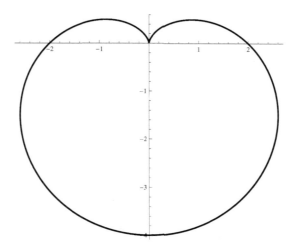

19. This is a limaçon with an inner loop. It is traced once for $\theta \in [0, 2\pi)$. It is symmetric around the y axis, since if (r, θ) is on the curve, then $2 - \sqrt{8}\sin(\pi - \theta) = 2 - \sqrt{8}\sin\theta$, so that $(r, \pi - \theta) = (-r, -\theta)$ is on the curve. So we need only plot the curve from 0 to π and reflect it across the y axis.

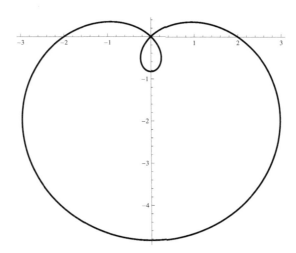

21. This is a lemniscate. It is traced once for $\theta \in \left[\frac{\pi}{2}, \pi\right) \cup \left[\frac{3\pi}{2}, 2\pi\right)$ since $\sin 2\theta$ is positive elsewhere. The curve is symmetric around the origin, since if (r, θ) is on the curve, then $r^2 = -\sin 2\theta$, so that also $(-r)^2 = -\sin 2\theta$ and $(-r, \theta)$ is on the curve. So we need only plot the curve from $\frac{\pi}{2}$ to π and reflect it through the origin.

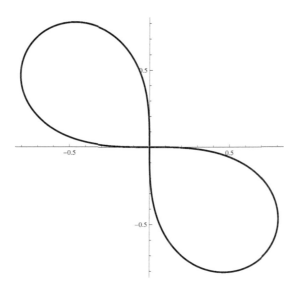

23. By Theorem 9.14, since $f(\theta) = a$ so that $f'(\theta) = 0$, we get for the arc length

$$\int_0^{2\pi} \sqrt{0^2 + a^2}\, d\theta = \int_0^{2\pi} a\, d\theta = 2\pi a.$$

25. This is a circle of radius $\frac{a}{2}$ that is traced once for $0 \le \theta < \pi$. With $f(\theta) = a\sin\theta$ we have $f'(\theta) = a\cos\theta$, so that the arc length is

$$\int_0^{\pi} \sqrt{(a\cos\theta)^2 + (a\sin\theta)^2}\, d\theta = \int_0^{\pi} \sqrt{a^2(\cos^2\theta + \sin^2\theta)}\, d\theta = \int_0^{\pi} a\, d\theta = a\pi.$$

27. This cardioid is symmetric around the y axis. The right half is traced from $-\frac{\pi}{2}$ to $\frac{\pi}{2}$, so we compute that arc length and double it. With $f(\theta) = 1 - \sin\theta$, we have $f'(\theta) = -\cos\theta$, so that the arc length is

$$2\int_{-\pi/2}^{\pi/2} \sqrt{(1-\sin\theta)^2 + (-\cos\theta)^2}\, d\theta = 2\int_{-\pi/2}^{\pi/2} \sqrt{1 - 2\sin\theta + \sin^2\theta + \cos^2\theta}\, d\theta$$

$$= 2\int_{-\pi/2}^{\pi/2} \sqrt{2 - 2\sin\theta}\, d\theta.$$

To evaluate this integral, make the substitution $u = 1 - \sin\theta$; then $du = -\cos\theta\, d\theta$. But

$$\cos\theta = \sqrt{1 - \sin^2\theta} = \sqrt{1 - (1-u)^2} = \sqrt{2u - u^2}$$

(note that $\cos\theta$ is nonnegative on $\left[-\frac{\pi}{2}, \frac{\pi}{2}\right]$, so we can use the positive square root), so that we get $-\frac{du}{\sqrt{2u-u^2}} = d\theta$. Then

$$2\int_{-\pi/2}^{\pi/2} \sqrt{2 - 2\sin\theta}\, d\theta = -2\int_{\theta=-\pi/2}^{\theta=\pi/2} \frac{\sqrt{2u}}{\sqrt{2u - u^2}}\, du$$

$$= -2\sqrt{2} \int_{\theta=-\pi/2}^{\theta=\pi/2} \frac{1}{\sqrt{2-u}}\, du$$

$$= -2\sqrt{2} \int_{\theta=-\pi/2}^{\theta=\pi/2} (2-u)^{-1/2}\, du$$

$$= 4\sqrt{2} \left[(2-u)^{1/2}\right]_{\theta=-\pi/2}^{\theta=\pi/2}$$

$$= 4\sqrt{2} \left[(1 + \sin\theta)^{1/2}\right]_{-\pi/2}^{\pi/2} = 8.$$

29. This is a segment of a vertical line. With $f(\theta) = \sec\theta$, we have $f'(\theta) = \sec\theta\tan\theta$, so that the arc length is

$$\int_{-\pi/6}^{\pi/6}\sqrt{\sec^2\theta\tan^2\theta + \sec^2\theta}\, d\theta = \int_{-\pi/6}^{\pi/6}\sqrt{\sec^2\theta(\tan^2\theta + 1)}\, d\theta$$

$$= \int_{-\pi/6}^{\pi/6}\sqrt{\sec^4\theta}\, d\theta = \int_{-\pi/6}^{\pi/6}\sec^2\theta\, d\theta = [\tan\theta]_{-\pi/6}^{\pi/6} = \frac{2\sqrt{3}}{3}.$$

31. Since $r = f(\theta) = a$, the area of the circle is

$$\frac{1}{2}\int_0^{2\pi} a^2\, d\theta = \frac{1}{2}\left[a^2\theta\right]_0^{2\pi} = \pi a^2.$$

33. $r = \cos 2\theta$ is a four-petaled rose; one rose is traced for $-\frac{\pi}{4} \le \theta \le \frac{\pi}{4}$, so we compute the area inside the function from 0 to $\frac{\pi}{4}$ and double it. This is

$$2\cdot\frac{1}{2}\int_0^{\pi/4}(\cos 2\theta)^2\, d\theta = \int_0^{\pi/4}\cos^2 2\theta\, d\theta = \left[\frac{\theta}{2} + \frac{1}{8}\sin 4\theta\right]_0^{\pi/4} = \frac{\pi}{8}.$$

35. This cardioid is traced once for $0 \le \theta \le 2\pi$, so the area inside the cardioid is

$$\frac{1}{2}\int_0^{2\pi}(1 - \sin\theta)^2\, d\theta = \frac{1}{2}\int_0^{2\pi}(1 - 2\sin\theta + \sin^2\theta)\, d\theta = \frac{1}{2}\left[\theta + 2\cos\theta + \frac{\theta}{2} - \frac{1}{4}\sin 2\theta\right]_0^{2\pi} = \frac{3\pi}{2}.$$

37. The lemniscate is traced completely for $\frac{\pi}{2} \le \theta \le \pi$ if we take both square roots of r^2. So taking just the positive square root will plot the loop of the lemniscate in the second quadrant. The area is then

$$\frac{1}{2}\int_{\pi/2}^{\pi}(\sqrt{-\sin 2\theta})^2\, d\theta = \frac{1}{2}\int_{\pi/2}^{\pi}(-\sin 2\theta)\, d\theta = \frac{1}{2}\left[\frac{1}{2}\cos 2\theta\right]_{\pi/2}^{\pi} = \frac{1}{2}.$$

39. Plots of both curves, with $r^2 = \cos 2\theta$ the dashed graph, are:

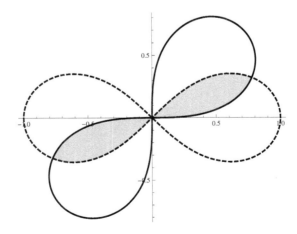

By symmetry, we can compute the area between the curves in the first quadrant and double it. The two curves intersect where $\cos 2\theta = \sin 2\theta$, or where $\tan 2\theta = 1$. This is at $\theta = \frac{\pi}{8}$. So we can compute the area between the curves by computing the area enclosed by $r^2 = \sin 2\theta$ for $\theta \in \left[0, \frac{\pi}{8}\right]$ and $r^2 = \cos^2 2\theta$ for $\theta \in \left[\frac{\pi}{8}, \frac{\pi}{4}\right]$ and doubling it:

$$2 \left(\frac{1}{2} \int_0^{\pi/8} \left(\sqrt{\sin 2\theta} \right)^2 d\theta + \frac{1}{2} \int_{\pi/8}^{\pi/4} \left(\sqrt{\cos 2\theta} \right)^2 d\theta \right) = \int_0^{\pi/8} \sin 2\theta \, d\theta + \int_{\pi/8}^{\pi/4} \cos 2\theta \, d\theta$$

$$= \left[-\frac{1}{2} \cos 2\theta \right]_0^{\pi/8} + \left[\frac{1}{2} \sin 2\theta \right]_{\pi/8}^{\pi/4}$$

$$= -\frac{\sqrt{2}}{4} + \frac{1}{2} + \left(\frac{1}{2} - \frac{\sqrt{2}}{4} \right)$$

$$= 1 - \frac{\sqrt{2}}{2}.$$

Chapter 10

Vectors

10.1 Cartesian Coordinates

Thinking Back

Lines parallel to the coordinate axes

▷ Since the line must be parallel to the x axis, its equation is of the form $y = a$. Since (π, e) is on the line, the equation must be $y = e$.

▷ Since the line is parallel to the y axis, its equation must be of the form $x = a$. Since (π, e) is on the line, the equation must be $x = \pi$.

▷ Since the line is parallel to the y axis, its equation must be of the form $x = a$. Since $(0, 5)$ is on the line, the equation must be $x = 0$. Thus the line is the y axis itself.

Circles in $\mathbb{R}^2$

▷ Since the circle is tangent to the x axis, its radius is 6 (the distance from $(3, 6)$ to the x axis), so its equation is $(x - 3)^2 + (y - 6)^2 = 36$.

▷ Since the circle is tangent to the y axis, its radius is 3 (the distance from $(3, 6)$ to the y axis), so its equation is $(x - 3)^2 + (y - 6)^2 = 9$.

▷ Since the circle contains $(5, -2)$, the distance from $(5, -2)$ to $(-4, 3)$ must be the radius; this distance is

$$\sqrt{(5 - (-4))^2 + (-2 - 3)^2} = \sqrt{81 + 25} = \sqrt{106}.$$

Thus the equation of the circle is $(x + 4)^2 + (y - 3)^2 = 106$.

Concepts

1. (a) False. If (a, b) and (c, d) are distinct, then either $a \neq b$ or $c \neq d$, so that either $(a - b)^2 > 0$ or $(c - d)^2 > 0$. Thus $\sqrt{(a - b)^2 + (c - d)^2} > 0$.

   (b) True. It is the vertical line through $(5, 0)$.

   (c) False. It is a plane consisting of all points $(5, y, z)$.

   (d) False. It is a hyperplane consisting of all points $(5, y, z, w)$.

(e) False. Complete the square:

$$x^2 + y^2 + z^2 + 2x - 4y - 10z + 50$$
$$= (x^2 + 2x + 1) + (y^2 - 4y + 4) + (z^2 - 10z + 25) + 50 - 30$$
$$= (x+1)^2 + (y-2)^2 + (z-5)^2 + 20 = 0.$$

Thus this is the equation $(x+1)^2 + (y-2)^2 + (z-5)^2 = -20$, which has no solutions.

(f) True. Since it is tangent to each of the coordinate planes, the distance from its center to each of the coordinate planes is the same. But the distance from the center to the xy plane is the z coordinate of the center, and similarly for the other planes. Thus the three coordinates of the center are the same, so it must be (c, c, c) for some constant c.

(g) True. If the spheres are tangent, they intersect in a point. Otherwise, they intersect in a circle.

(h) True. Two noncollinear points determine a line, and thus an infinite number of planes containing that line. One and only one of those planes contains the third point.

3. (a) In two dimensions, $y = 5$ is a line parallel to the x axis and $x = -3$ is a line parallel to the y axis; these two lines determine the single point $(-3, 5)$.

(b) In three dimensions, $y = 5$ is a plane parallel to the xz plane and $x = -3$ is a plane parallel to the yz axis; these two planes intersect in the line consisting of the points $(-3, 5, z)$.

5. Since the cube has side length 2 and center at the origin, each vertex is 1 unit from each of the coordinate planes. Thus the vertices are at $(\pm 1, \pm 1, \pm 1)$.

7. See Definition 10.2. A sphere is the set of points in 3-space at a fixed distance from a given point.

9. (a) In two dimensions, this is a circle of radius 2 centered at the origin.

(b) In three dimensions, this consists of all points (x, y, z) such that $x^2 + y^2 = 4$, so it is a right circular cylinder lying over the circle of radius 2 centered at the origin in the xy plane.

11. If the points are symmetric about the origin, then 0 must be the sum of the corresponding coordinates on the two points, since $(0, 0, 0)$ is the midpoint of the segment joining them. Thus the other point is $(-5, 6, -7)$.

13. The symmetric point is $(3, -7, 6)$, since the midpoint of the segment joining the two points is then

$$\left(\frac{3+3}{2}, \frac{-7-7}{2}, \frac{6-4}{2} \right) = (3, -7, 1),$$

which lies on the plane $z = 1$.

15. Examining the six axis labelings shown in the preceding exercise, it is easily seen that interchanging any two labelings in one row gives a labeling in the other row. Thus exchanging two labels in a right-handed system creates a left handed system; similarly, exchanging two labels in a left-handed system gives a right-handed system. Thus if you exchange two pairs of labels, you return to a right-handed system.

17. The distance between (x_1, y_1, z_1, w_1) and (x_2, y_2, z_2, w_2) in four-space is

$$\sqrt{(x_2 - x_1)^2 + (y_2 - y_1)^2 + (z_2 - z_1)^2 + (w_2 - w_1)^2}.$$

The distance between $(x_1, y_1, z_1, w_1, v_1)$ and $(x_2, y_2, z_2, w_2, v_2)$ in five-space is

$$\sqrt{(x_2 - x_1)^2 + (y_2 - y_1)^2 + (z_2 - z_1)^2 + (w_2 - w_1)^2 + (v_2 - v_1)^2}.$$

In general, given two points $(x_1, x_2, \ldots, x_n)$ and $(y_1, y_2, \ldots, y_n)$ in n-space, the distance between them is

$$\sqrt{(y_1 - x_1)^2 + (y_2 - x_2)^2 + \cdots + (y_n - x_n)^2}.$$

19. (a) This is the region $x \geq 0$, $y \geq 0$ (so it is in the first quadrant), $z > 3$ (above the plane $z = 3$):

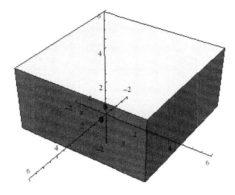

(b) This is the region closer than 5 to the point $(1, 2, 4)$, so it is the region $(x - 1)^2 + (y - 2)^2 + (z - 4)^2 < 25$:

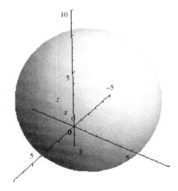

(c) Completing the square gives $(x - 2)^2 + (y + 3)^2 = 25$, so this is the region above the disk of radius 5 centered at $(2, -3)$:

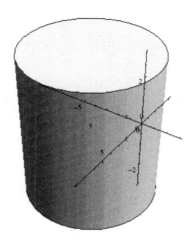

Skills

21. The distance is $\sqrt{(5-(-2))^2+(-6-3)^2} = \sqrt{49+81} = \sqrt{130}$.

23. The distance is $\sqrt{(1-(-2))^2+(4-3)^2+(7-5)^2} = \sqrt{9+1+4} = \sqrt{14}$.

25. The distance is $\sqrt{(-1-(-4))^2+(4-3)^2+(-3-1)^2} = \sqrt{9+1+16} = \sqrt{26}$.

27. The distance is

$$\sqrt{(-1-0)^2+(3-6)^2+(5-1)^2+(2-(-2))^2+(0-3)^2} = \sqrt{1+9+16+16+9} = \sqrt{51}.$$

29. By Definition 10.2, the equation is $(x-3)^2+(y+2)^2+(z-5)^2 = 25$.

31. Since the sphere is tangent to the xy plane, which is the plane $z = 0$, it must have radius equal to the magnitude of the z coordinate of the center, which is 7. Thus by Definition 10.2, the equation is $(x-2)^2+(y-5)^2+(z+7)^2 = 49$.

33. Since the origin is on the sphere, the radius of the sphere must be the distance from the origin to the center, which is $\sqrt{2^2+5^2+7^2} = \sqrt{78}$. Thus by Definition 10.2, the equation of the sphere is $(x-2)^2+(y-5)^2+(z+7)^2 = 78$.

35. The radius of the sphere is the distance between the two points, which is

$$\sqrt{(3-2)^2+(0-(-8))^2+(-1-0)^2} = \sqrt{1+64+1} = \sqrt{66}.$$

Then by Definition 10.2, the equation is $(x-2)^2+(y+8)^2+z^2 = 66$.

37. Since the given segment is a diameter, its midpoint must be the center of the sphere. Its midpoint is

$$\left(\frac{6-3}{2}, \frac{-1+3}{2}, \frac{4+1}{2}\right) = \left(\frac{3}{2}, 1, \frac{5}{2}\right).$$

The radius of the sphere must be the distance from this point to either of the two given points, which is (using the first point)

$$\sqrt{\left(\frac{3}{2}-6\right)^2 + (1-(-1))^2 + \left(\frac{5}{2}-4\right)^2} = \sqrt{\frac{81+16+9}{4}} = \frac{\sqrt{106}}{2}.$$

Thus by Definition 10.2, the equation of the sphere is $\left(x-\frac{3}{2}\right)^2 + (y-1)^2 + \left(z-\frac{5}{2}\right)^2 = \frac{53}{2}$.

39. Complete the square:

$$x^2 + y^2 + z^2 + 3y - 5z + 3 = (x^2) + \left(y^2 + 3y + \frac{9}{4}\right) + \left(z^2 - 5z + \frac{25}{4}\right) + 3 - \left(\frac{9}{4} + \frac{25}{4}\right)$$
$$= x^2 + \left(y+\frac{3}{2}\right)^2 + \left(z-\frac{5}{2}\right)^2 - \frac{11}{2} = 0$$

Rearranging terms gives the sphere $x^2 + \left(y+\frac{3}{2}\right)^2 + \left(z-\frac{5}{2}\right)^2 = \frac{11}{2}$, so this is a sphere of radius $\sqrt{\frac{11}{2}}$ centered at $\left(0, -\frac{3}{2}, \frac{5}{2}\right)$.

41. This surface is a circular cone with axis of symmetry the x axis; its level curves for x are circles $y^2 + z^2 = a^2$. Its trace in the xy plane is the pair of lines $x = \pm\frac{y}{3}$, or $y = \pm 3x$. Its trace in the xz plane is the pair of lines $z = \pm 3x$. Finally, its trace in the yz plane is the curve $\frac{x^2}{9} + \frac{y^2}{9} = 0$, so it consists only of the origin.

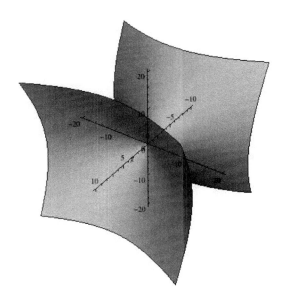

43. This surface is a hyperboloid of two sheets (it may be written $x^2 + y^2 - z^2 = -1$). Level curves are of the form $x^2 + y^2 = a^2 - 1$, so they are circles for $|a| > 1$, points when $a = \pm 1$, and empty for $|a| < 1$. Its trace in the xy plane is empty. Its trace in the xz plane is the hyperbola $z^2 - x^2 = 1$, and in the yz plane it is the hyperbola $z^2 - y^2 = 1$.

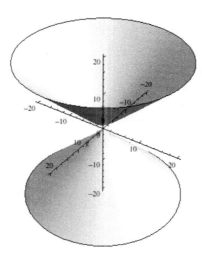

45. This surface is an elliptic paraboloid. Its level curves are $\frac{x^2}{9} + \frac{y^2}{25} = a$, so they are ellipses for $a > 0$, the origin for $a = 0$, and empty for $a < 0$. Its trace in the xy plane consists only of the origin. Its trace in the xz plane is the parabola $z = \frac{x^2}{9}$, and in the yz plane is the parabola $z = \frac{y^2}{25}$.

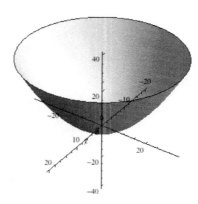

47. This is a hyperbolic paraboloid. Its level curves are the hyperbolas $x^2 - y^2 = a$, except where $a = 0$, when it consists of the pair of lines $x = \pm y$. The trace in the xy plane is the pair of lines $x = \pm y$; in the xz plane the trace is the parabola $z = x^2$. In the yz plane the trace is the parabola $z = -y^2$.

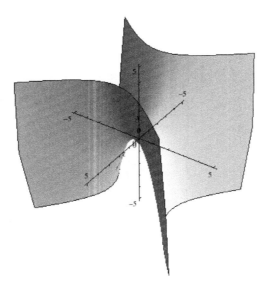

49. A labeled version of the diagram is below, assuming that the side length of the square is 2 so that the radius of the larger circle is 1:

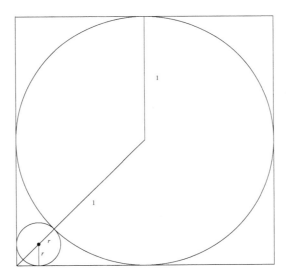

Since the radius of the smaller circle is r, the Pythagorean Theorem tells us that the distance from the center of that circle to the corner of the square is $r\sqrt{2}$. Thus the total length of the diagonal line is $1 + r + r\sqrt{2}$. But the length of that diagonal line is equal to half the diagonal of the square, so that $1 + r + r\sqrt{2} = \sqrt{2}$ and thus $r = \frac{\sqrt{2}-1}{\sqrt{2}+1}$.

51. We need to show that the distances between the three pairs of points are all equal. But:

$$\sqrt{(5-3)^2 + (4-6)^2 + (-1-(-1))^2} = \sqrt{4+4} = \sqrt{8}$$
$$\sqrt{(5-3)^2 + (4-4)^2 + (-1-1)^2} = \sqrt{4+4} = \sqrt{8}$$
$$\sqrt{(3-3)^2 + (6-4)^2 + (-1-1)^2} = \sqrt{4+4} = \sqrt{8}.$$

53. We need to show that the distances between the three pairs of points satisfy the Pythagorean Theorem. But:

$$\sqrt{(1-3)^2 + (5-8)^2 + (0-6)^2} = \sqrt{4+9+36} = \sqrt{49} = 7$$

$$\sqrt{(1-7)^2 + (5-(-7))^2 + (0-4)^2} = \sqrt{36+144+16} = \sqrt{196} = 14$$

$$\sqrt{(3-7)^2 + (8-(-7))^2 + (6-4)^2} = \sqrt{16+225+4} = \sqrt{245},$$

and $7^2 + 14^2 = 49 + 196 = 245 = \left(\sqrt{245}\right)^2$. Since it is a right triangle, its base and height are given by its legs, 7 and 14, so its area is $\frac{1}{2}bh = 49$.

55. For the tetrahedron whose fourth vertex is at $(1,1,1)$ (see previous exercise), the center of the sphere is at

$$\left(\frac{1+0+0+1}{4}, \frac{0+1+0+1}{4}, \frac{0+0+1+1}{4}\right) = \left(\frac{1}{2}, \frac{1}{2}, \frac{1}{2}\right).$$

The distance from this point to, say $(1,0,0)$ is

$$\sqrt{\left(1-\frac{1}{2}\right)^2 + \left(\frac{1}{2}\right)^2 + \left(\frac{1}{2}\right)^2} = \frac{\sqrt{3}}{2},$$

so this is the radius of the sphere. Thus the equation of this sphere is

$$\left(x-\frac{1}{2}\right)^2 + \left(y-\frac{1}{2}\right)^2 + \left(z-\frac{1}{2}\right)^2 = \frac{3}{4}.$$

For the tetrahedron whose fourth vertex is at $\left(-\frac{1}{3}, -\frac{1}{3}, -\frac{1}{3}\right)$, the center of the sphere is at

$$\left(\frac{1+0+0-\frac{1}{3}}{4}, \frac{0+1+0-\frac{1}{3}}{4}, \frac{0+0+1-\frac{1}{3}}{4}\right) = \left(\frac{1}{6}, \frac{1}{6}, \frac{1}{6}\right).$$

The distance from this point to, say $(1,0,0)$ is

$$\sqrt{\left(1-\frac{1}{6}\right)^2 + \left(\frac{1}{6}\right)^2 + \left(\frac{1}{6}\right)^2} = \sqrt{\frac{3}{4}},$$

so this is the radius of the sphere. Thus the equation of this sphere is

$$\left(x-\frac{1}{6}\right)^2 + \left(y-\frac{1}{6}\right)^2 + \left(z-\frac{1}{6}\right)^2 = \frac{3}{4}.$$

57. The octahedron consists of two square pyramids, one constructed from the five points $(1,0,0)$, $(0,1,0)$, $(0,-1,0)$, $(0,0,1)$, and $(0,0,-1)$ and the other from the five points $(-1,0,0)$, $(0,1,0)$, $(0,-1,0)$, $(0,0,1)$, and $(0,0,-1)$. For each of these pyramids, the base is the square formed by the latter four points, which has side length $\sqrt{2}$ and thus area 2. The height of the pyramid is 1, so the volume of each pyramid is $\frac{2}{3}$ and thus the volume of the octahedron is $\frac{4}{3}$.

Applications

59. Note that in the usual coordinate system, where we regard an angle of 0 as being due east, $15°$ east of north is an angle of $75°$ and $100°$ east of north is an angle of $-10°$. Suppose Annie's point is (x, y). Then the line from (x, y) to the summit of Constitution Peak makes an angle of $75°$ with due east, and the line from (x, y) to the summit of Blakely Peak makes an angle of $-10°$ with due east. We thus get the two equations

$$\tan 75° = \frac{4.0 - y}{7.9 - x}, \qquad \tan(-10°) = \frac{-3.4 - y}{9.3 - x}.$$

This gives the two equations (after approximating the two tangent functions)

$$y = 3.73x - 25.47, \qquad y = -0.176x - 1.76.$$

Solving gives $x \approx 6.07$ and $y \approx -2.83$, so that Annie is at location $(6.07, -2.83)$, which is

$$\sqrt{6.07^2 + (-2.83)^2} \approx 6.697 \text{ miles from Deer Harbor.}$$

61. The portion of the hyperbolic paraboloid constituting the chip achieves its maximum z value at $(\pm 1.3, 0)$, and its minimum z value at $(0, \pm 0.8)$. The difference of these two z values should be the height of the chip, which is 0.16. This gives the equation

$$\left(\frac{1.3^2}{a^2} - \frac{0^2}{a^2} \right) - \left(\frac{0^2}{a^2} - \frac{0.8^2}{a^2} \right) = \frac{1.3^2}{a^2} + \frac{0.8^2}{a^2} = 0.16.$$

Simplifying gives $0.16a^2 = 2.33$. Then $a^2 = 14.5625$, so that the equation of the chip is

$$z = \frac{x^2}{14.5625} - \frac{y^2}{14.5625}.$$

Proofs

63. We may as well assume that the regular tetrahedron is the one we found in Exercise 54(a), with vertices $(1, 0, 0)$, $(0, 1, 0)$, $(0, 0, 1)$, and $(1, 1, 1)$, since any other regular tetrahedron may be obtained from that one by stretching all the sides by some constant and then rotating it into place. Neither of those operations changes the relative length of segments, so if the property holds for this regular tetrahedron, it will hold for all of them.

Consider the two edges $\mathcal{E}$ joining $(1, 0, 0)$ and $(0, 1, 0)$ and $\mathcal{E}'$ joining $(0, 0, 1)$ and $(1, 1, 1)$. The midpoints of those edges are $\left(\frac{1}{2}, \frac{1}{2}, 0 \right)$ and $\left(\frac{1}{2}, \frac{1}{2}, 1 \right)$, and the midpoint of the segment joining those two points is $\left(\frac{1}{2}, \frac{1}{2}, \frac{1}{2} \right)$.

Another pair of edges are the edge joining $(1, 0, 0)$ and $(0, 0, 1)$ and the edge joining $(0, 1, 0)$ and $(1, 1, 1)$. The midpoint of the first of these is $\left(\frac{1}{2}, 0, \frac{1}{2} \right)$ and the midpoint of the second is $\left(\frac{1}{2}, 1, \frac{1}{2} \right)$. Thus the midpoint of the segment joining these two midpoints is again $\left(\frac{1}{2}, \frac{1}{2}, \frac{1}{2} \right)$.

The third pair of edges works the same way: one edge joins $(1, 0, 0)$ and $(1, 1, 1)$ and the other joins $(0, 1, 0)$ and $(0, 0, 1)$. The midpoint of the first edge is $\left(1, \frac{1}{2}, \frac{1}{2} \right)$ and the midpoint of the second edge is $\left(0, \frac{1}{2}, \frac{1}{2} \right)$. The midpoint of the segment joining these two midpoints is again $\left(\frac{1}{2}, \frac{1}{2}, \frac{1}{2} \right)$. Thus all three midpoints coincide, so that the line segments connecting the midpoints of opposite sides bisect one another.

Thinking Forward

Hyperspheres

▷ The distance between the two points is the radius; this is

$$\sqrt{(1-2)^2 + (2-5)^2 + (3-3)^2 + (4-(-4))^2} = \sqrt{26},$$

so that the equation of the hypersphere is $(x-1)^2 + (y-2)^2 + (z-3)^2 + (w-4)^2 = 26$.

▷ Since the two given points are endpoints of a diameter, the midpoint of the segment joining them must be the center of the hypersphere; this point is

$$\left(\frac{1+2}{2}, \frac{2+5}{2}, \frac{3+3}{2}, \frac{4-4}{2} \right) = \left(\frac{3}{2}, \frac{7}{2}, 3, 0 \right).$$

The radius is half the distance between the given points, so it is

$$\frac{1}{2} \sqrt{(1-2)^2 + (2-5)^2 + (3-3)^2 + (4-(-4))^2} = \frac{\sqrt{26}}{2}.$$

Thus the equation of the hypersphere is $\left(x - \frac{3}{2}\right)^2 + \left(y - \frac{7}{2}\right)^2 + (z-3)^2 + w^2 = \frac{13}{2}$.

▷ A hypersphere in $\mathbb{R}^n$ is the set of points in $\mathbb{R}^n$ that are at a fixed distance from a given point in $\mathbb{R}^n$.

10.2　Vectors

Thinking Back

Properties of addition:

▷ The commutative property of addition says that if a and b are real numbers, then $a + b = b + a$.

▷ The associative property of addition says that if a, b, and c are real numbers, then $(a + b) + c = a + (b + c)$.

▷ The distributive property of multiplication over addition says that if a, b, and c are real numbers, then $a(b + c) = ab + ac$.

Computing distances

▷ The distance is $\sqrt{(-3-6)^2 + (4-(-5))^2} = \sqrt{81 + 81} = 9\sqrt{2}$.

▷ The distance is $\sqrt{(11-11)^2 + (12-12)^2 + (13-(-2))^2} = \sqrt{15^2} = 15$.

▷ The distance is $\sqrt{(5-3)^2 + (-2-0)^2 + (-1-(-4))^2} = \sqrt{4 + 4 + 9} = \sqrt{17}$.

Concepts

1.　(a) True. The norm of any vector is the square root of the sum of the squares of its coordinates. See Definition 10.5.

　　(b) False. This is true unless $\mathbf{v}$ is the zero vector.

　　(c) False. If the norm of $\mathbf{v}$ is less than 1, then the norm of the unit vector in the direction of $\mathbf{v}$, which is 1, exceeds the norm of $\mathbf{v}$.

(d) False. It is denoted by $\overrightarrow{AB}$.

(e) True. If you consider $\mathbb{R}^2$ as sitting inside $\mathbb{R}^3$ in the xy plane, then the point $(x, y) \in \mathbb{R}^2$ corresponds to the point $(x, y, 0) \in \mathbb{R}^3$.

(f) True. The standard basis vectors are vectors in which all coordinates but one are zero, and the nonzero coordinate is 1. The norm of such a vector is 1.

(g) True. Since each coordinate of $-\mathbf{v}$ is the negative of the corresponding coordinate of $\mathbf{v}$, the point represented by the tip of $\mathbf{v}$ is the negative of the point represented by the tip of $-\mathbf{v}$, so it points in the opposite direction.

(h) True. This follows from the previous item, since $\overrightarrow{BA}$ has the same length as $\overrightarrow{AB}$ but points in the opposite direction.

3. Since the terminal point of $\overrightarrow{AB}$ is the initial point of $\overrightarrow{BC}$, we know that $\overrightarrow{AB} + \overrightarrow{BC}$ is the vector from the initial point of $\overrightarrow{AB}$ to the terminal point of $\overrightarrow{BC}$, which is $\overrightarrow{AC}$.

5. The terminal point of $\overrightarrow{AB}$ is the initial point of $\overrightarrow{BA}$, so $\overrightarrow{AB} + \overrightarrow{BA} = \overrightarrow{AA} = \mathbf{0}$.

7. From Exercise 3, $\overrightarrow{AB} + \overrightarrow{BC} = \overrightarrow{AC}$, so the sum in question is equal to $\overrightarrow{AC} + \overrightarrow{CA}$. By Exercise 5, this sum is $\mathbf{0}$. (This sum amounts to tracing the three edges of a triangle with vertices A, B, and C, so it returns you to where you started).

9. To add two vectors algebraically, simply add each of their coordinates, the first to the first, the second to the second, and so on. The result is the sum of the vectors. To add $\mathbf{u}$ and $\mathbf{v}$ geometrically, position $\mathbf{v}$ so that its initial point is at the terminal point of $\mathbf{u}$; then $\mathbf{u} + \mathbf{v}$ is the vector from the initial point of $\mathbf{u}$ to the terminal point of $\mathbf{v}$.

11. $\langle 2a, 2b, 2c \rangle$ is such a vector. It is parallel to $\langle a, b, c \rangle$ since it is a scalar multiple of $\langle a, b, c \rangle$, and its norm is

$$\|\langle 2a, 2b, 2c \rangle\| = \sqrt{(2a)^2 + (2b)^2 + (2c)^2} = \sqrt{4(a^2 + b^2 + c^2)} = 2\sqrt{a^2 + b^2 + c^2} = 2\|\langle a, b, c \rangle\|.$$

13. Let (x, y, z) denote the terminal point of the vector. Then $\langle 2, 3, -5 \rangle = \langle x - (-3), y - 2, z - 4 \rangle$, so that $x + 3 = 2$, $y - 2 = 3$, and $z - 4 = -5$. It follows that $(x, y, z) = (-1, 5, -1)$ is the terminal point of the vector.

15. The magnitude of $\mathbf{u} = \langle 1, 2, 3 \rangle$ is $\|\langle 1, 2, 3 \rangle\| = \sqrt{1^2 + 2^2 + 3^2} = \sqrt{14}$, so that $\left\langle \frac{1}{\sqrt{14}}, \frac{2}{\sqrt{14}}, \frac{3}{\sqrt{14}} \right\rangle$ is a unit vector in the direction of $\mathbf{u}$. Thus a vector of magnitude 5 parallel to $\mathbf{u}$ is

$$5\left\langle \frac{1}{\sqrt{14}}, \frac{2}{\sqrt{14}}, \frac{3}{\sqrt{14}} \right\rangle = \left\langle \frac{5}{\sqrt{14}}, \frac{10}{\sqrt{14}}, \frac{15}{\sqrt{14}} \right\rangle.$$

If the initial point is $\langle 0, 3, -2 \rangle$, then the terminal point is at

$$\left(\frac{5}{\sqrt{14}}, \frac{10}{\sqrt{14}} + 3, \frac{15}{\sqrt{14}} - 2 \right).$$

17. (a) The set of points $\mathbf{v} = (x, y, z) \in \mathbb{R}^3$ satisfying $\|\mathbf{v}\| = 4$ is the set of points such that $\|\mathbf{v}\| = \sqrt{x^2 + y^2 + z^2} = 4$, which is the sphere of radius 4 centered at the origin.

(b) The set of points $\mathbf{v} = (x, y, z) \in \mathbb{R}^3$ satisfying $\|\mathbf{v}\| \leq 4$ is the set of points such that $\|\mathbf{v}\| = \sqrt{x^2 + y^2 + z^2} \leq 4$, which is the sphere of radius 4 centered at the origin together with its interior (that is, it is the closed ball of radius 4 centered at the origin).

(c) The set of points $\mathbf{v} = (x, y, z) \in \mathbb{R}^3$ satisfying $\|\mathbf{v} - \mathbf{v}_0\| = 4$ is the set of points such that $\|\mathbf{v} - \mathbf{v}_0\| = \sqrt{(x-a)^2 + (y-b)^2 + (z-c)^2} = 4$, i.e., such that $(x-a)^2 + (y-b)^2 + (z-c)^2 = 4^2$. This is the sphere of radius 4 centered at $\mathbf{v}_0 = (a, b, c)$.

19. All of the ideas generalize in a straightforward manner. Adding and subtracting vectors geometrically is identical, except you must work in a higher-dimensional space. Adding and subtracting vectors algebraically is also identical, adding or subtracting each of the coordinates. The concept of norm generalizes as well; if $\mathbf{u} = (x_1, x_2, \ldots, x_n)$, then $\|\mathbf{u}\| = \sqrt{x_1^2 + x_2^2 + \cdots + x_n^2}$. The concepts of equality, the zero vector, unit vectors, and parallel vectors also generalize in a straightforward manner.

21. The set of position vectors of magnitude (norm) 5 in $\mathbb{R}^3$ is

$$\{(x, y, z) \in \mathbb{R}^3 \mid \sqrt{x^2 + y^2 + z^2} = 5\} = \{(x, y, z) \in \mathbb{R}^3 \mid x^2 + y^2 + z^2 = 5^2\},$$

which is the sphere of radius 5 centered at the origin.

Skills

23. We have

$$\mathbf{u} + \mathbf{v} = \langle 2, -6 \rangle + \langle 6, 2 \rangle = \langle 2+6, -6+2 \rangle = \langle 8, -4 \rangle$$
$$\mathbf{u} - \mathbf{v} = \langle 2, -6 \rangle - \langle 6, 2 \rangle = \langle 2-6, -6-2 \rangle = \langle -4, -8 \rangle.$$

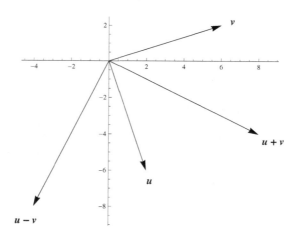

25. We have

$$\mathbf{u} + \mathbf{v} = \langle 1, -2 \rangle + \langle -3, 6 \rangle = \langle 1-3, -2+6 \rangle = \langle -2, 4 \rangle$$
$$\mathbf{u} - \mathbf{v} = \langle 1, -2 \rangle - \langle -3, 6 \rangle = \langle 1-(-3), -2-6 \rangle = \langle 4, -8 \rangle.$$

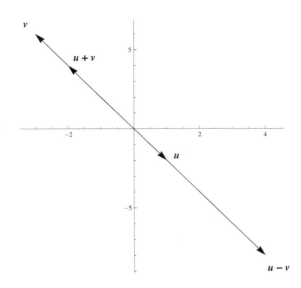

27. We have

$$\mathbf{u} + \mathbf{v} = \langle 1, -4, 6 \rangle + \langle 2, -4, 7 \rangle = \langle 1 + 2, -4 + (-4), 6 + 7 \rangle = \langle 3, -8, 13 \rangle$$
$$\mathbf{u} - \mathbf{v} = \langle 1, -4, 6 \rangle - \langle 2, -4, 7 \rangle = \langle 1 - 2, -4 - (-4), 6 - 7 \rangle = \langle -1, 0, -1 \rangle .$$

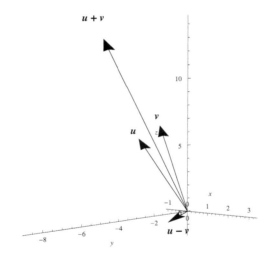

29. $\overrightarrow{PQ} = Q - P = \langle -3 - 3, -2 - 6 \rangle = \langle -6, -8 \rangle.$

31. $\overrightarrow{PQ} = Q - P = \langle 1 - 1, -7 - (-2), -2 - 5 \rangle = \langle 0, -5, -7 \rangle.$

33. The norm of $\mathbf{v}$ is $\| \langle 3, -4 \rangle \| = \sqrt{3^2 + (-4)^2} = \sqrt{9 + 16} = \sqrt{25} = 5.$

35. The norm of $\mathbf{v}$ is $\left\| \left\langle \frac{1}{3}, \frac{1}{3}, -\frac{1}{3} \right\rangle \right\| = \sqrt{\left(\frac{1}{3} \right)^2 + \left(\frac{1}{3} \right)^2 + \left(-\frac{1}{3} \right)^2} = \sqrt{\frac{1}{9} + \frac{1}{9} + \frac{1}{9}} = \sqrt{\frac{3}{9}} = \frac{\sqrt{3}}{3}.$

37. The norm of $\mathbf{v}$ is $\| \mathbf{v} \| = \sqrt{3^2 + (-4)^2} = \sqrt{25} = 5$, so that the unit vector in the direction of $\mathbf{v}$ is

$$\frac{\mathbf{v}}{\| \mathbf{v} \|} = \left\langle \frac{3}{5}, -\frac{4}{5} \right\rangle .$$

39. The norm of $\mathbf{v}$ is $\|\mathbf{v}\| = \sqrt{\left(\frac{1}{5}\right)^2 + \left(\frac{1}{3}\right)^2} = \sqrt{\frac{1}{25} + \frac{1}{9}} = \sqrt{\frac{34}{225}} = \frac{\sqrt{34}}{15}$, so that the unit vector in the direction of $\mathbf{v}$ is

$$\frac{\mathbf{v}}{\|\mathbf{v}\|} = \left\langle \frac{1}{5} \cdot \frac{15}{\sqrt{34}}, \frac{1}{3} \cdot \frac{15}{\sqrt{34}} \right\rangle = \left\langle \frac{3}{\sqrt{34}}, \frac{5}{\sqrt{34}} \right\rangle = \left\langle \frac{3\sqrt{34}}{34}, \frac{5\sqrt{34}}{34} \right\rangle.$$

41. The norm of $\mathbf{v}$ is $\|\mathbf{v}\| = \sqrt{1^2 + 1^2 + 1^2} = \sqrt{3}$, so that the unit vector in the direction of $\mathbf{v}$ is

$$\frac{\mathbf{v}}{\|\mathbf{v}\|} = \left\langle \frac{1}{\sqrt{3}}, \frac{1}{\sqrt{3}}, \frac{1}{\sqrt{3}} \right\rangle = \left\langle \frac{\sqrt{3}}{3}, \frac{\sqrt{3}}{3}, \frac{\sqrt{3}}{3} \right\rangle.$$

43. The norm of the vector is $\|\langle 3, 1, 2 \rangle\| = \sqrt{3^1 + 1^2 + 2^2} = \sqrt{14}$, so that a unit vector in the same direction is $\left\langle \frac{3}{\sqrt{14}}, \frac{1}{\sqrt{14}}, \frac{2}{\sqrt{14}} \right\rangle$. Thus a vector of magnitude 5 in that direction is

$$5 \left\langle \frac{3}{\sqrt{14}}, \frac{1}{\sqrt{14}}, \frac{2}{\sqrt{14}} \right\rangle = \left\langle \frac{15}{\sqrt{14}}, \frac{5}{\sqrt{14}}, \frac{10}{\sqrt{14}} \right\rangle.$$

45. The norm of the vector is $\|\langle 8, -7, 2 \rangle\| = \sqrt{8^2 + 7^2 + 2^2} = \sqrt{117} = 3\sqrt{13}$, so that a unit vector in the same direction is $\left\langle \frac{8}{3\sqrt{13}}, -\frac{7}{3\sqrt{13}}, \frac{2}{3\sqrt{13}} \right\rangle$. Thus a vector of magnitude 2 in that direction is

$$2 \left\langle \frac{8}{3\sqrt{13}}, -\frac{7}{3\sqrt{13}}, \frac{2}{3\sqrt{13}} \right\rangle = \left\langle \frac{16}{3\sqrt{13}}, -\frac{14}{3\sqrt{13}}, \frac{4}{3\sqrt{13}} \right\rangle.$$

47. The norm of the vector is $\|\langle -1, -4, -6 \rangle\| = \sqrt{(-1)^2 + (-4)^2 + (-6)^2} = \sqrt{53}$, so that a unit vector in the opposite direction is the negative of the given vector divided by its norm, which is $\left\langle \frac{1}{\sqrt{53}}, \frac{4}{\sqrt{53}}, \frac{6}{\sqrt{53}} \right\rangle$. Thus a vector in the opposite direction with magnitude 7 is

$$7 \left\langle \frac{1}{\sqrt{53}}, \frac{4}{\sqrt{53}}, \frac{6}{\sqrt{53}} \right\rangle = \left\langle \frac{7}{\sqrt{53}}, \frac{28}{\sqrt{53}}, \frac{42}{\sqrt{53}} \right\rangle.$$

49. A vector in the opposite direction to $\langle 0, -3, 4 \rangle$ is $\langle 0, 3, -4 \rangle$ (negate each of the components). The magnitude of this vector is $\sqrt{0^2 + 3^2 + 4^2} = 5$, so that to make the vector have magnitude 10, we must multiply it by 2. The requested vector is $\langle 0, 6, -8 \rangle$.

51. A vector opposite to $\langle 1, -2, 3 \rangle$ is $\langle -1, 2, -3 \rangle$. The norm of that vector is $\sqrt{(-1)^2 + 2^2 + 3^2} = \sqrt{14}$, so that a unit vector in the direction of $\langle -1, 2, -3 \rangle$ is $\left\langle -\frac{1}{\sqrt{14}}, \frac{2}{\sqrt{14}}, -\frac{3}{\sqrt{14}} \right\rangle$. Thus a vector of length 3 in the same direction is

$$3 \left\langle -\frac{1}{\sqrt{14}}, \frac{2}{\sqrt{14}}, -\frac{3}{\sqrt{14}} \right\rangle = \left\langle -\frac{3}{\sqrt{14}}, \frac{6}{\sqrt{14}}, -\frac{9}{\sqrt{14}} \right\rangle.$$

Applications

53. As in Example 3, let $\mathbf{u}$ be the force on the left rope, $\mathbf{v}$ the force on the right rope, and $\mathbf{w}$ the gravitational force. Then $\mathbf{u} + \mathbf{v} + \mathbf{w} = \mathbf{0}$. Since the angle on the left rope is $45°$, and that on the right rope is $30°$, we get (taking signs of forces into account)

$$\mathbf{u} = -\cos 45° \cdot \|\mathbf{u}\|\mathbf{i} + \sin 45° \cdot \|\mathbf{u}\|\mathbf{j} = -\frac{\sqrt{2}}{2} \cdot \|\mathbf{u}\|\mathbf{i} + \frac{\sqrt{2}}{2} \cdot \|\mathbf{u}\|\mathbf{j}$$

$$\mathbf{v} = \cos 30° \cdot \|\mathbf{v}\|\mathbf{i} + \sin 30° \cdot \|\mathbf{v}\|\mathbf{j} = \frac{\sqrt{3}}{2} \cdot \|\mathbf{v}\|\mathbf{i} + \frac{1}{2} \cdot \|\mathbf{v}\|\mathbf{j}$$

$$\mathbf{w} = -100\mathbf{j}.$$

Then

$$\mathbf{0} = \mathbf{u} + \mathbf{v} + \mathbf{w} = \left(-\frac{\sqrt{2}}{2} \cdot \|\mathbf{u}\| + \frac{\sqrt{3}}{2} \cdot \|\mathbf{v}\| \right) \mathbf{i} + \left(\frac{\sqrt{2}}{2} \cdot \|\mathbf{u}\| + \frac{1}{2} \cdot \|\mathbf{v}\| - 100 \right) \mathbf{j}.$$

The first equation gives $\|\mathbf{u}\| = \frac{\sqrt{3}}{\sqrt{2}} \|\mathbf{v}\|$; substituting into the second equation gives

$$\frac{\sqrt{2}}{2} \cdot \frac{\sqrt{3}}{\sqrt{2}} \|\mathbf{v}\| + \frac{1}{2} \cdot \|\mathbf{v}\| = 100,$$

so that $\|\mathbf{v}\| = \frac{200}{1+\sqrt{3}} \approx 73.2051$ pounds and thus $\|\mathbf{u}\| = \frac{\sqrt{3}}{\sqrt{2}} \|\mathbf{v}\| \approx 89.6575$ pounds.

55. (a) Assuming the positive x axis points east and the positive y axis points north, the vector is $\langle 0, -2 \rangle$.

(b) Assuming the positive x axis points east and the positive y axis points north, the vector is in the direction $\langle 1, -1 \rangle$ with a length of 2, so it is $\left\langle \sqrt{2}, -\sqrt{2} \right\rangle$.

(c) With the same axes as in the previous two parts, the tide is $\langle -2, 0 \rangle$, and her velocity due to paddling is in the direction $\langle 1, -1 \rangle$ with a length of 3, so it $\left\langle \sqrt{3}, -\sqrt{3} \right\rangle$. The sum of these two vectors is $\left\langle -2 + \sqrt{3}, -\sqrt{3} \right\rangle$.

Proofs

57. Using the commutative property for addition of real numbers, we get

$$\mathbf{u} + \mathbf{v} = \langle u_1, u_2, u_3 \rangle + \langle v_1, v_2, v_3 \rangle = \langle u_1 + v_1, u_2 + v_2, u_3 + v_3 \rangle = \langle v_1 + u_1, v_2 + u_2, v_3 + u_3 \rangle$$
$$= \langle v_1, v_2, v_3 \rangle + \langle u_1, u_2, u_3 \rangle = \mathbf{v} + \mathbf{u}.$$

59. Using the distributive property of multiplication over addition for real numbers gives

$$c(\mathbf{u} + \mathbf{v}) = c \left(\langle u_1, u_2, u_3 \rangle + \langle v_1, v_2, v_3 \rangle \right)$$
$$= c \langle u_1 + v_1, u_2 + v_2, u_3 + v_3 \rangle$$
$$= \langle c(u_1 + v_1), c(u_2 + v_2), c(u_3 + v_3) \rangle$$
$$= \langle cu_1 + cv_1, cu_2 + cv_2, cu_3 + cv_3 \rangle$$
$$= \langle cu_1, cu_2, cu_3 \rangle + \langle cv_1, cv_2, cv_3 \rangle$$
$$= c \langle u_1, u_2, u_3 \rangle + c \langle v_1, v_2, v_3 \rangle$$
$$= c\mathbf{u} + c\mathbf{v}.$$

61. Suppose $\mathbf{v} = \langle v_1, v_2, v_3 \rangle$. Then using the distributive property of multiplication over addition for reals, we have

$$(c + d)\mathbf{v} = (c + d) \langle v_1, v_2, v_3 \rangle$$
$$= \langle (c+d)v_1, (c+d)v_2, (c+d)v_3 \rangle$$
$$= \langle cv_1 + dv_1, cv_2 + dv_2, cv_3 + dv_3 \rangle$$
$$= \langle cv_1, cv_2, cv_3 \rangle + \langle dv_1, dv_2, dv_2 \rangle$$
$$= c \langle v_1, v_2, v_3 \rangle + d \langle v_1, v_2, v_3 \rangle$$
$$= c\mathbf{v} + d\mathbf{v}.$$

63. Consider the following diagram of the situation:

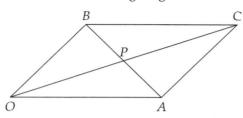

Place the parallelogram so that one vertex O is at the origin and let $\mathbf{u}$ and $\mathbf{v}$ be the vectors describing the two sides emanating from the origin, so that their terminal points are vertices A and B of the parallelogram. The fourth vertex is then at $\mathbf{u} + \mathbf{v}$; call that vertex C. Call the point of intersection of the diagonals P.

The midpoints of the diagonals are

$$\text{midpoint of } OC = \frac{\mathbf{0} + (\mathbf{u} + \mathbf{v})}{2} = \frac{\mathbf{u} + \mathbf{v}}{2} \text{ and}$$
$$\text{midpoint of } AB = \frac{\mathbf{u} + \mathbf{v}}{2}.$$

The two midpoints are identical vectors, so the midpoints of the diagonals coincide. Thus the two diagonals bisect each other at P.

65. Consider the following diagram of the situation:

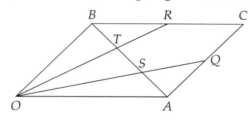

Place the parallelogram so that one vertex O is at the origin and let $\mathbf{u}$ and $\mathbf{v}$ be the vectors describing the two sides emanating from the origin, so that their terminal points are vertices A and B of the parallelogram. The fourth vertex is then at $\mathbf{u} + \mathbf{v}$; call that vertex C. The other intersections are as labeled.

Since $\overrightarrow{BA} = \mathbf{v} - \mathbf{u}$, one third of the way from B to A is $\frac{1}{3}(\mathbf{v} - \mathbf{u})$, so that the vector between O and that point is $K = \mathbf{u} + \frac{1}{3}(\mathbf{v} - \mathbf{u}) = \frac{\mathbf{v}+2\mathbf{u}}{3}$. Now, $\overrightarrow{OR}$ is the midpoint of BC, so it is the vector $\frac{\mathbf{u}+(\mathbf{u}+\mathbf{v})}{2} = \frac{2\mathbf{u}+\mathbf{v}}{2}$. Then K is a scalar multiple (namely $\frac{2}{3}$) of that vector, so it lies on the segment OR. Hence OR and AB meet at a point that is one third of the way from B to A.

Similarly, the point two thirds of the way from B to A is $\frac{2}{3}(\mathbf{v} - \mathbf{u})$, so that the vector between O and that point is $L = \mathbf{u} + \frac{2}{3}(\mathbf{v} - \mathbf{u}) = \frac{2\mathbf{v}+\mathbf{u}}{3}$. Now, $\overrightarrow{OQ}$ is the midpoint of AC, so it is the vector $\frac{\mathbf{v}+(\mathbf{u}+\mathbf{v})}{2} = \frac{\mathbf{u}+2\mathbf{v}}{2}$. Then L is a scalar multiple (namely $\frac{2}{3}$) of that vector, so it lies on the segment OQ. Hence OQ and AB meet at a point that is one third of the way from A to B. Thus the two lines OR and OQ divide AB into thirds.

Thinking Forward

Perpendicular vectors in $\mathbb{R}^2$

▷ There are an infinite number. Suppose the unit vector in the direction of $\langle a, b \rangle$ is $\langle \cos \theta, \sin \theta \rangle$. Then $\langle \cos \left(\theta + \frac{\pi}{2} \right), \sin \left(\theta + \frac{\pi}{2} \right) \rangle = \langle -\sin \theta, \cos \theta \rangle$ is a perpendicular unit vector, so that any scalar multiple of that vector is perpendicular to $\langle a, b \rangle$.

▷ There are two. One, as above, is $\langle -\sin \theta, \cos \theta \rangle$; the other is the unit vector in the opposite direction, which is $\langle \sin \theta, -\cos \theta \rangle$.

▷ Since $\| \langle 1, 3 \rangle \| = \sqrt{1^2 + 3^2} = \sqrt{10}$, the unit vector in the direction of $\langle 1, 3 \rangle$ is $\langle \frac{1}{\sqrt{10}}, \frac{3}{\sqrt{10}} \rangle$. By the first part of this problem, the vector $\langle -\frac{3}{\sqrt{10}}, \frac{1}{\sqrt{10}} \rangle$ is a vector (in fact, a unit vector) perpendicular to $\langle 1, 3 \rangle$.

▷ See the previous part. $\left\langle -\frac{3}{\sqrt{10}}, \frac{1}{\sqrt{10}} \right\rangle$ is such a unit vector.

Perpendicular vectors in $\mathbb{R}^3$

▷ Any plane in $\mathbb{R}^3$ that is perpendicular to a given vector contains an infinite number of lines that are perpendicular to the given vector. Each of these lines provides an infinite number of vectors (scalar multiples of a direction vector) that are perpendicular to the given vector.

▷ Each line in the previous part that is perpendicular to the given vector provides two unit vectors, one in each direction, that are perpendicular to the given vector.

▷ **j** is perpendicular to both **i** and **k**, since it points in the direction of the y axis, which is perpendicular to both the x and z axes; these axes are the directions of **i** and **j**.

10.3 Dot Product

Thinking Back

Law of Cosines

Let A be the angle between the sides of length 2 and 3, B the angle between the sides of length 3 and 4, and C the remaining angle. Then by the Law of Cosines,

$$4^2 = 2^2 + 3^2 - 2 \cdot 2 \cdot 3 \cos A = 13 - 12 \cos A,$$
$$2^2 = 3^2 + 4^2 - 2 \cdot 3 \cdot 4 \cos B = 25 - 24 \cos B,$$
$$3^2 = 2^2 + 4^2 - 2 \cdot 2 \cdot 4 \cos C = 20 - 16 \cos C.$$

Thus

$$A = \cos^{-1} \frac{16 - 13}{-12} = \cos^{-1} \frac{-1}{4} \approx 104.48°,$$
$$B = \cos^{-1} \frac{4 - 25}{-24} = \cos^{-1} \frac{7}{8} \approx 29.96°,$$
$$C = \cos^{-1} \frac{9 - 20}{-16} = \cos^{-1} \frac{11}{16} \approx 46.57°.$$

The distance between a point and a sphere

The shortest distance from $(11, -3, 5)$ to the given sphere is the point where the line segment from $(11, -3, 5)$ to the center of the sphere, at $(-2, 3, 0)$, intersects the sphere. The length of this segment is 4, the radius of the sphere, plus the distance from the point to the sphere. The length of the segment is

$$\sqrt{(11 - (-2))^2 + (-3 - 3)^2 + (5 - 0)^2} = \sqrt{169 + 36 + 25} = \sqrt{230}.$$

Thus the distance from the $(11, -3, 5)$ to the surface of the sphere is $\sqrt{230} - 4$.

The Triangle Inequality

We have
$$|a + b| = |5 + (-3)| = |2| = 2, \qquad |a| + |b| = |5| + |-3| = 5 + 3 = 8,$$
and indeed $2 \le 8$.

Concepts

1. (a) True. See Theorem 10.14(a).

   (b) False. The component of the projection is the *norm* of the projection vector $\text{proj}_{\mathbf{u}}\,\mathbf{v}$, and is therefore a scalar.

   (c) False. $\text{proj}_{\mathbf{u}}\,\mathbf{v}$ is parallel to $\mathbf{u}$.

   (d) False. Two curves are orthogonal at a point if they intersect there and if their tangent lines at that point are perpendicular. See Definition 10.18.

   (e) True. By Theorem 10.15, $\mathbf{u}\cdot\mathbf{v} = \|\mathbf{u}\|\,\|\mathbf{v}\|\cos\theta$, so that $\cos\theta = \frac{\mathbf{u}\cdot\mathbf{v}}{\|\mathbf{u}\|\|\mathbf{v}\|}$.

   (f) True. Since

   $$\text{proj}_{\mathbf{u}}\,\mathbf{v} = \frac{\mathbf{u}\cdot\mathbf{v}}{\|\mathbf{u}\|^2}\mathbf{u}, \qquad \text{proj}_{\mathbf{v}}\,\mathbf{u} = \frac{\mathbf{v}\cdot\mathbf{u}}{\|\mathbf{v}\|^2}\mathbf{v},$$

   if the two are equal, then $\frac{\mathbf{u}\cdot\mathbf{v}}{\|\mathbf{u}\|^2}\mathbf{u} = \frac{\mathbf{v}\cdot\mathbf{u}}{\|\mathbf{v}\|^2}\mathbf{v}$. This holds if $\mathbf{u}$ and $\mathbf{v}$ are orthogonal since then $\mathbf{u}\cdot\mathbf{v} = \mathbf{v}\cdot\mathbf{u} = 0$. If they are not orthogonal, divide through by $\mathbf{u}\cdot\mathbf{v} = \mathbf{v}\cdot\mathbf{u}$ and rewrite the equation as

   $$\frac{1}{\|\mathbf{u}\|}\cdot\frac{\mathbf{u}}{\|\mathbf{u}\|} = \frac{1}{\|\mathbf{v}\|}\cdot\frac{\mathbf{v}}{\|\mathbf{v}\|}.$$

   Since each of $\frac{\mathbf{u}}{\|\mathbf{u}\|}$ and $\frac{\mathbf{v}}{\|\mathbf{v}\|}$ are unit vectors, and since both scalars are positive, it follows that if the equality is to hold, then the scalars must be equal, so that $\|\mathbf{u}\| = \|\mathbf{v}\|$. Replacing $\|\mathbf{v}\|$ by $\|\mathbf{u}\|$ in the above equation and simplifying gives $\mathbf{u} = \mathbf{v}$. So if the two projections are equal, the vectors are either orthogonal or equal.

   (g) True. If $\mathbf{u}\cdot\mathbf{v} = \|\mathbf{u}\|\,\|\mathbf{v}\|$, then the angle θ between $\mathbf{u}$ and $\mathbf{v}$ is given by $\cos\theta = \frac{\mathbf{u}\cdot\mathbf{v}}{\|\mathbf{u}\|\|\mathbf{v}\|} = 1$, so that $\mathbf{u}$ and $\mathbf{v}$ are parallel. Thus $\mathbf{u} = k\mathbf{v}$ where k is a constant. However, note also that since $\mathbf{u}\cdot\mathbf{v}$ is equal to a product of norms, it is positive, so that the angle between $\mathbf{u}$ and $\mathbf{v}$ must be acute. Since the vectors are parallel, they must therefore point in the same direction, so that $k > 0$.

   (h) False. Since $\mathbf{u}\cdot\mathbf{v}$ is a scalar, the dot product of that scalar with $\mathbf{w}$ is not defined — dot products are products of two vectors.

3. See Definition 10.13. In $\mathbb{R}^2$, if $\mathbf{u} = \langle u_1, u_2\rangle$ and $\mathbf{v} = \langle v_1, v_2\rangle$, then $\mathbf{u}\cdot\mathbf{v} = u_1 v_1 + u_2 v_2$.

5. In the formula $a^2 = b^2 + c^2 - 2bc\cos A$, the angle A is the angle where the sides of length b and c meet. If A is a right angle, then $\cos A = 0$ and the formula becomes $a^2 = b^2 + c^2$, so that a is the hypotenuse of a right triangle with legs b and c.

7. (a) The vectors are parallel if one is a multiple of the other. Thus if none of a, b, c, α, β, γ is zero, then we need $\frac{\alpha}{a} = \frac{\beta}{b} = \frac{\gamma}{c}$. If one of the coefficients is zero, then the corresponding coefficient in the other formula must also be zero, and all remaining nonzero coefficients must satisfy the same ratio equality as above. Thus in either case the condition is $a\mathbf{i} + b\mathbf{j} + c\mathbf{k} = k(\alpha\mathbf{i} + \beta\mathbf{j} + \gamma\mathbf{k})$ where k is a scalar.

   (b) The vectors are perpendicular if their dot product is zero, so the condition is

   $$(a\mathbf{i} + b\mathbf{j} + c\mathbf{k})\cdot(\alpha\mathbf{i} + \beta\mathbf{j} + \gamma\mathbf{k}) = a\alpha + b\beta + c\gamma = 0.$$

9. (a) For example, let $\mathbf{u} = \langle 1, 0\rangle$, $\mathbf{v} = \langle 0, 1\rangle$, and $\mathbf{w} = \langle 0, 2\rangle$. Then $\mathbf{u}\cdot\mathbf{v} = \mathbf{u}\cdot\mathbf{w} = 0$, but $\mathbf{v} \neq \mathbf{w}$.

   (b) If $\mathbf{u}\cdot\mathbf{v} = \mathbf{u}\cdot\mathbf{w}$, then $\mathbf{u}\cdot\mathbf{v} - \mathbf{u}\cdot\mathbf{w} = \mathbf{u}\cdot(\mathbf{v} - \mathbf{w}) = 0$, so that $\mathbf{u}$ must be perpendicular to $\mathbf{v} - \mathbf{w}$.

11. Since

$$\|\mathbf{u} + \mathbf{v}\|^2 = (\mathbf{u} + \mathbf{v}) \cdot (\mathbf{u} + \mathbf{v})$$
$$= \mathbf{u} \cdot \mathbf{u} + \mathbf{u} \cdot \mathbf{v} + \mathbf{v} \cdot \mathbf{u} + \mathbf{v} \cdot \mathbf{v}$$
$$= \|\mathbf{u}\|^2 + 2\mathbf{u} \cdot \mathbf{v} + \|\mathbf{v}\|^2$$
$$= \|\mathbf{u}\|^2 + 2\|\mathbf{u}\|\,\|\mathbf{v}\| \cos\theta + \|\mathbf{v}\|^2 .$$

This last expression is equal to $(\|\mathbf{u}\| + \|\mathbf{v}\|)^2$ if and only if $\cos\theta = 1$ (and thus $\theta = 0$), where θ is the angle between $\mathbf{u}$ and $\mathbf{v}$. So $\|\mathbf{u} + \mathbf{v}\|^2 = (\|\mathbf{u}\| + \|\mathbf{v}\|)^2$ (and thus $\|\mathbf{u} + \mathbf{v}\| = \|\mathbf{u}\| + \|\mathbf{v}\|$) if and only if the vectors point in the same direction.

13. Let $\mathbf{v} = \langle v_1, v_2, v_3 \rangle$. Then $\mathbf{v} \cdot \mathbf{i} = v_1 \cdot 1 + v_2 \cdot 0 + v_3 \cdot 0 = v_1$, so that $\mathbf{v} \cdot \mathbf{i} = 0$ if and only if $v_1 = 0$, so that $\mathbf{v} = \langle 0, v_2, v_3 \rangle$ where v_2 and v_3 are arbitrary.

15. Let $\mathbf{v} = \langle v_1, v_2, v_3 \rangle$. Then

$$\text{proj}_{\mathbf{i}}\, \mathbf{v} = \frac{\mathbf{v} \cdot \mathbf{i}}{\|\mathbf{i}\|^2}\mathbf{i} = \frac{v_1 \cdot 1 + v_2 \cdot 0 + v_3 \cdot 0}{1}\mathbf{i} = v_1\mathbf{i}.$$

Thus $\text{proj}_{\mathbf{i}}\, \mathbf{v} = \mathbf{i}$ if and only if $\mathbf{v} = \langle 1, v_2, v_3 \rangle$ where v_2 and v_3 are arbitrary.

17. (a) The slope of the line can be seen as a vector parallel to the line, so for example $\langle 1, m \rangle$ is parallel to the line. (Another way to see this is that $\langle 1, m \rangle$ has the same "rise-over-run" as $y = mx + b$).

    (b) If $m \neq 0$, then the slope of a line perpendicular to $y = mx + b$ is $-\frac{1}{m}$, so that $\left\langle 1, -\frac{1}{m} \right\rangle$, or (multiplying through by the scalar $-m$) $\langle -m, 1 \rangle$ is parallel to that perpendicular line.

    (c) $\langle 1, m \rangle \cdot \langle -m, 1 \rangle = 1 \cdot (-m) + m \cdot 1 = 0$.

19. (a) If they were, say $\mathbf{v}_2 = k\mathbf{v}_1$, then we would have $3 = k \cdot 1$ so that $k = 3$, but also $5 = k \cdot (-2)$, so that $k = -\frac{5}{2}$. This is plainly impossible.

    (b) The equations in the previous exercise become

$$1c_1 + 3c_2 = 5$$
$$-2c_1 + 5c_2 = 1.$$

Solving gives $c_1 = 2$ and $c_2 = 1$, so that

$$\langle 5, 1 \rangle = 2 \langle 1, -2 \rangle + \langle 3, 5 \rangle .$$

Skills

21. We have $\langle 2, 0, -5 \rangle \cdot \langle -3, 7, -1 \rangle = 2 \cdot (-3) + 0 \cdot 7 - 5 \cdot (-1) = -1$. Then the angle θ between the vectors satisfies

$$\cos\theta = -\frac{1}{\|\langle 2, 0, -5 \rangle\| \, \|\langle -3, 7, -1 \rangle\|} = -\frac{1}{\sqrt{29} \cdot \sqrt{59}} = -\frac{1}{\sqrt{1711}}.$$

Then $\theta = \cos^{-1}\left(-\frac{1}{\sqrt{1711}}\right) \approx 91.4°$.

23. We have $\langle -5, 1, 3 \rangle \cdot \langle -3, 2, 7 \rangle = -5 \cdot (-3) + 1 \cdot 2 + 3 \cdot 7 = 38$. Then the angle θ between the vectors satisfies

$$\cos\theta = \frac{38}{\|\langle -5, 1, 3 \rangle\| \, \|\langle -3, 2, 7 \rangle\|} = \frac{38}{\sqrt{35} \cdot \sqrt{62}} = \frac{38}{\sqrt{2170}}.$$

Then $\theta = \cos^{-1}\frac{38}{\sqrt{2170}} \approx 35.3°$.

25. We have $\mathbf{u} \cdot \mathbf{v} = \langle 3, -1, 2 \rangle \cdot \langle -4, -6, 3 \rangle = 3 \cdot (-4) - 1 \cdot (-6) + 2 \cdot 3 = 0$, so that $\mathbf{u}$ and $\mathbf{v}$ are perpendicular. Then

$$\text{comp}_{\mathbf{u}} \mathbf{v} = \frac{\mathbf{u} \cdot \mathbf{v}}{\|\mathbf{u}\|} = 0, \quad \text{and} \quad \text{proj}_{\mathbf{u}} \mathbf{v} = \frac{\mathbf{u} \cdot \mathbf{v}}{\|\mathbf{u}\|^2} \mathbf{u} = \mathbf{0},$$

so that the component of $\mathbf{v}$ orthogonal to $\mathbf{u}$ is

$$\mathbf{v} - \text{proj}_{\mathbf{u}} \mathbf{v} = \langle -4, -6, 3 \rangle - \mathbf{0} = \langle -4, -6, 3 \rangle .$$

27. We have $\|\mathbf{u}\| = \sqrt{3^2 + 1^2 + (-2)^2} = \sqrt{14}$, and

$$\text{comp}_{\mathbf{u}} \mathbf{v} = \frac{\mathbf{u} \cdot \mathbf{v}}{\|\mathbf{u}\|} = \frac{\langle 3, 1, -2 \rangle \cdot \langle -6, -2, 4 \rangle}{\sqrt{3^2 + 1^2 + (-2)^2}} = -\frac{28}{\sqrt{14}} = -2\sqrt{14}.$$

Then

$$\text{proj}_{\mathbf{u}} \mathbf{v} = \frac{\mathbf{u} \cdot \mathbf{v}}{\|\mathbf{u}\|^2} \mathbf{u} = \text{comp}_{\mathbf{u}} \mathbf{v} \frac{\mathbf{u}}{\|\mathbf{u}\|} = -2\mathbf{u} = \langle -6, -2, 4 \rangle ,$$

so that the component of $\mathbf{v}$ orthogonal to $\mathbf{u}$ is

$$\mathbf{v} - \text{proj}_{\mathbf{u}} \mathbf{v} = \langle -6, -2, 4 \rangle - \langle -6, -2, 4 \rangle = \mathbf{0}.$$

Since the orthogonal component is zero, the two vectors are parallel — in fact, $\mathbf{v} = -2\mathbf{u}$.

29. With $\mathbf{u} = \langle 3, -4 \rangle$ and $\mathbf{v} = \langle 16, 12 \rangle$ we have $\mathbf{u} \cdot \mathbf{v} = 3 \cdot 16 - 4 \cdot 12 = 0$, so that $\text{proj}_{\mathbf{u}} \mathbf{v} = \text{proj}_{\mathbf{v}} \mathbf{u} = \mathbf{0}$.

31. With $\mathbf{u} = \langle 1, -5, -1 \rangle$ and $\mathbf{v} = \langle 0, 1, 0 \rangle$ we have

$$\text{proj}_{\mathbf{u}} \mathbf{v} = \frac{\mathbf{u} \cdot \mathbf{v}}{\|\mathbf{u}\|^2} \mathbf{u} = \frac{1 \cdot 0 - 5 \cdot 1 - 1 \cdot 0}{1^2 + (-5)^2 + (-1)^2} \langle 1, -5, -1 \rangle = -\frac{5}{27} \langle 1, -5, -1 \rangle$$

$$\text{proj}_{\mathbf{v}} \mathbf{u} = \frac{\mathbf{u} \cdot \mathbf{v}}{\|\mathbf{v}\|^2} \mathbf{v} = \frac{1 \cdot 0 - 5 \cdot 1 - 1 \cdot 0}{0^2 + 1^2 + 0^2} \langle 0, 1, 0 \rangle = -5 \langle 0, 1, 0 \rangle = \langle 0, -5, 0 \rangle .$$

33. With $\mathbf{u} = \langle -2, 3, 5 \rangle$ and $\mathbf{v} = \langle 13, -5, 8 \rangle$, we have

$$\mathbf{u} \cdot \mathbf{v} = -2 \cdot 13 + 3 \cdot (-5) + 5 \cdot 8 = -1$$

$$\|\mathbf{u}\| = \sqrt{(-2)^2 + 3^2 + 5^2} = \sqrt{38}, \quad \|\mathbf{v}\| = \sqrt{13^2 + (-5)^2 + 8^2} = \sqrt{258},$$

so that

(a) $\mathbf{u} \cdot \mathbf{v} = -1$.

(b) If θ is the angle between $\mathbf{u}$ and $\mathbf{v}$, then

$$\cos \theta = \frac{\mathbf{u} \cdot \mathbf{v}}{\|\mathbf{u}\| \|\mathbf{v}\|} = \frac{-1}{\sqrt{38}\sqrt{258}} \approx -0.0101,$$

so that $\theta \approx 90.58°$.

(c) $\text{proj}_{\mathbf{u}} \mathbf{v} = \frac{\mathbf{u} \cdot \mathbf{v}}{\|\mathbf{u}\|^2} \mathbf{u} = \frac{-1}{38} \langle -2, 3, 5 \rangle = \frac{1}{38} \langle 2, -3, -5 \rangle$.

35. With $\mathbf{u} = \langle 2, 4, -1, 2 \rangle$ and $\mathbf{v} = \langle -1, 3, -2, 6 \rangle$, we have

$$\mathbf{u} \cdot \mathbf{v} = 2 \cdot (-1) + 4 \cdot 3 - 1 \cdot (-2) + 2 \cdot 6 = 24,$$

$$\|\mathbf{u}\| = \sqrt{2^2 + 4^2 + (-1)^2 + 2^2} = 5, \quad \|\mathbf{v}\| = \sqrt{(-1)^2 + 3^2 + (-2)^2 + 6^2} = 5\sqrt{2},$$

so that

(a) $\mathbf{u} \cdot \mathbf{v} = 24$.

(b) If θ is the angle between $\mathbf{u}$ and $\mathbf{v}$, then

$$\cos\theta = \frac{\mathbf{u} \cdot \mathbf{v}}{\|\mathbf{u}\| \, \|\mathbf{v}\|} = \frac{24}{25\sqrt{2}} \approx 0.6788,$$

so that $\theta \approx 47.25°$.

(c) $\operatorname{proj}_{\mathbf{u}} \mathbf{v} = \frac{\mathbf{u} \cdot \mathbf{v}}{\|\mathbf{u}\|^2} \mathbf{u} = \frac{24}{25} \langle 2, 4, -1, 2 \rangle$.

37. Following the method of Example 4, the required distance is the length of the vector component of $\overrightarrow{QP} = \langle 4, 4, 12 \rangle$ orthogonal to the vector $\overrightarrow{QR} = \langle -1, -1, 9 \rangle$. This component is

$$\langle 4, 4, 12 \rangle - \operatorname{proj}_{\langle -1, -1, 9 \rangle} \langle 4, 4, 12 \rangle = \langle 4, 4, 12 \rangle - \frac{-1 \cdot 4 - 1 \cdot 4 + 9 \cdot 12}{(-1)^2 + (-1)^2 + 9^2} \langle -1, -1, 9 \rangle$$

$$= \langle 4, 4, 12 \rangle - \frac{100}{83} \langle -1, -1, 9 \rangle$$

$$= \langle 4, 4, 12 \rangle + \left\langle \frac{100}{83}, \frac{100}{83}, -\frac{900}{83} \right\rangle$$

$$= \left\langle \frac{432}{83}, \frac{432}{83}, \frac{96}{83} \right\rangle.$$

Its length is $\frac{\sqrt{432^2 + 432^2 + 96^2}}{83} = \frac{\sqrt{(48 \cdot 9)^2 + (48 \cdot 9)^2 + (48 \cdot 2)^2}}{83} = \frac{48}{83}\sqrt{166}$.

39. Since the cube is symmetric, we may as well assume that the vertex where the face diagonal and the edge meet is the origin, that the face diagonal is the vector $\langle r, r, 0 \rangle$, and that the edge is $\langle 0, 0, r \rangle$. Then the angle θ between these two vectors satisfies

$$\cos\theta = \frac{\langle r, r, 0 \rangle \cdot \langle 0, 0, r \rangle}{\|\langle r, r, 0 \rangle\| \, \|\langle 0, 0, r \rangle\|} = \frac{0}{\sqrt{2r^2}\sqrt{r^2}} = 0,$$

so that $\theta = \cos^{-1} 0 = 90°$.

41. Since the cube is symmetric, we may as well assume that the vertex where the cube diagonal and the face diagonal meet is the origin, so that the cube diagonal is the vector $\langle r, r, r \rangle$; assume the face diagonal is $\langle r, r, 0 \rangle$. Then the angle θ between these two vectors satisfies

$$\cos\theta = \frac{\langle r, r, r \rangle \cdot \langle r, r, 0 \rangle}{\|\langle r, r, r \rangle\| \, \|\langle r, r, 0 \rangle\|} = \frac{2r^2}{\sqrt{3r^2}\sqrt{2r^2}} = \frac{2}{\sqrt{6}} = \frac{\sqrt{6}}{3}.$$

so that $\theta = \cos^{-1} \frac{\sqrt{6}}{3} \approx 35.26°$.

43. By the formula for θ, the angle between two vectors, we get

$$\alpha = \cos^{-1}\left(\frac{\langle 1, 2, 3 \rangle \cdot \mathbf{i}}{\|\langle 1, 2, 3 \rangle\| \, \|\mathbf{i}\|} \right) = \cos^{-1} \frac{1}{\sqrt{14}} \approx 74.5°$$

$$\beta = \cos^{-1}\left(\frac{\langle 1, 2, 3 \rangle \cdot \mathbf{j}}{\|\langle 1, 2, 3 \rangle\| \, \|\mathbf{j}\|} \right) = \cos^{-1} \frac{2}{\sqrt{14}} \approx 57.69°$$

$$\gamma = \cos^{-1}\left(\frac{\langle 1, 2, 3 \rangle \cdot \mathbf{k}}{\|\langle 1, 2, 3 \rangle\| \, \|\mathbf{k}\|} \right) = \cos^{-1} \frac{3}{\sqrt{14}} \approx 36.7°,$$

and the direction cosines are the arguments of $\cos^{-1}$ above.

45. By the formula for θ, the angle between two vectors, we get

$$\alpha = \cos^{-1}\left(\frac{\langle -1,1,-4\rangle \cdot \mathbf{i}}{\|\langle -1,1,-4\rangle\|\,\|\mathbf{i}\|}\right) = \cos^{-1}\frac{-1}{3\sqrt{2}} \approx 103.63°$$

$$\beta = \cos^{-1}\left(\frac{\langle -1,1,-4\rangle \cdot \mathbf{j}}{\|\langle -1,1,-4\rangle\|\,\|\mathbf{j}\|}\right) = \cos^{-1}\frac{1}{3\sqrt{2}} \approx 76.37°$$

$$\gamma = \cos^{-1}\left(\frac{\langle -1,1,-4\rangle \cdot \mathbf{k}}{\|\langle -1,1,-4\rangle\|\,\|\mathbf{k}\|}\right) = \cos^{-1}\frac{-4}{3\sqrt{2}} \approx 160.53°.$$

and the direction cosines are the arguments of $\cos^{-1}$ above.

47. If $\mathbf{v} = \langle a,b,c\rangle$, then the direction cosines are

$$\cos\alpha = \frac{\langle a,b,c\rangle \cdot \mathbf{i}}{\|\mathbf{v}\|} = \frac{a}{\|\mathbf{v}\|},$$

$$\cos\beta = \frac{\langle a,b,c\rangle \cdot \mathbf{j}}{\|\mathbf{v}\|} = \frac{b}{\|\mathbf{v}\|},$$

$$\cos\gamma = \frac{\langle a,b,c\rangle \cdot \mathbf{k}}{\|\mathbf{v}\|} = \frac{c}{\|\mathbf{v}\|},$$

so that

$$\|\mathbf{v}\|((\cos\alpha)\mathbf{i}+(\cos\beta)\mathbf{j}+(\cos\gamma)\mathbf{k}) = \|\mathbf{v}\|\cdot\frac{a}{\|\mathbf{v}\|}\mathbf{i}+\|\mathbf{v}\|\cdot\frac{b}{\|\mathbf{v}\|}\mathbf{j}+\|\mathbf{v}\|\cdot\frac{c}{\|\mathbf{v}\|}\mathbf{k}$$
$$= a\mathbf{i}+b\mathbf{j}+c\mathbf{k} = \langle a,b,c\rangle = \mathbf{v}.$$

49. $1 = \cos^2\alpha + \cos^2\beta + \cos^2\gamma = \frac{1}{4}+\frac{1}{4}+\cos^2\gamma$, so that $\cos^2\gamma = \frac{1}{2}$ and then $\cos\gamma = \pm\frac{\sqrt{2}}{2}$.

51. $1 = \cos^2\alpha + \cos^2\beta + \cos^2\gamma = \frac{1}{4}+\cos^2\beta+\frac{3}{16}$, so that $\cos^2\beta = \frac{9}{16}$ and then $\cos\beta = \pm\frac{3}{4}$.

53. (a) Note that $\|\mathbf{v}\| = \sqrt{1^2+1^2+\cdots+1^2} = \sqrt{n}$. Now, using the notation from the previous exercise, we get

$$\alpha_i = \cos^{-1}\frac{v_i}{\|\mathbf{v}\|} = \cos^{-1}\frac{1}{\sqrt{n}}.$$

This is independent of i, so that all the direction angles are equal to $\cos^{-1}\frac{1}{\sqrt{n}}$.

(b) We have $\lim_{n\to\infty}\alpha_i = \lim_{n\to\infty}\cos^{-1}\frac{1}{\sqrt{n}} = \cos^{-1}\lim_{n\to\infty}\frac{1}{\sqrt{n}} = \cos^{-1}0 = \frac{\pi}{2}.$

Applications

55. (a) By Theorem 10.15, the angle θ between Annie's heading and directly south (which is $\langle 0,-1\rangle$) is given by

$$\cos\theta = \frac{\langle 1,-4\rangle \cdot \langle 0,-1\rangle}{\|\langle 1,-4\rangle\|\cdot\|\langle 0,-1\rangle\|} = \frac{4}{\sqrt{17}}.$$

Thus the angle is $\theta = \cos^{-1}\frac{4}{\sqrt{17}} \approx 14.04°.$

(b) Since she is traveling at 2 miles per hour, her velocity vector is

$$2\cdot\frac{\langle 1,-4\rangle}{\|\langle 1,-4\rangle\|} = \frac{2}{\sqrt{17}}\langle 1,-4\rangle,$$

so that the southward component of her velocity is $\frac{-8}{\sqrt{17}}.$

(c) Since the crossing is two miles south, it will take her $\frac{2}{8/\sqrt{17}} = \frac{\sqrt{17}}{4}$ hours, or just over an hour.

Proofs

57. Let $\mathbf{v} = \langle v_1, v_2, v_3 \rangle$. Then

$$(\mathbf{v} \cdot \mathbf{i})\mathbf{i} + (\mathbf{v} \cdot \mathbf{j})\mathbf{j} + (\mathbf{v} \cdot \mathbf{k})\mathbf{k}$$
$$= (\langle v_1, v_2, v_3 \rangle \cdot \langle 1, 0, 0 \rangle)\mathbf{i} + (\langle v_1, v_2, v_3 \rangle \cdot \langle 0, 1, 0 \rangle)\mathbf{j} + (\langle v_1, v_2, v_3 \rangle \cdot \langle 0, 0, 1 \rangle)\mathbf{k}$$
$$= v_1\mathbf{i} + v_2\mathbf{j} + v_3\mathbf{k} = \langle v_1, v_2, v_3 \rangle$$
$$= \mathbf{v}.$$

59. Since $|\cos\theta| \leq 1$ for any θ, we have

$$|\mathbf{u} \cdot \mathbf{v}| = |\|\mathbf{u}\| \, \|\mathbf{v}\| \cos\theta| = \|\mathbf{u}\| \, \|\mathbf{v}\| \, |\cos\theta| \leq \|\mathbf{u}\| \, \|\mathbf{v}\| \,.$$

For the two expressions to be equal, we must have $|\cos\theta| = 1$, so that $\theta = 0$ or $\theta = \pi$. The two vectors must be parallel in order for the inequality to be an equality.

61. We have $\mathbf{u} \cdot \mathbf{v} = 0 \iff \frac{\mathbf{u} \cdot \mathbf{v}}{\|\mathbf{u}\|\|\mathbf{v}\|} = 0 \iff \cos\theta = 0 \iff \theta$ is a right angle $\iff \mathbf{u}$ and $\mathbf{v}$ are orthogonal.

63. Suppose $\mathbf{v} = \mathbf{t}_\perp + \mathbf{t}_\| = \mathbf{s}_\perp + \mathbf{s}_\|$, where $\mathbf{t}_\perp$ and $\mathbf{s}_\perp$ are orthogonal to $\mathbf{u}$ and $\mathbf{t}_\|$ and $\mathbf{s}_\|$ are parallel to $\mathbf{u}$. Then $\mathbf{t}_\| = k\mathbf{u}$ and $\mathbf{s}_\| = l\mathbf{u}$ for some scalars k and l. Then $\mathbf{t}_\perp = \mathbf{v} - k\mathbf{u}$ and $\mathbf{s}_\perp = \mathbf{v} - l\mathbf{u}$. It follows that

$$0 = \mathbf{t}_\perp \cdot \mathbf{u} = (\mathbf{v} - k\mathbf{u}) \cdot \mathbf{u} = \mathbf{v} \cdot \mathbf{u} - k\mathbf{u} \cdot \mathbf{u}$$
$$0 = \mathbf{s}_\perp \cdot \mathbf{u} = (\mathbf{v} - l\mathbf{u}) \cdot \mathbf{u} = \mathbf{v} \cdot \mathbf{u} - l\mathbf{u} \cdot \mathbf{u}.$$

Since these expressions are both zero, they are equal to each other, so that $k\mathbf{u} \cdot \mathbf{u} = l\mathbf{u} \cdot \mathbf{u}$ and thus $k = l$ since $\mathbf{u} \neq \mathbf{0}$. Thus $\mathbf{t}_\| = \mathbf{s}_\|$, so that $\mathbf{t}_\perp = \mathbf{v} - \mathbf{t}_\| = \mathbf{v} - \mathbf{s}_\| = \mathbf{s}_\perp$. Thus the $\mathbf{s}$ and $\mathbf{t}$ decompositions are equal.

65. Let $ABCD$ be a parallelogram with $\overrightarrow{AB} = \mathbf{u}$ and $\overrightarrow{AD} = \mathbf{v}$, so that $\overrightarrow{AC} = \mathbf{u} + \mathbf{v}$ and $\overrightarrow{DB} = \mathbf{u} - \mathbf{v}$. Then

$$\overrightarrow{AC} \cdot \overrightarrow{DB} = (\mathbf{u} + \mathbf{v})(\mathbf{u} - \mathbf{v}) = \mathbf{u} \cdot \mathbf{u} + \mathbf{v} \cdot \mathbf{u} - \mathbf{u}\mathbf{v} - \mathbf{v} \cdot \mathbf{v} = \|\mathbf{u}\|^2 - \|\mathbf{v}\|^2 \,.$$

Now, AC and DB are perpendicular if and only if $\overrightarrow{AC} \cdot \overrightarrow{DB} = 0$. From the above equation, this is equivalent to the statement that $\|\mathbf{u}\|^2 - \|\mathbf{v}\|^2 = 0$, which holds if and only if $\|\mathbf{u}\| = \|\mathbf{v}\|$. But this is equivalent to saying that $\mathbf{u}$ and $\mathbf{v}$ have the same length, which is equivalent to saying that the parallelogram is a rhombus.

Thinking Forward

Lines and vectors

▷ The slope of this line is $\frac{3}{5}$, so the vector $\langle 5, 3 \rangle$ is parallel to the line. The norm of this vector is $\sqrt{25 + 9} = \sqrt{34}$, so that two unit vectors parallel to the line are $\left\langle \frac{5}{\sqrt{34}}, \frac{3}{\sqrt{34}} \right\rangle$ and $\left\langle -\frac{5}{\sqrt{34}}, -\frac{3}{\sqrt{34}} \right\rangle$.

▷ The slope of this line is m, so the vector $\langle 1, m \rangle$ is parallel to the line. The norm of this vector is $\sqrt{1 + m^2}$, so that two unit vectors parallel to the line are the vectors $\left\langle \frac{1}{\sqrt{1+m^2}}, \frac{m}{\sqrt{1+m^2}} \right\rangle$ and $\left\langle -\frac{1}{\sqrt{1+m^2}}, -\frac{m}{\sqrt{1+m^2}} \right\rangle$.

▷ The slope of this line is $\frac{3}{5}$, so the slope of a perpendicular line is $-\frac{5}{3}$ and thus the vector $\langle -3, 5 \rangle$ is perpendicular to the line. The norm of this vector is $\sqrt{25 + 9} = \sqrt{34}$, so that two unit vectors parallel to the line are $\left\langle -\frac{3}{\sqrt{34}}, \frac{5}{\sqrt{34}} \right\rangle$ and $\left\langle \frac{3}{\sqrt{34}}, -\frac{5}{\sqrt{34}} \right\rangle$.

▷ If $m \neq 0$, then the slope of a perpendicular line is $-\frac{1}{m}$, and thus the vector $\langle -m, 1 \rangle$ is perpendicular to the line. The norm of this vector is $\sqrt{(-m)^2 + 1^2} = \sqrt{1 + m^2}$, so that two unit vectors parallel to the line are $\left\langle -\frac{m}{\sqrt{1+m^2}}, \frac{1}{\sqrt{1+m^2}} \right\rangle$ and $\left\langle \frac{m}{\sqrt{1+m^2}}, -\frac{1}{\sqrt{1+m^2}} \right\rangle$. If $m = 0$, then the line is $y = b$, which is a horizontal line. The two vertical unit vectors $\pm\mathbf{j}$ are perpendicular to the line $y = b$.

10.4 Cross Product

Thinking Back

Coordinate system

See the beginning of Section 10.1. If the index finger of the right hand points in the positive x direction and the middle finger of the right hand points in the positive y direction, then if the thumb points in the positive z direction, it is a right handed coordinate system; if it points in the negative z direction, it is a left handed coordinate system.

Orthogonal vectors

Since both vectors lie in the xy plane, they are both of the form $\langle x, y, 0 \rangle$, so that $\mathbf{k}$ and $-\mathbf{k}$ are two unit vectors perpendicular to both of them.

Concepts

1. (a) False. In general, $\mathbf{u} \times \mathbf{v} = -\mathbf{v} \times \mathbf{u}$; see Theorem 10.27.

    (b) False. $\mathbf{u} \times \mathbf{v}$ is a vector, while $\mathbf{u} \cdot \mathbf{v}$ is a scalar.

    (c) True. $\mathbf{u} \times \mathbf{v} = -\mathbf{v} \times \mathbf{u}$, so if this is equal to $\mathbf{v} \times \mathbf{u}$, then $\mathbf{v} \times \mathbf{u} = \mathbf{0}$. If $\mathbf{u}$ and $\mathbf{v}$ are not parallel, then by Theorem 10.33(a), $0 = \|\mathbf{u} \times \mathbf{v}\| = \|\mathbf{u}\| \|\mathbf{v}\| \sin \theta$. Then either $\theta = 0$ or $\theta = \pi$, which implies that the vectors are parallel, which is a contradiction, or $\mathbf{u} = \mathbf{0}$ or $\mathbf{v} = \mathbf{0}$, which again implies that they are parallel, a contradiction. Thus $\mathbf{u}$ and $\mathbf{v}$ are parallel.

    (d) False. For example, $(\mathbf{i} \times \mathbf{j}) \times \mathbf{j} = \mathbf{k} \times \mathbf{j} = -\mathbf{i}$ while $\mathbf{i} \times (\mathbf{j} \times \mathbf{j}) = \mathbf{i} \times \mathbf{0} = \mathbf{0}$.

    (e) True. See Theorem 10.36.

    (f) True. By Corollary 10.37(a), $\mathbf{u} \cdot (\mathbf{v} \times \mathbf{w}) = \mathbf{v} \cdot (\mathbf{w} \times \mathbf{u})$; since the cross product is anticommutative, this is equal to $\mathbf{v} \cdot (-\mathbf{u} \times \mathbf{w}) = -\mathbf{v} \cdot (\mathbf{u} \times \mathbf{w})$.

    (g) True. Since $\mathbf{u} \cdot \mathbf{v} = \|\mathbf{u}\| \|\mathbf{v}\| \cos \theta$ and $\|\mathbf{u} \times \mathbf{v}\| = \|\mathbf{u}\| \|\mathbf{v}\| \sin \theta$, dividing the two equations gives $\frac{\mathbf{u} \cdot \mathbf{v}}{\|\mathbf{u} \times \mathbf{v}\|} = \cot \theta$. (Note that $\|\mathbf{u}\| \neq 0$ and $\|\mathbf{v}\| \neq 0$ since $\mathbf{u}$ and $\mathbf{v}$ are nonparallel, so we can cancel these expressions in the quotient. Also note that $\sin \theta \neq 0$ for the same reason, so the denominator is nonzero).

    (h) False. The length of $\mathbf{u} \times \mathbf{v}$ is $\|\mathbf{u} \times \mathbf{v}\| = \|\mathbf{u}\| \|\mathbf{v}\| \sin \theta = \sin \theta$, so if $\mathbf{u}$ and $\mathbf{v}$ are not perpendicular, the length of $\mathbf{u} \times \mathbf{v}$ will be less than 1.

3. See Definition 10.25. The cross product $\mathbf{u} \times \mathbf{v}$ can be thought of as the determinant of a 3×3 matrix whose first row is $\mathbf{i}, \mathbf{j}, \mathbf{k}$, whose second row is $\mathbf{u}$, and whose third row is $\mathbf{v}$. Its value is

$$\mathbf{u} \times \mathbf{v} = (u_2 v_3 - u_3 v_2)\mathbf{i} + (u_3 v_1 - u_1 v_3)\mathbf{j} + (u_1 v_2 - u_2 v_1)\mathbf{k}.$$

5. $\mathbf{u} \times \mathbf{v}$ is perpendicular to $\mathbf{u}$ and $\mathbf{v}$. See Theorem 10.31. Further, the magnitude of $\mathbf{u} \times \mathbf{v}$ is $\|\mathbf{u} \times \mathbf{v}\| = \|\mathbf{u}\| \|\mathbf{v}\| \sin \theta$ where θ is the angle between $\mathbf{u}$ and $\mathbf{v}$. Finally, the vectors $\mathbf{u}, \mathbf{v},$ and $\mathbf{u} \times \mathbf{v}$ form a right-hand triple.

7. The dot product of two vectors is zero if and only if the vectors are orthogonal. Thus $\mathbf{u} \cdot (\mathbf{u} \times \mathbf{v}) = \mathbf{0}$ and $\mathbf{v} \cdot (\mathbf{u} \times \mathbf{v}) = \mathbf{0}$ tell us that $\mathbf{u} \times \mathbf{v}$ is orthogonal to $\mathbf{u}$ and to $\mathbf{v}$.

9. The parallelogram is

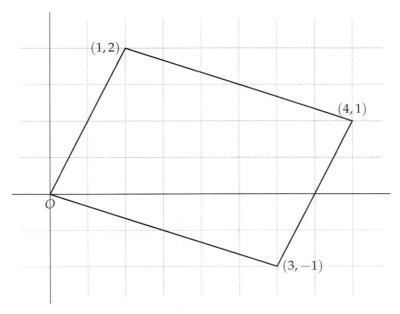

Think of the parallelogram as lying in the xy plane in $\mathbb{R}^3$, so that $\mathbf{u} = \langle 3, -1, 0 \rangle$ and $\mathbf{v} = \langle 1, 2, 0 \rangle$. Then by Corollary 10.33, the area is

$$\|\mathbf{u} \times \mathbf{v}\| = \|\langle -1 \cdot 0 - 0 \cdot 2, 0 \cdot 1 - 3 \cdot 0, 3 \cdot 2 - (-1) \cdot 1 \rangle\| = \|\langle 0, 0, 7 \rangle\| = 7.$$

11. Since $\|\mathbf{u} \times \mathbf{v}\| = \|\mathbf{u}\| \, \|\mathbf{v}\| \sin \theta$, if this is equal to $\|\mathbf{u}\| \, \|\mathbf{v}\|$, then $\sin \theta = 1$, so that $\mathbf{u}$ and $\mathbf{v}$ must be perpendicular.

13. For example,

$$\langle 1, 1, 0 \rangle \times \langle 2, 3, 0 \rangle = \langle 1 \cdot 0 - 0 \cdot 3, 0 \cdot 2 - 1 \cdot 0, 1 \cdot 3 - 1 \cdot 2 \rangle = \langle 0, 0, 1 \rangle$$
$$\langle 1, 1, 0 \rangle \times \langle 3, 4, 0 \rangle = \langle 1 \cdot 0 - 0 \cdot 4, 0 \cdot 3 - 1 \cdot 0, 1 \cdot 4 - 1 \cdot 3 \rangle = \langle 0, 0, 1 \rangle .$$

So the two cross products are equal, but $\langle 2, 3, 0 \rangle \neq \langle 3, 4, 0 \rangle$. If $\mathbf{u} \times \mathbf{v} = \mathbf{u} \times \mathbf{w}$, then by Theorem 10.29(a), we have $\mathbf{u} \times (\mathbf{v} - \mathbf{w}) = 0$, so that $\mathbf{u}$ must be parallel to $\mathbf{v} - \mathbf{w}$. Note that this holds in the example above: $\mathbf{v} - \mathbf{w} = \langle 2, 3, 0 \rangle - \langle 3, 4, 0 \rangle = \langle -1, -1, 0 \rangle = -1 \cdot \langle 1, 1, 0 \rangle$.

15. If $\mathbf{u} \cdot (\mathbf{v} \times \mathbf{w})$ is 0, then $\mathbf{v} \times \mathbf{w}$ is perpendicular to $\mathbf{u}$. But $\mathbf{v} \times \mathbf{w}$ is also perpendicular to $\mathbf{v}$ and $\mathbf{w}$, so it is perpendicular to all three vectors. Thus all three vectors lie in a plane, and $\mathbf{v} \times \mathbf{w}$ is perpendicular to that plane.

17. Grouping must be specified, since $(\mathbf{u} \times \mathbf{v}) \times \mathbf{w} \neq \mathbf{u} \times (\mathbf{v} \times \mathbf{w})$ in general. Thus the expression as given does not have a well-defined meaning.

19. If $\mathbf{u} \cdot \mathbf{v} = 0$, then $\mathbf{u}$ and $\mathbf{v}$ are perpendicular. If $\mathbf{u} \times \mathbf{v} = 0$, then $\mathbf{u}$ and $\mathbf{v}$ are parallel. The only way both of these statements can be true is for $\mathbf{u}$, $\mathbf{v}$, or both to be $\mathbf{0}$.

21. Since $\mathbf{u} \times \mathbf{v} = -\mathbf{v} \times \mathbf{u}$, if this is also equal to $\mathbf{v} \times \mathbf{u}$, then $-\mathbf{v} \times \mathbf{u} = \mathbf{v} \times \mathbf{u}$ so that $\mathbf{v} \times \mathbf{u} = \mathbf{0}$ and thus $\mathbf{u}$ and $\mathbf{v}$ are parallel.

Skills

23.

$$\mathbf{u} \times \mathbf{w} = \langle 2,1,-3 \rangle \times \langle -2,6,5 \rangle = \langle 1 \cdot 5 - (-3) \cdot 6, (-3) \cdot (-2) - 2 \cdot 5, 2 \cdot 6 - 1 \cdot (-2) \rangle$$
$$= \langle 23, -4, 14 \rangle$$
$$\mathbf{w} \times \mathbf{u} = \langle -2,6,5 \rangle \times \langle 2,1,-3 \rangle = \langle 6 \cdot (-3) - 5 \cdot 1, 5 \cdot 2 - (-2) \cdot (-3), (-2) \cdot 1 - 6 \cdot 2 \rangle$$
$$= \langle -23, 4, -14 \rangle .$$

25. Using the previous exercises,

$$(\mathbf{u} \times \mathbf{v}) \times \mathbf{w} = \langle 1,-14,-4 \rangle \times \langle -2,6,5 \rangle$$
$$= \langle (-14) \cdot 5 - (-4) \cdot 6, (-4) \cdot (-2) - 1 \cdot 5, 1 \cdot 6 - (-14) \cdot (-2) \rangle$$
$$= \langle -46, 3, -22 \rangle$$
$$\mathbf{u} \times (\mathbf{v} \times \mathbf{w}) = \langle 2,1,-3 \rangle \times \langle -6,-22,24 \rangle$$
$$= \langle 1 \cdot 24 - (-3) \cdot (-22), (-3) \cdot (-6) - 2 \cdot 24, 2 \cdot (-22) - 1 \cdot (-6) \rangle$$
$$= \langle -42, -30, -38 \rangle .$$

27. By Corollary 10.3, and using Exercise 22, the area of the parallelogram determined by $\mathbf{u}$ and $\mathbf{v}$ is

$$\|\mathbf{u} \times \mathbf{v}\| = \|\langle 1,-14,-4 \rangle\| = \sqrt{1^2 + (-14)^2 + (-4)^2} = \sqrt{213}.$$

29. By Theorem 10.36, and using Exercise 24, the volume of the parallelepiped is

$$|\mathbf{u} \cdot (\mathbf{v} \times \mathbf{w})| = |\langle 2,1,-3 \rangle \cdot \langle -6,-22,24 \rangle| = |-12 - 22 - 72| = |-106| = 106.$$

Since $\mathbf{u} \cdot (\mathbf{v} \times \mathbf{w}) < 0$, Theorem 10.36 tells us that $\mathbf{u}, \mathbf{v}, \mathbf{w}$ do not form a right-handed triple.

31.

$$\mathbf{u} \times \mathbf{w} = \langle -3,1,-4 \rangle \times \langle 1,3,13 \rangle = \langle 1 \cdot 13 - (-4) \cdot 3, (-4) \cdot 1 - (-3) \cdot 13, (-3) \cdot 3 - 1 \cdot 1 \rangle$$
$$= \langle 25, 35, -10 \rangle$$
$$\mathbf{w} \times \mathbf{u} = \langle 1,3,13 \rangle \times \langle -3,1,-4 \rangle = \langle 3 \cdot (-4) - 13 \cdot 1, 13 \cdot (-3) - 1 \cdot (-4), 1 \cdot 1 - 3 \cdot (-3) \rangle$$
$$= \langle -25, -35, 10 \rangle .$$

33. Using the previous exercises,

$$(\mathbf{u} \times \mathbf{v}) \cdot \mathbf{w} = \langle 5,7,-2 \rangle \cdot \langle 1,3,13 \rangle = 5 \cdot 1 + 7 \cdot 3 - 2 \cdot 13 = 0$$
$$\mathbf{u} \cdot (\mathbf{v} \times \mathbf{w}) = \langle -3,1,-4 \rangle \cdot \langle -15,-21,6 \rangle = -3 \cdot (-15) + 1 \cdot (-21) - 4 \cdot 6 = 0.$$

35. By Theorem 10.36, and using Exercise 33, the volume of the parallelepiped is

$$|\mathbf{u} \cdot (\mathbf{v} \times \mathbf{w})| = |0| = 0.$$

Since the volume of the parallelepiped is zero, the three vectors $\mathbf{u}, \mathbf{v},$ and $\mathbf{w}$ must lie in a plane.

37. The three points give the two vectors $\mathbf{u} = \overrightarrow{PQ} = \langle -4,-3,5 \rangle$ and $\mathbf{v} = \overrightarrow{PR} = \langle -5,0,8 \rangle$. Then

(a) A vector perpendicular to the plane determined by P, Q, and R must also be perpendicular to $\mathbf{u}$ and $\mathbf{v}$, so that $\mathbf{u} \times \mathbf{v}$ is such a vector:

$$\mathbf{u} \times \mathbf{v} = \langle -4,-3,5 \rangle \times \langle -5,0,8 \rangle$$
$$= \langle (-3) \cdot 8 - 5 \cdot 0, 5 \cdot (-5) - (-4) \cdot 8, (-4) \cdot 0 - (-3) \cdot (-5) \rangle \quad = \langle -24, 7, -15 \rangle .$$

(b) We have $\|\mathbf{u} \times \mathbf{v}\| = \sqrt{(-24)^2 + 7^2 + (-15)^2} = \sqrt{850} = 5\sqrt{34}$, so that two unit vectors perpendicular to the plane are

$$\left\langle -\frac{24}{5\sqrt{34}}, \frac{7}{5\sqrt{34}}, -\frac{3}{\sqrt{34}} \right\rangle \text{ and } \left\langle \frac{24}{5\sqrt{34}}, -\frac{7}{5\sqrt{34}}, \frac{3}{\sqrt{34}} \right\rangle.$$

(c) The area of the triangle determined by the points is

$$\frac{1}{2} \|\mathbf{u} \times \mathbf{v}\| = \frac{5\sqrt{34}}{2}.$$

39. The three points give the two vectors $\mathbf{u} = \overrightarrow{PQ} = \langle -6, 10, 7 \rangle$ and $\mathbf{v} = \overrightarrow{PR} = \langle -3, 0, 2 \rangle$. Then

(a) A vector perpendicular to the plane determined by P, Q, and R must also be perpendicular to $\mathbf{u}$ and $\mathbf{v}$, so that $\mathbf{u} \times \mathbf{v}$ is such a vector:

$$\begin{aligned} \mathbf{u} \times \mathbf{v} &= \langle -6, 10, 7 \rangle \times \langle -3, 0, 2 \rangle \\ &= \langle 10 \cdot 2 - 7 \cdot 0, 7 \cdot (-3) - (-6) \cdot 2, (-6) \cdot 0 - 10 \cdot (-3) \rangle \\ &= \langle 20, -9, 30 \rangle. \end{aligned}$$

(b) We have $\|\mathbf{u} \times \mathbf{v}\| = \sqrt{20^2 + (-9)^2 + 30^2} = \sqrt{1381}$, so that two unit vectors perpendicular to the plane are

$$\left\langle \frac{20}{\sqrt{1381}}, -\frac{9}{\sqrt{1381}}, \frac{30}{\sqrt{1381}} \right\rangle \text{ and } \left\langle -\frac{20}{\sqrt{1381}}, \frac{9}{\sqrt{1381}}, -\frac{30}{\sqrt{1381}} \right\rangle.$$

(c) The area of the triangle determined by the points is

$$\frac{1}{2} \|\mathbf{u} \times \mathbf{v}\| = \frac{\sqrt{1381}}{2}.$$

41. Thinking of the three points as lying in the xy plane in $\mathbb{R}^3$, with zero z coordinate, gives $P = (1, 6, 0)$, $Q = (0, -3, 0)$, and $R = (-5, 4, 0)$. These three points give the two vectors $\mathbf{u} = \overrightarrow{PQ} = \langle -1, -9, 0 \rangle$ and $\mathbf{v} = \overrightarrow{PR} = \langle -6, -2, 0 \rangle$. Then

(a) A vector perpendicular to the plane determined by P, Q, and R must also be perpendicular to $\mathbf{u}$ and $\mathbf{v}$, so that $\mathbf{u} \times \mathbf{v}$ is such a vector:

$$\begin{aligned} \mathbf{u} \times \mathbf{v} &= \langle -1, -9, 0 \rangle \times \langle -6, -2, 0 \rangle \\ &= \langle (-9) \cdot 0 - 0 \cdot (-2), 0 \cdot (-6) - (-1) \cdot 0, -1 \cdot (-2) - (-9) \cdot (-6) \rangle \\ &= \langle 0, 0, -52 \rangle. \end{aligned}$$

(b) We have $\|\mathbf{u} \times \mathbf{v}\| = \sqrt{0^2 + 0^2 + (-52)^2} = 52$, so that two unit vectors perpendicular to the plane are

$$\langle 0, 0, -1 \rangle \text{ and } \langle 0, 0, 1 \rangle.$$

(c) The area of the triangle determined by the points is

$$\frac{1}{2} \|\mathbf{u} \times \mathbf{v}\| = 26.$$

43. We have

$$a = \left\|\overrightarrow{PQ}\right\| = \sqrt{(0-4)^2 + (0-3)^2 + (3-(-2))^2} = \sqrt{16+9+25} = 5\sqrt{2},$$

$$b = \left\|\overrightarrow{PR}\right\| = \sqrt{(-1-4)^2 + (3-3)^2 + (6-(-2))^2} = \sqrt{25+64} = \sqrt{89},$$

$$c = \left\|\overrightarrow{QR}\right\| = \sqrt{(-1-0)^2 + (3-0)^2 + (6-3)^2} = \sqrt{1+9+9} = \sqrt{19},$$

$$s = \frac{1}{2}(a+b+c) = \frac{5\sqrt{2} + \sqrt{89} + \sqrt{19}}{2}.$$

Then the area of the triangle is

$$\sqrt{s(s-a)(s-b)(s-c)}$$

$$= \sqrt{\frac{5\sqrt{2}+\sqrt{89}+\sqrt{19}}{2} \cdot \frac{-5\sqrt{2}+\sqrt{89}+\sqrt{19}}{2} \cdot \frac{5\sqrt{2}-\sqrt{89}+\sqrt{19}}{2} \cdot \frac{5\sqrt{2}+\sqrt{89}-\sqrt{19}}{2}}$$

$$= \frac{5\sqrt{34}}{2}.$$

45. (a) For example, suppose all four sides of the quadrilateral are equal to 1. Let the angle at one vertex be θ and the corresponding vectors of the sides (both of length 1) be $\mathbf{u}$ and $\mathbf{v}$. Then the area of the parallelogram is

$$\|\mathbf{u} \times \mathbf{v}\| = \|\mathbf{u}\| \|\mathbf{v}\| \sin\theta = \sin\theta.$$

So by varying the angle at which the sides meet, the area can be chosen to be anything in the range $(0, 1]$. Thus the area is not determined by the side lengths.

(b) If we are given the side lengths plus one diagonal, then two of the sides plus the diagonal determine a triangle, and the other two sides plus the diagonal determine another triangle. We can find the areas of each of these triangles using Heron's formula, and the area of the quadrilateral is their sum.

(c) If we know the side lengths, and the angle θ at one vertex, we can compute the vectors $\mathbf{u}$ and $\mathbf{v}$ emanating from the vertex with a known angle. For example, place one side along the x axis; then the other side is at an angle θ, and we know the lengths of both sides. Then $\mathbf{u} - \mathbf{v}$ is a diagonal of the quadrilateral, whose length we now know. From part (b), we can compute the area of the quadrilateral.

47. We have two triangles: $\triangle PQS$ has sides 6, 9, and 9, while $\triangle QRS$ has sides 7, 8, and 9. For $\triangle PQS$, we have $s = \frac{6+9+9}{2} = 12$, while for $\triangle QRS$ we have $s = \frac{7+8+9}{2} = 12$. Thus by Heron's formula, the area of the quadrilateral is

$$\text{Area}(\triangle PQS) + \text{Area}(\triangle QRS) = \sqrt{12 \cdot 6 \cdot 3 \cdot 3} + \sqrt{12 \cdot 5 \cdot 4 \cdot 3}$$

$$= \sqrt{648} + \sqrt{720}$$

$$= 18\sqrt{2} + 12\sqrt{5}.$$

49. Since $\angle R = 60°$, $\triangle QRS$ has two sides each equal to 8 and the included angle $60°$, so it is an equilateral triangle. Thus the third side, which is the diagonal QS, is 8 as well. Then for $\triangle PQS$ we have $s = \frac{7+8+9}{2} = 12$, while for $\triangle QRS$ we have $s = \frac{8+8+8}{2} = 12$. Hence by Heron's formula the

area of the quadrilateral is

$$\text{Area}(\triangle PQS) + \text{Area}(\triangle QRS) = \sqrt{12 \cdot 5 \cdot 4 \cdot 3} + \sqrt{12 \cdot 4 \cdot 4 \cdot 4}$$
$$= \sqrt{720} + \sqrt{768}$$
$$= 12\sqrt{5} + 16\sqrt{3}$$

51. (a) Consider the three vectors $\overrightarrow{PQ}$, $\overrightarrow{PR}$, $\overrightarrow{PS}$. If $PQRS$ is a parallelogram, then two of these are sides, which we may call $\mathbf{u}$ and $\mathbf{v}$, and the third is a diagonal, which must be $\mathbf{u} + \mathbf{v}$. These three vectors are $\overrightarrow{PQ} = \langle 3, 2 \rangle$, $\overrightarrow{PR} = \langle 5, -2 \rangle$, and $\overrightarrow{PS} = \langle 2, -4 \rangle$. Since $\overrightarrow{PQ} + \overrightarrow{PS} = \overrightarrow{PR}$, it follows that PR is a diagonal, and PQ and PS are adjacent sides of a parallelogram.

   (b) We have $\mathbf{u} = \overrightarrow{PQ} = \langle 3, 2 \rangle$ and $\mathbf{v} = \overrightarrow{PS} = \langle 2, -4 \rangle$. Regarding these vectors as lying in the xy plane with zero z coordinate, we get for the area of the parallelogram

   $$\|\mathbf{u} \times \mathbf{v}\| = \|\langle 2 \cdot 0 - 0 \cdot (-4), 0 \cdot 2 - 3 \cdot 0, 3 \cdot (-4) - 2 \cdot 2 \rangle\| = \|-16\| = 16.$$

53. (a) Consider the three vectors $\overrightarrow{PQ}$, $\overrightarrow{PR}$, $\overrightarrow{PS}$. If $PQRS$ is a parallelogram, then two of these are sides, which we may call $\mathbf{u}$ and $\mathbf{v}$, and the third is a diagonal, which must be $\mathbf{u} + \mathbf{v}$. However, these three vectors are $\langle 3, 2 \rangle$, $\langle 7, 0 \rangle$, and $\langle 5, -5 \rangle$. Since none of these is the sum of the other two, this is not a parallelogram.

   (b) The quadrilateral is

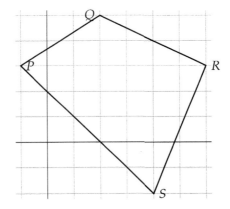

   So it is the sum of two triangles, $\triangle PQR$ and $\triangle PRS$. For $\triangle PQR$, the sides and semiperimeter are

   $$a = \left\| \overrightarrow{PQ} \right\| = \sqrt{(2 - (-1))^2 + (5 - 3)^2} = \sqrt{13},$$

   $$b = \left\| \overrightarrow{QR} \right\| = \sqrt{(6 - 2)^2 + (3 - 5)^2} = \sqrt{20},$$

   $$c = \left\| \overrightarrow{PR} \right\| = \sqrt{(6 - (-1))^2 + (3 - 3)^2} = 7,$$

   $$s = \frac{1}{2}(a + b + c) = \frac{7 + \sqrt{13} + \sqrt{20}}{2},$$

   so the area of $\triangle PQR$ is

   $$\sqrt{s(s-a)(s-b)(s-c)}$$
   $$= \sqrt{\frac{7 + \sqrt{13} + \sqrt{20}}{2} \cdot \frac{-7 + \sqrt{13} + \sqrt{20}}{2} \cdot \frac{7 - \sqrt{13} + \sqrt{20}}{2} \cdot \frac{7 + \sqrt{13} - \sqrt{20}}{2}} = 7.$$

For $\triangle PRS$, the sides and semiperimeter are

$$a = \left\| \overrightarrow{PR} \right\| = \sqrt{(6 - (-1))^2 + (3 - 3)^2} = 7,$$

$$b = \left\| \overrightarrow{RS} \right\| = \sqrt{(4 - 6)^2 + (-2 - 3)^2} = \sqrt{29},$$

$$c = \left\| \overrightarrow{PS} \right\| = \sqrt{(4 - (-1))^2 + (-2 - 3)^2} = \sqrt{50},$$

$$s = \frac{1}{2}(a + b + c) = \frac{7 + \sqrt{29} + \sqrt{50}}{2},$$

so the area of $\triangle PRS$ is

$$\sqrt{s(s-a)(s-b)(s-c)}$$

$$= \sqrt{\frac{7 + \sqrt{29} + \sqrt{50}}{2} \cdot \frac{-7 + \sqrt{29} + \sqrt{50}}{2} \cdot \frac{7 - \sqrt{29} + \sqrt{50}}{2} \cdot \frac{7 + \sqrt{29} - \sqrt{50}}{2}} = \frac{35}{2}.$$

Then the area of the quadrilateral is $7 + \frac{35}{2} = \frac{49}{2}$.

55. (a) Let $\mathbf{u} = \overrightarrow{PQ} = \langle -4, 7, -2 \rangle$, $\mathbf{v} = \overrightarrow{PR} = \langle 5, 21, -15 \rangle$, and $\mathbf{w} = \overrightarrow{PS} = \langle 13, 7, -11 \rangle$. The four points are coplanar if these three vectors are coplanar, which happens if and only if their scalar triple product is zero. But

$$\mathbf{u} \cdot (\mathbf{v} \times \mathbf{w}) = \langle -4, 7, -2 \rangle \cdot \langle 21 \cdot (-11) - (-15) \cdot 7, -15 \cdot 13 - 5 \cdot (-11), 5 \cdot 7 - 21 \cdot 13 \rangle$$

$$= \langle -4, 7, -2 \rangle \cdot \langle -126, -140, -238 \rangle$$

$$= 504 - 980 + 476 = 0.$$

Thus the vectors are coplanar.

(b) Since $\overrightarrow{RS} = \langle 8, -14, 4 \rangle = -2\overrightarrow{PQ}$, these two sides are parallel. However, $\overrightarrow{QR} = \langle 9, 14, -13 \rangle$, which is not a multiple of $\overrightarrow{PS}$, so that these two sides are not parallel and this is not a parallelogram. However, it is a trapezoid.

(c) The area of a trapezoid is $\frac{1}{2}(b_1 + b_2)h$, where b_1 and b_2 are the lengths of the parallel bases and h is the height. Here, $b_1 = \left\| \overrightarrow{PQ} \right\| = \sqrt{69}$ and $b_2 = \left\| \overrightarrow{RS} \right\| = 2\sqrt{69}$. To determine h, we must find the distance between the parallel lines $\overrightarrow{PQ}$ and $\overrightarrow{RS}$. This is just the norm of the vector component of $\overrightarrow{PR}$ orthogonal to $\overrightarrow{PQ}$; that component is

$$\overrightarrow{PR} - \text{proj}_{\overrightarrow{PQ}} \overrightarrow{PR} = \langle 5, 21, -15 \rangle - \frac{\langle -4, 7, -2 \rangle \cdot \langle 5, 21, -15 \rangle}{\left\| \langle -4, 7, -2 \rangle \right\|^2} \langle -4, 7, -2 \rangle$$

$$= \langle 5, 21, -15 \rangle - \frac{157}{69} \langle -4, 7, -2 \rangle$$

$$= \frac{1}{69} \langle 973, 350, -721 \rangle.$$

Then the length of that orthogonal component is

$$\frac{1}{69} \sqrt{973^2 + 350^2 + 721^2} = 7\sqrt{\frac{470}{69}}.$$

Thus the area of the trapezoid is

$$\frac{1}{2}(\sqrt{69} + 2\sqrt{69}) \cdot 7\sqrt{\frac{470}{69}} = \frac{21}{2}\sqrt{470}.$$

57. (a) Let $\mathbf{u} = \overrightarrow{PQ} = \langle 4, -4, 8 \rangle$, $\mathbf{v} = \overrightarrow{PR} = \langle -1, -3, 9 \rangle$, and $\mathbf{w} = \overrightarrow{PS} = \langle 2, -6, 15 \rangle$. The four points are coplanar if these three vectors are coplanar, which happens if and only if their scalar triple product is zero. But

$$\begin{aligned}
\mathbf{u} \cdot (\mathbf{v} \times \mathbf{w}) &= \langle 4, -4, 8 \rangle \cdot \langle -3 \cdot 15 - 9 \cdot (-6), 9 \cdot 2 - (-1) \cdot 15, -1 \cdot (-6) - (-3) \cdot 2 \rangle \\
&= \langle 4, -4, 8 \rangle \cdot \langle 9, 33, 12 \rangle \\
&= 36 - 132 + 96 = 0.
\end{aligned}$$

(b) If $PQRS$ is a parallelogram, then two of the three of PQ, PR, and PS must be sides adjacent to P and the other must be a diagonal, which must equal the sum of the sides. However, no two of these three sum to the third, so this is not a parallelogram.

(c) The quadrilateral is the sum of two triangles, $\triangle PQR$ and $\triangle QRS$. For $\triangle PQR$, the sides and semiperimeter are

$$a = \left\| \overrightarrow{PQ} \right\| = \sqrt{(7-3)^2 + (0-4)^2 + (6-(-2))^2} = \sqrt{96},$$

$$b = \left\| \overrightarrow{QR} \right\| = \sqrt{(2-7)^2 + (1-0)^2 + (7-6)^2} = \sqrt{27},$$

$$c = \left\| \overrightarrow{PR} \right\| = \sqrt{(2-3)^2 + (1-4)^2 + (7-(-2))^2} = \sqrt{91},$$

$$s = \frac{1}{2}(a+b+c) = \frac{\sqrt{27} + \sqrt{91} + \sqrt{96}}{2},$$

so the area of $\triangle PQR$ is

$$\begin{aligned}
&\sqrt{s(s-a)(s-b)(s-c)} \\
&= \sqrt{\frac{\sqrt{27} + \sqrt{91} + \sqrt{96}}{2} \cdot \frac{-\sqrt{27} + \sqrt{91} + \sqrt{96}}{2} \cdot \frac{\sqrt{27} - \sqrt{91} + \sqrt{96}}{2} \cdot \frac{\sqrt{27} + \sqrt{91} - \sqrt{96}}{2}} \\
&\hspace{10cm} = 2\sqrt{146}.
\end{aligned}$$

For $\triangle QRS$, the sides and semiperimeter are

$$a = \left\| \overrightarrow{QR} \right\| = \sqrt{(2-7)^2 + (1-0)^2 + (7-6)^2} = \sqrt{27},$$

$$b = \left\| \overrightarrow{RS} \right\| = \sqrt{(5-2)^2 + (-2-1)^2 + (13-7)^2} = \sqrt{54},$$

$$c = \left\| \overrightarrow{QS} \right\| = \sqrt{(5-7)^2 + (-2-0)^2 + (13-6)^2} = \sqrt{57},$$

$$s = \frac{1}{2}(a+b+c) = \frac{\sqrt{27} + \sqrt{54} + \sqrt{57}}{2},$$

so the area of $\triangle QRS$ is

$$\sqrt{s(s-a)(s-b)(s-c)}$$

$$= \sqrt{\frac{\sqrt{27}+\sqrt{54}+\sqrt{57}}{2} \cdot \frac{-\sqrt{27}+\sqrt{54}+\sqrt{57}}{2} \cdot \frac{\sqrt{27}-\sqrt{54}+\sqrt{57}}{2} \cdot \frac{\sqrt{27}+\sqrt{54}-\sqrt{57}}{2}}$$

$$= \frac{3\sqrt{146}}{2}.$$

Then the area of the quadrilateral is $2\sqrt{146} + \frac{3\sqrt{146}}{2} = \frac{7\sqrt{146}}{2}$.

Applications

59. As the hint suggests, let one edge lie along the x axis, so that it is $\langle 2,0,0 \rangle$. If a second edge lies in the first quadrant, forming an angle of $45°$ with the x axis, then it is $2\langle \cos 45°, \sin 45°, 0 \rangle = \langle \sqrt{2}, \sqrt{2}, 0 \rangle$. If the third vector is $\langle a, b, c \rangle$, then it forms a $45°$ angle with both of the other sides. Since if $\mathbf{u}$ and $\mathbf{v}$ are vectors of norm 2 with an angle of $45°$ between them

$$\mathbf{u} \cdot \mathbf{v} = \|\mathbf{u}\|\,\|\mathbf{v}\| \cos 45° = 2 \cdot 2 \cdot \frac{\sqrt{2}}{2} = 2\sqrt{2},$$

we have

$$\langle a, b, c \rangle \cdot \langle 2, 0, 0 \rangle = 2a = 2\sqrt{2},$$
$$\langle a, b, c \rangle \cdot \langle \sqrt{2}, \sqrt{2}, 0 \rangle = \sqrt{2}\,a + \sqrt{2}\,b = 2\sqrt{2}.$$

Solving gives $a = \sqrt{2}$ and $b = 2 - \sqrt{2}$. Finally, since $\|\langle a, b, c \rangle\| = 2$, we have

$$4 = \|\langle a, b, c \rangle\|^2 = \left(\sqrt{2}\right)^2 + \left(2 - \sqrt{2}\right)^2 + c^2 = 2 + 4 - 4\sqrt{2} + 2 + c^2$$

so that $c^2 = 4\sqrt{2} - 4$ and $c = 2\sqrt{\sqrt{2} - 1}$. Then the volume of the crystal is

$$\left\langle \sqrt{2}, 2 - \sqrt{2}, 2\sqrt{\sqrt{2}-1} \right\rangle \cdot \left(\langle 2,0,0 \rangle \times \left\langle \sqrt{2}, \sqrt{2}, 0 \right\rangle \right)$$

$$= \left\langle \sqrt{2}, 2 - \sqrt{2}, 2\sqrt{\sqrt{2}-1} \right\rangle \cdot$$

$$\left\langle 0 \cdot 0 - 0 \cdot \sqrt{2}, 0 \cdot \sqrt{2} - 2 \cdot 0, 2 \cdot \sqrt{2} - 0 \cdot \sqrt{2} \right\rangle$$

$$= \left\langle \sqrt{2}, 2 - \sqrt{2}, 2\sqrt{\sqrt{2}-1} \right\rangle \cdot \left\langle 0, 0, 2\sqrt{2} \right\rangle$$

$$= 4\sqrt{2\sqrt{2}-2}.$$

61. Since the magnitude of the torque is $\|\mathbf{r} \times \mathbf{F}\| = \|\mathbf{r}\|\,\|\mathbf{F}\| \sin \theta$, we have for the magnitude of the torque $40 \cdot \frac{9}{12} \cdot \sin 90° = 30$ foot-pounds.

Proofs

63. We have (by Definition 10.24)

$$\det \begin{bmatrix} a_1 & a_2 & a_3 \\ b_1 & b_2 & b_3 \\ c_1 & c_2 & c_3 \end{bmatrix} = a_1b_2c_3 - a_1b_3c_2 + a_2b_3c_1 - a_2b_1c_3 + a_3b_1c_2 - a_3b_2c_1$$

$$\det \begin{bmatrix} a_1 & a_2 & a_3 \\ c_1 & c_2 & c_3 \\ b_1 & b_2 & b_3 \end{bmatrix} = a_1c_2b_3 - a_1c_3b_2 + a_2c_3b_1 - a_2c_1b_3 + a_3c_1b_2 - a_3c_2b_1$$

$$= -a_1b_2c_3 + a_1b_3c_2 - a_2b_3c_1 + a_2b_1c_3 - a_3c_2b_1 + a_3b_2c_1$$

$$\det \begin{bmatrix} b_1 & b_2 & b_3 \\ a_1 & a_2 & a_3 \\ c_1 & c_2 & c_3 \end{bmatrix} = b_1a_2c_3 - b_1a_3c_2 + b_2a_3c_1 - b_2a_1c_3 + b_3a_1c_2 - b_3a_2c_1$$

$$= -a_1b_2c_3 + a_1c_2b_3 - a_2b_3c_1 + a_2b_1c_3 - a_3b_1c_2 + a_3b_2c_1.$$

Thus exchanging the second and third rows, or the first and second rows, negates the determinant. But exchanging the first and second rows, then the second and third rows, then the first and second rows again results in exchanging the first and third rows. Since each of these three exchanges negates the determinant, the end result is three inversions, so that exchanging the first and third rows also negates the determinant.

65. Suppose that $\mathbf{u} = \langle u_1, u_2, u_3 \rangle$ and $\mathbf{v} = k\mathbf{u} = \langle ku_1, ku_2, ku_3 \rangle$. Then

$$\mathbf{u} \times \mathbf{v} = \det \begin{bmatrix} \mathbf{i} & \mathbf{j} & \mathbf{k} \\ u_1 & u_2 & u_3 \\ ku_1 & ku_2 & ku_3 \end{bmatrix}$$

$$= u_2(ku_3)\mathbf{i} - u_3(ku_2)\mathbf{i} + u_3(ku_1)\mathbf{j} - u_1(ku_3)\mathbf{j} + u_1(ku_2)\mathbf{k} - u_2(ku_1)\mathbf{k}$$

$$= (ku_2u_3 - ku_2u_3)\mathbf{i} + (ku_1u_3 - ku_1u_3)\mathbf{j} + (ku_1u_2 - ku_1u_2)\mathbf{k}$$

$$= \mathbf{0}.$$

67. We have, if $\mathbf{u} = \langle u_1, u_2, u_3 \rangle$ and $\mathbf{v} = \langle v_1, v_2, v_3 \rangle$,

$$c(\mathbf{u} \times \mathbf{v}) = c\left((u_2v_3 - u_3v_2)\mathbf{i} + (u_3v_1 - u_1v_3)\mathbf{j} + (u_1v_2 - u_2v_1)\mathbf{k}\right)$$

$$= ((cu_2)v_3 - (cu_3)v_2)\mathbf{i} + ((cu_3)v_1 - (cu_1)v_3)\mathbf{j} + ((cu_1)v_2 - (cu_2)v_1)\mathbf{k})$$

$$= (c\mathbf{u}) \times \mathbf{v}.$$

To show that $c(\mathbf{u} \times \mathbf{v}) = \mathbf{u} \times (c\mathbf{v})$, note that

$$c(\mathbf{u} \times \mathbf{v}) = c(-\mathbf{v} \times \mathbf{u}) = -c(\mathbf{v} \times \mathbf{u}) = -(c\mathbf{v}) \times \mathbf{u} = \mathbf{u} \times (c\mathbf{v}).$$

69. Let $\mathbf{u} = \langle u_1, u_2, u_3 \rangle$ and $\mathbf{v} = \langle v_1, v_2, v_3 \rangle$. Then

$$\mathbf{v} \cdot (\mathbf{u} \times \mathbf{v}) = \langle v_1, v_2, v_3 \rangle \cdot \langle u_2v_3 - u_3v_2, u_3v_1 - u_1v_3, u_1v_2 - u_2v_1 \rangle$$

$$= v_1u_2v_3 - v_1u_3v_2 + v_2u_3v_1 - v_2u_1v_3 + v_3u_1v_2 - v_3u_2v_1$$

$$= v_1v_3u_2 - v_1v_3u_2 - v_1v_2u_3 + v_1v_2u_3 - v_2v_3u_1 + v_2v_3u_1$$

$$= 0.$$

71. (a) The volume of the parallelpiped is $\mathbf{u} \cdot (\mathbf{v} \times \mathbf{w})$. Now, $\mathbf{v} \times \mathbf{w}$ is, by Definition 10.25, in each component a difference of a product of a component of $\mathbf{v}$ and a component of $\mathbf{w}$. Thus $\mathbf{v} \times \mathbf{w}$ has integral components. But $\mathbf{u}$ also has integral components, so that $\mathbf{u} \cdot (\mathbf{v} \times \mathbf{w})$ is the dot product of two vectors with integral components, so it has integral components as well.

(b) For example, if $\mathbf{u} = \langle 1,0,0 \rangle = \mathbf{i}$ and $\mathbf{v} = \langle 0,1,0 \rangle = \mathbf{j}$, then the area of the parallelogram determined by $\mathbf{u}$ and $\mathbf{v}$ is

$$\|\mathbf{u} \times \mathbf{v}\| = \|\mathbf{i} \times \mathbf{j}\| = \|\mathbf{k}\| = 1,$$

which is an integer. However, if $\mathbf{u} = \langle 1,0,0 \rangle$ and $\mathbf{v} = \langle 1,1,1 \rangle$, then the area of the parallelogram determined by $\mathbf{u}$ and $\mathbf{v}$ is

$$\|\mathbf{u} \times \mathbf{v}\| = \|\langle 1,0,0 \rangle \times \langle 1,1,1 \rangle\| = \|\langle 0 \cdot 1 - 0 \cdot 1, 0 \cdot 1 - 1 \cdot 1, 1 \cdot -0 \cdot 1 \rangle\|$$

$$= \|\langle 0, -1, 1 \rangle\| = \sqrt{0^2 + (-1)^2 + 1^2} = \sqrt{2}.$$

73. Since $\|\mathbf{u} \times \mathbf{v}\| = \|\mathbf{u}\|\,\|\mathbf{v}\| \sin\theta$ and $\mathbf{u} \cdot \mathbf{v} = \|\mathbf{u}\|\,\|\mathbf{v}\| \cos\theta$, dividing the two equations (since $\mathbf{u} \cdot \mathbf{v} \neq 0$) gives $\frac{\|\mathbf{u} \times \mathbf{v}\|}{\mathbf{u} \cdot \mathbf{v}} = \tan\theta$.

75. By Exercise 74, if $\mathbf{u} \times \mathbf{v} = \mathbf{u} \times \mathbf{w}$, then $\mathbf{v}$ is parallel to $\mathbf{w}$, say $\mathbf{v} = k\mathbf{w}$. Then $\mathbf{u} \cdot \mathbf{v} = \mathbf{u} \cdot (k\mathbf{w}) = k(\mathbf{u} \cdot \mathbf{w})$, so if this is equal to $\mathbf{u} \cdot \mathbf{w}$, we must have $k = 1$, so that $\mathbf{v} = \mathbf{w}$.

77. Let $\mathbf{r} = \langle r_1, r_2, r_3 \rangle$, $\mathbf{s} = \langle s_1, s_2, s_3 \rangle$, $\mathbf{u} = \langle u_1, u_2, u_3 \rangle$, and $\mathbf{v} = \langle v_1, v_2, v_3 \rangle$. Then

$$(\mathbf{r} \cdot \mathbf{u})(\mathbf{s} \cdot \mathbf{v}) - (\mathbf{r} \cdot \mathbf{v})(\mathbf{s} \cdot \mathbf{u})$$

$$= (r_1 u_1 + r_2 u_2 + r_3 u_3)(s_1 v_1 + s_2 v_2 + s_3 v_3) - (r_1 v_1 + r_2 v_2 + r_3 v_3)(s_1 u_1 + s_2 u_2 + s_3 u_3)$$

$$= r_1 s_1 u_1 v_1 + r_1 s_2 u_1 v_2 + r_1 s_3 u_1 v_3 + r_2 s_1 u_2 v_1 + r_2 s_2 u_2 v_2$$

$$\quad + r_2 s_3 u_2 v_3 + r_3 s_1 u_3 v_1 + r_3 s_2 u_3 v_2 + r_3 s_3 u_3 v_3$$

$$\quad - r_1 s_1 u_1 v_1 - r_1 s_2 u_2 v_1 - r_1 s_3 u_3 v_1 - r_2 s_1 u_1 v_2 - r_2 s_2 u_2 v_2$$

$$\quad - r_2 s_3 u_3 v_3 - r_3 s_1 u_1 v_3 - r_3 s_2 u_2 v_3 - r_3 s_3 u_3 v_3$$

$$= r_1 s_2 u_1 v_2 + r_1 s_3 u_1 v_3 + r_2 s_1 u_2 v_1 + r_2 s_3 u_2 v_3 + r_3 s_1 u_3 v_1 + r_3 s_2 u_3 v_2$$

$$\quad - r_1 s_2 u_2 v_1 - r_1 s_3 u_3 v_1 - r_2 s_1 u_1 v_2 - r_2 s_3 u_3 v_3 - r_3 s_1 u_1 v_3 - r_3 s_2 u_2 v_3$$

$$= (r_1 s_2 - r_2 s_1) u_1 v_2 + (r_1 s_3 - r_3 s_1) u_1 v_3 + (r_2 s_1 - r_1 s_2) u_2 v_1$$

$$\quad + (r_2 s_3 - r_3 s_2) u_2 v_3 + (r_3 s_1 - r_1 s_3) u_3 v_1 + (r_3 s_2 - r_2 s_3) u_3 v_2$$

$$= (r_2 s_3 - r_3 s_2)(u_2 v_3 - u_3 v_2) + (r_3 s_1 - r_1 s_3)(u_3 v_1 - u_1 v_3) + (r_1 s_2 - r_2 s_1)(u_1 v_2 - u_2 v_1)$$

$$= \langle r_2 s_3 - r_3 s_2, r_3 s_1 - r_1 s_3, r_1 s_2 - r_2 s_1 \rangle \cdot \langle u_2 v_3 - u_3 v_2, u_3 v_1 - u_1 v_3, u_1 v_2 - u_2 v_1 \rangle$$

$$= (\mathbf{r} \times \mathbf{s}) \cdot (\mathbf{u} \times \mathbf{v}).$$

79. If the vectors can be transported so that they all lie in the same plane, then $\mathbf{r} \times \mathbf{s}$ and $\mathbf{u} \times \mathbf{v}$ are both perpendicular to that plane, so they are parallel. But the cross product of parallel vectors is zero, so that

$$(\mathbf{r} \times \mathbf{s}) \times (\mathbf{u} \times \mathbf{v}) = \mathbf{0}.$$

Thinking Forward

Planes in $\mathbb{R}^3$

▷ Translate the vectors so that their initial point is at the given point. The given point together with the terminal points of the vectors then give three points; these points are not collinear since the vectors are not parallel. Thus the three points determine a unique plane; that plane contains the given point as well as the two terminal points, so that it contains the vectors as well.

▷ Translate the vector so that its initial point is at the given point P. Then the plane consists of all points Q such that $\overrightarrow{PQ}$ is perpendicular to the given vector.

▷ For example, any collection of three noncollinear points determine a plane, and a position vector together with a point not on the line determined by the position vector determine a plane.

10.5   Lines in Three-Dimensional Space

Thinking Back

Lines in the plane

▷ The slope of this line is 2, so its direction vector is $(1, 2)$. Since $(0, 5)$ lies on the line, we let $\mathbf{P}_0 = \langle 0, 5 \rangle$ and $\mathbf{d} = \langle 1, 2 \rangle$. Then the line is $\langle 0, 5 \rangle + t \langle 1, 2 \rangle$ as a vector parametrization. In symmetric form, we have $y - 5 = 2x$, so that $\frac{y-5}{2} = \frac{x-0}{1}$.

▷ Since the line perpendicular has slope $-\frac{1}{2}$, it has equation $y = -\frac{1}{2}x + b$. Since $(2, -1)$ lies on the line, we see that $b = 0$, so that the equation is $y = -\frac{1}{2}x$. Since the slope is $-\frac{1}{2}$, a direction vector is $\langle -2, 1 \rangle$; since $\langle 0, 0 \rangle$ lies on the line, we let $\mathbf{P}_0 = \langle 0, 0 \rangle$, so that the line is $\langle 0, 0 \rangle + t \langle -2, 1 \rangle = t \langle -2, 1 \rangle$ as a vector parametrization. In symmetric form, we have $\frac{y-0}{1} = -\frac{x-0}{2}$.

▷ The slope of this line is m, so its direction vector is $\langle 1, m \rangle$. Since $(0, b)$ lies on the line, we let $\mathbf{P}_0 = \langle 0, b \rangle$ and $\mathbf{d} = \langle 1, m \rangle$. Then the line is $\langle 0, b \rangle + t \langle 1, m \rangle$ as a vector parametrization. In symmetric form, we have $y - b = mx$, so that (assuming $m \neq 0$) $\frac{y-b}{m} = \frac{x-0}{1}$. If $m = 0$, then the line is $y = b$ and there is no symmetric form since x does not appear in the equation of the line.

▷ Assuming $m \neq 0$, the perpendicular line has slope $-\frac{1}{m}$, so its equation is $y = -\frac{1}{m}x + C$. Since $(0, b)$ is on the line, we have $C = b$, so that the equation is $y = -\frac{1}{m}x + b$. Since the slope is $-\frac{1}{m}$, a direction vector is $\langle -m, 1 \rangle$; since $\langle 0, b \rangle$ lies on the line, we let $\mathbf{P}_0 = \langle 0, b \rangle$, so that the line is $\langle 0, b \rangle + t \langle -m, 1 \rangle$ as a vector parametrization. In symmetric form, we have $\frac{y-b}{1} = -\frac{x-0}{m}$. If on the other hand $m = 0$, then the line is horizontal, so the requested perpendicular line is the vertical line through $(0, b)$, which is the line $x = 0$. This has infinite slope, so its direction vector is $\langle 0, 1 \rangle$ and its vector parametrization is $\langle 0, b \rangle + t \langle 0, 1 \rangle$ (or, if you prefer, simply $t \langle 0, 1 \rangle$, since it passes through all points of the form $(0, y)$).

▷ The distance between the two points is $\sqrt{(x_2 - x_1)^2 + (y_2 - y_1)^2}$.

▷ The distance between the point $Q = (x_0, y_0)$ and the line $y = mx + b$ can be found, for example, by choosing two different points P and R on the line. Compute the vectors $\overrightarrow{PQ}$ and $\overrightarrow{PR}$. The distance from Q to the line is then $\overrightarrow{PQ} - \text{proj}_{\overrightarrow{PR}} \overrightarrow{PQ}$, the orthogonal component of $\overrightarrow{PQ}$ perpendicular to $\overrightarrow{PR}$. An alternative method is presented in the text in this section.

Concepts

1. (a) True. If they are parallel, then we can choose a common direction vector $\mathbf{d}$ for both lines. If they share a point $\mathbf{P}$, then both lines can be written in vector form as $\mathbf{P} + t\mathbf{d}$, so they are identical.

   (b) False. For example, the line $\langle 0, 0, 1 \rangle + t \langle 1, 1, 0 \rangle$ always has a z coordinate of 1, so it never intersects the xy plane.

   (c) True. If the line is parallel to the xz plane, then it is parallel to some line in the xz plane, which (since the y coordinate is zero in the xz plane) has direction vector $\mathbf{d} = \langle a_1, 0, c_1 \rangle$. But parallel lines have direction vectors that differ by a scalar multiple, so that $\mathbf{r}(t)$ has a direction vector of the form $\langle a, 0, c \rangle$.

   (d) False. For example, let $\mathbf{r}_1(t) = \langle 0, 0, 0 \rangle + t \langle 1, 0, 0 \rangle$ and $\mathbf{r}_2(t) = \langle 0, 0, 0 \rangle + t \langle 0, 1, 0 \rangle$. Then the direction vectors of the two lines are unequal, yet they intersect at the origin.

   (e) True. Since $\mathbf{P}_1 = \mathbf{P}_2$, the two lines intersect when $t = 0$ for each.

(f) False. For example, the line $\mathbf{r}_1(t) = \langle 0,0,0 \rangle + t\langle 1,0,0 \rangle$ and $\mathbf{r}_2(t) = \langle 1,0,0 \rangle + t\langle 2,0,0 \rangle$ both represent the x axis.

(g) False. See part (f); in fact, the lines can be identical, so they would be parallel.

(h) False. See the example in part (d) for two lines with this property that intersect.

3. A linear equation in three variables is an equation of the form $ax + by + cz = d$, where a, b, c, and d are real numbers. For example, $x + y + z = 12$ is such an equation. The graph of a linear equation in three variables is a plane. See the discussion at the start of this section.

5. P is a point on the line. A direction vector for the line is given by the vector $Q - P$. Thus from the discussion at the start of the section, an equation for the line is $\mathbf{r}(t) = P + t(Q - P)$.

7. If the slopes of the lines are equal, then they are either parallel or identical. (If they share a common point, they are identical). If they are neither parallel nor identical, then they intersect in a single point, which may be found by solving the pair of linear equations given by the lines.

9. For example, $\mathbf{r}(t) = \langle -1,3,7 \rangle + t\langle 2,-4,9 \rangle$. See the discussion in the text.

(a) Since the vector $\langle 2,-4,9 \rangle$ points in the same direction as $2\langle 2,-4,9 \rangle = \langle 4,-8,18 \rangle$, another vector equation for the same line is $\mathbf{r}_1(t) = \langle -1,3,7 \rangle + t\langle 4,-8,18 \rangle$. Many other answers are possible.

(b) From the equation for $\mathbf{r}(t)$, we see that for $t = 3$ we get $\mathbf{r}(3) = \langle -1,3,7 \rangle + 3\langle 2,-4,9 \rangle = \langle 5,-9,34 \rangle$, so that this point lies on the line. So if we let $y_0 = -9$ and $z_0 = 34$, and then let $\langle a,b,c \rangle = \langle 2,-4,9 \rangle$, we get an equation for the line $\mathbf{r}_2(t) = \langle 5,-9,34 \rangle + t\langle 2,-4,9 \rangle$. This is the same line since it has the same direction vector as $\mathbf{r}(t)$, and they share the point $\langle 5,-9,34 \rangle$. This vector equation can be rewritten as $x = 2t + 5$, $y = -4t - 9$, $z = 9t + 34$.

11. $\mathbf{r}(t)$ intersects the xy plane when its z coordinate is zero, which is when $5 + t = 0$, so that $t = -5$. At $t = -5$, $\mathbf{r}(t) = \langle 4 - 3(-5), 8 + 7(-5), 0 \rangle = \langle 19, -27, 0 \rangle$, so it intersects the xy plane at $(19, -27, 0)$. $\mathbf{r}(t)$ intersects the xz plane when its y coordinate is zero, which is when $8 + 7t = 0$, so when $t = -\frac{8}{7}$. At that value of t, $\mathbf{r}(t) = \langle 4 + 3\frac{8}{7}, 0, 5 - \frac{8}{7} \rangle = \langle \frac{52}{7}, 0, \frac{27}{7} \rangle$, so it intersects the xz plane at $\left(\frac{52}{7}, 0, \frac{27}{7} \right)$. $\mathbf{r}(t)$ intersects the yz plane when its x coordinate is zero, which is when $4 - 3t = 0$, or $t = \frac{4}{3}$. At that value of t, $\mathbf{r}(t) = \left\langle 0, 8 + 7 \cdot \frac{4}{3}, 5 + \frac{4}{3} \right\rangle = \left\langle 0, \frac{52}{3}, \frac{19}{3} \right\rangle$, so it intersects the yz plane at $\left(0, \frac{52}{3}, \frac{19}{3} \right)$.

13. This line intersects the xy plane when the z coordinate is zero, so when $-4t - 7 = 0$. This is for $t = -\frac{7}{4}$. When that happens, we have

$$(x,y,z) = \left(-\frac{7}{4} + 2,\ 2\left(-\frac{7}{4} \right) - 5,\ -4\left(-\frac{7}{4} \right) - 7 \right) = \left(\frac{1}{4}, -\frac{17}{2}, 0 \right).$$

It intersects the xz plane when the y coordinate is zero, so when $2t - 5 = 0$. This is for $t = \frac{5}{2}$. When that happens, we have

$$(x,y,z) = \left(\frac{5}{2} + 2,\ 2\left(\frac{5}{2} \right) - 5,\ -4\left(\frac{5}{2} \right) - 7 \right) = \left(\frac{9}{2}, 0, -17 \right).$$

It intersects the yz plane when the x coordinate is zero, so when $t + 2 = 0$. This is for $t = -2$. When that happens, we have

$$(x,y,z) = (-2 + 2, 2(-2) - 5, -4(-2) - 7) = (0, -9, 1).$$

15. This line intersects the xz plane when the y coordinate is zero, so when $t = 0$. When that happens, we have $(x, y, z) = (-7, 0, 5)$. The line intersects the xy plane when the z coordinate is zero, which never happens since $z = 5$ everywhere. It intersects the yz plane when the x coordinate is zero, which never happens since $x = -7$ throughout. Thus the line does not intersect either of those coordinate planes.

17. (a) Since the vector components give the x, y, and z coordinates, the parametric equation for $\mathcal{L}$ is
$$x = 2 + 7t, \qquad y = 3 - 5t, \qquad z = 2t, \text{ where } -\infty < t < \infty.$$

   (b) Solving the above equations for t gives
$$t = \frac{x-2}{7}, \qquad t = \frac{3-y}{5}, \qquad t = \frac{z}{2},$$
   so that the symmetric equation for the line is
$$\frac{x-2}{7} = \frac{3-y}{5} = \frac{z}{2}.$$

19. (a) Setting $t = 0$ gives $\langle 4, 3, 0 \rangle$ as a point on the line. From the parametric equations, we see that $\langle 0, -5, 1 \rangle$ is a direction vector. Thus a vector parametrization is $\langle 4, 3, 0 \rangle + t \langle 0, -5, 1 \rangle$.

   (b) Solving the equations given for t gives
$$x = 4, \qquad t = \frac{3-y}{5} = z,$$
   so that the equation in symmetric form is
$$x = 4, \qquad z = \frac{3-y}{5}.$$

Skills

21. The slope of the line is $\frac{-1-5}{2-0} = -3$, so the equation of the line is $y - 5 = -3(x - 0)$, or $y = -3x + 5$.

   To find a vector parametrization, set $\mathbf{P}_0 = \langle 0, 5 \rangle$; a direction vector is $\overrightarrow{PQ} = \langle 2, -1 \rangle - \langle 0, 5 \rangle = \langle 2, -6 \rangle$. Thus a vector parametrization is $\mathbf{r}(t) = \langle 0, 5 \rangle + t \langle 2, -6 \rangle$ for $-\infty < t < \infty$. To show that the two are equivalent, note that P and Q are both on $y = -3x + 5$, that $P = \mathbf{r}(0)$, and that $Q = \mathbf{r}(1)$, so that P and Q are also on $\mathbf{r}(t)$. Since two points determine a unique line, these two representations are of the same line.

23. (a) Let $\mathbf{P}_0 = P = \langle 0, 0, 0 \rangle$. Then a vector parametrization is $\mathbf{r}(t) = \langle 0, 0, 0 \rangle + t\mathbf{d} = t \langle 1, 2, -4 \rangle$ for $-\infty < t < \infty$.

   (b) Since the vector equation above provides formulas for the x, y, and z components, this gives for the parametric form
$$x = t, \qquad y = 2t, \qquad z = -4t \text{ for } -\infty < t < \infty.$$

   (c) Solving the parametric equations for t gives $t = x = \frac{y}{2} = -\frac{z}{4}$, so that an equation of the line in symmetric form is
$$x = \frac{y}{2} = -\frac{z}{4}.$$

25. (a) Let $\mathbf{P}_0 = P = \langle -1, 3, 7 \rangle$. Then a vector parametrization is $\mathbf{r}(t) = \langle -1, 3, 7 \rangle + t\mathbf{d} = \langle -1, 3, 7 \rangle + t \langle 2, 0, 4 \rangle$ for $-\infty < t < \infty$.

(b) Since the vector equation above provides formulas for the x, y, and z components, this gives for the parametric form

$$x = 2t - 1, \qquad y = 3, \qquad z = 4t + 7 \text{ for } -\infty < t < \infty.$$

(c) Solving the parametric equations for t gives $t = \frac{x+1}{2} = \frac{z-7}{4}$, so that an equation of the line in symmetric form is

$$\frac{x+1}{2} = \frac{z-7}{4}, \quad y = 3.$$

27. (a) Let $\mathbf{P}_0 = P = \langle 3, 1 \rangle$. Then a vector parametrization is $\mathbf{r}(t) = \langle 3, 1 \rangle + t\mathbf{d} = \langle 3, 1 \rangle + t \langle 2, 5 \rangle$ for $-\infty < t < \infty$.

(b) Since the vector equation above provides formulas for the x and y components, this gives for the parametric form

$$x = 2t + 3, \qquad y = 5t + 1 \text{ for } -\infty < t < \infty.$$

(c) Solving the parametric equations for t gives $t = \frac{x-3}{2} = \frac{y-1}{5}$, so that an equation of the line in symmetric form is

$$\frac{x-3}{2} = \frac{y-1}{5}.$$

29. (a) Let $\mathbf{P}_0 = P = \langle 0, 0, 0 \rangle$. Then a direction vector $\mathbf{d}$ is $\mathbf{d} = \overrightarrow{PQ} = \langle 4, -1, 6 \rangle$, so that a vector parametrization is $\mathbf{r}(t) = \langle 0, 0, 0 \rangle + t \langle 4, -1, 6 \rangle = t \langle 4, -1, 6 \rangle$ for $-\infty < t < \infty$.

(b) Since the vector equation above provides formulas for the x, y, and z components, this gives for the parametric form

$$x = 4t, \qquad y = -t, \qquad z = 6t \text{ for } -\infty < t < \infty.$$

(c) Solving the parametric equations for t gives $t = \frac{x}{4} = -y = \frac{z}{6}$, so that an equation of the line in symmetric form is

$$\frac{x}{4} = -y = \frac{z}{6}.$$

31. (a) Let $\mathbf{P}_0 = P = \langle -4, 11, 0 \rangle$. Then a direction vector $\mathbf{d}$ is $\mathbf{d} = \overrightarrow{PQ} = \langle 8, 0, 2 \rangle$, so that a vector parametrization is $\mathbf{r}(t) = \langle -4, 11, 0 \rangle + t \langle 8, 0, 2 \rangle$ for $-\infty < t < \infty$.

(b) Since the vector equation above provides formulas for the x, y, and z components, this gives for the parametric form

$$x = 8t - 4, \qquad y = 11, \qquad z = 2t \text{ for } -\infty < t < \infty.$$

(c) Solving the parametric equations for t gives $t = \frac{x+4}{8} = \frac{z}{2}$, so that an equation of the line in symmetric form is

$$\frac{x+4}{8} = \frac{z}{2}, \quad y = 11.$$

33. (a) Let $\mathbf{P}_0 = P = \langle 1, 6 \rangle$. Then a direction vector $\mathbf{d}$ is $\mathbf{d} = \overrightarrow{PQ} = \langle 3, -1 \rangle$, so that a vector parametrization is $\mathbf{r}(t) = \langle 1, 6 \rangle + t \langle 3, -1 \rangle$ for $-\infty < t < \infty$.

(b) Since the vector equation above provides formulas for the x and y components, this gives for the parametric form

$$x = 3t + 1, \qquad y = -t + 6 \text{ for } -\infty < t < \infty.$$

(c) Solving the parametric equations for t gives $t = \frac{x-1}{3} = 6 - y$, so that an equation of the line in symmetric form is

$$\frac{x-1}{3} = -y + 6.$$

35. (a) Let $\mathbf{P}_0 = P = \langle x_0, y_0, z_0 \rangle$. Then a direction vector $\mathbf{d}$ is $\mathbf{d} = \overrightarrow{PQ} = \langle x_1 - x_0, y_1 - y_0, z_1 - z_0 \rangle$, so that a vector parametrization is $\mathbf{r}(t) = \langle x_0, y_0, z_0 \rangle + t \langle x_1 - x_0, y_1 - y_0, z_1 - z_0 \rangle$ for $-\infty < t < \infty$.

(b) Since

$$\mathbf{r}(0) = \langle x_0, y_0, z_0 \rangle = P$$
$$\mathbf{r}(1) = \langle x_0, y_0, z_0 \rangle + \langle x_1 - x_0, y_1 - y_0, z_1 - z_0 \rangle = \langle x_1, y_1, z_1 \rangle = Q,$$

restricting t to $0 \leq t \leq 1$ parametrizes the line segment from P to Q.

37. Let $\mathbf{P}_0 = P = \langle 1, 7, 3 \rangle$. Then a direction vector $\mathbf{d}$ is $\mathbf{d} = \overrightarrow{PQ} = \langle -2, -9, 2 \rangle$, so that a vector parametrization is $\mathbf{r}(t) = \langle 1, 7, 3 \rangle + t \langle -2, -9, 2 \rangle$ for $0 \leq t \leq 1$. Since the vector equation above provides formulas for the x, y, and z components, this gives for the parametric form

$$x = -2t + 1, \qquad y = -9t + 7, \qquad z = 2t + 3 \text{ for } 0 \leq t \leq 1.$$

39. Let $\mathbf{P}_0 = P = \langle 3, -1, 4 \rangle$. Then a direction vector $\mathbf{d}$ is $\mathbf{d} = \overrightarrow{PQ} = \langle -4, 6, 5 \rangle$, so that a vector parametrization is $\mathbf{r}(t) = \langle 3, -1, 4 \rangle + t \langle -4, 6, 5 \rangle$ for $0 \leq t \leq 1$. Since the vector equation above provides formulas for the x, y, and z components, this gives for the parametric form

$$x = -4t + 3, \qquad y = 6t - 1, \qquad z = 5t + 4 \text{ for } 0 \leq t \leq 1.$$

41. The direction vectors are $\mathbf{d}_1 = \langle 2, -1, 3 \rangle$ and $\mathbf{d}_2 = \langle -4, 2, -6 \rangle = -2 \langle 2, -1, 3 \rangle$. Thus the lines are parallel or identical. Now, $\mathbf{r}_1(0) = \langle 6, 1, 0 \rangle$. If $\mathbf{r}_2(t) = \langle 6, 1, 0 \rangle$, then looking at the z component shows that we must have $t = \frac{1}{3}$, in which case $\mathbf{r}_2(t) = \left\langle -4 \left(\frac{1}{3} \right) + 3, 2 \left(\frac{1}{3} \right), -6 \left(\frac{1}{3} \right) + 2 \right\rangle = \langle \frac{5}{3}, \frac{2}{3}, 0 \rangle \neq \langle 6, 1, 0 \rangle$. Thus the lines are not identical, so they must be parallel. To find the distance between the lines, it suffices to find the distance from any point on $\mathbf{r}_1$, say $\langle 6, 1, 0 \rangle$, to $\mathbf{r}_2$. Using Theorem 10.38, with $\mathbf{r}_2(t) = \langle 3, -1, 2 \rangle + t \langle -4, 2, -6 \rangle$, we get

$$\frac{\|\langle -4, 2, -6 \rangle \times (\langle 6, 1, 0 \rangle - \langle 3, -1, 2 \rangle)\|}{\|\langle -4, 2, -6 \rangle\|} = \frac{\|\langle -4, 2, -6 \rangle \times \langle 3, 2, -2 \rangle\|}{\sqrt{4^2 + 2^2 + 6^2}}$$
$$= \frac{\|\langle 2 \cdot (-2) + 6 \cdot 2, -6 \cdot 3 + 4 \cdot (-2), -4 \cdot 2 - 2 \cdot 3 \rangle\|}{\sqrt{56}}$$
$$= \frac{\|\langle 8, -26, -14 \rangle\|}{2\sqrt{14}}$$
$$= \frac{\sqrt{64 + 676 + 196}}{2\sqrt{14}} = \sqrt{\frac{117}{7}}.$$

43. The direction vectors are $\mathbf{d}_1 = \langle 5, -4, 1 \rangle$ and $\mathbf{d}_2 = \langle -3, -1, -2 \rangle$. Since these are not scalar multiples of each other, the lines are not parallel or identical. To see if they intersect, reparametrize $\mathbf{r}_2$ using the variable u rather than t; then we want to see if there are values of t and u such that all three coordinates are equal, i.e., such that

$$5t + 2 = -3u + 4, \qquad -4t = -u + 12, \qquad t - 7 = -2u - 1.$$

The third equation gives $t = 6 - 2u$; substituting into the second equation gives $-4(6 - 2u) = -u + 12$, or $-24 + 8u = 12 - u$. This gives $u = 4$ and thus $t = 6 - 2 \cdot 4 = -2$. Substituting these values into the first equation gives $5(-2) + 2 = -3(4) + 4$, or $-8 = -8$, which is true. Thus the lines intersect when $t = -2$ in $\mathbf{r}_1(t)$ and $u = 4$ in $\mathbf{r}_2(t)$. This point is $\mathbf{r}_1(-2) = \langle -8, 8, -9 \rangle$. The angle between the two lines may be found by looking at their direction vectors:

$$
\begin{aligned}
\sin\theta &= \frac{\|\mathbf{d}_1 \times \mathbf{d}_2\|}{\|\mathbf{d}_1\| \, \|\mathbf{d}_2\|} \\
&= \frac{\|\langle 5, -4, 1 \rangle \times \langle -3, -1, -2 \rangle\|}{\sqrt{5^2 + 4^2 + 1^2}\sqrt{3^2 + 1^2 + 2^2}} \\
&= \frac{\|\langle -4 \cdot (-2) - 1 \cdot (-1), 1 \cdot (-3) - 5 \cdot (-2), 5 \cdot (-1) - (-4) \cdot (-3) \rangle\|}{\sqrt{42}\sqrt{14}} \\
&= \frac{\|\langle 9, 7, -17 \rangle\|}{14\sqrt{3}} \\
&= \frac{\sqrt{9^2 + 7^2 + 17^2}}{14\sqrt{3}} \\
&= \frac{\sqrt{419}}{14\sqrt{3}} \approx 0.8441.
\end{aligned}
$$

Thus $\theta \approx \sin^{-1} 0.8441 \approx 57.58°$.

45. The direction vectors are $\mathbf{d}_1 = \langle 5, -4, 1 \rangle$ and $\mathbf{d}_2 = \langle -10, 8, -2 \rangle = -2 \langle 5, -4, 1 \rangle$. Since the direction vectors are multiples of one another, the lines are either parallel or identical. With $t = 0$, we get the point $(2, 6, 8)$ on $\mathbf{r}_1$. In order for $\mathbf{r}_2(u) = \langle 2, 6, 8 \rangle$, we must have (from the z component) $8 = -2u$ so that $u = -4$. But then the x component is $2 - 10(-4) = 42$. Thus the lines are parallel but not identical. To find the distance between the lines, it suffices to find the distance from any point on $\mathbf{r}_1$, say $\langle 2, 6, 8 \rangle$, to $\mathbf{r}_2$. Using Theorem 10.38, with $\mathbf{r}_2(t) = \langle 2, 3, 0 \rangle + t \langle -10, 8, -2 \rangle$, we get

$$
\begin{aligned}
\frac{\|\langle -10, 8, -2 \rangle \times (\langle 2, 6, 8 \rangle - \langle 2, 3, 0 \rangle)\|}{\|\langle -10, 8, -2 \rangle\|} &= \frac{\|\langle -10, 8, -2 \rangle \times \langle 0, 3, 8 \rangle\|}{\sqrt{10^2 + 8^2 + 2^2}} \\
&= \frac{\|\langle 8 \cdot 8 - (-2) \cdot 3, -2 \cdot 0 - (-10) \cdot 8, -10 \cdot 3 - 0 \cdot 8 \rangle\|}{\sqrt{168}} \\
&= \frac{\|\langle 70, 80, -30 \rangle\|}{2\sqrt{42}} \\
&= \frac{10\sqrt{7^2 + 8^2 + 3^2}}{2\sqrt{42}} = \frac{5\sqrt{122}}{\sqrt{42}} = \frac{5\sqrt{61}}{\sqrt{21}}.
\end{aligned}
$$

47. (a) Let $Q = \mathbf{r}(0) = \langle 1, -6, 0 \rangle$ and $R = \mathbf{r}(1) = \langle 6, -5, -4 \rangle$. Then the component of $\overrightarrow{QP} = \langle -1, 7, -2 \rangle$ parallel to $\overrightarrow{QR} = \langle 5, 1, -4 \rangle$ is

$$
\text{proj}_{\langle 5,1,-4 \rangle} \langle -1, 7, -2 \rangle = \frac{\langle -1, 7, -2 \rangle \cdot \langle 5, 1, -4 \rangle}{\|\langle 5, 1, -4 \rangle\|^2} \langle 5, 1, -4 \rangle = \frac{10}{42} \langle 5, 1, -4 \rangle = \frac{5}{21} \langle 5, 1, -4 \rangle.
$$

Thus the perpendicular component, whose length is the distance from P to the line, is

$$
\overrightarrow{QP} - \frac{5}{21} \langle 5, 1, -4 \rangle = \langle -1, 7, -2 \rangle - \left\langle \frac{25}{21}, \frac{5}{21}, -\frac{20}{21} \right\rangle = \left\langle -\frac{46}{21}, \frac{142}{21}, -\frac{22}{41} \right\rangle,
$$

so that the distance is

$$
\left\| \left\langle -\frac{46}{21}, \frac{142}{21}, -\frac{22}{41} \right\rangle \right\| = \frac{2}{21}\sqrt{5691} = 2\sqrt{\frac{271}{21}}.
$$

(b) Using Theorem 10.38, we choose $P_0 = \mathbf{r}(0) = \langle 1, -6, 0 \rangle$ and $d = \langle 5, 1, -4 \rangle$. Then the distance from P to the line is

$$\frac{\| \langle 5, 1, -4 \rangle \times \langle -1, 7, -2 \rangle \|}{\| \langle 5, 1, -4 \rangle \|} = \frac{\| \langle 26, 14, 36 \rangle \|}{\sqrt{42}} = \frac{2\sqrt{542}}{\sqrt{42}} = 2\sqrt{\frac{271}{21}}.$$

49. Let $P_0 = (0, -3)$. Since the slope is 2, we can choose $d = \langle 1, 2 \rangle$, so that a parametric equation of the line is $\mathbf{r}(t) = \langle 0, -3 \rangle + t \langle 1, 2 \rangle$.

(a) Let $Q = \mathbf{r}(0) = \langle 0, -3 \rangle$ and $R = \mathbf{r}(1) = \langle 1, -1 \rangle$. Then the component of $\overrightarrow{QP} = \langle 2, 8 \rangle$ parallel to $\overrightarrow{QR} = \langle 1, 2 \rangle$ is

$$\mathrm{proj}_{\langle 1,2 \rangle} \langle 2, 8 \rangle = \frac{\langle 2, 8 \rangle \cdot \langle 1, 2 \rangle}{\| \langle 1, 2 \rangle \|^2} \langle 1, 2 \rangle = \frac{18}{5} \langle 1, 2 \rangle = \left\langle \frac{18}{5}, \frac{36}{5} \right\rangle.$$

Thus the perpendicular component, whose length is the distance from P to the line, is

$$\overrightarrow{QP} - \left\langle \frac{18}{5}, \frac{36}{5} \right\rangle = \langle 2, 8 \rangle - \left\langle \frac{18}{5}, \frac{36}{5} \right\rangle = \left\langle -\frac{8}{5}, \frac{4}{5} \right\rangle,$$

so that the distance is

$$\left\| \left\langle -\frac{8}{5}, \frac{4}{5} \right\rangle \right\| = \sqrt{\frac{80}{25}} = \sqrt{\frac{16}{5}} = \frac{4}{\sqrt{5}}.$$

(b) Using Theorem 10.38, pretend the situation is actually in 3-space, with third components all equal to zero, so that we can use the cross-product. Then the distance is

$$\frac{\| \langle 1, 2, 0 \rangle \times \langle 2, 8, 0 \rangle \|}{\| \langle 1, 2, 0 \rangle \|} = \frac{\| \langle 0, 0, 4 \rangle \|}{\sqrt{5}} = \frac{4}{\sqrt{5}}.$$

51. (a) The lines are parallel when their direction vectors are parallel, which means that their cross product is zero. Thus we want

$$\langle -1, 2, 5 \rangle \times \langle \alpha, -2\alpha, 15 \rangle = \langle 30 + 10\alpha, 15 + 5\alpha, 0 \rangle = \langle 0, 0, 0 \rangle,$$

so that $\alpha = -3$.

(b) The lines are perpendicular when their direction vectors are perpendicular, which means that their dot products must be zero. Thus we want

$$\langle -1, 2, 5 \rangle \cdot \langle \alpha, -2\alpha, 15 \rangle = -\alpha - 4\alpha + 75 = 0,$$

so that $\alpha = 15$.

Applications

53. Since Ian is due east of the summit, which is at $(0, 0, 5.96)$, his coordinates are $(x, 0, z)$. Since his position and the two given points form a straight line, the cross product of any two of the resulting three vectors should be zero. The vector between the two summits is

$$\langle 11.2 - 5.4, -5.6 - (-2.5), 4.2 - 4.5 \rangle = \langle 5.8, -3.1, -0.3 \rangle,$$

and the vector between Ian's position and the minor point is $\langle 5.4 - x, -2.5, 4.5 - z \rangle$. The cross product of these two vectors is $\langle -14.7 + 3.1z, -27.72 + 0.3x + 5.8z, 2.24 - 3.1x \rangle$. Setting this equal to zero and solving gives $x \approx 0.7226$ and $z \approx 4.742$. Thus Ian is at height 4.742, and his distance from the summit as the crow flies is

$$\sqrt{(0.7226 - 0)^2 + (0 - 0)^2 + (4.742 - 5.96)^2} \approx 1.42.$$

Proofs

55. Let Q be the point on $\mathbf{r}(t)$ that is closest to P, so that $\|PQ\|$ is the distance we wish to find. Then (look at the diagram for Exercise 20) P_0P is the hypotenuse of a right triangle, and if θ is the angle between the line and P_0P, then since $\mathbf{d}$ is the direction vector for the line, we know that $\dfrac{\left\|\overrightarrow{PQ}\right\|}{\left\|\overrightarrow{P_0P}\right\|} = \sin\theta$ (from the definition of $\sin\theta$). Clearing fractions gives $\left\|\overrightarrow{PQ}\right\| = \left\|\overrightarrow{P_0P}\right\|\sin\theta$. But also from Theorem 10.32, know that the angle between two vectors $\mathbf{d}$ and $\overrightarrow{P_0P}$ is given by

$$\sin\theta = \frac{\left\|\mathbf{d} \times \overrightarrow{P_0P}\right\|}{\|\mathbf{d}\|\left\|\overrightarrow{P_0P}\right\|}.$$

Substitute this value for $\sin\theta$ into the first equation to get

$$\left\|\overrightarrow{PQ}\right\| = \left\|\overrightarrow{P_0P}\right\|\left(\frac{\left\|\mathbf{d} \times \overrightarrow{P_0P}\right\|}{\|\mathbf{d}\|\left\|\overrightarrow{P_0P}\right\|}\right) = \frac{\left\|\mathbf{d} \times \overrightarrow{P_0P}\right\|}{\|\mathbf{d}\|}.$$

Thinking Forward

Lines tangent to spheres

Since there is at least one line tangent to the sphere at each point on the sphere, there are an infinite number of lines that are tangent to the sphere (those lines are all different, since each one touches the sphere at a different point, so cannot coincide with any other tangent line). At any given point, since the sphere has a tangent plane at that point, any line in that tangent plane is tangent to the sphere there, so the sphere has an infinite number of tangents at any given point. To have a tangent parallel to the xy plane means that the tangent plane to the sphere must be parallel to the xy plane. This happens only at the north and south poles $(0, 0, \pm 1)$. At each of those points, there are an infinite number of tangents each of which is parallel to the xy plane. By symmetry, the same holds for the yz plane, with tangents at $(\pm 1, 0, 0)$ and for the xz plane, with tangents at $(0, \pm 1, 0)$. Finally, the vertical tangent line at any point on the equator of the sphere (the circle $x^2 + y^2 = 1$, $z = 0$) is parallel to the z axis; similar statements hold for the other two axes.

Planes tangent to spheres

Since each point on the sphere has a tangent plane, there are an infinite number (note that no two of those planes are the same since they each touch the sphere at only one point, so cannot coincide). At any given point on the sphere, however, there is a unique tangent plane. To have a tangent plane parallel to the xy plane means that the tangent plane must be horizontal; this happens only at the north and south poles $(0, 0, \pm 1)$, so there are exactly two tangent planes parallel to the xy plane. Similarly, there are two tangent planes parallel to the yz plane, at $(\pm 1, 0, 0)$, and two parallel to the xz plane, at $(0, \pm 1, 0)$. Finally, the tangent plane at any point on the equator of the sphere (the circle $x^2 + y^2 = 1$, $z = 0$) is parallel to the z axis; similar statements hold for the other two axes.

10.6   Planes

Thinking Back

Linear equations

$2x - 3y = 5$ represents a line in the xy plane since for any value of x there is exactly one corresponding value of y that satisfies that equation; the points lie in a line because this can be written in vector form as $\langle -2, -3 \rangle + t \langle 3, 2 \rangle$, which is the equation of a line. In $\mathbb{R}^3$, this is a plane, since although it is a line in the xy plane, the z coordinate is arbitrary, so this is a vertical plane lying over the line $2x - 3y - 5$.

Orthogonal vectors

Using properties of the dot product from Theorem 10.14, we hve

$$\mathbf{u} \cdot (a\mathbf{v} + b\mathbf{w}) = \mathbf{u} \cdot (a\mathbf{v}) + \mathbf{u} \cdot (b\mathbf{w}) = a(\mathbf{u} \cdot \mathbf{v}) + b(\mathbf{u} \cdot \mathbf{w}).$$

With the vectors given, we have

$$\mathbf{u} \cdot \mathbf{v} = \langle 1, 2, -3 \rangle \cdot \langle 1, 1, 1 \rangle = 1 \cdot 1 + 2 \cdot 1 - 3 \cdot 1 = 0,$$
$$\mathbf{u} \cdot \mathbf{w} = \langle 1, 2, -3 \rangle \cdot \langle -1, 2, 1 \rangle = -1 \cdot 1 + 2 \cdot 2 - 3 \cdot 1 = 0.$$

Thus $\mathbf{u}$ is orthogonal to $a\mathbf{v} + b\mathbf{w}$ for any real numbers a and b. Geometrically, this means that if $\mathbf{u}$ is orthogonal to two vectors that point in different directions, it is orthogonal to any combination of those two vectors; i.e., it is orthogonal to the plane determined by the vectors.

Concepts

1. (a) False. The graph of a linear equation in $\mathbb{R}^3$ is a plane. For example, $x = 0$ is the yz plane.

   (b) False. They determine a plane only if they are not collinear. If they are collinear, they determine a line, which lies in an infinite number of possible planes.

   (c) False. If the lines are skew, there is no plane that contains both lines.

   (d) False. For example, the three coordinate axes are concurrent, but there is no plane containing all three of them.

   (e) True. If the lines intersect, we can use the direction vectors of each to determine a common normal; the resulting plane contains both lines. If the lines are parallel, they have the same direction vector. Use that direction vector together with the vector from any point on one line to any point on the other to determine a normal vector; the resulting plane through the chosen point with that normal will contain the first line, and therefore also the second since they have the same direction vector.

   (f) True. See the discussion in the text. The direction vector of the line of intersection can be found by taking the cross product of the normal vectors to the two planes.

   (g) False. If the direction vector is *orthogonal* to the normal vector of the plane, the line and plane are parallel. Consider the plane $x = 0$, with normal vector $\mathbf{i}$, and the line with direction vector $\mathbf{i}$ that is $\langle 1, 0, 0 \rangle + t\mathbf{i}$. When $t = -1$, this line goes through the origin, which lies on $x = 0$, so the line and the plane are not parallel.

   (h) True. Since $\mathbf{N}_1 \times \mathbf{N}_2 = \mathbf{0}$, the normal vectors are parallel and thus the planes are parallel.

3. The plane intersects the x axis when $y = z = 0$, so the equation becomes $ax = d$, so that $x = \frac{d}{a}$. Similarly, the plane intersects the y axis when $x = z = 0$, so when $by = d$ and $y = \frac{d}{b}$. Finally, the plane intersects the z axis when $z = \frac{d}{c}$. So the three points of intersection are $\left(\frac{d}{a}, 0, 0 \right)$, $\left(0, \frac{d}{b}, 0 \right)$, and $\left(0, 0, \frac{d}{c} \right)$. Since any three noncollinear points determine a plane, these three points may be used to sketch the plane.

5. Since $\langle a, b, c \rangle$ and $\langle \alpha, \beta, \gamma \rangle$ are not parallel, neither is a multiple of the other. Then at least one of the following conditions holds: $\langle a, b \rangle$ is not a multiple of $\langle \alpha, \beta \rangle$, $\langle a, c \rangle$ is not a multiple of $\langle \alpha, \gamma \rangle$, or $\langle b, c \rangle$ is not a multiple of $\langle \beta, \gamma \rangle$, for if none of those three held, then the two normal vectors would be parallel. So assume that $\langle a, b \rangle$ is not a multiple of $\langle \alpha, \beta \rangle$. Then setting $z = 0$ in the original equations, we get the pair of equations $ax + by = d$ and $\alpha x + \beta y = \delta$. Since $\langle a, b \rangle$ is not a multiple of $\langle \alpha, \beta \rangle$, this pair of equations has a solution (x_0, y_0). Then the point $(x_0, y_0, 0)$ lies on both planes. The argument is identical if either of the other conditions holds.

7. Three points are collinear if they lie on the same line. To tell if three points P, Q, R are collinear, it suffices to show that $\overrightarrow{PQ}$ and $\overrightarrow{PR}$ are parallel, i.e., to show that their cross product is zero. Three points are not collinear if they do not lie on the same line; in this case, the three points determine a unique plane.

9. Given a line $\mathcal{L}$ and a point P not on $\mathcal{L}$, choose two distinct points Q and R on $\mathcal{L}$. Since P is not on $\mathcal{L}$, the three points are not collinear. Then the procedure in the previous exercise results in the equation of the plane containing P, Q, and R, and thus containing P and $\mathcal{L}$. If P lies on $\mathcal{L}$, then since there are an infinite number of planes containing $\mathcal{L}$ (see the previous exercise, or Exercise 6), any of those planes also contains P. So there is not a unique plane containing P and $\mathcal{L}$.

11. If two lines are parallel but distinct, choose two different points P and Q on one of the lines and a third point R on the other line. Since the lines are distinct, P, Q, and R are not collinear, so they determine a plane using the method in Exercise 8. That plane contains both P and Q, so it contains the line determined by the direction vector $\overrightarrow{PQ}$, which is the first line. Since the plane also contains R, and the second line has the same direction vector, it also contains the second line.

13. Let the lines be $\mathcal{L}_1$ and $\mathcal{L}_2$, and choose points P_1 on $\mathcal{L}_1$ and P_2 on $\mathcal{L}_2$. Let the direction vectors of the lines be $\mathbf{d}_1$ and $\mathbf{d}_2$; note that $\mathbf{d}_1$ and $\mathbf{d}_2$ are not parallel. Now the two lines $\mathcal{L}_1$ and $P_1 + t\mathbf{d}_2$ are intersecting distinct lines, so they determine a plane containing P_1. Similarly, $\mathcal{L}_2$ and $P_2 + t\mathbf{d}_1$ are intersecting distinct lines, so they determine a unique plane containing P_2. The normal vector to each plane is $\mathbf{d}_1 \times \mathbf{d}_2$, so that the planes are parallel. To see that the pair of planes is unique, note that any other such pair of planes through P_1 and P_2 have a common normal, since they are parallel, so that normal is orthogonal to the line $P_1 + t\mathbf{d}_2$ and to $P_2 + t\mathbf{d}_1$, so those lines lie in the first and second planes respectively. Thus the two planes are the pair of planes we determined above.

15. A normal vector to $\mathcal{P}$ is $\langle \alpha, \beta, \gamma \rangle$. The given line $\mathcal{L}$, which has vector parametrization $\langle x_0, y_0, z_0 \rangle + t \langle a, b, c \rangle$, has direction vector $\langle a, b, c \rangle$. Then $\mathcal{L}$ is orthogonal to $\mathcal{P}$ if and only if $\mathcal{L}$ is parallel to the normal vector to $\mathcal{P}$ if and only if $\langle a, b, c \rangle$ is parallel to $\langle \alpha, \beta, \gamma \rangle$ if and only if $\langle a, b, c \rangle$ is a scalar multiple of $\langle \alpha, \beta, \gamma \rangle$.

17. A diagram of the situation is

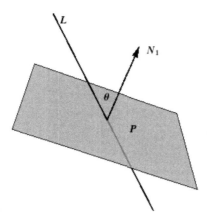

where **L** is the line and $\mathbf{N}_1$ is the normal to the plane. Since $\mathbf{N}_1$ forms a right angle with **P**, the angle between **L** and **P** is clearly $90° - \theta$.

19. The idea of this derivation is that the distance between the skew lines is the distance between the unique pair of parallel planes each of which contains one of the lines (see Exercise 13). If the direction vectors of the two lines are $\mathbf{d}_1$ and $\mathbf{d}_2$, then a normal to each of the planes is $\mathbf{d}_1 \times \mathbf{d}_2$; if P is a point on the first line in the first plane, and Q is a point on the second line in the second plane, then the formula for the distance from a point to a plane gives for the distance from P to the plane containing Q (since $\mathbf{d}_1 \times \mathbf{d}_2$ is a normal to the plane)

$$\frac{\left| (\mathbf{d}_1 \times \mathbf{d}_2) \cdot \overrightarrow{QP} \right|}{\| \mathbf{d}_1 \times \mathbf{d}_2 \|}.$$

Skills

21. Since the normal vector is $\langle 4, -1, 5 \rangle$ and the plane contains $\langle 0, 0, 0 \rangle$, its equation is

$$4(x - 0) - (y - 0) + 5(z - 0) = 0, \text{ or } 4x - y + 5z = 0.$$

See the discussion at the beginning of the section.

23. Since a normal vector is $\langle 2, -1, 6 \rangle$ and the point $\langle 2, -1, 6 \rangle$ is on the plane, its equation is

$$2(x - 2) - (y - (-1)) + 6(z - 6) = 0, \text{ or } 2(x - 2) - (y + 1) + 6(z - 6) = 0, \text{ or } 2x - y + 6z = 41.$$

25. Since the plane contains $P = (2, 4, 3)$, $Q = (3, -5, 0)$, and $R = (-4, 1, 6)$, a normal vector is

$$\begin{aligned}
\mathbf{N} = \overrightarrow{PQ} \times \overrightarrow{PR} &= \langle 1, -9, -3 \rangle \times \langle -6, -3, 3 \rangle \\
&= \langle -9 \cdot 3 - (-3) \cdot (-3), -3 \cdot (-6) - 1 \cdot 3, 1 \cdot (-3) - (-9) \cdot (-6) \rangle = \langle -36, 15, -57 \rangle.
\end{aligned}$$

Then since P is on the plane, its equation is

$$-36(x - 2) + 15(y - 4) - 57(z - 3) = 0, \text{ or } -12(x - 2) + 5(y - 4) - 19(z - 3) = 0, \text{ or }$$
$$-12x + 5y - 19z = -61.$$

27. Since the plane contains $P = (x_0, 0, 0)$, $Q = (0, y_0, 0)$, and $R = (0, 0, z_0)$, a normal vector is

$$\mathbf{N} = \overrightarrow{PQ} \times \overrightarrow{PR} = \langle -x_0, y_0, 0 \rangle \times \langle -x_0, 0, z_0 \rangle$$
$$= \langle y_0 \cdot z_0 - 0 \cdot 0, 0 \cdot (-x_0) - (-x_0) \cdot z_0, (-x_0) \cdot 0 - y_0 \cdot (-x_0) \rangle = \langle y_0 z_0, x_0 z_0, x_0 y_0 \rangle.$$

Then since P is on the plane, its equation is

$$y_0 z_0 (x - x_0) + x_0 z_0 (y - 0) + x_0 y_0 (z - 0) = 0, \text{ or } y_0 z_0 (x - x_0) + x_0 z_0 y + x_0 y_0 z = 0, \text{ or }$$
$$y_0 z_0 x + x_0 z_0 y + x_0 y_0 z = x_0 y_0 z_0.$$

29. Since $(3, -2, 3)$ is on the line $\mathbf{r}(t)$, and $\langle 7, 1, 4 \rangle$ is its direction vector, a normal to the plane is

$$\mathbf{N} = ((\langle -4, 1, 3 \rangle - \langle 3, -2, 3 \rangle) \times \langle 7, 1, 4 \rangle$$
$$= \langle -7, 3, 0 \rangle \times \langle 7, 1, 4 \rangle$$
$$= \langle 3 \cdot 4 - 0 \cdot 1, 0 \cdot 7 - (-7) \cdot 4, -7 \cdot 1 - 3 \cdot 7 \rangle$$
$$= \langle 12, 28, -28 \rangle.$$

Dividing through by 4, another normal to the plane is $\langle 3, 7, -7 \rangle$. Using $(3, -2, 3)$, which is on the plane, we get for an equation

$$3(x - 3) + 7(y - (-2)) - 7(z - 3) = 0, \text{ or } 3(x - 3) + 7(y + 2) - 7(z - 3) = 0, \text{ or }$$
$$3x + 7y - 7z = -26.$$

31. The direction vector of $\mathbf{r}_1(t)$ is $\langle -5, 2, 4 \rangle$, and the direction vector of $\mathbf{r}_2(t)$ is $\langle 15, -6, -12 \rangle = -3 \langle -5, 2, 4 \rangle$. Since the direction vectors are scalar multiples of each other, the lines are parallel. Then three noncollinear points on the plane determined by the lines are $\mathbf{r}_1(0) = \langle -2, 3, 0 \rangle$, $\mathbf{r}_1(1) = \langle -7, 5, 4 \rangle$, and $\mathbf{r}_2(0) = \langle 8, 1, 3 \rangle$. Thus a normal vector to the plane is

$$\mathbf{N} = (\mathbf{r}_1(1) - \mathbf{r}_1(0)) \times (\mathbf{r}_2(0) - \mathbf{r}_1(0))$$
$$= ((\langle -7, 5, 4 \rangle - \langle -2, 3, 0 \rangle) \times ((\langle 8, 1, 3 \rangle - \langle -2, 3, 0 \rangle)$$
$$= \langle 5, -2, 4 \rangle \times \langle 10, -2, 3 \rangle$$
$$= \langle -2 \cdot 3 - 4 \cdot (-2), 4 \cdot 10 - 5 \cdot 3, 5 \cdot (-2) - (-2) \cdot 10 \rangle$$
$$= \langle 2, 25, 10 \rangle.$$

Since $\langle -2, 3, 0 \rangle$ lies on the plane, its equation is

$$2(x - (-2)) + 25(y - 3) + 10(z - 0) = 0, \text{ or } 2(x + 2) + 25(y - 3) + 10z = 0, \text{ or }$$
$$2x + 25y + 10z = 71.$$

33. Reparametrize $\mathbf{r}_2$ using the variable u, so that $\mathbf{r}_2(u) = \langle 6 - u, 3 + 8u, 9 - 5u \rangle$. Then if the two lines intersect at a point, at that point we must have corresponding components of $\mathbf{r}_1$ and $\mathbf{r}_2$ equal, so that $7 = 6 - u$, $3 - 4t = 3 + 8u$, and $2 + 6t = 9 - 5u$. From the first equation we get $u = -1$; substituting into the second equation gives $3 - 4t = 3 + 8(-1) = -5$ so that $4t = 8$ and $t = 2$. The third equation now becomes $2 + 6(2) = 9 - 5(-1)$, which is a true equation. Thus the two lines intersect when $t = 2$ and $u = -1$, which is at the point $(7, -5, 14)$. The normal to the resulting plane is thus perpendicular to the direction vectors of both lines, so it is

$$\mathbf{N} = \langle 0, -4, 6 \rangle \times \langle -1, 8, -5 \rangle$$
$$= \langle -4 \cdot (-5) - 6 \cdot 8, 6 \cdot (-1) - 0 \cdot (-5), 0 \cdot 8 - (-4) \cdot (-1) \rangle$$
$$= \langle -28, -6, -4 \rangle.$$

Using the point $(7, -5, 14)$ on the plane, we get for an equation

$$-28(x - 7) - 6(y - (-5)) - 4(z - 14) = 0, \text{ or } 14(x - 7) + 3(y + 5) + 2(z - 14) = 0, \text{ or }$$
$$14x + 3y + 2z = 111.$$

35. The normal vectors for the planes are $\langle 1,2,3 \rangle$ and $\langle -2,1,-4 \rangle$, so the direction vector for the line is orthogonal to both of them. A direction vector for the line is thus

$$\langle 1,2,3 \rangle \times \langle -2,1,-4 \rangle = \langle 2 \cdot (-4) - 3 \cdot 1, 3 \cdot (-2) - 1 \cdot (-4), 1 \cdot 1 - 2 \cdot (-2) \rangle = \langle -11, -2, 5 \rangle.$$

Since the z component of the direction vector is nonzero, we know that the line of intersection crosses the plane $z = 0$ (by choosing an appropriate value for t, we can make $at + b = 0$ since $a \neq 0$). So to find a point on the plane, set $z = 0$ in both equations to get $x + 2y = 4$ and $-2x + y = 6$. Solving this pair of equations gives $y = \frac{14}{5}$ and thus $x = -\frac{8}{5}$. So an equation for the line of intersection is

$$\left\langle -\frac{8}{5}, \frac{14}{5}, 0 \right\rangle + t \langle -11, -2, 5 \rangle.$$

37. The normal vectors for the planes are $\langle 1,0,0 \rangle$ and $\langle 3,-5,2 \rangle$, so the direction vector for the line is orthogonal to both of them. A direction vector for the line is thus

$$\langle 1,0,0 \rangle \times \langle 3,-5,2 \rangle = \langle 0 \cdot 2 - 0 \cdot (-5), 0 \cdot 3 - 1 \cdot 2, 1 \cdot (-5) - 0 \cdot 3 \rangle = \langle 0, -2, -5 \rangle.$$

Since the z component of the direction vector is nonzero, we know that the line of intersection crosses the plane $z = 0$ (by choosing an appropriate value for t, we can make $at + b = 0$ since $a \neq 0$). So to find a point on the plane, set $z = 0$ in both equations to get $x = 4$ and $3x - 5y = -3$. This gives $y = 3$. So an equation for the line of intersection is

$$\langle 4,3,0 \rangle + t \langle 0,-2,-5 \rangle.$$

39. First find a point on the plane. For example, $R = (3,2,0)$ is on the plane, since $3 \cdot 3 - 4 \cdot 2 + 5 \cdot 0 = 1$. With $P = (2,0,-3)$, we have $\overrightarrow{RP} = \langle 2,0,-3 \rangle - \langle 3,2,0 \rangle = \langle -1,-2,-3 \rangle$. A normal vector to the plane is $\langle 3,-4,5 \rangle$. So by Theorem 10.39, the distance from P to the plane is

$$\left| \text{comp}_{\langle 3,-4,5 \rangle} \overrightarrow{RP} \right| = \frac{|\langle 3,-4,5 \rangle \cdot \langle -1,-2,-3 \rangle|}{\|\langle 3,-4,5 \rangle\|} = \frac{10}{5\sqrt{2}} = \sqrt{2}.$$

41. Normal vectors to the planes are $\langle 2,-3,5 \rangle$ and $\langle -6,9,-15 \rangle$; these are scalar multiples of each other. Since the normal vectors are parallel, so are the planes. $R_1 = (-7,-7,0)$ lies on $2x - 3y + 5z = 7$, and $R_2 = \left(-\frac{4}{3}, 0, 0 \right)$ lies on $-6x + 9y - 15z = 8$. So by Corollary 10.41, the distance between the planes is

$$\frac{\left| \langle 2,-3,5 \rangle \cdot \overrightarrow{R_1 R_2} \right|}{\|\langle 2,-3,-5 \rangle\|} = \frac{\left| \langle 2,-3,-5 \rangle \cdot \left\langle \frac{17}{3}, 7, 0 \right\rangle \right|}{\|\langle 2,-3,-5 \rangle\|} = \frac{|-29|}{3\sqrt{2^2 + 3^2 + 5^2}} = \frac{29}{3\sqrt{38}}.$$

43. Putting the two equations into parametric form gives

$$\mathbf{r}_1(t) = \langle -1,3,2 \rangle + t \langle 2,-4,5 \rangle \qquad \text{and} \qquad \mathbf{r}_2(u) = \langle 4,-1,0 \rangle + u \langle 3,2,3 \rangle.$$

Since the direction vectors are not parallel, the lines are not parallel. Thus they either intersect or are skew. If they intersect, then there are t and u so that corresponding components of $\mathbf{r}_1$ and $\mathbf{r}_2$ are equal, i.e., so that $2t - 1 = 3u + 4$, $-4t + 3 = 2u - 1$, and $5t + 2 = 3u$. Using the third equation, substitute $5t + 2$ for $3u$ in the first equation to get $2t - 1 = 5t + 6$, so that $t = -\frac{7}{3}$. Substituting that value into the second and third equations gives $-4 \left(-\frac{7}{3} \right) + 3 = 2u - 1$ and $5 \left(-\frac{7}{3} \right) + 2 = 3u$. These have solutions $u = \frac{20}{3}$ and $u = -\frac{29}{9}$. So there are no values of t and u that satisfy all three equations; thus, the lines are skew. Then by Corollary 10.42, the distance between the lines is

$$\frac{\left| (\mathbf{d}_1 \times \mathbf{d}_2) \cdot \overrightarrow{P_1 P_2} \right|}{\|\mathbf{d}_1 \times \mathbf{d}_2\|}.$$

Now, $\mathbf{d}_1 \times \mathbf{d}_2 = \langle 2, -4, 5 \rangle \times \langle 3, 2, 3 \rangle = \langle -4 \cdot 3 - 5 \cdot 2, 5 \cdot 3 - 2 \cdot 3, 2 \cdot 2 - (-4) \cdot 3 \rangle = \langle -22, 9, 16 \rangle$.
Thus the distance between the lines is

$$\frac{|\langle -22, 9, 16 \rangle \cdot \langle 5, -4, -2 \rangle|}{\|\langle -22, 9, 16 \rangle\|} = \frac{178}{\sqrt{22^2 + 9^2 + 16^2}} = \frac{178}{\sqrt{821}}.$$

45. The two normal vectors are $\langle -1, 7, -2 \rangle$ and $\langle 3, 5, -4 \rangle$, so that by Exercise 14, the angle between the planes is

$$\begin{aligned} \theta &= \cos^{-1} \frac{\langle -1, 7, -2 \rangle \cdot \langle 3, 5, -4 \rangle}{\|\langle -1, 7, -2 \rangle\| \, \|\langle 3, 5, -4 \rangle\|} \\ &= \cos^{-1} \frac{40}{\sqrt{1^2 + 7^2 + 2^2} \sqrt{3^2 + 5^2 + 4^2}} \\ &= \cos^{-1} \frac{40}{\sqrt{54}\sqrt{50}} = \cos^{-1} \frac{40}{3\sqrt{6} \cdot 5\sqrt{2}} \\ &= \cos^{-1} \frac{4}{3\sqrt{3}} \approx 39.66°. \end{aligned}$$

47. The angle θ between the direction vector $\langle 5, -1 - 3 \rangle$ for the line and the normal $\langle -10, 2, 6 \rangle$ to the plane is given by

$$\cos\theta = \frac{\langle 5, -1, -3 \rangle \cdot \langle -10, 2, 6 \rangle}{\|\langle 5, -1, -3 \rangle\| \, \|\langle -10, 2, 6 \rangle\|} = \frac{-70}{\sqrt{5^2 + 1^2 + 3^2}\sqrt{10^2 + 2^2 + 6^2}} = \frac{-70}{\sqrt{35}\sqrt{140}} = -1.$$

So the angle between the direction vector and the normal is $180°$, or, as an acute angle, $0°$. Thus the line is perpendicular to the plane (the angle is $90° - 0° = 90°$). Note that this is also obvious from the initial calculations, since the normal vector to the plane is a scalar multiple of the direction vector to the line, so the two are parallel and thus the line is perpendicular to the plane.

49. The direction vector for the line is $\mathbf{d} = \langle -2, 0, -4 \rangle$, and the normal vector for the plane is $\mathbf{n} = \langle 2, 5, -1 \rangle$. Since $\mathbf{d} \cdot \mathbf{n} = -4 + 0 + 4 = 0$, the line is perpendicular to the normal vector, so it is parallel to the plane. Since $P = (3, -4, 5)$ does not satisfy the equation of the plane, the line does not lie in the plane. Now, $R = (1, 1, 0)$ is a point on the plane, so by Theorem 10.39, the distance from P to the plane, which is the distance from the line to the plane, is

$$\left| \text{comp}_{\langle 2,5,-1 \rangle} \overrightarrow{RP} \right| = \frac{\left| \langle 2, 5, -1 \rangle \cdot \overrightarrow{RP} \right|}{\|\langle 2, 5, -1 \rangle\|} = \frac{|\langle 2, 5, -1 \rangle \cdot \langle 2, -5, 5 \rangle|}{\sqrt{2^2 + 5^2 + 1^2}} = \frac{|4 - 25 - 5|}{\sqrt{30}} = \frac{26}{\sqrt{30}}.$$

51. The direction vector for the line is $\mathbf{d} = \langle 6, -1, 0 \rangle$, and the normal vector for the plane is $\mathbf{n} = \langle 0, 7, -5 \rangle$. Since $\mathbf{d} \cdot \mathbf{n} = 0 - 7 + 0 \neq 0$, the line is not perpendicular to the normal vector, so is not parallel to the plane, so it intersects the plane. To find the point of intersection, plug the parametric formula for the line into the equation for the plane and solve for t:

$$7(3 - t) - 5(0) = 3, \text{ or } 21 - 7t = 3,$$

so that $t = \frac{18}{7}$. Hence the point of intersection is

$$\left\langle 5 + 6\left(\frac{18}{7}\right), 3 - \frac{18}{7}, 0 \right\rangle = \left\langle \frac{143}{7}, \frac{3}{7}, 0 \right\rangle.$$

To find the angle at which they intersect, by Exercise 17, compute the angle θ between the normal and the direction vector of the line:

$$\cos\theta = \frac{\mathbf{d} \cdot \mathbf{n}}{\|\mathbf{d}\| \, \|\mathbf{n}\|} = \frac{0 - 7 + 0}{\sqrt{6^2 + 1^2 + 0^2}\sqrt{0^2 + 7^2 + 5^2}} = \frac{-7}{\sqrt{37}\sqrt{74}} = -\frac{7}{37\sqrt{2}}.$$

Then $\theta \approx 97.69°$. Using the acute angle instead, we get $\theta = 180° - 97.69° = 82.31°$. Then the angle at which the line intersects the plane is $90° - 82.31° = 7.69°$.

53. (a) Applying the formula from Exercise 52 with $a = b = c = 0$ gives

$$\frac{1}{2}x + \frac{1}{4}y + \frac{\sqrt{11}}{4}z = \frac{1}{4} + \frac{1}{16} + \frac{11}{16}, \text{ or}$$

$$\frac{1}{2}x + \frac{1}{4}y + \frac{\sqrt{11}}{4}z = 1, \text{ which can be further simplified if desired to}$$

$$2x + y + \sqrt{11}\,z = 4.$$

(b) Applying the formula from Exercise 52 with $a = 1$, $b = -2$, and $c = 0$, and $(x_0, y_0, z_0) = (3, 0, 1)$ gives

$$(3 - 1)x + (0 - (-2))y + (1 - 0)z = 3(3 - 1) + 0(0 - (-2)) + 1(1 - 0), \text{ or } 2x + 2y + z = 7.$$

Applications

55. The origin of the coordinate system lies where the buoy meets the water. The three given points, since they all lie on the beach, lie on a plane. The two direction vectors $\langle 150, 30, 5 \rangle - \langle 120, 40, 0 \rangle = \langle 30, -10, 5 \rangle$ and $\langle 150, 30, 5 \rangle - \langle 140, 60, 5 \rangle = \langle 10, -30, 0 \rangle$ are not multiples of each other. So a vector perpendicular to them, and to any plane they lie in, is their cross product, which is

$$\langle 30, -10, 5 \rangle \times \langle 10, -30, 0 \rangle = \langle 150, 50, -800 \rangle.$$

Using the point $(120, 40, 0)$ to lie in the plane gives for the equation of the plane

$$150(x - 120) + 50(y - 40) - 800z = 0.$$

Divide through by 50 to get $3(x - 120) + y - 40 - 16z = 0$, or $3x + y - 16z = 400$. Since the buoy sits at $x = y = 0$, the equation becomes $-16z = 400$, so that $z = -25$. The buoy is in 25 feet of water.

Proofs

57. Two lines are perpendicular if and only if their normal vectors are perpendicular as well. The normal vectors to the two given planes are $\langle a, b, c \rangle$ and $\langle \alpha, \beta, \gamma \rangle$. Those vectors are perpendicular if their dot product is zero; that dot product is $a\alpha + b\beta + c\gamma$.

59. (a) Note that the point $R = \left(-\frac{d}{3a}, -\frac{d}{3b}, -\frac{d}{3c} \right)$ lies on the plane, and that $\langle a, b, c \rangle$ is a normal vector, so that by Theorem 10.39, the distance from P to the plane is

$$\left| \text{comp}_\mathbf{n} \overrightarrow{RP} \right| = \frac{\left| \mathbf{n} \cdot \overrightarrow{RP} \right|}{\|\mathbf{n}\|}$$

$$= \frac{\left| \langle a, b, c \rangle \cdot \left\langle x_0 + \frac{d}{3a}, y_0 + \frac{d}{3b}, z_0 + \frac{d}{3c} \right\rangle \right|}{\| \langle a, b, c \rangle \|}$$

$$= \frac{\left| ax_0 + \frac{d}{3} + by_0 + \frac{d}{3} + cz_0 + \frac{d}{3} \right|}{\sqrt{a^2 + b^2 + c^2}}$$

$$= \frac{|ax_0 + by_0 + cz_0 + d|}{\sqrt{a^2 + b^2 + c^2}}.$$

(b) Here $P = (2, 0, -3)$ and the plane is $3x - 4y + 5z - 1 = 0$, so by part (a) the distance from P to the plane is

$$\frac{|3 \cdot 2 - 4 \cdot 0 + 5 \cdot (-3) - 1|}{\sqrt{3^2 + 4^2 + 5^2}} = \frac{10}{\sqrt{50}} = \frac{10}{5\sqrt{2}} = \sqrt{2}.$$

61. If $\mathbf{d}_1$ and $\mathbf{d}_2$ are parallel, then the lines are parallel and hence lie in the same plane (choose a point on $\mathbf{r}_2$ and use that point together with $\mathbf{r}_1$ to find a plane containing the point and $\mathbf{r}_1$; since $\mathbf{r}_1$ and $\mathbf{r}_2$ are parallel and the plane contains a point of $\mathbf{r}_2$, the entire line $\mathbf{r}_2$ must lie on the plane as well). In that case, $\mathbf{d}_1 \times \mathbf{d}_2 = 0$ so that $(\mathbf{P}_0 - \mathbf{Q}_0) \cdot (\mathbf{d}_1 \times \mathbf{d}_2) = 0$. Thus the assertion holds if $\mathbf{d}_1$ and $\mathbf{d}_2$ are parallel. So assume they are not. Then by Exercise 60, $\mathcal{L}_1$ and $\mathcal{L}_2$ intersect if and only if $\mathbf{P}_0 - \mathbf{P}_1$, $\mathbf{d}_1$, and $\mathbf{d}_2$ are coplanar, which is the case if and only if their scalar triple product is zero: $(\mathbf{P}_0 - \mathbf{Q}_0) \cdot (\mathbf{d}_1 \times \mathbf{d}_2) = 0$. But since the lines are not parallel, they intersect if and only if they lie in the same plane.

Thinking Forward

A plane tangent to a surface

Since $\mathbf{v}_x$ and $\mathbf{v}_y$ are tangent to the surface, their cross product must be normal to the surface; that cross product is

$$\mathbf{v}_x \times \mathbf{v}_y = \langle 1, -3, 0 \rangle \times \langle 1, 0, 4 \rangle = \langle -3 \cdot 4 - 0 \cdot 0, 0 \cdot 1 - 1 \cdot 4, 1 \cdot 0 - (-3) \cdot 1 \rangle = \langle -12, -4, 3 \rangle.$$

Since the tangent plane touches the surface at $(2, -3, 4)$, this point lies on the plane, and its normal vector is $\langle -12, -4, 3 \rangle$, so its equation is

$$-12(x - 2) - 4(y + 3) + 3(z - 4) = 0, \quad \text{or} \quad 12x + 4y - 3z = 0.$$

A plane tangent to a surface

Since $\mathbf{v}_x$ and $\mathbf{v}_y$ are tangent to the surface, their cross product must be normal to the surface; that cross product is

$$\mathbf{v}_x \times \mathbf{v}_y = \langle 1, \alpha, 0 \rangle \times \langle 1, 0, \beta \rangle = \langle \alpha \cdot \beta - 0 \cdot 0, 0 \cdot 1 - 1 \cdot \beta, 1 \cdot 0 - \alpha \cdot 1 \rangle = \langle \alpha\beta, -\beta, -\alpha \rangle.$$

Since the tangent plane touches the surface at (x_0, y_0, z_0), this point lies on the plane, and its normal vector is $\langle \alpha\beta, -\beta, -\alpha \rangle$, so its equation is

$$\alpha\beta(x - x_0) - \beta(y - y_0) - \alpha(z - z_0) = 0.$$

Chapter Review and Self-Test

1. The distance is $\sqrt{(4 - 1)^2 + (7 - 2)^2 + (-3 - (-3))^2} = \sqrt{3^2 + 5^2} = \sqrt{34}$.

3. The distance is $\sqrt{(3 - 1)^2 + (9 - 5)^2 + (-1 - (-2))^2} = \sqrt{2^2 + 4^2 + 1^2} = \sqrt{21}$.

5. The equation is $(x - 2)^2 + (y - (-3))^2 + (z - 4)^2 = 6^2$, or $(x - 2)^2 + (y + 3)^2 + (z - 4)^2 = 36$.

7. The midpoint of that segment must be the center of the sphere; that midpoint is

$$\left(\frac{1 + 3}{2}, \frac{5 + 9}{2}, \frac{-2 - 1}{2} \right) = \left(2, 7, -\frac{3}{2} \right).$$

The radius of the sphere is half the distance between the given points, or

$$\frac{1}{2} \sqrt{(3 - 1)^2 + (9 - 5)^2 + (-1 - (-2))^2} = \frac{1}{2} \sqrt{2^2 + 4^2 + 1^2} = \frac{\sqrt{21}}{2}.$$

Thus the equation of the sphere is $(x - 2)^2 + (y - 7)^2 + \left(z + \frac{3}{2} \right)^2 = \frac{21}{4}$.

9.

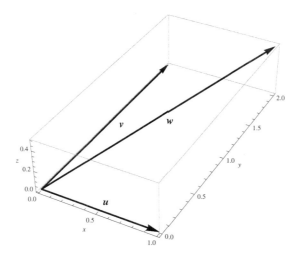

11.

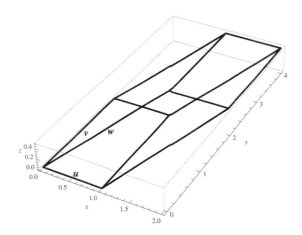

13. $\|\mathbf{v}\| = \sqrt{0^2 + 2^2 + 0^2} = 2$. The graphs above already have scales on them.

15. $\mathbf{u} \cdot \mathbf{v} = \langle 1,0,0 \rangle \cdot \langle 0,2,0 \rangle = 1 \cdot 0 + 0 \cdot 2 + 0 \cdot 0 = 0$.

17. The angle θ between $\mathbf{u}$ and $\mathbf{v}$, by Theorem 10.15, is given by $\cos \theta = \frac{\mathbf{u} \cdot \mathbf{v}}{\|\mathbf{u}\|\|\mathbf{v}\|} = 0$, since by Exercise 15 $\mathbf{u} \cdot \mathbf{v} = 0$. Since $\cos \theta = 0$, it follows that $\theta = 90°$.

19. The diagonals are $\mathbf{u} + \mathbf{v}$ and $\mathbf{u} - \mathbf{v}$; their lengths are

$$\|\mathbf{u} + \mathbf{v}\| = \|\langle 1,2,0 \rangle\| = \sqrt{1^2 + 2^2 + 0^2} = \sqrt{5},$$
$$\|\mathbf{u} - \mathbf{v}\| = \|\langle 1,-2,0 \rangle\| = \sqrt{1^2 + (-2)^2 + 0^2} = \sqrt{5}.$$

21. Since $\langle 0,0,2 \rangle$ is orthogonal to both $\mathbf{u}$ and $\mathbf{v}$ and has norm 2, it follows that $\frac{1}{2} \langle 0,0,2 \rangle = \langle 0,0,1 \rangle$ is a unit vector orthogonal to both $\mathbf{u}$ and $\mathbf{v}$.

23. Each face is a parallelogram, and opposite faces have equal areas since the solid is a parallelepiped. The bottom and top faces are parallelograms with sides $\mathbf{u}$ and $\mathbf{v}$; their areas were computed in

Exercise 18 to be 2. The left and right faces have sides $\mathbf{v}$ and $\mathbf{w}$, and the front and back faces have sides $\mathbf{u}$ and $\mathbf{w}$; their areas are

$$\|\mathbf{v} \times \mathbf{w}\| = \left\|\langle 0, 2, 0 \rangle \times \left\langle 1, 2, \frac{1}{2} \right\rangle\right\| = \left\|\left\langle 2 \cdot \frac{1}{2} - 0 \cdot 2, 0 \cdot 1 - 0 \cdot \frac{1}{2}, 0 \cdot 2 - 2 \cdot 1 \right\rangle\right\|$$

$$= \|\langle 1, 0, -2 \rangle\| = \sqrt{1^2 + 0^2 + (-2)^2} = \sqrt{5},$$

$$\|\mathbf{u} \times \mathbf{w}\| = \left\|\langle 1, 0, 0 \rangle \times \left\langle 1, 2, \frac{1}{2} \right\rangle\right\| = \left\|\left\langle 0 \cdot \frac{1}{2} - 0 \cdot 2, 0 \cdot 1 - 1 \cdot \frac{1}{2}, 1 \cdot 2 - 0 \cdot 1 \right\rangle\right\|$$

$$= \left\|\left\langle 0, -\frac{1}{2}, 2 \right\rangle\right\| = \sqrt{0^2 + \left(-\frac{1}{2}\right)^2 + 2^2} = \frac{\sqrt{17}}{2}.$$

25. $\|\mathbf{u}\| = \sqrt{2^2 + 4^2 + (-1)^2} = \sqrt{21}.$

27. $\mathbf{u} \cdot \mathbf{v} = \langle 2, 4, -1 \rangle \cdot \langle 0, -3, 2 \rangle = 2 \cdot 0 + 4 \cdot (-3) + (-1) \cdot 2 = -14.$

29. If θ is the angle between $\mathbf{u}$ and $\mathbf{v}$, then

$$\cos\theta = \frac{\mathbf{u} \cdot \mathbf{v}}{\|\mathbf{u}\|\,\|\mathbf{v}\|} = -\frac{14}{\sqrt{21}\sqrt{13}}.$$

Then $\theta \approx 147.92°$.

31. The diagonals of the parallelogram are $\mathbf{u} + \mathbf{v}$ and $\mathbf{u} - \mathbf{v}$, so their lengths are

$$\|\mathbf{u} + \mathbf{v}\| = \|\langle 2, 4, -1 \rangle + \langle 0, -3, 2 \rangle\| = \|\langle 2, 1, 1 \rangle\| = \sqrt{2^2 + 1^2 + 1^2} = \sqrt{6}$$

$$\|\mathbf{u} - \mathbf{v}\| = \|\langle 2, 4, -1 \rangle - \langle 0, -3, 2 \rangle\| = \|\langle 2, 7, -3 \rangle\| = \sqrt{2^2 + 7^2 + (-3)^2} = \sqrt{62}.$$

33. Since $\langle 5, -4, -6 \rangle$ is orthogonal to both $\mathbf{u}$ and $\mathbf{v}$, it follows that

$$\frac{1}{\|\langle 5, -4, -6 \rangle\|} \langle 5, -4, -6 \rangle = \frac{1}{\sqrt{77}} \langle 5, -4, -6 \rangle$$

is a unit vector orthogonal to both $\mathbf{u}$ and $\mathbf{v}$.

35. Each face is a parallelogram, and opposite faces have equal areas since the solid is a parallelepiped. The bottom and top faces are parallelograms with sides $\mathbf{u}$ and $\mathbf{v}$; their areas were computed in Exercise 30 to be $\sqrt{77}$. The left and right faces have sides $\mathbf{v}$ and $\mathbf{w}$, and the front and back faces have sides $\mathbf{u}$ and $\mathbf{w}$; their areas are

$$\|\mathbf{v} \times \mathbf{w}\| = \|\langle 0, -3, 2 \rangle \times \langle -1, 1, 5 \rangle\| = \|\langle -3 \cdot 5 - 2 \cdot 1, 2 \cdot (-1) - 0 \cdot 5, 0 \cdot 1 - (-3) \cdot (-1) \rangle\|$$

$$= \|\langle -17, -2, -3 \rangle\| = \sqrt{17^2 + 2^2 + 3^2} = \sqrt{302}$$

$$\|\mathbf{u} \times \mathbf{w}\| = \|\langle 2, 4, -1 \rangle \times \langle -1, 1, 5 \rangle\| = \|\langle 4 \cdot 5 - (-1) \cdot 1, -1 \cdot (-1) - 2 \cdot 5, 2 \cdot 1 - 4 \cdot (-1) \rangle\|$$

$$= \|\langle 21, -9, 6 \rangle\| = \sqrt{21^2 + 9^2 + 6^2} = \sqrt{558} = 3\sqrt{62}.$$

37. A direction vector for the line containing P and Q is $\overrightarrow{PQ} = \langle 1, 4, -5 \rangle$; using the point $P = (2, -3, 5)$, we get the equation

$$\mathbf{r}(t) = \langle 2, -3, 5 \rangle + t \langle 1, 4, -5 \rangle.$$

39. Using the results of the previous two exercises, we see that the vectors $\langle 1, 4, -5 \rangle$ and $\langle -3, 3, 2 \rangle$ both lie in the plane, so their cross product is a normal to the plane; that cross product is

$$\langle 4 \cdot 2 - (-5) \cdot 3, -5 \cdot (-3) - 1 \cdot 2, 1 \cdot 3 - 4 \cdot (-3) \rangle = \langle 23, 13, 15 \rangle .$$

Using the point $P = (2, -3, 5)$, the equation for the plane containing all three points is

$$23(x - 2) + 13(y + 3) + 15(z - 5) = 0, \quad \text{or} \quad 23x + 13y + 15z = 82.$$

41. By Theorem 10.38, the distance from P to $\mathcal{L} = \langle 2, -3, 0 \rangle + t \langle 5, 1, -4 \rangle$ is

$$
\begin{aligned}
\frac{\left\| \mathbf{d} \times \overrightarrow{P_0 P} \right\|}{\|\mathbf{d}\|} &= \frac{\|\langle 5, 1, -4 \rangle \times (\langle 1, 2, -3 \rangle - \langle 2, -3, 0 \rangle)\|}{\|\langle 5, 1, -4 \rangle\|} \\
&= \frac{\|\langle 5, 1, -4 \rangle \times \langle -1, 5, -3 \rangle\|}{\sqrt{5^2 + 1^2 + 4^2}} \\
&= \frac{\|\langle 1 \cdot (-3) - (-4) \cdot 5, -4 \cdot (-1) - 5 \cdot (-3), 5 \cdot 5 - 1 \cdot (-1) \rangle\|}{\sqrt{42}} \\
&= \frac{\|\langle 17, 19, 26 \rangle\|}{\sqrt{42}} \\
&= \frac{\sqrt{17^2 + 19^2 + 26^2}}{\sqrt{42}} = \frac{\sqrt{1326}}{\sqrt{42}} = \sqrt{\frac{221}{7}}.
\end{aligned}
$$

43. To see where $\mathcal{L}$ intersects $\mathcal{P}$, substitute the parametric equations for $\mathcal{L}$ into the equation for $\mathcal{P}$ and solve for t:
$$(2 + 5t) - 3(-3 + t) + 4(-4t) = 5, \quad \text{or} \quad -14t + 11 = 5.$$

Solving gives $t = \frac{3}{7}$, so the point of intersection is

$$\left\langle 2 + 5 \cdot \frac{3}{7}, -3 + \frac{3}{7}, -4 \cdot \frac{3}{7} \right\rangle = \left\langle \frac{29}{7}, -\frac{18}{7}, -\frac{12}{7} \right\rangle.$$

Chapter 11

Vector Functions

11.1 Vector-Valued Functions

Thinking Back

Parametric equations for the unit circle

▷ This is the usual parametrization of the circle, counterclockwise from $(1,0)$; it is given by $\mathbf{r}(t) = \langle \cos t, \sin t \rangle$.

▷ This is the same as the parametrization in the previous exercise, except it is clockwise. We can model that by $\mathbf{r}(t) = \langle \cos(-t), \sin(-t) \rangle = \langle \cos t, -\sin t \rangle$.

▷ Since the graph is traced twice from 0 to 2π, we must have $\mathbf{r}(t) = \langle \cos 2t, \sin 2t \rangle$, so that when $t = \pi$ the graph returns to $(1,0)$ since $2t = 2\pi$.

▷ Generalizing the previous part, the equation of a unit circle centered at the origin that is traced counterclockwise k times on $[0, 2\pi]$ is $\langle \cos kt, \sin kt \rangle$. To place this on a circle of radius ρ, starting at $(\rho, 0)$, multiply through by ρ to get $\langle \rho \cos kt, \rho \sin kt \rangle$. Finally, translate the curve to a center of (a, b) by adding that vector, to get $\mathbf{r}(t) = \langle a + \rho \cos kt, b + \rho \sin kt \rangle$.

Concepts

1. (a) False. For example, both $\cos t\mathbf{i} + \sin t\mathbf{j}$ and $\cos t\mathbf{i} - \sin t\mathbf{j}$ for $0 \le t < 2\pi$ parametrize the unit circle.

    (b) False. For example, the curve $\cos t\mathbf{i} + \sin t\mathbf{j} + t\mathbf{k}$ for $-\infty < t < \infty$ can also be parametrized as $\cos 2t\mathbf{i} + \sin 2t\mathbf{j} + 2t\mathbf{k}$.

    (c) True. This is simply a change in notation. Given the parametric equations $x = x(t)$, $y = y(t)$, this can be written as a vector-valued function $\mathbf{r}(t) = \langle x(t), y(t) \rangle = x(t)\mathbf{i} + y(t)\mathbf{j}$.

    (d) True. This is simply a change in notation. Given the vector valued function

    $$\mathbf{r}(t) = \langle x(t), y(t), z(t) \rangle,$$

    we get the parametric equations $x = x(t)$, $y = y(t)$, $z = z(t)$.

    (e) True. This is simply a change in notation. Given the parametric equations $x = x(t)$, $y = y(t)$, $z = z(t)$, this can be written as a vector-valued function $\mathbf{r}(t) = \langle x(t), y(t), z(t) \rangle = x(t)\mathbf{i} + y(t)\mathbf{j} + z(t)\mathbf{k}$, so the two have the same graph.

(f) False. It exists if and only if the limit exists for each component. For example, if $x(t)$ is the sign of t, so that it is -1 for $t < 0$, 0 for $t = 0$, and 1 for $t > 0$, then no matter what $y(t)$ is,

$$\lim_{t \to 0} \langle x(t), y(t) \rangle = \left\langle \lim_{t \to 0} x(t), \lim_{t \to 0} y(t) \right\rangle \text{ does not exist.}$$

(g) False. For example, let $x(t)$ be 1 if $t \neq 0$ and 0 if $t = 0$, and let $y(t) = 1$ for all t. Then if

$$\mathbf{r}(t) = \langle x(t), y(t) \rangle, \text{ we have } \mathbf{r}(0) = \langle 0, 1 \rangle \text{ but } \lim_{t \to 0} \langle x(t), y(t) \rangle = \left\langle \lim_{t \to 0} x(t), \lim_{t \to 0} y(t) \right\rangle = \langle 1, 1 \rangle.$$

(h) False. For example, the function in part (g) is discontinuous at $t = 0$.

3. See Definition 1.5. $\lim_{x \to c} f(x) = L$ if for every $\epsilon > 0$ there exists $\delta > 0$ such that if $x \in (c - \delta, c) \cup (c, c + \delta)$ then $f(x) \in (L - \epsilon, L + \epsilon)$.

5. An epsilon-delta definition is not needed because the limit of a vector valued function is defined in terms of limits of real-valued functions. Evaluating each of those limits in the end relies on an epsilon-delta computation.

7. See Definition 11.4. $\mathbf{r}(t)$ is continuous at c if $\lim_{t \to c} \mathbf{r}(t) = \mathbf{r}(c)$, using the definition of limit from Definition 11.3.

9. Since $x(t)$ is continuous, it follows from the extreme value theorem that there is some $c \in [a, b]$ such that $x(c) = M_x$ is a maximum for $x(t)$ on $[a, b]$. Similarly, since $y(t)$ is continuous, there is some $d \in [a, b]$ such that $y(d) = M_y$ is a maximum for $y(t)$ on $[a, b]$. Then for any $t \in [a, b]$, we have $\|\mathbf{r}(t)\| \leq \sqrt{x(t)^2 + y(t)^2} \leq \sqrt{M_x^2 + M_y^2}$. Thus $\mathbf{r}(t)$ is contained in a circle of radius $\sqrt{M_x^2 + M_y^2}$ centered at the origin.

11. Since $x(t)$ is continuous, it follows from the extreme value theorem that there is some $c \in [a, b]$ such that $x(c) = M_x$ is a maximum for $x(t)$ on $[a, b]$. Similarly, since $y(t)$ is continuous, there is some $d \in [a, b]$ such that $y(d) = M_y$ is a maximum for $y(t)$ on $[a, b]$. A similar statement holds for $z(t)$, showing that for some M_z, $z(e) = M_z$ is a maximum for some $e \in [a, b]$. Then for any $t \in [a, b]$, we have $\|\mathbf{r}(t)\| \leq \sqrt{x(t)^2 + y(t)^2 + z(t)^2} \leq \sqrt{M_x^2 + M_y^2 + M_z^2}$. Thus $\mathbf{r}(t)$ is contained in a circle of radius $\sqrt{M_x^2 + M_y^2 + M_z^2}$ centered at the origin.

13. Consider $\mathbf{r}_1(t) = \langle \cos 2t, \sin 2t, 2t \rangle$. For a given value of t, we have

$$\mathbf{r}_1(t) = \langle \cos 2t, \sin 2t, 2t \rangle = \mathbf{r}(2t),$$

so that $\mathbf{r}_1$ is twice as far along the curve as $\mathbf{r}$ is for a given value of t ($\mathbf{r}$ does not get to that point until time $2t$).

15. A horizontal asymptote means that the y coordinate of the graph approaches a finite value as the x coordinate approaches either ∞ or $-\infty$, while $t \to \infty$. This is equivalent to saying that $\lim_{t \to \infty} y(t)$ exists but that $\lim_{t \to \infty} x(t) = \pm\infty$. For example, the curve $\mathbf{r}(t) = \langle t^2, 1 + e^{-t} \rangle$ has a horizontal asymptote at $y = 1$ since $\lim_{t \to \infty} (1 + e^{-t}) = 1$ while $\lim_{t \to \infty} t^2 = \infty$:

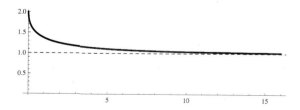

17. Dot products of functions are just like dot products of vectors:

$$\mathbf{r}_1(t) \cdot \mathbf{r}_2(t) = x_1(t)x_2(t) + y_1(t)y_2(t).$$

19. For example,

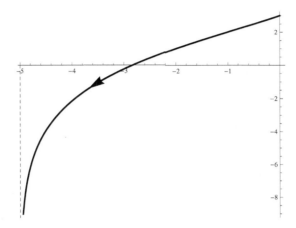

21. For example,

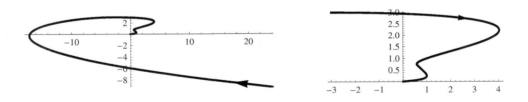

The graph on the left shows the trace for $t \geq -3$, while the graph on the right zooms in on the origin to show the behavior as t gets large positively.

23. The graphs of the two functions coincide. To see this, let $\mathbf{r}_1(t) = \mathbf{r}(kt)$. Then $\mathbf{r}(t) = \mathbf{r}_1\left(\frac{t}{k}\right)$, and $\mathbf{r}_1(t) = \mathbf{r}(kt)$, so that every point on one graph is also on the other graph. The difference is that $\mathbf{r}_1$ is traversed more rapidly: $\mathbf{r}_1(1) = \mathbf{r}(k)$, so that $\mathbf{r}_1$ is at the same point when $t = 1$ that $\mathbf{r}$ is when $t = k > 1$.

25. A point on the graph of $\mathbf{r}(t)$ has Cartesian coordinates $(\cos t, \sin t, \cos t)$. Such a point lies on $x^2 + y^2 = 1$ since substituting the x and y coordinates gives $\cos^2 t + \sin^2 t = 1$. It also lies on the cylinder $y^2 + z^2 = 1$ since substituting the y and z coordinates gives $\sin^2 t + \cos^2 t = 1$.

Skills

27. Parametric equations for this curve are $x = 2 - \sin t, y = 4 + \cos t$ for $0 \le t \le 2\pi$; a graph is

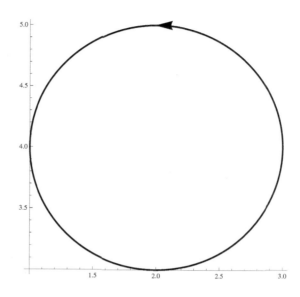

At $t = 0$ the trace starts at $(2, 5)$; from 0 to 2π it traces the entire graph counterclockwise once.

29. Parametric equations for this curve are $x = 1 + \sin t, y = 3 - \cos 2t$ for $0 \le t \le 2\pi$; a graph is

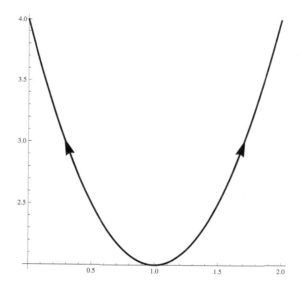

At $t = 0$ the trace starts at $(1, 2)$; from 0 to $\frac{\pi}{2}$ it traces the graph up and to the right, to $(2, 4)$. From $\frac{\pi}{2}$ the trace is back down to $(1, 2)$, then up and to the left, to $(0, 4)$ at $t = \frac{3\pi}{2}$. Finally, the graph is traced back down to $(1, 2)$ at $t = 2\pi$.

31. Parametric equations for this curve are $x = t$, $y = t^2$, and $z = t^3$ for $0 \leq t \leq 2$; a graph is

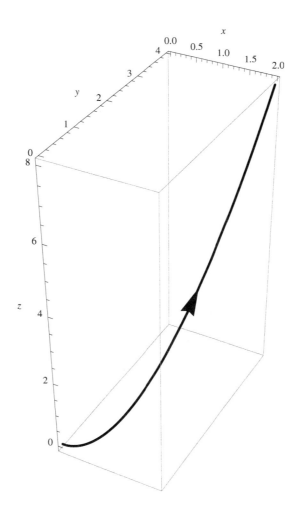

The graph is traced in the direction of increasing z.

33. Parametric equations for this curve are $x = \cos^2 t$, $y = \sin 2t$ for $0 \le t \le 2\pi$; a graph is

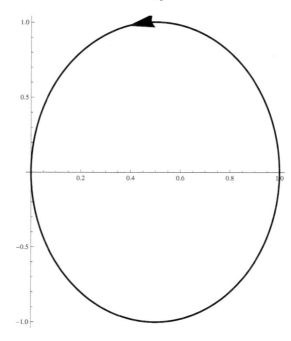

The graph is traced counterclockwise starting from $(1,0)$ at $t = 0$. It is traced twice on $[0, 2\pi]$, returning to $(1,0)$ at $t = \pi$.

35. $\langle 1, 3t, t^3 \rangle + \langle t, t^2, t^3 \rangle = \langle 1 + t, 3t + t^2, t^3 + t^3 \rangle = \langle 1 + t, 3t + t^2, 2t^3 \rangle.$

37. $5 \langle \cos t, \sin t \rangle = \langle 5 \cos t, 5 \sin t \rangle.$

39. $\langle \sin t, \cos t \rangle \cdot \langle \cos t, -\sin t \rangle = \sin t \cdot \cos t + \cos t \cdot (-\sin t) = 0.$

41. $\langle 1, t, t^2 \rangle \times \langle t, t^2, t^3 \rangle = \langle t \cdot t^3 - t^2 \cdot t^2, t^2 \cdot t - 1 \cdot t^3, 1 \cdot t^2 - t \cdot t \rangle = \langle 0, 0, 0 \rangle.$ Note that for any particular value of t, the vector $\langle t, t^2, t^3 \rangle = t \langle 1, t, t^2 \rangle$, so that the two vectors are always parallel, and their cross product is zero.

43. By Definition 11.3,

$$\lim_{t \to \pi} \langle \sin t, \cos t, \sec t \rangle = \left\langle \lim_{t \to \pi} \sin t, \lim_{t \to \pi} \cos t, \lim_{t \to \pi} \sec t \right\rangle = \langle \sin \pi, \cos \pi, \sec \pi \rangle = \langle 0, -1, -1 \rangle.$$

45. By Definition 11.3,

$$\lim_{t \to 0^+} \left\langle \frac{\sin t}{t}, \frac{1 - \cos t}{t}, \left(1 + \frac{1}{t}\right)^t \right\rangle = \left\langle \lim_{t \to 0^+} \frac{\sin t}{t}, \lim_{t \to 0^+} \frac{1 - \cos t}{t}, \lim_{t \to 0^+} \left(1 + \frac{1}{t}\right)^t \right\rangle = \langle 1, 0, 1 \rangle,$$

using results about these three limits from earlier chapters.

47. The initial value problem for x is $\frac{dx}{dt} = 1 + x^2$, $x(0) = 0$. This gives $\frac{dx}{1 + x^2} = dt$; integrating both sides gives $\tan^{-1} x = t + C$. Then the initial condition gives $\tan^{-1} x(0) = 0 + C = 0$, so that $C = 0$. Thus the equation is $\tan^{-1} x = t$, or $x(t) = \tan t$. For the second equation, we have $y'(t) = x^2 = \tan^2 t$. Use the Pythagorean identity $\tan^2 t = \sec^2 t - 1$ to integrate:

$$y(t) = \int \tan^2 t \, dt = \int (\sec^2 t - 1) \, dt = \tan t - t + C.$$

Then $y(0) = \tan 0 - 0 + C = 1$ gives $C = 1$, so that $y(t) = \tan t - t + 1$. Thus $\mathbf{r}(t) = \langle x(t), y(t) \rangle = \langle \tan t, \tan t - t + 1 \rangle$. A graph of this function for $t \in \left(-\frac{\pi}{2}, \frac{\pi}{2} \right)$ is

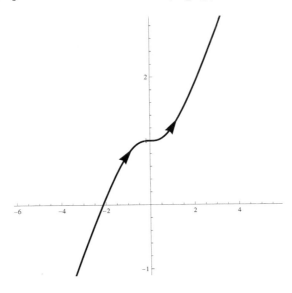

49. Clearly the points lie on a sphere of radius 5, since

$$\| \langle 3 \sin t, 5 \cos t, 4 \sin t \rangle \| = \sqrt{9 \sin^2 t + 25 \cos^2 t + 16 \sin^2 t} = \sqrt{25(\cos^2 t + \sin^2 t)} = 5.$$

If for each t and u the three points

$$P = (0,0,0), \qquad Q = (3 \sin t, 5 \cos t, 4 \sin t), \qquad R = (3 \sin u, 5 \cos u, 4 \sin u)$$

are coplanar, then all points on the curve lie on the same plane. But these three points are coplanar if and only if the scalar triple product $\overrightarrow{QR} \cdot (\overrightarrow{PQ} \times \overrightarrow{PR})$ is zero.

$$\begin{aligned}
\overrightarrow{QR} \cdot (\overrightarrow{PQ} \times \overrightarrow{PR}) &= \overrightarrow{QR} \cdot (\langle 3 \sin t, 5 \cos t, 4 \sin t \rangle \cdot \langle 3 \sin u, 5 \cos u, 4 \sin u \rangle) \\
&= \overrightarrow{QR} \cdot \langle 20 \cos t \sin u - 20 \sin t \cos u, 12 \sin t \sin u - 12 \sin t \sin u, \\
& \qquad\quad 15 \sin t \cos u - 15 \cos t \sin u \rangle \\
&= \overrightarrow{QR} \cdot \langle 20 \sin(u - t), 0, 15 \sin(t - u) \rangle \\
&= \langle 3(\sin u - \sin t), 5(\cos u - \cos t), 4(\sin u - \sin t) \rangle \cdot \langle 20 \sin(u - t), 0, 15 \sin(t - u) \rangle \\
&= 60 \sin u \sin(u - t) - 60 \sin t \sin(u - t) + 60 \sin u \sin(t - u) - 60 \sin t \sin(t - u) \\
&= 0.
\end{aligned}$$

Since the scalar triple product is zero, the points all lie on a plane. Since the points lie on a sphere, they are all equidistant from the center of the sphere. Since they also line in a plane, they form a circle.

Applications

51. Whatever Arne does, he wants his height to be at least 300 meters when he reaches the peak. Since his vertical position is $1200 - 14t$ meters per second, he will reach a height of 300 meters when $1200 - 14t = 300$, or $t = \frac{900}{14} = \frac{450}{7}$ seconds. At that time, his y coordinate should be at least 2000 if he is to clear the cliff, so we want $\frac{A}{2} \cdot \frac{450}{7} \geq 2000$, or $A \geq \frac{28000}{450} \approx 62.2$. So A must be at least 62.2.

53. (a) A sketch of the two curves, with h_1 in black and h_2 in gray, is (with $\alpha = 1$)

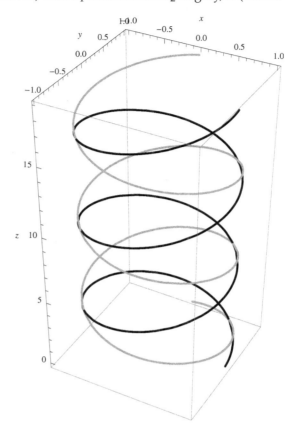

Clearly h_1 traces counterclockwise when viewed from above, while h_2 traces clockwise.

(b) The two helices intersect if $h_1(t) = h_2(u)$; that is, if $\cos t = \sin u$, $\sin t = \cos u$, and $\alpha t = \alpha u$. The third equation implies that $t = u$, and the first two then hold if $t = u = \frac{\pi}{4} + n\pi$. Thus the two curves always intersect.

Proofs

55. To see that it lies on the cone, substitute the parametrized values for x and y into the right-hand side to get

$$\sqrt{x^2 + y^2} = \sqrt{(t\sin t)^2 + (t\cos t)^2} = \sqrt{t^2(\sin^2 t + \cos^2 t)} = \sqrt{t^2} = t = z,$$

where we are justified in saying $\sqrt{t^2} = t$ since we are looking at the range $t \geq 0$.

57. If $\mathbf{r}_1(t) = \langle x_1(t), y_1(t) \rangle$ and $\mathbf{r}_2(t) = \langle x_2(t), y_2(t) \rangle$, then

$$\lim_{t \to t_0} (\mathbf{r}_1(t) \cdot \mathbf{r}_2(t)) = \lim_{t \to t_0} (\langle x_1(t), y_1(t) \rangle \cdot \langle x_2(t), y_2(t) \rangle) = \lim_{t \to t_0} (x_1(t)x_2(t) + y_1(t)y_2(t)).$$

But this is just the limit of a sum of products of continuous real-valued functions, which must itself be continuous. Thus the limit is just

$$x_1(t_0)x_2(t_0) + y_1(t_0)y_2(t_0) = \langle x_1(t_0), y_1(t_0) \rangle \cdot \langle x_2(t_0), y_2(t_0) \rangle = \mathbf{r}_1(t_0) \cdot \mathbf{r}_2(t_0).$$

59. Choose t_0 in the domain of t. Showing that the dot product is continuous means showing that if we take the limit of $\mathbf{r}_1(t) \cdot \mathbf{r}_2(t)$ for $t \to t_0$, we get the same value as we get by taking $\mathbf{r}_1(t) \cdot \mathbf{r}_2(t)$ and evaluating it at t_0. But $\mathbf{r}_1(t) \cdot \mathbf{r}_2(t)$ evaluated at t_0 is $\mathbf{r}_1(t_0) \cdot \mathbf{r}_2(t_0)$. So we want to show that

$$\lim_{t \to t_0} (\mathbf{r}_1(t) \cdot \mathbf{r}_2(t)) = \mathbf{r}_1(t_0) \cdot \mathbf{r}_2(t_0),$$

and this is just Exercise 57.

61. The dot product is $\mathbf{r}_1(t) \cdot \mathbf{r}_2(t) = x_1(t)y_1(t) + x_2(t)y_2(t)$. Since the $x_i(t)$ and $y_i(t)$ are differentiable, so are $x_1(t)y_1(t)$ and $x_2(t)y_2(t)$, by the product rule (Theorem 2.11). Then $x_1(t)y_1(t) + x_2(t)y_2(t)$ is also differentiable, by the sum rule (Theorem 2.10).

Thinking Forward

The derivative of a vector function

A vector-valued function $\mathbf{r}(t) = \langle x(t), y(t), z(t) \rangle$ is differentiable if and only if $x(t)$, $y(t)$, and $z(t)$ are all differentiable, and in that case the derivative is given by $\mathbf{r}'(t) = \langle x'(t), y'(t), z'(t) \rangle$.

The integral of a vector function

A vector-valued function $\mathbf{r}(t) = \langle x(t), y(t), z(t) \rangle$ is integrable on an interval $t \in [a, b]$ if and only if each component function is integrable there, and in that case the integral is

$$\int_a^b \mathbf{r}(t)\, dt = \left\langle \int_a^b x(t)\, dt, \ \int_a^b y(t)\, dt, \ \int_a^b z(t)\, dt \right\rangle.$$

11.2 The Calculus of Vector Functions

Thinking Back

Differentiability

▷ We have

$$f'(x) = \lim_{h \to 0} \frac{f(x+h) - f(x)}{h} = \lim_{h \to 0} \frac{(x+h)^2 - x^2}{h} = \lim_{h \to 0} \frac{2xh + h^2}{h} = \lim_{h \to 0} (2x + h) = 2x.$$

▷ We have

$$f'(x) = \lim_{h \to 0} \frac{f(x+h) - f(x)}{h} = \lim_{h \to 0} \frac{(x+h)^3 - 1 - (x^3 - 1)}{h}$$
$$= \lim_{h \to 0} \frac{3x^2 h + 3xh^2 + h^3}{h} = \lim_{h \to 0} (3x^2 + 3xh + h^2) = 3x^2.$$

▷ We have

$$
\begin{aligned}
f'(x) &= \lim_{h \to 0} \lim_{h \to 0} \frac{f(x+h) - f(x)}{h} \\
&= \lim_{h \to 0} \frac{\sqrt{x+h+1} - \sqrt{x+1}}{h} \\
&= \lim_{h \to 0} \frac{(\sqrt{x+h+1} - \sqrt{x+1})(\sqrt{x+h+1} + \sqrt{x+1})}{h(\sqrt{x+h+1} + \sqrt{x+1})} \\
&= \lim_{h \to 0} \frac{x+h+1 - (x+1)}{h(\sqrt{x+h+1} + \sqrt{x+1})} \\
&= \lim_{h \to 0} \frac{h}{h(\sqrt{x+h+1} + \sqrt{x+1})} \\
&= \lim_{h \to 0} \frac{1}{\sqrt{x+h+1} + \sqrt{x+1}} \\
&= \frac{1}{2\sqrt{x+1}}.
\end{aligned}
$$

Integrability

▷ Use a right approximation. With n subintervals, we have $\Delta x = \frac{3}{n}$, and $x_k = 1 + \frac{3}{n}k$ for $k = 0, 1, \ldots, n$. Then the integral is the limit of the corresponding Riemann sum:

$$
\begin{aligned}
\int_1^4 (x+3)\, dx &= \lim_{n \to \infty} \sum_{k=1}^n (x_k + 3)\Delta x \\
&= \lim_{n \to \infty} \sum_{k=1}^n \left(4 + \frac{3}{n}k \right) \cdot \frac{3}{n} \\
&= \lim_{n \to \infty} \sum_{k=1}^n \left(\frac{12}{n} + \frac{9}{n^2}k \right) \\
&= \lim_{n \to \infty} \left(\frac{12}{n} \sum_{k=1}^n 1 + \frac{9}{n^2} \sum_{k=1}^n k \right) \\
&= \lim_{n \to \infty} \left(12 + \frac{9}{n^2} \cdot \frac{n(n+1)}{2} \right) \\
&= 12 + \lim_{n \to \infty} \frac{9n^2 + 9n}{2n^2} \\
&= 12 + \lim_{n \to \infty} \frac{9 + 9/n}{2} \\
&= 12 + \frac{9}{2} = \frac{33}{2}.
\end{aligned}
$$

▷ Use a right approximation. With n subintervals, we have $\Delta x = \frac{2}{n}$, and $x_k = 0 + \frac{2}{n}k = \frac{2}{n}k$ for $k = 0, 1, \ldots, n$. Then the integral is the limit of the corresponding Riemann sum:

$$
\begin{aligned}
\int_0^2 x^2 \, dx &= \lim_{n \to \infty} \sum_{k=1}^n x_k^2 \Delta x \\
&= \lim_{n \to \infty} \sum_{k=1}^n \left(\frac{2}{n}k \right)^2 \cdot \frac{2}{n} \\
&= \lim_{n \to \infty} \left(\frac{8}{n^3} \sum_{k=1}^n k^2 \right) \\
&= \lim_{n \to \infty} \left(\frac{8}{n^3} \cdot \frac{n(n+1)(2n+1)}{6} \right) \\
&= \lim_{n \to \infty} \frac{4(n+1)(2n+1)}{3n^2} \\
&= \lim_{n \to \infty} \frac{8n^2 + 12n + 4}{3n^2} \\
&= \lim_{n \to \infty} \frac{8 + 12/n + 4/n^2}{3} \\
&= \frac{8}{3}.
\end{aligned}
$$

▷ Use a right approximation. With n subintervals, we have $\Delta x = \frac{2}{n}$, and $x_k = 1 + \frac{2}{n}k$ for $k = 0, 1, \ldots, n$. Then the integral is the limit of the corresponding Riemann sum:

$$
\begin{aligned}
\int_1^3 (x^2 - 3) \, dx &= \lim_{n \to \infty} \sum_{k=1}^n (x_k^2 - 3)\Delta x \\
&= \lim_{n \to \infty} \sum_{k=1}^n \left(\left(1 + \frac{2}{n}k \right)^2 - 3 \right) \cdot \frac{2}{n} \\
&= \lim_{n \to \infty} \sum_{k=1}^n \left(\frac{4}{n}k + \frac{4}{n^2}k^2 - 2 \right) \cdot \frac{2}{n} \\
&= \lim_{n \to \infty} \left(\frac{8}{n^2} \sum_{k=1}^n k + \frac{8}{n^3} \sum_{k=1}^n k^2 - \frac{4}{n} \sum_{k=1}^n 1 \right) \\
&= \lim_{n \to \infty} \left(\frac{8}{n^2} \cdot \frac{n(n+1)}{2} + \frac{8}{n^3} \cdot \frac{n(n+1)(2n+1)}{6} - \frac{4}{n} \cdot n \right) \\
&= \lim_{n \to \infty} \left(\frac{4n+4}{n} + \frac{4(n+1)(2n+1)}{3n^2} - 4 \right) \\
&= \lim_{n \to \infty} \left(\frac{4n+4}{n} + \frac{8n^2 + 12n + 4}{3n^2} - 4 \right) \\
&= \lim_{n \to \infty} \left(\frac{4 + 4/n}{1} + \frac{8 + 12/n + 4/n^2}{3} - 4 \right) \\
&= \frac{8}{3}.
\end{aligned}
$$

Concepts

1. (a) False. For example, if $\mathbf{r}(t) = \langle \ln|t|, 1 \rangle$ for $t \neq 0$ and $\mathbf{r}(0) = \langle 0, 1 \rangle$, then $\lim_{h \to 0} \mathbf{r}(t+h)$ is undefined at $t = 0$, since the x coordinate of $\mathbf{r}(t)$ goes to $-\infty$ as $t \to 0$.

   (b) False. This statement holds if and only if the component functions of $\mathbf{r}(t)$ are differentiable. For example, if $\mathbf{r}(t) = \langle |t|, 1 \rangle$, then $x(t) = |t|$ is not differentiable at $t = 0$, so the limit is not defined at $t = 0$.

   (c) True. See Theorem 11.12. The statement follows from the fact that $\|\mathbf{r}(t)\|^2 = \mathbf{r}(t) \cdot \mathbf{r}(t)$; taking derivatives of both sides, the result follows from the product rule.

   (d) True. If $\mathbf{r}(t) \cdot \mathbf{r}'(t) = 0$, then $\mathbf{r}(t) \cdot \mathbf{r}'(t) + \mathbf{r}'(t) \cdot \mathbf{r}(t) = 0$, so that $(\mathbf{r}(t) \cdot \mathbf{r}(t))' = 0$. But $\mathbf{r}(t) \cdot \mathbf{r}(t) = \|\mathbf{r}(t)\|^2$, so that $\left(\|\mathbf{r}(t)\|^2 \right)' = 0$ and thus $\|\mathbf{r}(t)\|^2$ is a constant. It follows that $\|\mathbf{r}(t)\|$ is also a constant.

   (e) False. For example, let $\mathbf{r}(t) = \langle \sin t \cos t, \sin^2 t, \cos t \rangle$. Then

$$\|\mathbf{r}(t)\| = \sqrt{\sin^2 t \cos^2 t + \sin^4 t + \cos^2 t} = \sqrt{\sin^2 t(\cos^2 t + \sin^2 t) + \cos^2 t}$$
$$= \sqrt{\sin^2 t + \cos^2 t} = 1,$$

   so that $\|\mathbf{r}(t)\|$ is constant. However,

$$\mathbf{r}'(t) = \left\langle \cos^2 t - \sin^2 t, 2 \sin t \cos t, -\sin t \right\rangle = \langle \cos 2t, \sin 2t, -\sin t \rangle,$$
$$\mathbf{r}''(t) = \langle -2 \sin 2t, 2 \cos 2t, -\cos t \rangle,$$

   so that

$$\mathbf{r}'(t) \cdot \mathbf{r}''(t) = \langle \cos 2t, \sin 2t, -\sin t \rangle \cdot \langle -2 \sin 2t, 2 \cos 2t, -\cos t \rangle$$
$$= -2 \cos 2t \sin 2t + 2 \sin 2t \cos 2t + \sin t \cos t = \sin t \cos t,$$

   which is not a constant.

   (f) False. If either $\mathbf{r}(t)$ or $\mathbf{r}'(t)$ is zero, then $\mathbf{r}(t) \times \mathbf{r}'(t) = 0$. This will happen any time $\mathbf{r}(t)$ is a constant vector. However, if neither $\mathbf{r}(t)$ nor $\mathbf{r}'(t)$ is zero, then since by part (c) $\mathbf{r}(t) \cdot \mathbf{r}'(t) = 0$, the two vectors are perpendicular, so their cross product will not be zero.

   (g) False. For example, if $\mathbf{r}(t)$ is a constant vector, then both the velocity and acceleration vectors are zero.

   (h) False. For example, let $[a, b] = [0, 1]$, and let $\mathbf{r}(t) = \langle x(t), 1 \rangle$, where

$$x(t) = \begin{cases} \frac{1}{t}, & t > 0, \\ 0, & t = 0. \end{cases}$$

   Then $\int_0^1 \mathbf{r}(t)\, dt = \left\langle \int_0^1 x(t)\, dt, \int_0^1 1\, dt \right\rangle$, but $\int_0^1 x(t)\, dt$ does not exist.

3. A scalar function $y = f(x)$ is differentiable at a point c in the domain of f if the limit

$$\lim_{h \to 0} \frac{f(c+h) - f(c)}{h}$$

   exists.

5. An explicit limit is not required since the derivative of a vector function $\mathbf{r}(t) = \langle x(t), y(t), z(t) \rangle$ is defined in terms of the derivatives of $x(t)$, $y(t)$, and $z(t)$. Each of those definitions involves a limit.

7. For example, let $\mathbf{r}(t) = \langle |t|, t \rangle$. A graph of this function for $-2 \leq t \leq 2$ is

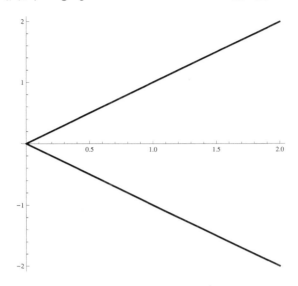

This function is not differentiable at $t = 0$ since $|t|$ is not differentiable at $t = 0$. As can be seen from the graph, either the line $t \langle 1, 1 \rangle$ or the line $t \langle -1, 1 \rangle$ could be thought of as a tangent line to the curve at $t = 0$. More precisely, we have

$$\lim_{h \to 0^+} \frac{\mathbf{r}(0+h) - \mathbf{r}(0)}{h} = \lim_{h \to 0^+} \frac{\langle |h|, h \rangle - \langle 0, 0 \rangle}{h} = \lim_{h \to 0^+} \frac{\langle h, h \rangle}{h} = \lim_{h \to 0^+} \langle 1, 1 \rangle = \langle 1, 1 \rangle$$

$$\lim_{h \to 0^-} \frac{\mathbf{r}(0+h) - \mathbf{r}(0)}{h} = \lim_{h \to 0^-} \frac{\langle |h|, h \rangle - \langle 0, 0 \rangle}{h} = \lim_{h \to 0^+} \frac{\langle -h, h \rangle}{h} = \lim_{h \to 0^+} \langle -1, 1 \rangle = \langle -1, 1 \rangle.$$

9. A scalar function $y = f(x)$ is integrable on $[a, b]$, if the Riemann sum

$$\lim_{n \to \infty} \sum_{k=1}^{n} f(x_k^*) \Delta x$$

exists, where the nth Riemann sum is obtained by dividing the interval $[a, b]$ into n equal subintervals, each of width Δx, and x_k^* is an arbitrary point in the kth subinterval.

11. An explicit Riemann sum, which involves a limit, is not required since the integral of a vector function $\mathbf{r}(t) = \langle x(t), y(t), z(t) \rangle$ is defined in terms of the integrals of $x(t)$, $y(t)$, and $z(t)$. Each of those definitions involves a limit.

13. If $\mathbf{r}(t) = \langle x(t), y(t), z(t) \rangle$ is a differentiable position function, then the **velocity** vector $\mathbf{v}(t)$ is $\mathbf{v}(t) = \mathbf{r}'(t) = \langle x'(t), y'(t), z'(t) \rangle$. See Definition 11.10.

15. By Theorem 11.8 (and Exercise 67),

$$\frac{d}{d\tau}\mathbf{r}(f(\tau)) = \frac{d}{d\tau}\mathbf{r}(t) = \frac{d}{dt}\mathbf{r}(t) \cdot \frac{dt}{d\tau} = \frac{d\mathbf{r}}{dt} \cdot \frac{df}{d\tau}.$$

17. By Definition 11.6, the derivative of $\mathbf{r}(t) = \langle t^3, 5t^3, -2t^3 \rangle$ is $\mathbf{r}'(t) = \langle 3t^2, 15t^2, -6t^2 \rangle$, found by taking the derivative of each of the components. Then if t_0 and t_1 are two points, we have (since $t > 0$)

$$\mathbf{r}'(t_1) = \left\langle 3t_1^2, 15t_1^2, -6t_1^2 \right\rangle = \frac{t_1^2}{t_0^2}\left\langle 3t_0^2, 15t_0^2, -2t_0^2 \right\rangle = \frac{t_1^2}{t_0^2}\mathbf{r}'(t_0).$$

Thus the derivative at any point is a constant multiple of the derivative at any other point. The tangent line to the graph of $\mathbf{r}(t)$ has direction vector $\mathbf{r}'(t)$, so all tangent lines have parallel direction vectors, so they have equal slopes. Thus the slope of the tangent line to the graph is constant, so that the graph must be a line. A graph of $\mathbf{r}(t)$ is

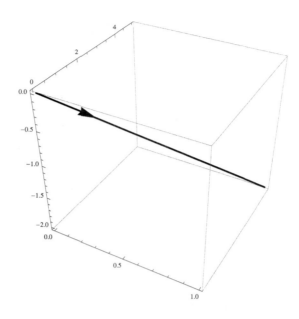

19. Since

$$\int \mathbf{r}'(t)\, dt = \left\langle \int x'(t)\, dt, \int y'(t)\, dt, \int z'(t)\, dt \right\rangle = \langle x(t) + C_1, y(t) + C_2, z(t) + C_3 \rangle$$
$$= \langle x(t), y(t), z(t) \rangle + \langle C_1, C_2, C_3 \rangle = \mathbf{r}(t) + \langle C_1, C_2, C_3 \rangle,$$

the two differ by a constant vector.

Skills

21. (a) Using the chain rule,

$$\frac{d\mathbf{r}}{d\tau} = \frac{d\mathbf{r}}{dt}\frac{dt}{d\tau} = \left\langle \frac{d}{dt}t, \frac{d}{dt}t^2, \frac{d}{dt}t^3 \right\rangle \cdot \frac{d}{d\tau}(\sin\tau) = \left\langle 1, 2t, 3t^2 \right\rangle \cdot \cos\tau$$
$$= \left\langle \cos\tau, 2\sin\tau\cos\tau, 3\sin^2\tau\cos\tau \right\rangle.$$

(b) Substituting $\sin\tau$ for t in $\mathbf{r}(t)$ gives $\mathbf{r}(\tau) = \langle \sin\tau, \sin^2\tau, \sin^3\tau \rangle$, so that

$$\frac{d\mathbf{r}}{d\tau} = \left\langle \frac{d}{d\tau}\sin\tau, \frac{d}{d\tau}\sin^2\tau, \frac{d}{d\tau}\sin^3\tau \right\rangle = \left\langle \cos\tau, 2\sin\tau\cos\tau, 3\sin^2\tau\cos\tau \right\rangle,$$

and the two are equal.

23. (a) Using the chain rule,

$$\frac{d\mathbf{r}}{d\tau} = \frac{d\mathbf{r}}{dt}\frac{dt}{d\tau} = \left\langle \frac{d}{dt}(t^2 \sin t^2), \frac{d}{dt}t, \frac{d}{dt}(t^2 \cos t^2) \right\rangle \cdot \frac{d}{d\tau}\sqrt{\tau}$$

$$= \left\langle 2t \sin t^2 + t^2 \cos(t^2) \cdot 2t, 1, 2t \cos t^2 - t^2 \sin(t^2) \cdot 2t \right\rangle \cdot \frac{1}{2\sqrt{\tau}}$$

$$= \left\langle 2t \sin t^2 + 2t^3 \cos t^2, 1, 2t \cos t^2 - 2t^3 \sin t^2 \right\rangle \cdot \frac{1}{2\sqrt{\tau}}$$

$$= \left\langle 2\sqrt{\tau} \sin \tau + 2\tau\sqrt{\tau} \cos \tau, 1, 2\sqrt{\tau} \cos \tau - 2\tau\sqrt{\tau} \sin \tau \right\rangle \cdot \frac{1}{2\sqrt{\tau}}$$

$$= \left\langle \sin \tau + \tau \cos \tau, \frac{1}{2\sqrt{\tau}}, \cos \tau - \tau \sin \tau \right\rangle.$$

(b) Substituting $\sqrt{\tau}$ for t in $\mathbf{r}(t)$ gives $\mathbf{r}(\tau) = \langle \tau \sin \tau, \sqrt{\tau}, \tau \cos \tau \rangle$, so that

$$\frac{d\mathbf{r}}{d\tau} = \left\langle \frac{d}{d\tau}(\tau \sin \tau), \frac{d}{d\tau}\sqrt{\tau}, \frac{d}{d\tau}(\tau \cos \tau) \right\rangle = \left\langle \sin \tau + \tau \cos \tau, \frac{1}{2\sqrt{\tau}}, \cos \tau - \tau \sin \tau \right\rangle,$$

and the two are equal.

25. With $\mathbf{r}(t) = te^t\mathbf{i} + t \ln t\mathbf{j} = \langle te^t, t \ln t, 0 \rangle$, we have

$$\mathbf{r}'(t) = \langle e^t + te^t, 1 + \ln t, 0 \rangle = \langle (t+1)e^t, 1 + \ln t, 0 \rangle.$$

The given point $\langle e, 0, 0 \rangle$ is $\mathbf{r}(1)$. Then by Definition 11.9, the line tangent to the curve at $(e, 0, 0)$ is

$$\mathcal{L}(t) = \mathbf{r}(1) + t\mathbf{r}'(1) = \langle e, 0, 0 \rangle + t\left\langle (1+1)e^1, 1 + \ln 1, 0 \right\rangle = \langle e, 0, 0 \rangle + t\langle 2e, 1, 0 \rangle.$$

The plane orthogonal to this line has normal vector $\langle 2e, 1, 0 \rangle$, so it is of the form $2ex + y = d$. Since $(e, 0, 0)$ lies on the plane, we have $d = 2e^2$, so the equation of the plane is $2ex + y = 2e^2$.

27. With $\mathbf{r}(t) = \langle \sin t, \cos t, 2 \sin 2t \rangle$, we have

$$\mathbf{r}'(t) = \langle \cos t, - \sin t, 4 \cos 2t \rangle.$$

The given point $\langle 1, 0, 0 \rangle$ is $\mathbf{r}\left(\frac{\pi}{2}\right)$. Then by Definition 11.9, the line tangent to the curve at $(1, 0, 0)$ is

$$\mathcal{L}(t) = \mathbf{r}\left(\frac{\pi}{2}\right) + t\mathbf{r}'\left(\frac{\pi}{2}\right) = \langle 1, 0, 0 \rangle + t\left\langle \cos \frac{\pi}{2}, - \sin \frac{\pi}{2}, 4 \cos \pi \right\rangle = \langle 1, 0, 0 \rangle + t\langle 0, -1, -4 \rangle.$$

The plane orthogonal to this line has normal vector $\langle 0, -1, -4 \rangle$, so it is of the form $-y - 4z = d$. Since $(1, 0, 0)$ lies on the plane, we have $d = 0$, so the equation of the plane is $-y - 4z = 0$, or $y + 4z = 0$.

29. With $\mathbf{r}(t) = \cos 3t\mathbf{i} + \sin 4t\mathbf{j} + t\mathbf{k} = \langle \cos 3t, \sin 4t, t \rangle$, we have

$$\mathbf{r}'(t) = \langle -3 \sin 3t, 4 \cos 4t, 1 \rangle.$$

The given point $\langle 0, 0, \frac{\pi}{2} \rangle$ is $\mathbf{r}\left(\frac{\pi}{2}\right)$. Then by Definition 11.9, the line tangent to the curve at $\left(0, 0, \frac{\pi}{2}\right)$ is

$$\mathcal{L}(t) = \mathbf{r}\left(\frac{\pi}{2}\right) + t\mathbf{r}'\left(\frac{\pi}{2}\right) = \left\langle 0, 0, \frac{\pi}{2} \right\rangle + t\left\langle -3 \sin \frac{3\pi}{2}, 4 \cos 2\pi, 1 \right\rangle = \left\langle 0, 0, \frac{\pi}{2} \right\rangle + t\langle 3, 4, 1 \rangle.$$

The plane orthogonal to this line has normal vector $\langle 3, 4, 1 \rangle$, so it is of the form $3x + 4y + z = d$. Since $\left(0, 0, \frac{\pi}{2}\right)$ lies on the plane, we have $d = \frac{\pi}{2}$, so the equation of the plane is $3x + 4y + z = \frac{\pi}{2}$.

31. Since $\mathbf{v}(t) = \mathbf{r}'(t)$ and $\mathbf{a}(t) = \mathbf{v}'(t) = \mathbf{r}''(t)$, we get

$$\mathbf{v}(t) = \left\langle e^t + te^t, \ln t + t \cdot \frac{1}{t} \right\rangle = \left\langle (t+1)e^t, 1 + \ln t \right\rangle$$

$$\mathbf{a}(t) = \left\langle (t+1)e^t + e^t, \frac{1}{t} \right\rangle = \left\langle (t+2)e^t, \frac{1}{t} \right\rangle.$$

33. Since $\mathbf{v}(t) = \mathbf{r}'(t)$ and $\mathbf{a}(t) = \mathbf{v}'(t) = \mathbf{r}''(t)$, we get

$$\mathbf{v}(t) = \left\langle \cos t, -\sin t, 4\cos 2t \right\rangle$$

$$\mathbf{a}(t) = \left\langle -\sin t, -\cos t, -8\sin 2t \right\rangle.$$

35. Substitute $\tau^3 + 1$ for t to get $\mathbf{r}(\tau) = \left\langle \tau^3 + 1, (\tau^3 + 1)^2, (\tau^3 + 1)^3 \right\rangle$. Then

$$\mathbf{r}'(\tau) = \left\langle 3\tau^2, 2(\tau^3 + 1)(3\tau^2), 3(\tau^3 + 1)^2(3\tau^2) \right\rangle = \left\langle 3\tau^2, 6\tau^2(\tau^3 + 1), 9\tau^2(\tau^3 + 1)^2 \right\rangle.$$

37. Use the chain rule:

$$\mathbf{r}'(\tau) = \frac{d\mathbf{r}}{dt}\frac{dt}{d\tau} = \left\langle \sec t \tan t, -\frac{1}{t^2}, e^t \ln t + e^t \cdot \frac{1}{t} \right\rangle \cdot \left(-\frac{1}{\tau^2} \right)$$

$$= \left\langle \sec t \tan t, -\frac{1}{t^2}, e^t \left(\ln t + \frac{1}{t} \right) \right\rangle \cdot \left(-\frac{1}{\tau^2} \right)$$

$$= \left\langle \sec \frac{1}{\tau} \tan \frac{1}{\tau}, -\tau^2, e^{1/\tau} \left(\ln \frac{1}{\tau} + \tau \right) \right\rangle \cdot \left(-\frac{1}{\tau^2} \right)$$

$$= \left\langle -\frac{1}{\tau^2} \sec \frac{1}{\tau} \tan \frac{1}{\tau}, 1, \frac{1}{\tau^2} e^{1/\tau} (\ln \tau - \tau) \right\rangle.$$

39. Substitute $5\tau - 2$ for t to get $\mathbf{r}(\tau) = \left\langle \cos(15\tau - 6), \sin(20\tau - 8), 5\tau - 2 \right\rangle$. Then

$$\mathbf{r}'(\tau) = \left\langle -15\sin(15\tau - 6), 20\cos(20\tau - 8), 5 \right\rangle.$$

41. By Definition 11.13,

$$\int \left\langle \sin t, \cos t, \tan t \right\rangle dt = \left\langle \int \sin t\, dt, \int \cos t\, dt, \int \tan t\, dt \right\rangle$$

$$= \left\langle -\cos t + C_1, \sin t + C_2, \ln|\sec t| + C_3 \right\rangle = \left\langle -\cos t, \sin t, \ln|\sec t| \right\rangle + \mathbf{C}.$$

43. By Definition 11.13,

$$\int_0^{2\pi} \left\langle \sin t, \cos t, t \right\rangle dt = \left\langle \int_0^{2\pi} \sin t\, dt, \int_0^{2\pi} \cos t\, dt, \int_0^{2\pi} t\, dt \right\rangle$$

$$= \left\langle [-\cos t]_0^{2\pi}, [\sin t]_0^{2\pi}, [\tfrac{1}{2}t^2]_0^{2\pi} \right\rangle$$

$$= \left\langle 0, 0, 2\pi^2 \right\rangle.$$

45. Since $\mathbf{r}'(t) = \mathbf{v}(t)$, we get

$$\mathbf{r}(t) = \int \mathbf{v}(t)\, dt = \left\langle \int t\, dt, \int t^2\, dt \right\rangle = \left\langle \frac{1}{2}t^2 + C_1, \frac{1}{3}t^3 + C_2 \right\rangle.$$

Then $\mathbf{r}(0) = \left\langle C_1, C_2 \right\rangle = \left\langle 3, -4 \right\rangle$, so that $C_1 = 3$ and $C_2 = -4$. Thus $\mathbf{r}(t) = \left\langle \frac{1}{2}t^2 + 3, \frac{1}{3}t^3 - 4 \right\rangle$.

47. Since $\mathbf{r}'(t) = \mathbf{v}(t)$, we get

$$\mathbf{r}(t) = \int \mathbf{v}(t)\,dt = \left\langle \int e^t\,dt, \int \ln t\,dt \right\rangle = \left\langle e^t + C_1, t\ln t - t + C_2 \right\rangle.$$

Then $\mathbf{r}(1) = \langle e + C_1, -1 + C_2 \rangle = \langle 1, -6 \rangle$, so that $C_1 = 1 - e$ and $C_2 = -5$. Thus

$$\mathbf{r}(t) = \left\langle e^t + 1 - e, t\ln t - t - 5 \right\rangle.$$

49. Integrate $\mathbf{a}(t)$ to get $\mathbf{v}(t) = \langle t^2 + C_1, t^3 + C_2 \rangle$. Since $\mathbf{v}(0) = \langle C_1, C_2 \rangle = \langle 1, -2 \rangle$, we get $C_1 = 1$ and $C_2 = -2$, so that $\mathbf{v}(t) = \langle t^2 + 1, t^3 - 2 \rangle$. Integrate again to get $\mathbf{r}(t) = \left\langle \frac{1}{3}t^3 + t + C_1, \frac{1}{4}t^4 - 2t + C_2 \right\rangle$. The initial condition gives $\mathbf{r}(0) = \langle C_1, C_2 \rangle = \langle 2, -3 \rangle$, so that $C_1 = 2$ and $C_2 = -3$. Thus $\mathbf{r}(t) = \left\langle \frac{1}{3}t^3 + t + 2, \frac{1}{4}t^4 - 2t - 3 \right\rangle$.

51. Integrate $\mathbf{a}(t) = \langle 0, -32, 0 \rangle$ to get $\mathbf{v}(t) = \langle C_1, -32t + C_2, C_3 \rangle$. Since $\mathbf{v}(0) = \langle 5, 5, 0 \rangle$, we get $C_1 = 5$, $C_2 = 5$, and $C_3 = 0$, so that $\mathbf{v}(t) = \langle 5, 5 - 32t, 0 \rangle$. Integrate again to get $\mathbf{r}(t) = \langle 5t + C_1, 5t - 16t^2 + C_2, C_3 \rangle$. The initial condition gives $\mathbf{r}(0) = \langle C_1, C_2, C_3 \rangle = \langle 0, 26, 0 \rangle$, so that $C_1 = C_3 = 0$ and $C_2 = 26$. Thus $\mathbf{r}(t) = \langle 5t, 5t - 16t^2 + 26, 0 \rangle$.

53. To find the intersection points, substitute t for x, t^2 for y, and t^3 for z and solve for t. This will give values of t for which a point of the form $\langle t, t^2, t^3 \rangle$ also satisfies the equation of the plane, i.e., the intersection points. We get

$$3t - 3t^2 + t^3 = 1, \quad \text{or} \quad t^3 - 3t^2 + 3t - 1 = 0, \quad \text{or} \quad (t-1)^3 = 0.$$

The only intersection point is when $t = 1$, which is at the point $(1, 1^2, 1^3) = (1, 1, 1)$.

55. The curves intersect for values of t and u where

$$\mathbf{r}_1(t) = \langle 3\cos t, 3\sin t \rangle = \mathbf{r}_2(u) = \langle 2 + 2\sin u, 2\cos u \rangle,$$

so when $3\cos t = 2 + 2\sin u$ and $3\sin t = 2\cos u$. To solve this pair of equations, square both of them and add to get

$$9(\sin^2 t + \cos^2 t) = 4 + 8\sin u + 4(\sin^2 u + \cos^2 u), \quad \text{so that} \quad \sin u = \frac{1}{8}.$$

Thus $u = \sin^{-1}\frac{1}{8}$ or $u = \pi - \sin^{-1}\frac{1}{8}$. Substituting $\sin u = \frac{1}{8}$ into the first equation gives

$$3\cos t = 2 + \frac{1}{4}, \quad \text{so that} \quad \cos t = \frac{3}{4}.$$

Thus $t = \cos^{-1}\frac{3}{4}$ or $t = -\cos^{-1}\frac{3}{4}$. Now,

$$\mathbf{r}_1\left(\cos^{-1}\frac{3}{4} \right) = \left\langle \frac{9}{4}, \frac{3\sqrt{7}}{4} \right\rangle, \quad \mathbf{r}_1\left(-\cos^{-1}\frac{3}{4} \right) = \left\langle \frac{9}{4}, -\frac{3\sqrt{7}}{4} \right\rangle$$

$$\mathbf{r}_2\left(\sin^{-1}\frac{1}{8} \right) = \left\langle \frac{9}{4}, \frac{3\sqrt{7}}{4} \right\rangle, \quad \mathbf{r}_2\left(\pi - \sin^{-1}\frac{1}{8} \right) = \left\langle \frac{9}{4}, -\frac{3\sqrt{7}}{4} \right\rangle,$$

so that the two points of intersection are

$$P_1 = \left\langle \frac{9}{4}, \frac{3\sqrt{7}}{4} \right\rangle \quad \text{for } t_1 = \sin^{-1}\frac{1}{8}, \; u_1 = \cos^{-1}\frac{3}{4},$$

$$P_2 = \left\langle \frac{9}{4}, -\frac{3\sqrt{7}}{4} \right\rangle \quad \text{for } t_2 = \pi - \sin^{-1}\frac{1}{8}, \; u_2 = -\cos^{-1}\frac{3}{4}.$$

Now, $\mathbf{r}_1'(t) = \langle -3\sin t, 3\cos t \rangle$ and $\mathbf{r}_2'(t) = \langle 2\cos t, -2\sin t \rangle$, so that at P_1 we have

$$\mathbf{r}_1'(t_1) = \left\langle -3 \cdot \frac{1}{8}, 3\cos\left(\sin^{-1}\frac{1}{8}\right) \right\rangle = \left\langle -\frac{3}{8}, \frac{9\sqrt{7}}{8} \right\rangle,$$

$$\mathbf{r}_2'(u_1) = \left\langle 2 \cdot \frac{3}{4}, -2\sin\left(\cos^{-1}\frac{3}{4}\right) \right\rangle = \left\langle \frac{3}{2}, -\frac{\sqrt{7}}{2} \right\rangle.$$

Then the angle θ between $\mathbf{r}_1(t)$ and $\mathbf{r}_2(t)$ at P_1 is given by

$$\cos\theta = \frac{\mathbf{r}_1'(t_1) \cdot \mathbf{r}_2'(u_1)}{\|\mathbf{r}_1'(t_1)\| \, \|\mathbf{r}_2'(u_1)\|} = \frac{-\frac{9}{16} - \frac{63}{16}}{\sqrt{\frac{9}{64} + \frac{567}{64}} \sqrt{\frac{9}{4} + \frac{7}{4}}} = -\frac{9}{2\,(3 \cdot 2)} = -\frac{3}{4}.$$

Thus $\theta = \cos^{-1}\left(-\frac{3}{4}\right) \approx 138.59°$, so the acute angle is $180 - 138.59 \approx 41.41°$. For P_2, we have

$$\mathbf{r}_1'(t_2) = \left\langle -3 \cdot \frac{1}{8}, 3\cos\left(\pi - \sin^{-1}\frac{1}{8}\right) \right\rangle = \left\langle -\frac{3}{8}, -\frac{9\sqrt{7}}{8} \right\rangle,$$

$$\mathbf{r}_2'(u_2) = \left\langle 2 \cdot \frac{3}{4}, -2\sin\left(-\cos^{-1}\frac{3}{4}\right) \right\rangle = \left\langle \frac{3}{2}, \frac{\sqrt{7}}{2} \right\rangle.$$

Then $\|\mathbf{r}_1'(t_2)\| = \|\mathbf{r}_1'(t_1)\|$ and $\|\mathbf{r}_2'(u_2)\| = \|\mathbf{r}_2'(u_1)\|$. Also, $\mathbf{r}_1'(t_2) \cdot \mathbf{r}_2'(u_2) = \mathbf{r}_1'(t_1) \cdot \mathbf{r}_2'(u_1)$. Thus we get the same angle θ.

So in summary, the two curves meet at the two points

$$\left(\frac{9}{4}, \pm\frac{3\sqrt{7}}{4} \right),$$

and the acute angle between the curves at each of those points is approximately $41.41°$.

57. Let $\mathbf{r}(t) = \langle x(t), y(t), z(t) \rangle$. Then $\mathbf{r}'(t) = \langle x'(t), y'(t), z'(t) \rangle$. Since $\mathbf{r}(t) = \mathbf{r}'(t)$, we have $x(t) = x'(t)$, $y(t) = y'(t)$, and $z(t) = z'(t)$. Now, $x'(t) = x(t)$ means $\frac{dx}{dt} = x$, so that $\frac{dx}{x} = dt$. Integrate both sides to get $\ln|x| = t + C_1$. Exponentiate to get $|x| = e^{t+C_1} = C_2 e^t$. We can remove the absolute value signs by negating the constant if required, so the solution is $x = A_1 e^t$. Similarly, $y = A_2 e^t$ and $z = A_3 e^t$ for constants A_1, A_2, and A_3. Thus $\mathbf{r}(t) = \langle A_1 e^t, A_2 e^t, A_3 e^t \rangle$. Since the curve passes through $(0, 1, 2)$, we can parametrize the curve so that this occurs at $t = 0$. Then

$$\mathbf{r}(0) = \langle A_1, A_2, A_3 \rangle = \langle 0, 1, 2 \rangle, \quad \text{so that} \quad A_1 = 0, \ A_2 = 1, \ A_3 = 2.$$

Then the equation of the curve is $\mathbf{r}(t) = \langle 0, e^t, 2e^t \rangle$. Its graph (drawn as a two-dimensional graph in the yz plane, since $x = 0$ always) is

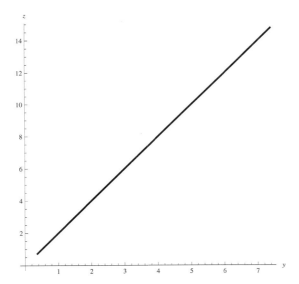

Note that since $\frac{2e^t}{e^t} = 2$ is a constant, this graph is a line with slope 2 extending into the first quadrant from $(0,0)$ (however, $(0,0)$ is not actually on the graph, since e^t is never zero).

Applications

59. (a) His velocity will be the integral of the acceleration vector, so that

$$\mathbf{v}(t) = \left\langle \int (-0.5d \cdot y'(t))\, dt, \int (-9.8 - d \cdot y'(t))\, dt \right\rangle$$
$$= \langle -0.5d \cdot y(t) + C_1, -9.8t - d \cdot y(t) + C_2 \rangle.$$

Since $\mathbf{v}(0) = \langle 0,0 \rangle$ and $y(0) = 0$, we get $C_1 = 0$ and $C_2 = 0$. Thus

$$\mathbf{v}(t) = \langle -0.5d \cdot y(t), -9.8t - d \cdot y(t) \rangle.$$

(b) Since $y'(t)$ is his vertical speed, we get $y'(t) = -9.8t - d \cdot y(t)$; solving for $y(t)$ gives $y(t) = -\frac{9.8t + y'(t)}{d}$.

(c) We have, using part (b), $x'(t) = -0.5d \cdot y(t) = -0.5d \cdot \left(-\frac{9.8t + y'(t)}{d} \right) = 0.5(9.8t + y'(t))$.

(d) $x(t) = \int (0.5(9.8t + y'(t)))\, dt = 0.5(4.9t^2 + y(t)) + C = 2.45t^2 + 0.5y(t) + C$. Since $y(0) = 0$, we get $C = 0$, so that his position vector at time t is $\langle x(t), y(t) \rangle = \langle 2.45t^2 + 0.5y(t), y(t) \rangle$.

(e) After 7.4 seconds, we know that $y(t) > -200$, so that

$$2.45 \cdot 7.4^2 + 0.5y(7.4) > 134.162 + 0.5(-200) = 34.162 \text{ m}.$$

After 7.4 seconds, Arne has only traveled about 34 m horizontally. Whether he makes it or not depends on how much longer it takes to drop 200 m, but things don't look good.

Proofs

61. If $\mathbf{a}(t) = \mathbf{0}$, then

$$\mathbf{v}(t) = \int \mathbf{a}(t)\, dt = \int \langle 0, 0, \ldots, 0 \rangle\, dt = \langle C_1, C_2, \ldots, C_n \rangle.$$

Next,

$$\mathbf{r}(t) = \int \mathbf{v}(t)\, dt = \int \langle C_1, C_2, \ldots, C_n \rangle \, dt$$

$$= \langle C_1 t + D_1, C_2 t + D_2, \ldots, C_n t + D_n \rangle$$

$$= \langle D_1, D_2, \ldots, D_n \rangle + t \langle C_1, C_2, \ldots, C_n \rangle,$$

and this is the graph of a line through the point $\langle D_1, D_2, \ldots, D_n \rangle$ that has the direction vector $\langle C_1, C_2, \ldots, C_n \rangle$.

63. Suppose $\mathbf{r}(t) = \langle x_1(t), x_2(t), \ldots, x_n(t) \rangle$. Then using Definition 11.6 as well as the Constant Multiple Rule (Theorem 2.10) for scalar functions, we have

$$\frac{d}{dt}(k\mathbf{r}(t)) = \frac{d}{dt}\left(k \langle x_1(t), x_2(t), \ldots, x_n(t) \rangle\right)$$

$$= \frac{d}{dt}\langle kx_1(t), kx_2(t), \ldots, kx_n(t) \rangle$$

$$= \left\langle \frac{d}{dt}(kx_1(t)), \frac{d}{dt}(kx_2(t)), \ldots, \frac{d}{dt}(kx_n(t)) \right\rangle$$

$$= \left\langle k\frac{d}{dt}x_1(t), k\frac{d}{dt}x_2(t), \ldots, k\frac{d}{dt}x_n(t) \right\rangle$$

$$= k \left\langle \frac{d}{dt}x_1(t), \frac{d}{dt}x_2(t), \ldots, \frac{d}{dt}x_n(t) \right\rangle$$

$$= k\frac{d}{dt}\langle x_1(t), x_2(t), \ldots, x_n(t) \rangle$$

$$= k\frac{d}{dt}\mathbf{r}(t)$$

$$= k\mathbf{r}'(t).$$

65. Suppose $\mathbf{r}_1(t) = \langle x_1(t), x_2(t), x_3(t) \rangle$ and $\mathbf{r}_2(t) = \langle y_1(t), y_2(t), y_3(t) \rangle$. Then using Definition 11.6 as well as the Sum and Product Rules (Theorems 2.10 and 2.11) for scalar functions, we have

$$\frac{d}{dt}\left(\mathbf{r}_1(t) \cdot \mathbf{r}_2(t)\right) = \frac{d}{dt}\left(\langle x_1(t), x_2(t), x_3(t) \rangle \cdot \langle y_1(t), y_2(t), y_3(t) \rangle\right)$$

$$= \frac{d}{dt}\left(x_1(t)y_1(t) + x_2(t)y_2(t) + x_3(t)y_3(t)\right)$$

$$= \frac{d}{dt}(x_1(t)y_1(t)) + \frac{d}{dt}(x_2(t)y_2(t)) + \frac{d}{dt}(x_3(t)y_3(t))$$

$$= x_1'(t)y_1(t) + x_1(t)y_1'(t) + x_2'(t)y_2(t) + x_2(t)y_2'(t) + x_3'(t)y_3(t) + x_3(t)y_3'(t)$$

$$= x_1'(t)y_1(t) + x_2'(t)y_2(t) + x_3'(t)y_3(t) + x_1(t)y_1'(t) + x_2(t)y_2'(t) + x_3(t)y_3'(t)$$

$$= \langle x_1'(t), x_2'(t), x_3'(t) \rangle \cdot \langle y_1(t), y_2(t), y_3(t) \rangle$$

$$\qquad + \langle x_1(t), x_2(t), x_3(t) \rangle \cdot \langle y_1'(t), y_2'(t), y_3'(t) \rangle$$

$$= \left(\frac{d}{dt}\langle x_1(t), x_2(t), x_3(t) \rangle\right) \cdot \langle y_1(t), y_2(t), y_3(t) \rangle$$

$$\qquad + \langle x_1(t), x_2(t), x_3(t) \rangle \cdot \left(\frac{d}{dt}\langle y_1(t), y_2(t), y_3(t) \rangle\right)$$

$$= \mathbf{r}_1'(t) \cdot \mathbf{r}_2(t) + \mathbf{r}_1(t) \cdot \mathbf{r}_2'(t).$$

67. Suppose $\mathbf{r}(t) = \langle x(t), y(t), z(t) \rangle$. Then using Definition 11.6 as well as the chain rule for scalar functions, we get

$$\begin{aligned}
\frac{d}{d\tau}(\mathbf{r}(t)) &= \frac{d}{d\tau}\langle x(t), y(t), z(t) \rangle \\
&= \left\langle \frac{d}{d\tau}x(t), \frac{d}{d\tau}y(t), \frac{d}{d\tau}z(t) \right\rangle \\
&= \left\langle \frac{dx}{dt}\frac{dt}{d\tau}, \frac{dy}{dt}\frac{dt}{d\tau}, \frac{dz}{dt}\frac{dt}{d\tau} \right\rangle \\
&= \left\langle \frac{dx}{dt}, \frac{dy}{dt}, \frac{dz}{dt} \right\rangle \frac{dt}{d\tau} \\
&= \frac{d\mathbf{r}}{dt}\frac{dt}{d\tau}.
\end{aligned}$$

69. If $\mathbf{r}(t) \cdot \mathbf{r}'(t) = 0$, then $\mathbf{r}(t) \cdot \mathbf{r}'(t) + \mathbf{r}'(t) \cdot \mathbf{r}(t) = 0$, so that $(\mathbf{r}(t) \cdot \mathbf{r}(t))' = 0$. But $\mathbf{r}(t) \cdot \mathbf{r}(t) = \|\mathbf{r}(t)\|^2$, so that $\left(\|\mathbf{r}(t)\|^2 \right)' = 0$ and thus $\|\mathbf{r}(t)\|^2$ is a constant. It follows that $\|\mathbf{r}(t)\|$ is also a constant since $\|\mathbf{r}\|$ is always nonnegative.

71. Since the speed is constant, i.e., $\|\mathbf{v}\|$ is constant, Theorem 11.12 tells us that $\mathbf{v} \cdot \mathbf{v}' = 0$, so that $\mathbf{v}$ and $\mathbf{v}' = \mathbf{a}$ are orthogonal.

Thinking Forward

Unit Tangent Vectors in $\mathbb{R}^2$

Since $\mathbf{r}'(t) = \langle 2t, 3t^2 \rangle$, we see that $\mathbf{r}'(1) = \langle 2, 3 \rangle$. Now, $\|\mathbf{r}'(1)\| = \sqrt{2^2 + 3^2} = \sqrt{13}$, so that a unit vector tangent to the curve at $\mathbf{r}(1)$ is $\left\langle \frac{2}{\sqrt{13}}, \frac{3}{\sqrt{13}} \right\rangle$. Of course, $\left\langle -\frac{2}{\sqrt{13}}, -\frac{3}{\sqrt{13}} \right\rangle$ is the other unit vector tangent to the curve at $\mathbf{r}(1)$, pointing in the opposite direction. In $\mathbb{R}^2$, vectors orthogonal to these tangent vectors lie on a single line, the line orthogonal to this direction through $\mathbf{r}(1)$. There are thus two unit vectors orthogonal to the tangent vector, namely the unit vectors pointing in opposite directions on this line. They are $\left\langle -\frac{2}{\sqrt{13}}, \frac{3}{\sqrt{13}} \right\rangle$ and $\left\langle \frac{2}{\sqrt{13}}, -\frac{3}{\sqrt{13}} \right\rangle$.

Unit Tangent Vectors in $\mathbb{R}^3$

Since $\mathbf{r}'(t) = \langle 1, 2t, 3t^2 \rangle$, we see that $\mathbf{r}'(1) = \langle 1, 2, 3 \rangle$. Now, $\|\mathbf{r}'(1)\| = \sqrt{1^2 + 2^2 + 3^2} = \sqrt{14}$, so that a unit vector tangent to the curve at $\mathbf{r}(1)$ is $\left\langle \frac{1}{\sqrt{14}}, \frac{2}{\sqrt{14}}, \frac{3}{\sqrt{14}} \right\rangle$. Of course, $\left\langle -\frac{1}{\sqrt{14}}, -\frac{2}{\sqrt{14}}, -\frac{3}{\sqrt{14}} \right\rangle$ is the other unit vector tangent to the curve at $\mathbf{r}(1)$, pointing in the opposite direction. In $\mathbb{R}^3$, the vectors orthogonal to this vector form a plane whose normal vector is the tangent vector to the curve at $\mathbf{r}(1)$. Any unit vector in that plane whose initial point is $\mathbf{r}(1)$ will be orthogonal to the tangent vector, so there are an infinite number of them.

Orthogonal Vectors

With $\mathbf{r}(t) = \langle a\sin bt, a\cos bt, ct \rangle$, we get

$$\begin{aligned}
\mathbf{v}(t) &= \mathbf{r}'(t) = \langle ab\cos bt, -ab\sin bt, c \rangle \\
\mathbf{a}(t) &= \mathbf{v}'(t) = \left\langle -ab^2\sin bt, -ab^2\cos bt, 0 \right\rangle.
\end{aligned}$$

Then

$$\mathbf{v}(t) \cdot \mathbf{a}(t) = \langle ab \cos bt, -ab \sin bt, c \rangle \cdot \left\langle -ab^2 \sin bt, -ab^2 \cos bt, 0 \right\rangle$$
$$= -a^2 b^2 \sin bt \cos bt + a^2 b^3 \sin bt \cos bt + c \cdot 0 = 0.$$

Horizontal Vectors

With $\mathbf{r}(t) = \langle a \sin bt, a \cos bt, ct \rangle$, we get

$$\mathbf{v}(t) = \mathbf{r}'(t) = \langle ab \cos bt, -ab \sin bt, c \rangle$$
$$\mathbf{a}(t) = \mathbf{v}'(t) = \left\langle -ab^2 \sin bt, -ab^2 \cos bt, 0 \right\rangle.$$

Since the z component of $\mathbf{a}(t)$ is zero, the vector is parallel to the xy plane, since in fact the position vector $\langle -ab^2 \sin bt, -ab^2 \cos bt, 0 \rangle$ lies in the xy plane.

11.3   Unit Tangent and Unit Normal Vectors

Thinking Back

Unit Vectors

Recall that $\|k\mathbf{v}\| = k \|\mathbf{v}\|$ is k if a scalar. Since $\frac{1}{\|\mathbf{v}\|}$ is a scalar, we have

$$\left\| \frac{\mathbf{v}}{\|\mathbf{v}\|} \right\| = \left\| \frac{1}{\|\mathbf{v}\|} \mathbf{v} \right\| = \frac{1}{\|\mathbf{v}\|} \|\mathbf{v}\| = 1,$$

so that $\frac{\mathbf{v}}{\|\mathbf{v}\|}$ is a unit vector.

Equation of a plane

A normal vector to the plane is the cross product of the given vectors:

$$\langle 1, -3, 5 \rangle \times \langle 2, 4, -1 \rangle = \langle -3 \cdot (-1) - 5 \cdot 4, \ 5 \cdot 2 - 1 \cdot (-1), \ 1 \cdot 4 - (-3) \cdot 2 \rangle = \langle -17, 11, 10 \rangle.$$

Thus the equation of the plane is $-17x + 11y + 10z = d$. Since $(0, 3, -2)$ lies in the plane, we get $-17(0) + 11(3) + 10(-2) = 13 = d$, so the equation of the plane is $-17x + 11y + 10z = 13$.

Concepts

1. (a) False. For example, let $\mathbf{r}(t) = \langle t, t^2 \rangle$. Then $\mathbf{r}'(t) = \langle 1, 2t \rangle$ and $\mathbf{r}''(t) = \langle 0, 2 \rangle$, so that $\mathbf{r}'(t) \cdot \mathbf{r}''(t) = 4t$.

    (b) True. Since $\mathbf{N}(t) = \frac{\mathbf{T}'(t)}{\|\mathbf{T}'(t)\|}$, $\mathbf{N}(t)$ is undefined if $\|\mathbf{T}'(t)\| = 0$. See Definition 11.15.

    (c) False. Since $\|\mathbf{u} \times \mathbf{v}\| = \|\mathbf{u}\| \|\mathbf{v}\| \sin \theta$, where θ is the angle between $\mathbf{u}$ and $\mathbf{v}$, if both $\mathbf{u}$ and $\mathbf{v}$ are unit vectors this gives $\|\mathbf{u} \times \mathbf{v}\| = \sin \theta$, so that depending on the angle between the vectors, the cross product may be any length from 0 to 1.

    (d) True. For a curve in the plane, we have $\mathbf{r}(t) = \langle x(t), y(t), 0 \rangle$, so that $\mathbf{T}(t) = \frac{\langle x'(t), y'(t), 0 \rangle}{\|\mathbf{r}'(t)\|}$, which also lies in the xy plane. Write $\mathbf{T}(t) = \langle t_1(t), t_2(t), 0 \rangle$. Then $\mathbf{N}(t) = \frac{\langle t_1'(t), t_2'(t), 0 \rangle}{\|\mathbf{T}'(t)\|}$, so this lies in the xy plane as well. Then $\mathbf{B} = \mathbf{T} \times \mathbf{N}$ is parallel to the z axis. But it is the normal vector to the osculating plane, so that the osculating plane at any point t is parallel to the xy plane. Since it contains at least one point of the curve ($\mathbf{r}(t)$), it is in fact equal to the xy plane.

(e) True, since $\mathbf{B}(t_0) = \mathbf{T}(t_0) \times \mathbf{N}(t_0)$. Note that since the unit normal vector exists, this implies that $\mathbf{T}'(t_0) \neq \mathbf{0}$, so that $\mathbf{B}(t_0)$ exists (see Definition 11.16).

(f) True. Since $\mathbf{T}$, $\mathbf{N}$, $\mathbf{B}$ form a right-handed system in that order, it follows that $\mathbf{N}(t_0) \times \mathbf{B}(t_0) = \mathbf{T}(t_0)$. Since the cross product is anticommutative, this means that $\mathbf{B}(t_0) \times \mathbf{N}(t_0) = -\mathbf{T}(t_0)$.

(g) True. The absolute value of this triple scalar product is the volume of the parallelepiped determined by $\mathbf{T}(t)$, $\mathbf{N}(t)$, and $\mathbf{B}(t)$. Since $\mathbf{T}$, $\mathbf{N}$, $\mathbf{B}$ form a right-handed system in that order, it follows that the triple scalar product is positive, so that in fact the triple scalar product is the volume of the parallelepiped. But the parallelepiped is in fact a cube, since $\mathbf{T}$, $\mathbf{N}$, and $\mathbf{B}$ are mutually perpendicular; since each of these is of unit length, the cube has side length 1, so that its volume is 1.

(h) True. The osculating plane may be different at different points. For instance, Example 2 shows that the helix $\mathbf{r}(t) = \langle \cos t, \sin t, t \rangle$ has osculating planes

$$x + z = \frac{\pi}{2} \quad \text{at } \mathbf{r}\left(\frac{\pi}{2}\right) = \left(0, 1, \frac{\pi}{2}\right)$$

$$-x + y + \sqrt{2}z = \frac{5\pi\sqrt{2}}{4} \quad \text{at } \mathbf{r}\left(\frac{5\pi}{4}\right) = \left(-\frac{\sqrt{2}}{2}, -\frac{\sqrt{2}}{2}, \frac{5\pi}{4}\right).$$

3. As you are ascending, the unit tangent vector points upwards and forward, and the principal unit normal vector points either up and backwards or down and forward depending on whether the slope of the road is increasing or decreasing (it always points in the direction of the bend in the curve). If the slope is not changing, the unit normal does not exist. The binormal, which exists if unit normal vector exists, points to the right if the slope is increasing and to the left if it is decreasing, since $\mathbf{TNB}$ is a right-handed system.

As you are descending, the unit tangent vector points downwards and forward, and the principal unit normal vector points either down and backwards or up and forward depending on whether the slope of the road is increasing negatively or decreasing negatively (it always points in the direction of the bend in the curve). If the slope is not changing, the unit normal does not exist. The binormal, which exists if unit normal vector exists, points to the left if the slope is increasing negatively and to the right if it is decreasing negatively, since $\mathbf{TNB}$ is a right-handed system.

As you are turning right, the unit tangent vector points ahead and to the right, and the unit normal points backwards and to the right. Since $\mathbf{TNB}$ is a right-handed system, the binormal vector points downwards.

Finally, as you are turning left, the unit tangent vector points ahead and to the left, and the unit normal points backwards and to the left. Since $\mathbf{TNB}$ is a right-handed system, the binormal vector points upwards.

5. By Definition 11.15, $\mathbf{r}(t)$ does not have a unit tangent vector at t_0 if $\mathbf{r}'(t_0) = \mathbf{0}$.

7. Assuming $\mathbf{r}'(t_0) \neq \mathbf{0}$, by Definition 11.14, $\mathbf{T}(t_0) = \frac{\mathbf{r}'(t_0)}{\|\mathbf{r}'(t_0)\|}$. That is, $\mathbf{T}(t)$ is the unit vector in the same direction as $\mathbf{r}'(t_0)$. If $\mathbf{r}'(t_0) = \mathbf{0}$, then $\mathbf{T}(t_0)$ does not exist.

9. Assuming $\mathbf{T}(t)$ exists and $\mathbf{T}'(t) \neq \mathbf{0}$, then $\mathbf{N}(t) = \frac{\mathbf{T}'(t)}{\|\mathbf{T}'(t)\|}$. That is, it is a unit vector pointing in the same direction as $\mathbf{T}'(t)$. If $\mathbf{T}(t)$ does not exist or has zero length, then $\mathbf{N}(t)$ does not exist.

11. Assuming that $\mathbf{T}(t_0)$ and $\mathbf{N}(t_0)$ exist, if either is zero, then they are certainly orthogonal. If neither is zero, then near t_0, $\mathbf{T}(t)$ is a unit vector, so it has constant length, so that by Theorem 11.12, $\mathbf{T}(t_0) \cdot \mathbf{T}'(t_0) = 0$. Since $\mathbf{N}(t_0)$ is a scalar multiple of $\mathbf{T}'(t_0)$, it follows that $\mathbf{T}(t_0) \cdot \mathbf{N}(t_0) = 0$. Thus $\mathbf{T}$ and $\mathbf{N}$ are orthogonal at t_0. Similarly, if either $\mathbf{N}(t_0)$ or $\frac{d\mathbf{N}}{dt}\big|_{t_0} = \mathbf{N}'(t_0)$ is zero, then they are orthogonal; otherwise, again by Theorem 11.12, since $\mathbf{N}(t_0)$ is of constant length, we get $\mathbf{N}(t_0) \cdot \mathbf{N}'(t_0) = 0$, so that these vectors are orthogonal as well. There is no particular relationship between $\mathbf{T}(t_0)$ and $\mathbf{N}'(t_0)$.

13. Assuming that $\mathbf{T}(t)$ and $\mathbf{N}(t)$ both exist at a point t, the binormal vector at t is $\mathbf{B}(t) = \mathbf{T}(t) \times \mathbf{N}(t)$. See Definition 11.16.

15. By Definition 11.16, $\mathbf{r}(t)$ does not have a principal binormal vector at t_0 if either $\mathbf{T}'(t_0) = \mathbf{0}$ or if $\mathbf{T}'(t_0)$ does not exist.

17. First, assuming $\mathbf{r}'(t_0) \neq \mathbf{0}$, compute

$$\mathbf{T}(t_0) = \frac{\mathbf{r}'(t_0)}{\|\mathbf{r}'(t_0)\|}.$$

(If $\mathbf{r}'(t_0) = \mathbf{0}$, then the osculating plane does not exist). Next, assuming $\mathbf{T}'(t_0) \neq \mathbf{0}$, compute

$$\mathbf{N}(t_0) = \frac{\mathbf{T}'(t_0)}{\|\mathbf{T}'(t_0)\|}.$$

(If $\mathbf{T}'(t_0) = \mathbf{0}$, then the osculating plane does not exist). Then

$$\mathbf{B}(t_0) = \mathbf{T}(t_0) \times \mathbf{N}(t_0)$$

is the binormal vector at t_0. This is a normal vector to the osculating plane, which must also pass through $\mathbf{r}(t_0) = \langle x(t_0), y(t_0), z(t_0) \rangle$. Thus the equation of the osculating plane is

$$\mathbf{B}(t_0) \cdot \langle x - x(t_0),\ y - y(t_0),\ z - z(t_0) \rangle = 0.$$

(See the discussion at the beginning of Section 10.6 if you do not recall why this is true).

19. Since the vectors are perpendicular, any two of them plus the point $\mathbf{r}(t_0)$ determine a unique plane. Thus there are three planes, one containing $\mathbf{N}(t_0)$ and $\mathbf{B}(t_0)$, one containing $\mathbf{N}(t_0)$ and $\mathbf{T}(t_0)$, and one containing $\mathbf{T}(t_0)$ and $B(t_0)$ as well as $\mathbf{r}(t_0)$.

21. Using the results of the preceding exercise, we have $\mathbf{T}'(t) = \frac{1}{\sqrt{2}} \langle -\cos t, -\sin t, 0 \rangle$, so that

$$\|\mathbf{T}'(t)\| = \frac{1}{\sqrt{2}} \sqrt{\cos^2 t + \sin^2 t + 0} = \frac{1}{\sqrt{2}}.$$

Thus

$$\mathbf{N}(t) = \frac{\mathbf{T}'(t)}{\|\mathbf{T}'(t)\|} = \langle -\cos t, -\sin t, 0 \rangle.$$

But the plane containing $\mathbf{T}$ and $\mathbf{B}$ has $\mathbf{B} \times \mathbf{T} = \mathbf{N}$ as a normal vector, and $\mathbf{N}(\pi) = \langle -1, 0, 0 \rangle$. Thus the equation of the rectifying plane is $-x = d$ for some d. Since the plane contains the point $\mathbf{r}(\pi) = \langle \cos \pi, \sin \pi, \pi \rangle = \langle -1, 0, \pi \rangle$, substituting gives $-1(-1) = d$ so that $d = 1$. Thus the equation of the rectifying plane is $-x = 1$ or $x = -1$.

Skills

23. Since $\mathbf{r}'(t) = \langle 2t, 5, 12t^2 \rangle$, we see that $\|\mathbf{r}'(t)\| = \sqrt{4t^2 + 25 + 144t^4}$, so that

$$\mathbf{T}(t) = \frac{\mathbf{r}'(t)}{\|\mathbf{r}'(t)\|} = \left\langle \frac{2t}{\sqrt{144t^4 + 4t^2 + 25}},\ \frac{5}{\sqrt{144t^4 + 4t^2 + 25}},\ \frac{12t^2}{\sqrt{144t^4 + 4t^2 + 25}} \right\rangle.$$

25. Since $\mathbf{r}'(t) = \langle -3\cos^2 t \sin t, 3\sin^2 t \cos t \rangle$, we see that

$$\|\mathbf{r}'(t)\| = \sqrt{9\cos^4 t \sin^2 t + 9\sin^4 t \cos^2 t} = 3\sqrt{(\cos^2 t + \sin^2 t) \sin^2 t \cos^2 t} = 3\,|\sin t \cos t|.$$

Then

$$\mathbf{T}(t) = \frac{\mathbf{r}'(t)}{\|\mathbf{r}'(t)\|} = \left\langle -\frac{\cos^2 t \sin t}{|\sin t \cos t|},\ \frac{\sin^2 t \cos t}{|\sin t \cos t|} \right\rangle.$$

27. Since $\mathbf{r}'(t) = \langle 3\cos t, -5\sin t, 4\cos t \rangle$, we see that

$$\|\mathbf{r}'(t)\| = \sqrt{9\cos^2 t + 25\sin^2 t + 16\cos^2 t} = \sqrt{25(\sin^2 t + \cos^2 t)} = 5,$$

so that

$$\mathbf{T}(t) = \frac{\mathbf{r}'(t)}{\|\mathbf{r}'(t)\|} = \left\langle \frac{3}{5}\cos t, -\sin t, \frac{4}{5}\cos t \right\rangle.$$

29. Since $\mathbf{r}'(t) = \langle 1, 2t \rangle$, we see that $\|\mathbf{r}'(t)\| = \sqrt{1 + 4t^2}$, so that

$$\mathbf{T}(t) = \frac{\mathbf{r}'(t)}{\|\mathbf{r}'(t)\|} = \left\langle \frac{1}{\sqrt{1 + 4t^2}}, \frac{2t}{\sqrt{1 + 4t^2}} \right\rangle.$$

Then

$$\mathbf{T}'(t) = \left\langle -\frac{1}{2}(1 + 4t^2)^{-3/2} \cdot 8t, \; \frac{\sqrt{1 + 4t^2} \cdot 2 - 2t \cdot \frac{1}{2}(1 + 4t^2)^{-1/2} \cdot 8t}{1 + 4t^2} \right\rangle$$

$$= \left\langle -4t(1 + 4t^2)^{-3/2}, \; \frac{2(1 + 4t^2)^{1/2} - 8t^2(1 + 4t^2)^{-1/2}}{1 + 4t^2} \right\rangle$$

$$= \left\langle -4t(1 + 4t^2)^{-3/2}, \; \frac{2 + 8t^2 - 8t^2}{(1 + 4t^2)^{3/2}} \right\rangle$$

$$= \left\langle -\frac{4t}{(1 + 4t^2)^{3/2}}, \; \frac{2}{(1 + 4t^2)^{3/2}} \right\rangle.$$

Then

$$\mathbf{T}'(1) = \left\langle -\frac{4}{5\sqrt{5}}, \frac{2}{5\sqrt{5}} \right\rangle$$

$$\|\mathbf{T}'(1)\| = \sqrt{\frac{16}{125} + \frac{4}{125}} = \sqrt{\frac{4}{25}} = \frac{2}{5},$$

and thus

$$\mathbf{T}(1) = \left\langle \frac{1}{\sqrt{5}}, \frac{2}{\sqrt{5}} \right\rangle,$$

$$\mathbf{N}(1) = \frac{\mathbf{T}'(1)}{\|\mathbf{T}'(1)\|} = \frac{5}{2}\left\langle -\frac{4}{5\sqrt{5}}, \frac{2}{5\sqrt{5}} \right\rangle = \left\langle -\frac{2}{\sqrt{5}}, \frac{1}{\sqrt{5}} \right\rangle.$$

31. Since $\mathbf{r}'(t) = \langle -\alpha\sin\alpha t, \alpha\cos\alpha t \rangle$, we get (assuming $\alpha > 0$) $\|\mathbf{r}'(t)\| = \sqrt{\alpha^2\sin^2\alpha t + \alpha^2\cos^2\alpha t} = \alpha$, so that

$$\mathbf{T}(t) = \frac{\mathbf{r}'(t)}{\|\mathbf{r}'(t)\|} = \langle -\sin\alpha t, \cos\alpha t \rangle$$

$$\mathbf{T}'(t) = \alpha\langle -\cos\alpha t, -\sin\alpha t \rangle.$$

Then

$$\mathbf{T}'(\pi) = \alpha\langle -\cos\alpha\pi, -\sin\alpha\pi \rangle$$

$$\|\mathbf{T}'(\pi)\| = \alpha\sqrt{\cos^2\alpha\pi + \sin^2\alpha\pi} = \alpha.$$

Thus

$$\mathbf{T}(\pi) = \langle -\sin\alpha\pi, \cos\alpha\pi \rangle,$$

$$\mathbf{N}(\pi) = \frac{\mathbf{T}'(\pi)}{\|\mathbf{T}'(\pi)\|} = \langle -\cos\alpha\pi, -\sin\alpha\pi \rangle.$$

33. Since $\mathbf{r}'(t) = \langle 3\cos t, -5\sin t, 4\cos t\rangle$, we get

$$\|\mathbf{r}'(t)\| = \sqrt{9\cos^2 t + 25\sin^2 t + 16\cos^2 t} = \sqrt{25(\cos^2 t + \sin^2 t)} = 5.$$

Then

$$\mathbf{T}(t) = \frac{\mathbf{r}'(t)}{\|\mathbf{r}'(t)\|} = \left\langle \frac{3}{5}\cos t, -\sin t, \frac{4}{5}\cos t\right\rangle$$

$$\mathbf{T}'(t) = \left\langle -\frac{3}{5}\sin t, -\cos t, -\frac{4}{5}\sin t\right\rangle.$$

This gives

$$\mathbf{T}'(\pi) = \left\langle -\frac{3}{5}\sin\pi, -\cos\pi, -\frac{4}{5}\sin\pi\right\rangle = \langle 0, 1, 0\rangle$$

$$\|\mathbf{T}'(\pi)\| = \sqrt{0 + 1 + 0} = 1.$$

Thus

$$\mathbf{T}(\pi) = \left\langle \frac{3}{5}\cos\pi, -\sin\pi, \frac{4}{5}\cos\pi\right\rangle = \left\langle -\frac{3}{5}, 0, -\frac{4}{5}\right\rangle$$

$$\mathbf{N}(\pi) = \frac{\mathbf{T}'(\pi)}{\|\mathbf{T}'(\pi)\|} = \langle 0, 1, 0\rangle.$$

35. Since $\mathbf{r}'(t) = \langle 1, 2t, 2t^2\rangle$, we get $\|\mathbf{r}'(t)\| = \sqrt{1 + 4t^2 + 4t^4} = 2t^2 + 1$. Then

$$\mathbf{T}(t) = \frac{\mathbf{r}'(t)}{\|\mathbf{r}'(t)\|} = \left\langle \frac{1}{2t^2 + 1}, \frac{2t}{2t^2 + 1}, \frac{2t^2}{2t^2 + 1}\right\rangle$$

$$\mathbf{T}'(t) = \left\langle -\frac{1}{(2t^2 + 1)^2} \cdot 4t, \frac{(2t^2 + 1)(2) - 2t(4t)}{(2t^2 + 1)^2}, \frac{(2t^2 + 1)(4t) - (2t^2)(4t)}{(2t^2 + 1)^2}\right\rangle$$

$$= \left\langle -\frac{4t}{(2t^2 + 1)^2}, \frac{2 - 4t^2}{(2t^2 + 1)^2}, \frac{4t}{(2t^2 + 1)^2}\right\rangle.$$

This gives

$$\mathbf{T}'(1) = \left\langle -\frac{4}{9}, -\frac{2}{9}, \frac{4}{9}\right\rangle, \qquad \|\mathbf{T}'(1)\| = \frac{1}{9}\sqrt{4^2 + 2^2 + 4^2} = \frac{2}{3}.$$

Thus

$$\mathbf{T}(1) = \left\langle \frac{1}{3}, \frac{2}{3}, \frac{2}{3}\right\rangle, \qquad \mathbf{N}(1) = \frac{\mathbf{T}'(1)}{\|\mathbf{T}'(1)\|} = \left\langle -\frac{2}{3}, -\frac{1}{3}, \frac{2}{3}\right\rangle.$$

Hence

$$\mathbf{B}(1) = \mathbf{T}(1) \times \mathbf{N}(1) = \left\langle \frac{1}{3}, \frac{2}{3}, \frac{2}{3}\right\rangle \times \left\langle -\frac{2}{3}, -\frac{1}{3}, \frac{2}{3}\right\rangle$$

$$= \frac{1}{9}\left(\langle 1, 2, 2\rangle \times \langle -2, -1, 2\rangle\right)$$

$$= \frac{1}{9}\langle 2\cdot 2 - 2\cdot(-1), 2\cdot(-2) - 1\cdot 2, 1\cdot(-1) - 2\cdot(-2)\rangle$$

$$= \frac{1}{9}\langle 6, -6, 3\rangle = \left\langle \frac{2}{3}, -\frac{2}{3}, \frac{1}{3}\right\rangle,$$

so that the equation of the osculating plane is

$$\mathbf{B}(1) \cdot \left\langle x - 1, y - 1, z - \frac{2}{3} \right\rangle = 0, \text{ or}$$

$$\frac{2}{3}(x - 1) - \frac{2}{3}(y - 1) + \frac{1}{3}\left(z - \frac{2}{3}\right) = 0.$$

Multiplying through by 3 gives the simpler answer $2(x - 1) - 2(y - 1) + (z - \frac{2}{3}) = 0$, or $2x - 2y + z = \frac{2}{3}$.

37. Since $\mathbf{r}'(t) = \left\langle e^t, -e^{-t}, \sqrt{2} \right\rangle$, we get $\|\mathbf{r}'(t)\| = \sqrt{e^{2t} + 2 + e^{-2t}} = \sqrt{(e^t + e^{-t})^2} = e^t + e^{-t}$. Then

$$\mathbf{T}(t) = \frac{\mathbf{r}'(t)}{\|\mathbf{r}'(t)\|} = \left\langle \frac{e^t}{e^t + e^{-t}}, -\frac{e^{-t}}{e^t + e^{-t}}, \frac{\sqrt{2}}{e^t + e^{-t}} \right\rangle$$

$$\mathbf{T}'(t) = \left\langle \frac{(e^t + e^{-t})e^t - e^t(e^t - e^{-t})}{(e^t + e^{-t})^2}, -\frac{(e^t + e^{-t})(-e^{-t}) - e^{-t}(e^t - e^{-t})}{(e^t + e^{-t})^2}, -\frac{\sqrt{2}(e^t - e^{-t})}{(e^t + e^{-t})^2} \right\rangle$$

$$= \left\langle \frac{2}{(e^t + e^{-t})^2}, \frac{2}{(e^t + e^{-t})^2}, \frac{\sqrt{2}(e^{-t} - e^t)}{(e^t + e^{-t})^2} \right\rangle.$$

This gives

$$\mathbf{T}'(0) = \left\langle \frac{1}{2}, \frac{1}{2}, 0 \right\rangle, \qquad \|\mathbf{T}'(0)\| = \sqrt{\frac{1}{2} + \frac{1}{2} + 0} = \frac{1}{\sqrt{2}}.$$

Thus

$$\mathbf{T}(0) = \left\langle \frac{1}{2}, -\frac{1}{2}, \frac{1}{\sqrt{2}} \right\rangle$$

$$\mathbf{N}(0) = \frac{\mathbf{T}'(0)}{\|\mathbf{T}'(0)\|} = \left\langle \frac{1}{\sqrt{2}}, \frac{1}{\sqrt{2}}, 0 \right\rangle.$$

Hence

$$\mathbf{B}(0) = \left\langle \frac{1}{2}, -\frac{1}{2}, \frac{1}{\sqrt{2}} \right\rangle \times \left\langle \frac{1}{\sqrt{2}}, \frac{1}{\sqrt{2}}, 0 \right\rangle = \left\langle -\frac{1}{2}, \frac{1}{2}, \frac{1}{\sqrt{2}} \right\rangle,$$

so that the equation of the osculating plane is

$$\mathbf{B}(0) \cdot \langle x - 1, y - 1, z \rangle = 0, \text{ or}$$

$$-\frac{1}{2}(x - 1) + \frac{1}{2}(y - 1) + \frac{1}{\sqrt{2}}z = 0.$$

Multiplying through by 2 gives the simpler answer $-(x - 1) + (y - 1) + \sqrt{2}z = 0$, or $-x + y + \sqrt{2}z = 0$.

39. Since $\mathbf{r}'(t) = \langle 2\cos 2t, -2\sin 2t, 1 \rangle$, we get $\|\mathbf{r}'(t)\| = \sqrt{4\cos^2 2t + 4\sin^2 2t + 1} = \sqrt{5}$, so that

$$\mathbf{T}(t) = \frac{\mathbf{r}'(t)}{\|\mathbf{r}'(t)\|} = \left\langle \frac{2}{\sqrt{5}}\cos 2t, -\frac{2}{\sqrt{5}}\sin 2t, \frac{1}{\sqrt{5}} \right\rangle$$

$$\mathbf{T}'(t) = \left\langle -\frac{4}{\sqrt{5}}\sin 2t, -\frac{4}{\sqrt{5}}\cos 2t, 0 \right\rangle.$$

This gives

$$\mathbf{T}'\left(\frac{\pi}{2}\right) = \left\langle -\frac{4}{\sqrt{5}}\sin\pi,\ -\frac{4}{\sqrt{5}}\cos\pi,\ 0 \right\rangle = \left\langle 0,\ \frac{4}{\sqrt{5}},\ 0 \right\rangle$$

$$\left\|\mathbf{T}'\left(\frac{\pi}{2}\right)\right\| = \frac{4}{\sqrt{5}}.$$

Thus

$$\mathbf{T}\left(\frac{\pi}{2}\right) = \left\langle \frac{2}{\sqrt{5}}\cos\pi,\ -\frac{2}{\sqrt{5}}\sin\pi,\ \frac{1}{\sqrt{5}} \right\rangle = \left\langle -\frac{2}{\sqrt{5}},\ 0,\ \frac{1}{\sqrt{5}} \right\rangle$$

$$\mathbf{N}\left(\frac{\pi}{2}\right) = \frac{\mathbf{T}'\left(\frac{\pi}{2}\right)}{\left\|\mathbf{T}'\left(\frac{\pi}{2}\right)\right\|} = \langle 0,1,0 \rangle.$$

Then

$$\mathbf{B}\left(\frac{\pi}{2}\right) = \left\langle -\frac{2}{\sqrt{5}},\ 0,\ \frac{1}{\sqrt{5}} \right\rangle \times \langle 0,1,0 \rangle = \left\langle -\frac{1}{\sqrt{5}},\ 0,\ -\frac{2}{\sqrt{5}} \right\rangle,$$

so that the equation of the osculating plane is

$$\mathbf{B}\left(\frac{\pi}{2}\right) \cdot \left\langle x, y+1, z-\frac{\pi}{2} \right\rangle,\ \text{or}$$

$$-\frac{1}{\sqrt{5}}x - \frac{2}{\sqrt{5}}\left(z - \frac{\pi}{2}\right) = 0,\ \text{or}$$

$$x + 2z = \pi.$$

41. From Exercise 37, we have

$$\mathbf{T}(0) = \left\langle \frac{1}{2},\ -\frac{1}{2},\ \frac{1}{\sqrt{2}} \right\rangle = \frac{1}{2}\left\langle 1, -1, \sqrt{2} \right\rangle$$

$$\mathbf{N}(0) = \left\langle \frac{1}{\sqrt{2}},\ \frac{1}{\sqrt{2}},\ 0 \right\rangle = \frac{1}{\sqrt{2}}\langle 1,1,0 \rangle$$

$$\mathbf{B}(0) = \left\langle -\frac{1}{2},\ \frac{1}{2},\ \frac{\sqrt{2}}{2} \right\rangle = \frac{1}{2}\left\langle -1, 1, \sqrt{2} \right\rangle.$$

The normal plane has normal vector $\mathbf{N}(0) \times \mathbf{B}(0) = \mathbf{T}(0)$, so its equation is, ignoring the scalar factor of $\frac{1}{\sqrt{2}}$,

$$\left\langle 1, -1, \sqrt{2} \right\rangle \cdot \langle x-1, y-1, z \rangle = (x-1) - (y-1) + \sqrt{2}z = 0.$$

Simplifying gives $x - y + \sqrt{2}z = 0$.

The rectifying plane has normal vector $\mathbf{B}(0) \times \mathbf{T}(0) = N(0)$, so its equation is, again ignoring the factor of $\frac{1}{\sqrt{2}}$,

$$\langle 1,1,0 \rangle \cdot \langle x-1, y-1, z \rangle = (x-1) + (y-1) = 0.$$

Simplifying gives $x + y = 2$.

Applications

43. (a) Their velocity vector is $\mathbf{r}'(t) = \left\langle 1.5t^{0.5}, 1.7 - 2t \right\rangle$, so that their speed when not paddling is $\|\mathbf{r}'(t)\| = \sqrt{(1.5t^{0.5})^2 + (1.7 - 2t)^2}$. To maximize $\|r'(t)\|$ all we need to do is to maximize its square, so we want to maximize $(1.5t^{0.5})^2 + (1.7 - 2t)^2 = 4t^2 - 4.55t + 2.89$. Its second derivative is 8, so that the curve is concave up and thus the maximum speed must be either at $t = 0$ or at $t = 1.7$. Substituting into $\|\mathbf{r}'(t)\|$ gives a speed of 2.89 at $t = 0$ and of ≈ 6.72 at $t = 1.7$, so they would move fastest at $t = 1.7$ hours.

(b) The direction they are paddling is a vector in the direction of the principal normal vector. To compute the principal normal vector, start with

$$\mathbf{T}(t) = \frac{\mathbf{r}'(t)}{\|\mathbf{r}'(t)\|}.$$

Now, $\mathbf{r}'(t) = \langle 1.5t^{0.5}, 1.7 - 2t \rangle$, so that its norm is

$$\sqrt{2.25t + 2.89 - 6.8t + 4t^2} = \sqrt{4t^2 - 4.55t + 2.89},$$

and thus

$$\mathbf{T}(t) = (4t^2 - 4.55t + 2.89)^{-1/2} \langle 1.5t^{0.5}, 1.7 - 2t \rangle.$$

At $t = 0.1$, this is $\mathbf{T}(0.1) \approx \langle 0.30, 0.95 \rangle$. The principal normal vector is a vector perpendicular to this one, so is either $\langle 0.95, -0.30 \rangle$ or $\langle -0.95, 0.30 \rangle$. A graph of the position over time is

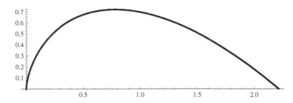

Since the graph opens downward, the inward-pointing normal is $\langle 0.95, -0.30 \rangle$.

Proofs

45. If $\mathbf{u}$ and $\mathbf{v}$ are two vectors, then $\|\mathbf{u} \times \mathbf{v}\| = \|\mathbf{u}\| \|\mathbf{v}\| \sin \theta$, where θ is the angle between the vectors (Theorem 10.35(a)). In this case, $\mathbf{u}$ and $\mathbf{v}$ are unit vectors, so that $\|\mathbf{u}\| = \|\mathbf{v}\| = 1$. Also, they are orthogonal, so that $\theta = \frac{\pi}{2}$ and thus $\sin \theta = 1$. The equation reduces to $\|\mathbf{u} \times \mathbf{v}\| = 1$, which means that $\mathbf{u} \times \mathbf{v}$ is a unit vector.

47. (a) Since both $\mathbf{T}(t_0)$ and $\mathbf{N}(t_0)$ are unit vectors by definition, and since they are orthogonal, it follows that (by Theorem 10.35(a))

$$\|\mathbf{B}(t_0)\| = \|\mathbf{T}(t_0) \times \mathbf{N}(t_0)\| = \|\mathbf{T}(t_0)\| \|\mathbf{N}(t_0)\| \sin \theta = \sin \frac{\pi}{2} = 1,$$

so that $\mathbf{B}(t_0)$ is a unit vector.

(b) Since the cross product of two vectors is orthogonal to both of the vectors, it follows that $\mathbf{B}(t_0)$ is orthogonal to both $\mathbf{T}(t_0)$ and $\mathbf{N}(t_0)$.

Thus $\mathbf{T}(t_0)$, $\mathbf{N}(t_0)$ and $\mathbf{B}(t_0)$ form an orthogonal set of three unit vectors, so they define a coordinate system. By Theorem 10.36, $\mathbf{u}$, $\mathbf{v}$, and $\mathbf{w}$ form a right-handed triple if and only if $\mathbf{u} \cdot (\mathbf{v} \times \mathbf{w}) > 0$. In this case, this means that $\mathbf{T}(t_0) \cdot (\mathbf{N}(t_0) \times \mathbf{B}(t_0)) > 0$. But in general, $\mathbf{u} \cdot (\mathbf{v} \times \mathbf{w}) = (\mathbf{u} \times \mathbf{v}) \cdot \mathbf{w}$ by Corollary 10.37(b), so the three vectors form a right-handed triple if and only if $(\mathbf{T}(t_0) \times \mathbf{N}(t_0)) \cdot \mathbf{B}(t_0) > 0$. But

$$(\mathbf{T}(t_0) \times \mathbf{N}(t_0)) \cdot \mathbf{B}(t_0) = \mathbf{B}(t_0) \cdot \mathbf{B}(t_0) = \|\mathbf{B}(t_0)\|^2 = 1 > 0.$$

Thus $\mathbf{T}(t_0)$, $\mathbf{N}(t_0)$ and $\mathbf{B}(t_0)$ form a right-handed coordinate system.

Thinking Forward

Equation of a circle

Consider $\mathbf{r}(t) = \langle a, b \rangle + t \langle \alpha, \beta \rangle$. The point on $\mathbf{r}(t)$ that is ρ units from $\langle a, b \rangle$ corresponds to a value of t for which $\|\mathbf{r}(t) - \mathbf{r}(0)\| = \rho$. But

$$\|\mathbf{r}(t) - \mathbf{r}(0)\| = \|t \langle \alpha, \beta \rangle\| = t\sqrt{\alpha^2 + \beta^2},$$

so that $t = \frac{\rho}{\sqrt{\alpha^2 + \beta^2}}$. Thus the center of the circle is at

$$\left(a + \frac{\rho\alpha}{\sqrt{\alpha^2 + \beta^2}}, \; b + \frac{\rho\beta}{\sqrt{\alpha^2 + \beta^2}} \right),$$

and it obviously has radius ρ, so its equation is

$$\left(x - \left(a + \frac{\rho\alpha}{\sqrt{\alpha^2 + \beta^2}} \right) \right)^2 + \left(y - \left(b + \frac{\rho\beta}{\sqrt{\alpha^2 + \beta^2}} \right) \right)^2 = \rho^2.$$

Equation of a sphere

Consider $\mathbf{r}(t) = \langle a, b, c \rangle + t \langle \alpha, \beta, \gamma \rangle$. The point on $\mathbf{r}(t)$ that is ρ units from $\langle a, b, c \rangle$ corresponds to a value of t for which $\|\mathbf{r}(t) - \mathbf{r}(0)\| = \rho$. But

$$\|\mathbf{r}(t) - \mathbf{r}(0)\| = \|t \langle \alpha, \beta, \gamma \rangle\| = t\sqrt{\alpha^2 + \beta^2 + \gamma^2},$$

so that $t = \frac{\rho}{\sqrt{\alpha^2 + \beta^2 + \gamma^2}}$. Thus the center of the sphere is at

$$\left(a + \frac{\rho\alpha}{\sqrt{\alpha^2 + \beta^2 + \gamma^2}}, \; b + \frac{\rho\beta}{\sqrt{\alpha^2 + \beta^2 + \gamma^2}}, \; c + \frac{\rho\gamma}{\sqrt{\alpha^2 + \beta^2 + \gamma^2}} \right),$$

and it obviously has radius ρ, so its equation is

$$\left(x - \left(a + \frac{\rho\alpha}{\sqrt{\alpha^2 + \beta^2 + \gamma^2}} \right) \right)^2 + \left(y - \left(b + \frac{\rho\beta}{\sqrt{\alpha^2 + \beta^2 + \gamma^2}} \right) \right)^2$$

$$+ \left(z - \left(c + \frac{\rho\gamma}{\sqrt{\alpha^2 + \beta^2 + \gamma^2}} \right) \right)^2 = \rho^2.$$

11.4 Arc Length Parametrizations and Curvature

Thinking Back

The arc length of an exponential function

With $f(x) = e^{x/2} + e^{-x/2}$, we have $f'(x) = \frac{1}{2}\left(e^{x/2} - e^{-x/2} \right)$, so that the arc length from 0 to 1 is

$$\int_0^1 \sqrt{1 + f'(x)^2}\, dx = \int_0^1 \sqrt{1 + \frac{1}{4}\left(e^x - 2 + e^{-x} \right)}\, dx = \frac{1}{2}\int_0^1 \sqrt{e^x + 2 + e^{-x}}\, dx$$

$$= \frac{1}{2}\int_0^1 \sqrt{(e^{x/2} + e^{-x/2})^2}\, dx = \frac{1}{2}\int_0^1 (e^{x/2} + e^{-x/2})\, dx$$

$$= \left[e^{x/2} - e^{-x/2} \right]_0^1 = e^{1/2} - e^{-1/2}.$$

The arc length of the cycloid

With the parametric equations as given, the arc length is

$$\int_0^{2\pi} \sqrt{(x'(\theta))^2 + (y'(\theta))^2}\, d\theta = \int_0^{2\pi} \sqrt{(1 - \cos\theta)^2 + (\sin\theta)^2}\, d\theta$$

$$= \int_0^{2\pi} \sqrt{2 - 2\cos\theta}\, d\theta$$

$$= \sqrt{2} \int_0^{2\pi} \sqrt{1 - \cos\theta}\, d\theta$$

To integrate, note that by the double-angle formula, $\sin^2\frac{\theta}{2} = \frac{1}{2}(1 - \cos\theta)$, so that $1 - \cos\theta = 2\sin^2\frac{\theta}{2}$. As θ varies from 0 to 2π, we see that $\frac{\theta}{2}$ varies from 0 to π, and $\sin\frac{\theta}{2}$ is thus positive over the range of integration, so we can write $\sqrt{1 - \cos\theta} = \sqrt{2}\sin\frac{\theta}{2}$. Then the integral becomes

$$\sqrt{2} \int_0^{2\pi} \sqrt{2}\sin\frac{\theta}{2}\, d\theta = 2\left[-2\cos\frac{\theta}{2}\right]_0^{2\pi} = 8.$$

Concepts

1. (a) False. For example, let $\mathbf{r}(t) = \langle \frac{t}{2}, 0, 0 \rangle$; then $\|\mathbf{r}'(t)\| = \left\|\langle \frac{1}{2}, 0, 0 \rangle\right\| = \frac{1}{2}$, so that $\int_a^b \|\mathbf{r}'(t)\|\, dt = \frac{1}{2}(b - a) < b - a$ assuming $a > b$.

    (b) False. Since the arc length parametrization depends on $\mathbf{r}'(t)$, the function must at least be differentiable.

    (c) True. A straight line is parametrized as $\mathbf{r}(t) = \mathbf{r}_0 + t\mathbf{d}$, so that $\mathbf{r}'(t) = \mathbf{d}$ and then $\mathbf{T}(t) = \frac{\mathbf{d}}{\|\mathbf{d}\|}$, which is a constant. Thus $\frac{d\mathbf{T}}{dt} = \mathbf{0}$, so the curvature is zero by Definition 11.21.

    (d) True. This is Definition 11.25(a).

    (e) False. A straight line is parametrized as $\mathbf{r}(t) = \mathbf{r}_0 + t\mathbf{d}$, so that $\mathbf{r}'(t) = \mathbf{d}$ and then $\mathbf{T}(t) = \frac{\mathbf{d}}{\|\mathbf{d}\|}$, which is a constant. Thus $\frac{d\mathbf{T}}{dt} = \mathbf{0}$, so the curvature is zero by Definition 11.21. This means that the radius of curvature, which is the reciprocal of the curvature, is infinite.

    (f) False. By Definition 11.21, the curvature is a norm, so it cannot be negative. But it can be zero at a particular point (for example, if the curve is a straight line; see parts (c) or (e)).

    (g) False. The osculating circle depends for its definition on the fact that $\mathbf{N}(t_0)$ is a unit vector. But even for a twice-differentiable function, $\mathbf{N}(t_0)$ could be zero, or it could be undefined if $\mathbf{T}(t_0) = \mathbf{0}$.

    (h) True. It is (by Definition 11.25(b)) the circle of radius $\rho = \frac{1}{\kappa}$, the radius of curvature, with center $\mathbf{r}(t_0) + \rho\mathbf{N}(t_0)$.

3. A curve is parametrized by arc length if $\mathbf{r}(t)$ moves along the curve at the same speed as t moves along the real line. This means that the arc length of the curve between $\mathbf{r}(a)$ and $\mathbf{r}(b)$ is just $b - a$.

5. Since a space curve has three dimensions to work with, it could be turning left and right, or up and down, or some combination. So there are many different "axes" to measure concavity or convexity against.

7. The definition is easy to understand both because it is simple and because it corresponds to the intuition that if the unit tangent vector is changing rapidly, then the curve is turning rapidly. However, it is difficult to use because it relies on an arc length parametrization, which may be difficult or impossible to obtain in terms of elementary functions.

9. The formula in the definition is easy to use if one happens to have an arc length parametrization in hand. If not, however, in the case of a space curve, the formula in Theorem 11.23(a) can be used if $\mathbf{T}(t)$ is reasonably easy to differentiate and then find the norm of. However, there are certainly cases where $\mathbf{T}'(t)$ is substantially more complicated than $\mathbf{r}''(t)$, in which case the formula in 11.23(b) might be simpler. If the curve happens to be a space curve, then the formula in Theorem 11.24 (a) is easier than those in Theorem 11.23, as it involves only differentiation of a scalar function, so less computation is involved in computing this formula. Theorem 11.24 (b) is best applied for a planar curve where only a parametric set of equations is given and Theorem 11.24(a) cannot easily be applied (for example, if an explicit form for the function cannot easily be derived from the parametric equations). Theorems 11.24(b) and 11.23(b) are actually the same formula; if $\mathbf{r}(t)$ is a planar curve, say $\mathbf{r}(t) = \langle x(t), y(t), 0 \rangle$ and write out the explicit computation in Theorem 11.23(b), you get the formula in 11.24(b).

11. (a) With $\mathbf{r}(t) = \langle a + \alpha t, b + \beta t, c + \gamma t \rangle$, we get $\mathbf{r}'(t) = \langle \alpha, \beta, \gamma \rangle$ and $\mathbf{r}''(t) = \mathbf{0}$. So by Theorem 11.23(b), $\kappa = 0$ at every point on the curve. Since the radius of curvature is the reciprocal of the curvature, it is infinite everywhere on the curve.

(b) Since the curve is a straight line, it does not curve, so it makes sense that a concept named curvature should be zero on this graph. Since the radius of curvature is the radius of a circle that best fits the curvature of the graph at a given point, it is clear that a circle of larger radius always fits the straight line better than one of smaller radius, so in some sense a circle of "infinite" radius, which is a straight line, best fits the graph.

13. First compute the curvature κ, using either the definition or the appropriate part of Theorem 11.24. The specifics may vary depending on which of these is selected, but in general the procedure will involve taking one or more derivatives of $\mathbf{r}(t)$ and several norms. Once κ is known, the radius of curvature is $\rho = \frac{1}{\kappa}$, since we are assuming $\kappa \neq 0$. Then the osculating circle must contain $\mathbf{r}(t_0)$, where $\mathbf{r}(t_0)$ is the point at which we are computing the osculating circle, and it has radius ρ. The center of the circle is ρ units away from $\mathbf{r}(t_0)$, in the direction of the principal unit normal. Thus the center of the osculating circle is $\mathbf{r}(t_0) + \rho \mathbf{N}(t_0)$. Note that the osculating circle will lie in the same plane as the curve when the curve is planar.

15. If $f(x) = x^2$, then $f'(x) = 2x$ and $f''(x) = 2$, a constant. By Theorem 11.24(a), however, its curvature is

$$\kappa = \frac{|f''(x)|}{(1 + f'(x)^2)^{3/2}} = \frac{2}{(1 + 4x^2)^{3/2}},$$

which is not a constant. Thus the curvature varies with x.

17. Since $\kappa(1) = 3$, use a piece of a circle with radius $\rho = \frac{1}{\kappa} = \frac{1}{3}$. Thus it will be its own osculating circle at $\mathbf{r}(1) = \langle 1, 2 \rangle$, so by Definition 11.25 the center of the circle will be at $\mathbf{r}(1) + \rho \mathbf{N}(1) = \langle 1, 2 \rangle + \frac{1}{3} \left\langle \frac{3}{5}, \frac{4}{5} \right\rangle = \left\langle \frac{6}{5}, \frac{34}{15} \right\rangle$. Similarly, at $(-3, 5)$, use a piece of a circle with radius $\frac{1}{3}$. The center of the circle will be at $\mathbf{r}(3) + \rho \mathbf{N}(3) = \langle -3, 5 \rangle + \frac{1}{3} \left\langle \frac{1}{\sqrt{2}}, \frac{1}{\sqrt{2}} \right\rangle$. Connect these two pieces with a straight line. The graph is below (of course, many other graphs are possible):

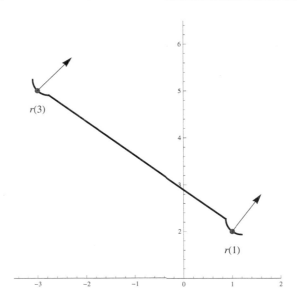

The two given points and a vector in the direction of the normal are marked.

19. Since $\kappa(1) = 3$, use a piece of a circle with radius $\rho = 3$. Thus it will be its own osculating circle at $\mathbf{r}(1) = \langle 3, 5 \rangle$, so by Definition 11.25 the center of the circle will be at $\mathbf{r}(1) + \rho \mathbf{N}(1) = \langle 3, 5 \rangle + 3 \langle 1, 0 \rangle = \langle 6, 5 \rangle$. Similarly, at $(-2, 4)$, use a piece of a circle with radius $\rho = 2$. The center of the circle will be at

$$\mathbf{r}(4) + \rho \mathbf{N}(4) = \langle -2, 4 \rangle + 2 \left\langle \frac{\sqrt{3}}{3}, \frac{\sqrt{6}}{3} \right\rangle = \left\langle -2 + \frac{2\sqrt{3}}{3}, 4 + \frac{2\sqrt{6}}{3} \right\rangle.$$

Connect these two pieces with an arbitrary curve. One such graph is below (of course, many others are possible):

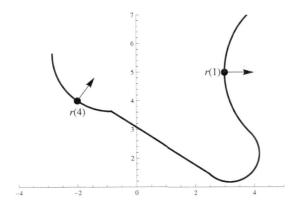

The two given points and a vectors in the directions of the normals are marked.

21. Since κ is always $\frac{1}{2}$, the radius of curvature is always 2, so we think of a circle of radius 2. Conveniently, the circle of radius 2 with center $(0, 0)$ passes through the two points $\langle 2, 0 \rangle$ and $\langle 0, 2 \rangle$; an equation for that circle is $\mathbf{r}(t) = \langle 2 \cos \left(\frac{\pi}{2} t \right), 2 \sin \left(\frac{\pi}{2} t \right) \rangle$. A plot is:

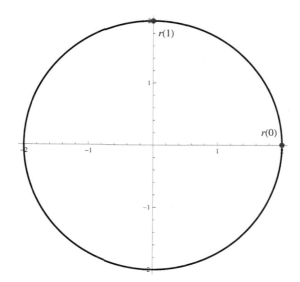

The two given points are marked.

Skills

23. Using Theorem 11.18, with $\mathbf{r}(t) = \langle 3\cos 4t, 3\sin 4t \rangle$ we have

$$\mathbf{r}'(t) = \langle -12\sin 4t, 12\cos 4t \rangle \qquad \text{so} \qquad \|\mathbf{r}'(t)\| = \sqrt{144\sin^2 4t + 144\cos^2 4t} = \sqrt{144} = 12.$$

Thus the arc length from $t = 0$ to $t = \frac{\pi}{2}$ is

$$\int_0^{\pi/2} \|\mathbf{r}'(t)\| \, dt = \int_0^{\pi/2} 12 \, dt = 6\pi.$$

25. Using Theorem 11.18, with $\mathbf{r}(t) = \left\langle 4\sin t, t^{3/2}, -4\cos t \right\rangle$, we have

$$\mathbf{r}'(t) = \left\langle 4\cos t, \frac{3}{2}t^{1/2}, 4\sin t \right\rangle$$

$$\|\mathbf{r}'(t)\| = \sqrt{16\cos^2 t + \frac{9}{4}t + 16\sin^2 t} = \frac{1}{2}\sqrt{64 + 9t}.$$

Thus the arc length from $t = 0$ to $t = 4$ is

$$\int_0^4 \|\mathbf{r}'(t)\| \, dt = \int_0^4 \frac{1}{2}\sqrt{64 + 9t} \, dt = \frac{1}{2}\int_0^4 \sqrt{64 + 9t} \, dt.$$

Using the substitution $u = 64 + 9t$, we get $du = 9\,dt$ so that $dt = \frac{1}{9}\,du$. Then the integral becomes

$$\frac{1}{2}\int_0^4 \sqrt{64 + 9t} \, dt = \frac{1}{18}\int_{t=0}^{t=4} \sqrt{u}\, du = \frac{1}{18}\left[\frac{2}{3}u^{3/2}\right]_{t=0}^{t=4} = \frac{1}{27}\left[(64 + 9t)^{3/2}\right]_0^4$$

$$= \frac{1}{27}(1000 - 512) = \frac{488}{27}.$$

27. Using Theorem 11.18, with $\mathbf{r}(t) = \langle e^t \sin t, e^t \cos t, e^t \rangle$, we have

$$\mathbf{r}'(t) = \langle e^t \sin t + e^t \cos t, e^t \cos t - e^t \sin t, e^t \rangle = e^t \langle \sin t + \cos t, \cos t - \sin t, 1 \rangle$$

$$\|\mathbf{r}'(t)\| = e^t \sqrt{(\sin t + \cos t)^2 + (\cos t - \sin t)^2 + 1}$$

$$= e^t \sqrt{\sin^2 t + 2 \sin t \cos t + \cos^2 t + \cos^2 t - 2 \sin t \cos t + \sin^2 t + 1}$$

$$= e^t \sqrt{3}.$$

Thus the arc length from $t = 0$ to $t = \pi$ is

$$\int_0^\pi \|\mathbf{r}'(t)\| \, dt = \sqrt{3} \int_0^\pi e^t \, dt = \sqrt{3} \left[e^t \right]_0^\pi = \sqrt{3} \left(e^\pi - 1 \right).$$

29. If $\mathbf{r}(t)$ were an arc length parametrization, then we would have $\|\mathbf{r}'(t)\| = 1$. But

$$\mathbf{r}'(t) = \langle 2, -1, 5 \rangle \quad \text{so that} \quad \|\mathbf{r}'(t)\| = \sqrt{2^2 + 1^2 + 5^2} = \sqrt{30}.$$

Following Theorem 11.20, let

$$s(t) = \int_0^t \|\mathbf{r}'(t)\| \, dt = t\sqrt{30}.$$

Then $t = \frac{s}{\sqrt{30}}$, so that an arc length parametrization is

$$\mathbf{r}(s) = \left\langle 3 + \cdot \frac{2}{\sqrt{30}} s, \ 4 - \frac{1}{\sqrt{30}} s, \ -1 + \frac{5}{\sqrt{30}} s \right\rangle,$$

and in fact

$$\|\mathbf{r}'(s)\| = \left\| \left\langle \frac{2}{\sqrt{30}}, \ -\frac{1}{\sqrt{30}}, \ \frac{5}{\sqrt{30}} \right\rangle \right\| = \sqrt{\frac{4}{30} + \frac{1}{30} + \frac{25}{30}} = 1.$$

31. With $f(x) = e^x$, we have $f'(x) = f''(x) = e^x$, so that by Theorem 11.24(a) the curvature at $x = 0$ is

$$\kappa = \frac{|f''(0)|}{(1 + f'(0)^2)^{3/2}} = \frac{|e^0|}{(1 + e^0)^{3/2}} = \frac{1}{2\sqrt{2}}.$$

To find the center of the osculating circle we must find $\mathbf{N}(0)$. A tangent vector at $x = 0$ is $\langle 1, f'(0) \rangle = \langle 1, 1 \rangle$. Since the magnitude of this vector is $\sqrt{2}$, the unit tangent vector at $x = 0$ is $\mathbf{T}(0) = \left\langle \frac{1}{\sqrt{2}}, \frac{1}{\sqrt{2}} \right\rangle$. Then the two unit normals at $x = 0$ are $\left\langle -\frac{1}{\sqrt{2}}, \frac{1}{\sqrt{2}} \right\rangle$ and $\left\langle \frac{1}{\sqrt{2}}, -\frac{1}{\sqrt{2}} \right\rangle$. Since $f''(0) = 1 > 0$, the curve is concave up at $x = 0$, so we choose the normal with the positive y coordinate. Thus

$$\mathbf{N}(0) = \left\langle -\frac{1}{\sqrt{2}}, \frac{1}{\sqrt{2}} \right\rangle.$$

So by Definition 11.25, the radius of the osculating circle is $\rho = \frac{1}{\kappa} = 2\sqrt{2}$, and the center is at

$$\langle 0, f(0) \rangle + \rho \mathbf{N}(0) = \langle 0, 1 \rangle + 2\sqrt{2} \left\langle -\frac{\sqrt{2}}{2}, \frac{\sqrt{2}}{2} \right\rangle = \langle -2, 3 \rangle.$$

Hence the equation of the osculating circle is $(x + 2)^2 + (y - 3)^2 = \rho^2 = 8$. Plots of $f(x)$ and a portion of the osculating circle are below, with the osculating circle in gray and its center and the point of intersection marked:

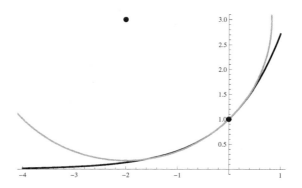

33. With $f(x) = \csc x$, we have $f'(x) = -\csc x \cot x$ and $f''(x) = \csc^3 x + \csc x \cot^2 x$, so that by Theorem 11.24(a) the curvature at $x = \frac{\pi}{2}$ is

$$\kappa = \frac{\left|f''\left(\frac{\pi}{2}\right)\right|}{\left(1 + f'\left(\frac{\pi}{2}\right)^2\right)^{3/2}} = \frac{\left|1^3 + 1^2 \cdot 0\right|}{\left(1 + (-1 \cdot 0)^2\right)^{3/2}} = 1.$$

To find the center of the osculating circle we must find $\mathbf{N}\left(\frac{\pi}{2}\right)$. A tangent vector at $x = \frac{\pi}{2}$ is $\left\langle 1, f'\left(\frac{\pi}{2}\right) \right\rangle = \langle 1, 0 \rangle$; since this is a unit vector, it is also equal to $\mathbf{T}\left(\frac{\pi}{2}\right)$. Then the two unit normals at $x = \frac{\pi}{2}$ are $\langle 0, \pm 1 \rangle$. Since $f''\left(\frac{\pi}{2}\right) = 1 > 0$, the curve is concave up at $x = \frac{\pi}{2}$, so we choose the normal that points upwards. Thus

$$\mathbf{N}\left(\frac{\pi}{2}\right) = \langle 0, 1 \rangle.$$

So by Definition 11.25, the radius of the osculating circle is $\rho = \frac{1}{\kappa} = 1$, and the center is at

$$\left\langle \frac{\pi}{2}, f\left(\frac{\pi}{2}\right) \right\rangle + \rho \mathbf{N}\left(\frac{\pi}{2}\right) = \left\langle \frac{\pi}{2}, 1 \right\rangle + 1 \langle 0, 1 \rangle = \left\langle \frac{\pi}{2}, 2 \right\rangle.$$

Hence the equation of the osculating circle is $\left(x - \frac{\pi}{2}\right)^2 + (y - 2)^2 = 1$. Plots of $f(x)$ and the osculating circle are below, with the osculating circle in gray and its center and the point of intersection marked:

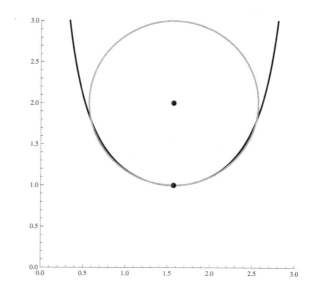

35. With $f(x) = \frac{1}{3}(x^2 + 2)^{3/2}$, we have

$$f'(x) = \frac{1}{2}(x^2 + 2)^{1/2} \cdot 2x = x(x^2 + 2)^{1/2}$$

$$f''(x) = (x^2 + 2)^{1/2} + \frac{1}{2}x(x^2 + 2)^{-1/2} \cdot 2x = \frac{x^2 + 2 + x^2}{(x^2 + 2)^{1/2}} = \frac{2(x^2 + 1)}{(x^2 + 2)^{1/2}}.$$

Then by Theorem 11.24(a), the curvature is

$$\kappa = \frac{|f''(1)|}{(1 + f'(1)^2)^{3/2}} = \frac{\left|\frac{4}{\sqrt{3}}\right|}{(1 + 3)^{3/2}} = \frac{1}{2\sqrt{3}}.$$

To find the center of the osculating circle, we must find $\mathbf{N}(1)$. A tangent vector at $x = 1$ is $\langle 1, f'(1) \rangle = \langle 1, \sqrt{3} \rangle$. Since the magnitude of this vector is $\sqrt{1 + 3} = 2$, the unit tangent vector is $\mathbf{T}(1) = \langle \frac{1}{2}, \frac{\sqrt{3}}{2} \rangle$. Then the two unit normals at $x(1)$ are $\langle -\frac{\sqrt{3}}{2}, \frac{1}{2} \rangle$ and $\langle \frac{\sqrt{3}}{2}, -\frac{1}{2} \rangle$. Since $f''(1) = \frac{4}{\sqrt{3}} > 0$, the curve is concave up near $x = 1$, so we choose the normal with a positive y coordinate. Thus

$$\mathbf{N}(1) = \left\langle -\frac{\sqrt{3}}{2}, \frac{1}{2} \right\rangle.$$

So by Definition 11.25, the radius of the osculating circle is $\rho = \frac{1}{\kappa} = 2\sqrt{3}$, and the center is at

$$\langle 1, f(1) \rangle + \rho \mathbf{N}(1) = \langle 1, \sqrt{3} \rangle + 2\sqrt{3} \left\langle -\frac{\sqrt{3}}{2}, \frac{1}{2} \right\rangle = \langle -2, 2\sqrt{3} \rangle.$$

Hence the equation of the osculating circle is

$$(x + 2)^2 + (y - 2\sqrt{3})^2 = (2\sqrt{3})^2 = 12.$$

Plots of $f(x)$ and the osculating circle are below, with the osculating circle in gray and the point of intersection marked (the center of the circle is not marked since it is too far away from the rest of the graph and it would be more difficult to see the graph features):

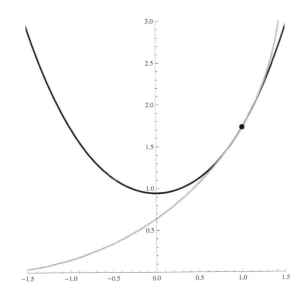

37. Using Theorem 11.24(b), since

$$x'(t) = -a\sin t, \quad x''(t) = -a\cos t, \quad y'(t) = b\cos t, \quad y''(t) = -b\sin t,$$

we get

$$\kappa = \frac{|x'(t)y''(t) - x''(t)y'(t)|}{((x'(t))^2 + (y'(t))^2))^{3/2}} = \frac{|(-a\sin t)(-b\sin t) - (-a\cos t)(b\cos t)|}{(a^2\sin^2 t + b^2\cos^2 t)^{3/2}}$$

$$= \frac{|ab\sin^2 t + ab\cos^2 t|}{(a^2\sin^2 t + b^2\cos^2 t)^{3/2}} = \frac{|ab|}{(a^2\sin^2 t + b^2\cos^2 t)^{3/2}}.$$

39. Convert to the parametric equations $x = t - \sin t$, $y = 1 - \cos t$. Then

$$x'(t) = 1 - \cos t, \quad x''(t) = \sin t, \quad y'(t) = \sin t, \quad y''(t) = \cos t,$$

so that by Theorem 11.24(b),

$$\kappa = \frac{|x'(t)y''(t) - x''(t)y'(t)|}{((x'(t))^2 + (y'(t))^2))^{3/2}} = \frac{|(1 - \cos t)(\cos t) - (\sin t)(\sin t)|}{((1 - \cos t)^2 + \sin^2 t)^{3/2}}$$

$$= \frac{|\cos t - 1|}{(2 - 2\cos t)^{3/2}} = \frac{1 - \cos t}{2^{3/2}(1 - \cos t)^{3/2}} = \frac{1}{2\sqrt{2 - 2\cos t}}.$$

41. We get

$$\mathbf{r}'(t) = \langle \sin t + t\cos t, \cos t - t\sin t, 1 \rangle$$

$$\mathbf{r}''(t) = \langle \cos t + \cos t - t\sin t, -\sin t - \sin t - t\cos t, 0 \rangle$$

$$= \langle 2\cos t - t\sin t, -2\sin t - t\cos t, 0 \rangle$$

$$\mathbf{r}'(t) \times \mathbf{r}''(t) = \langle \sin t + t\cos t, \cos t - t\sin t, 1 \rangle \times \langle 2\cos t - t\sin t, -2\sin t - t\cos t, 0 \rangle$$

$$= \langle (\cos t - t\sin t)(0) - 1(-2\sin t - t\cos t),$$

$$1(2\cos t - t\sin t) - (\sin t + t\cos t)(0),$$

$$(\sin t + t\cos t)(-2\sin t - t\cos t) - (\cos t - t\sin t)(2\cos t - t\sin t) \rangle$$

$$= \langle 2\sin t + t\cos t, \ 2\cos t - t\sin t,$$

$$-2\sin^2 t - t\sin t\cos t - 2t\sin t\cos t - t^2\cos^2 t$$

$$-2\cos^2 t + t\sin t\cos t + 2t\sin t\cos t - t^2\sin^2 t \rangle$$

$$= \langle 2\sin t + t\cos t, 2\cos t - t\sin t, -t^2 - 2 \rangle$$

$$\|\mathbf{r}'(t)\| = \sqrt{(\sin t + t\cos t)^2 + (\cos t - t\sin t)^2 + 1}$$

$$= \sqrt{\sin^2 t + 2t\sin t\cos t + t^2\cos^2 t + \cos^2 t - 2t\sin t\cos t + t^2\sin^2 t + 1}$$

$$= \sqrt{2 + t^2}$$

$$\|\mathbf{r}'(t) \times \mathbf{r}''(t)\| = \sqrt{(2\sin t + t\cos t)^2 + (2\cos t - t\sin t)^2 + (-t^2 - 2)^2}$$

$$= \sqrt{4\sin^2 t + 4t\sin t\cos t + t^2\cos^2 t + 4\cos^2 t - 4t\sin t\cos t + t^2\sin^2 t + t^4 + 4t^2 + 4}$$

$$= \sqrt{t^4 + 5t^2 + 8}.$$

Then by Theorem 11.23(b),

$$\kappa = \frac{\|\mathbf{r}'(t) \times \mathbf{r}''(t)\|}{\|\mathbf{r}'(t)\|^3} = \frac{\sqrt{t^4 + 5t^2 + 8}}{(t^2 + 2)^{3/2}}.$$

43. We get

$$\mathbf{r}'(t) = \langle \sin t + t \cos t, \cos t - t \sin t, \alpha \rangle$$
$$\mathbf{r}''(t) = \langle \cos t + \cos t - t \sin t, -\sin t - \sin t - t \cos t, 0 \rangle$$
$$= \langle 2 \cos t - t \sin t, -2 \sin t - t \cos t, 0 \rangle$$
$$\mathbf{r}'(t) \times \mathbf{r}''(t) = \langle \sin t + t \cos t, \cos t - t \sin t, \alpha \rangle \times \langle 2 \cos t - t \sin t, -2 \sin t - t \cos t, 0 \rangle$$
$$= \langle (\cos t - t \sin t)(0) - \alpha(-2 \sin t - t \cos t),$$
$$\alpha(2 \cos t - t \sin t) - (\sin t + t \cos t)(0),$$
$$(\sin t + t \cos t)(-2 \sin t - t \cos t) - (\cos t - t \sin t)(2 \cos t - t \sin t) \rangle$$
$$= \langle 2\alpha \sin t + \alpha t \cos t, \; 2\alpha \cos t - \alpha t \sin t,$$
$$-2 \sin^2 t - t \sin t \cos t - 2t \sin t \cos t - t^2 \cos^2 t$$
$$-2 \cos^2 t + t \sin t \cos t + 2t \sin t \cos t - t^2 \sin^2 t \rangle$$
$$= \left\langle 2\alpha \sin t + \alpha t \cos t, \; 2\alpha \cos t - \alpha t \sin t, \; -t^2 - 2 \right\rangle$$
$$\|\mathbf{r}'(t)\| = \sqrt{(\sin t + t \cos t)^2 + (\cos t - t \sin t)^2 + \alpha^2}$$
$$= \sqrt{\sin^2 t + 2t \sin t \cos t + t^2 \cos^2 t + \cos^2 t - 2t \sin t \cos t + t^2 \sin^2 t + \alpha^2}$$
$$= \sqrt{t^2 + \alpha^2 + 1}$$
$$\|\mathbf{r}'(t) \times \mathbf{r}''(t)\| = \sqrt{(2\alpha \sin t + \alpha t \cos t)^2 + (2\alpha \cos t - \alpha t \sin t)^2 + (-t^2 - 2)^2}$$
$$= \sqrt{4\alpha^2 \sin^2 t + 4\alpha^2 t \sin t \cos t + \alpha^2 t^2 \cos^2 t + 4\alpha^2 \cos^2 t - 4\alpha^2 t \sin t \cos t + \alpha^2 t^2 \sin^2 t + t^4 + 4t^2 + 4}$$
$$= \sqrt{t^4 + (4 + \alpha^2)t^2 + (4 + 4\alpha^2)}.$$

Then by Theorem 11.23(b),

$$\kappa = \frac{\|\mathbf{r}'(t) \times \mathbf{r}''(t)\|}{\|\mathbf{r}'(t)\|^3} = \frac{\sqrt{t^4 + (4 + \alpha^2)t^2 + (4 + 4\alpha^2)}}{(t^2 + \alpha^2 + 1)^{3/2}}.$$

45. With $f(x) = x^3$, we get $f'(x) = 3x^2$ and $f''(x) = 6x$, so that by Theorem 11.24(a) the curvature is

$$\kappa = \frac{|f''(x)|}{(1 + f'(x)^2)^{3/2}} = \frac{|6x|}{(9x^4 + 1)^{3/2}}.$$

This is an even function since both numerator and denominator are even, so to find the maxima it suffices to find the maxima for $x > 0$, where the function is

$$\kappa(x) = \frac{6x}{(9x^4 + 1)^{3/2}}.$$

Taking the derivative gives

$$\kappa'(x) = \frac{(9x^4 + 1)^{3/2}(6) - 6x\left(\frac{3}{2}(9x^4 + 1)^{1/2} \cdot 36x^3\right)}{(9x^4 + 1)^3} = \frac{(9x^4 + 1)(6) - 324x^4}{(9x^4 + 1)^{5/2}} = \frac{-270x^4 + 6}{(9x^4 + 1)^{5/2}}.$$

A local extremum may occur where the first derivative vanishes, which is where $6 - 270x^4 = 0$, or $x = \pm\sqrt[4]{\frac{1}{45}}$. Since we are assuming $x > 0$, we get $x = \sqrt[4]{\frac{1}{45}}$. By symmetry of κ, the value $x = -\sqrt[4]{\frac{1}{45}}$ is also a local extremum. Finally, since $\kappa'(x)$ changes sign from $+$ to $-$ at each of these points, they are both local maxima for the curvature.

47. Use Theorem 11.23(b). We have

$$
\begin{aligned}
\mathbf{r}'(t) &= \langle -a\sin t, a\cos t, b \rangle \\
\mathbf{r}''(t) &= \langle -a\cos t, -a\sin t, 0 \rangle \\
\mathbf{r}'(t) \times \mathbf{r}''(t) &= \langle -a\sin t, a\cos t, b \rangle \times \langle -a\cos t, -a\sin t, 0 \rangle \\
&= \langle (a\cos t)(0) - b(-a\sin t), b(-a\cos t) - (-a\sin t)(0), \\
&\qquad (-a\sin t)(-a\sin t) - (a\cos t)(-a\cos t) \rangle \\
&= \left\langle ab\sin t, -ab\cos t, a^2 \right\rangle \\
\|\mathbf{r}'(t)\| &= \sqrt{a^2\sin^2 t + a^2\cos^2 t + b^2} = \sqrt{a^2 + b^2} \\
\|\mathbf{r}'(t) \times \mathbf{r}''(t)\| &= \sqrt{a^2 b^2 \sin^2 t + a^2 b^2 \cos^2 t + a^4} = \sqrt{a^2}\sqrt{b^2 + a^2} = a\sqrt{a^2 + b^2}.
\end{aligned}
$$

Then

$$
\kappa = \frac{\|\mathbf{r}'(t) \times \mathbf{r}''(t)\|}{\|\mathbf{r}'(t)\|^3} = \frac{a(a^2 + b^2)^{1/2}}{(a^2 + b^2)^{3/2}} = \frac{a}{a^2 + b^2},
$$

which is a constant independent of t.

49. We have

$$
\begin{aligned}
\mathbf{r}'(t) &= \left\langle -\sin t, \cos t, -\frac{1}{t^2} \right\rangle, \\
\mathbf{r}''(t) &= \left\langle -\cos t, -\sin t, \frac{2}{t^3} \right\rangle, \\
\mathbf{r}'(t) \times \mathbf{r}''(t) &= \left\langle -\sin t, \cos t, -\frac{1}{t^2} \right\rangle \times \left\langle -\cos t, -\sin t, \frac{2}{t^3} \right\rangle \\
&= \left\langle \frac{2}{t^3}\cos t - \frac{1}{t^2}\sin t, \frac{1}{t^2}\cos t + \frac{2}{t^3}\sin t, \sin^2 t + \cos^2 t \right\rangle \\
&= \left\langle \frac{2}{t^3}\cos t - \frac{1}{t^2}\sin t, \frac{1}{t^2}\cos t + \frac{2}{t^3}\sin t, 1 \right\rangle \\
\|\mathbf{r}'(t)\| &= \sqrt{\sin^2 t + \cos^2 t + \frac{1}{t^4}} = \sqrt{1 + \frac{1}{t^4}} = \frac{\sqrt{t^4 + 1}}{t^2} \\
\|\mathbf{r}'(t) \times \mathbf{r}''(t)\| &= \sqrt{\left(\frac{2}{t^3}\cos t - \frac{1}{t^2}\sin t\right)^2 + \left(\frac{1}{t^2}\cos t + \frac{2}{t^3}\sin t\right)^2 + 1} \\
&= \sqrt{\frac{4}{t^6}\cos^2 t - \frac{4}{t^5}\sin t\cos t + \frac{1}{t^4}\sin^2 t + \frac{1}{t^4}\cos^2 t + \frac{4}{t^5}\sin t\cos t + \frac{4}{t^3}\sin^2 t + 1} \\
&= \sqrt{\frac{4}{t^6}(\sin^2 t + \cos^2 t) + \frac{1}{t^4}(\sin^2 t + \cos^2 t) + 1} \\
&= \sqrt{\frac{4}{t^6} + \frac{1}{t^4} + 1} \\
&= \frac{\sqrt{t^6 + t^2 + 4}}{t^3}.
\end{aligned}
$$

So by Theorem 11.23(a),

$$\kappa = \frac{\|\mathbf{r}'(t) \times \mathbf{r}''(t)\|}{\|\mathbf{r}'(t)\|^3} = \frac{\frac{\sqrt{t^6 + t^2 + 4}}{t^3}}{\frac{(t^4+1)^{3/2}}{t^6}} = \frac{t^3\sqrt{t^6 + t^2 + 4}}{(t^4+1)^{3/2}}.$$

Then clearly $\lim\limits_{t \to 0^+} \kappa = 0$ since the numerator goes to zero while the denominator goes to 1. We also have

$$\begin{aligned}
\lim_{t \to \infty} \kappa &= \lim_{t \to \infty} \frac{t^3 \sqrt{t^6 + t^2 + 4}}{(t^4+1)^{3/2}} \\
&= \lim_{t \to \infty} \frac{t^{-6} \cdot t^3 \sqrt{t^6 + t^2 + 4}}{t^{-6}(t^4+1)^{3/2}} \\
&= \lim_{t \to \infty} \frac{\sqrt{1 + 1/t^4 + 4/t^6}}{(1 + 1/t^4)^{3/2}} \\
&= 1.
\end{aligned}$$

51. Differentiating $\mathbf{B} = \mathbf{T} \times \mathbf{N}$ with respect to s and using Theorem 11.11(d) gives

$$\frac{d\mathbf{B}}{ds} = \frac{d}{ds}(\mathbf{T} \times \mathbf{N}) = \frac{d\mathbf{T}}{ds} \times \mathbf{N} + \mathbf{T} \times \frac{d\mathbf{N}}{ds}.$$

Since by assumption both $\frac{d\mathbf{T}}{ds}$ and $\frac{d\mathbf{N}}{ds}$ exist everywhere, so does $\frac{d\mathbf{B}}{ds}$.

53. Differentiating $\mathbf{N} = \mathbf{B} \times \mathbf{T}$ with respect to s using Theorem 11.11 gives

$$\frac{d\mathbf{N}}{ds} = \frac{d\mathbf{B}}{ds} \times \mathbf{T} + \mathbf{B} \times \frac{d\mathbf{T}}{ds} = -\tau \mathbf{N} \times \mathbf{T} + \mathbf{B} \times \kappa \mathbf{N} = -\tau \mathbf{N} \times \mathbf{T} + \kappa \mathbf{B} \times \mathbf{N}.$$

Now, since $\mathbf{BTN}$ is a right-handed system, we get $\mathbf{N} \times \mathbf{T} = -\mathbf{B}$ and $\mathbf{B} \times \mathbf{N} = -\mathbf{T}$. Making these substitutions gives

$$\frac{d\mathbf{N}}{ds} = \tau \mathbf{B} - \kappa \mathbf{T}.$$

55. With $\mathbf{r}(t) = \langle \cosh t, \sinh t, t \rangle$, we have $\mathbf{r}'(t) = \langle \sinh t, \cosh t, 1 \rangle$, so that

$$\|\mathbf{r}'(t)\| = \sqrt{\sinh^2 t + \cosh^2 t + 1} = \sqrt{2\cosh^2 t} = \sqrt{2}\,\cosh t.$$

Thus

$$\begin{aligned}
\mathbf{T}(t) &= \frac{\mathbf{r}'(t)}{\|\mathbf{r}'(t)\|} = \frac{1}{\sqrt{2}}\langle \tanh t, 1, \operatorname{sech} t \rangle \\
\mathbf{T}'(t) &= \frac{1}{\sqrt{2}}\left\langle \operatorname{sech}^2 t, 0, -\operatorname{sech} t \tanh t \right\rangle \\
\|\mathbf{T}'(t)\| &= \frac{1}{\sqrt{2}}\sqrt{\operatorname{sech}^4 t + \operatorname{sech}^2 t \tanh^2 t} \\
&= \frac{1}{\sqrt{2}}\sqrt{\operatorname{sech}^2 t(\operatorname{sech}^2 t + \tanh^2 t)} = \frac{1}{\sqrt{2}}\operatorname{sech} t
\end{aligned}$$

since $\mathrm{sech}^2 t + \tanh^2 t = 1$. Then

$$\mathbf{N}(t) = \frac{\mathbf{T}'(t)}{\|\mathbf{T}'(t)\|} = \langle \mathrm{sech}\, t, 0, -\tanh t \rangle$$

$$\mathbf{B}(t) = \mathbf{T}(t) \times \mathbf{N}(t) = \frac{1}{\sqrt{2}} \langle \tanh t, 1, \mathrm{sech}\, t \rangle \times \langle \mathrm{sech}\, t, 0, -\tanh t \rangle$$

$$= \frac{1}{\sqrt{2}} \langle 1 \cdot (-\tanh t) - \mathrm{sech}\, t \cdot 0, \mathrm{sech}^2 t + \tanh^2 t, \tanh t \cdot 0 - 1 \cdot \mathrm{sech}\, t \rangle$$

$$= \frac{1}{\sqrt{2}} \langle -\tanh t, 1, -\mathrm{sech}\, t \rangle .$$

Differentiating gives

$$\mathbf{B}'(t) = \frac{1}{\sqrt{2}} \langle -\mathrm{sech}^2 t, 0, \mathrm{sech}\, t \tanh t \rangle = \left(-\frac{1}{\sqrt{2}} \mathrm{sech}\, t \right) \mathbf{N}(t).$$

Thus $\tau = -\frac{1}{\sqrt{2}} \mathrm{sech}\, t$.

57. If the curve lies in a plane, then all the vectors $\mathbf{T}(t)$ and $\mathbf{N}(t)$ lie in the plane as well, so that all the vectors $\mathbf{B}(t)$ are parallel to the normal vector to that plane. But they all point in the same direction, by the right-hand rule, and they are all unit vectors, so they are all the same vector. Thus $\mathbf{B}(t)$ is constant, so its derivative is zero. But since $\mathbf{0} = \frac{d\mathbf{B}}{ds} = -\tau\mathbf{N}$, it must be the case that $\tau = 0$ (we do not have $\mathbf{N}(t) = 0$ if $\mathbf{N}(t)$ exists).

Applications

59. (a) A graph of his path is given by

   (b) Graphs of the x and y components of his path are

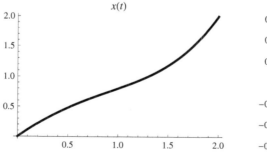 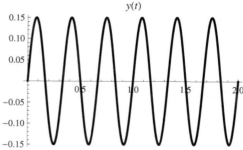

   Ian is walking at a roughly constant speed from west to east, since the graph of $x(t)$ is close to a straight line (although he slows down a bit at the start and speeds up a bit towards the end), but his path is oscillating north and south. It looks like the greatest curvature is at the top and bottom of his north-south traversal.

   (c) The greatest difficulties will occur at the peaks and valleys of the curve.

Proofs

61. By Theorem 11.24(a), the curvature of a function f is

$$\kappa = \frac{|f''(x)|}{(1 + f'(x)^2)^{3/2}}.$$

But if f has an inflection point at $x = a$, then $f''(a)$ vanishes, so that $\kappa = 0$ at a.

63. Suppose $\mathcal{C}$ is parametrized by $\mathbf{r}(t)$, and assume that the particle is moving with constant speed $\|\mathbf{r}'(t)\| = c$. Then the velocity of the particle is $\mathbf{r}'(t)$, and its acceleration is $\mathbf{r}''(t)$. But $\mathbf{T}(t) = \frac{\mathbf{r}'(t)}{\|\mathbf{r}'(t)\|}$, so that $\mathbf{r}'(t) = c\mathbf{T}(t)$; then $\mathbf{r}''(t) = c\mathbf{T}'(t)$. By Theorem 11.23(a), $\|\mathbf{T}'(t)\| = \kappa \|\mathbf{r}'(t)\|$. Thus the magnitude of the acceleration is

$$\|\mathbf{r}''(t)\| = \|c\mathbf{T}'(t)\| = c\|\mathbf{T}'(t)\| = c\kappa \|\mathbf{r}'(t)\| = c^2\kappa.$$

That is, the acceleration is a constant, c^2, times the curvature.

65. Use the parametrization $x = t$, $y = f(t)$. Then $\mathbf{r}(t) = \langle t, f(t), 0\rangle$, so that $\mathbf{r}'(t) = \langle 1, f'(t), 0\rangle$ and $\mathbf{r}''(t) = \langle 0, f''(t), 0\rangle$. Then by Theorem 11.23(b), the curvature is given by

$$\kappa = \frac{\|\mathbf{r}'(t) \times \mathbf{r}''(t)\|}{\|\mathbf{r}'(t)\|^3} = \frac{\|\langle 1, f'(t), 0\rangle \times \langle 0, f''(t), 0\rangle\|}{\left(\sqrt{1 + f'(t)^2}\right)^3} = \frac{\|\langle 0, 0, f''(t)\rangle\|}{(1 + f'(t)^2)^{3/2}} = \frac{|f''(t)|}{(1 + f'(t)^2)^{3/2}}.$$

Thinking Forward

A decomposition of the acceleration vector

Recall that in general, $\text{comp}_\mathbf{v} \mathbf{u} = \frac{\mathbf{v} \cdot \mathbf{u}}{\|\mathbf{v}\|}$ (see Definition 10.21). Then

▷ With $\mathbf{r}(t) = \langle t, t^2, t^3\rangle$, we get $\mathbf{v}(t) = \mathbf{r}'(t) = \langle 1, 2t, 3t^2\rangle$ and $\mathbf{a}(t) = \mathbf{r}''(t) = \langle 0, 2, 6t\rangle$. Then

$$\text{comp}_{\mathbf{v}(t)} \mathbf{a}(t) = \frac{\mathbf{v}(t) \cdot \mathbf{a}(t)}{\|\mathbf{v}(t)\|} = \frac{4t + 18t^3}{\sqrt{9t^4 + 4t^2 + 1}}.$$

▷ With $\mathbf{r}(t) = \langle \cos t, \sin t, t\rangle$, we get

$$\mathbf{v}(t) = \mathbf{r}'(t) = \langle -\sin t, \cos t, 1\rangle \qquad \text{and} \qquad \mathbf{a}(t) = \mathbf{r}''(t) = \langle -\cos t, -\sin t, 0\rangle.$$

Then

$$\text{comp}_{\mathbf{v}(t)} \mathbf{a}(t) = \frac{\mathbf{v}(t) \cdot \mathbf{a}(t)}{\|\mathbf{v}(t)\|} = \frac{\sin t \cos t - \sin t \cos t + 0}{\sqrt{\sin^2 t + \cos^2 t + 1}} = 0.$$

The acceleration and velocity are perpendicular.

▷ With $\mathbf{r}(t) = \langle e^t \sin t, e^t \cos t, e^t\rangle$, we get

$$\mathbf{v}(t) = \mathbf{r}'(t) = \langle e^t \sin t + e^t \cos t, e^t \cos t - e^t \sin t, e^t\rangle = \langle e^t(\sin t + \cos t), e^t(\cos t - \sin t), e^t\rangle$$

$$\mathbf{a}(t) = \mathbf{v}'(t) = \langle e^t(\sin t + \cos t) + e^t(\cos t - \sin t), e^t(\cos t - \sin t) + e^t(-\sin t - \cos t), e^t\rangle$$

$$= \langle 2e^t \cos t, -2e^t \sin t, e^t\rangle.$$

Then

$$\text{comp}_{\mathbf{v}(t)} \mathbf{a}(t) = \frac{\mathbf{v}(t) \cdot \mathbf{a}(t)}{\|\mathbf{v}(t)\|}$$

$$= \frac{2e^{2t} \cos^2 t + 2e^{2t} \sin t \cos t - 2e^{2t} \sin t \cos t + 2e^{2t} \sin t + e^{2t}}{\sqrt{e^{2t}(\cos t + \sin t)^2 + e^{2t}(\cos t - \sin t)^2 + e^{2t}}}$$

$$= \frac{3e^{2t}}{\sqrt{3e^{2t}}} = \sqrt{3e^{2t}} = \sqrt{3}\,e^t.$$

11.5 Motion

Thinking Back

Projecting one vector onto another

Suppose the angle between **u** and **v** is θ. Then **v** forms the hypotenuse of a right triangle one leg of which is $\text{proj}_\mathbf{u}\,\mathbf{v}$ and the other of which is the perpendicular from the terminal point of **v** to **u**. Then

$$\frac{\|\text{proj}_\mathbf{u}\,\mathbf{v}\|}{\|\mathbf{v}\|} = \cos\theta = \frac{\mathbf{u}\cdot\mathbf{v}}{\|\mathbf{u}\|\,\|\mathbf{v}\|}.$$

Clearing fractions gives $\text{comp}_\mathbf{u}\,\mathbf{v} = \|\text{proj}_\mathbf{u}\,\mathbf{v}\| = \frac{\mathbf{u}\cdot\mathbf{v}}{\|\mathbf{u}\|}$, and then $\text{proj}_\mathbf{u}\,\mathbf{v}$ is a vector of length $\text{comp}_\mathbf{u}\,\mathbf{v}$ in the direction of **u**, so it is

$$\text{proj}_\mathbf{u}\,\mathbf{v} = (\text{comp}_\mathbf{u}\,\mathbf{v})\,\frac{\mathbf{u}}{\|\mathbf{u}\|} = \frac{\mathbf{u}\cdot\mathbf{v}}{\|\mathbf{u}\|^2}\mathbf{u}.$$

The difference between a vector and its projection onto another vector

v forms the hypotenuse of a right triangle one leg of which is $\text{proj}_\mathbf{u}\,\mathbf{v}$ and the other of which is $\mathbf{v} - \text{proj}_\mathbf{u}\,\mathbf{v}$. See Section 10.3.

Concepts

1. (a) True. Since the acceleration due to gravity is a constant, the vertical position, which is the integral of the integral of acceleration, will be quadratic in time, while the horizontal velocity is constant, so the horizontal position varies linearly with time. The result is that vertical position varies quadratically in time, so is a parabola.

   (b) False. The vertical component of its velocity is zero, but it may still be moving horizontally (and will be unless it was fired straight up), since the horizontal component of velocity is $\|\mathbf{v}_0\|\cos\theta$.

   (c) True. The horizontal component is $\|\mathbf{v}_0\|\cos\theta$ where $\mathbf{v}_0$ is the initial velocity and θ is the angle at which the projectile is fired.

   (d) True. The acceleration due to gravity is a constant, in the downward direction, so is $-g\mathbf{k}$.

   (e) False. The *speed* is defined to be the magnitude of its *velocity*.

   (f) True. This is the definition of centripetal acceleration. See Theorem 11.27 and the discussion which follows.

   (g) False. The tangential component of acceleration and the normal component of acceleration are both scalars, while acceleration is a vector. If you add a vector in the tangential direction whose length is the tangential component of acceleration to one in the normal direction whose length is the normal component of acceleration, their sum will be the acceleration.

   (h) False. A parametrization is an arc length parametrization if for every a and b in the domain of **r** we have $\int_a^b \|\mathbf{r}'(t)\|\,dt = b - a$. This may or may not equal the distance between $\mathbf{r}(b)$ and $\mathbf{r}(a)$, which is the displacement between times a and b, as **r** may certainly twist around. See Definition 11.19.

3. The intuitive explanation is that the force due to gravity is smaller, so that it takes longer for gravity to drag the object down to the surface. During that time, the object will travel further horizontally. The equation of the parabola describing the motion is

$$y = -\frac{g}{2\|\mathbf{v}_0\|^2\cos^2\theta}x^2 + (\tan\theta)x + h$$

where $\mathbf{v}_0$ is the initial velocity, θ the angle of firing, h the initial height, and g the force due to gravity. If g is smaller, this "flattens out" the parabola, so that its x intercept is further from the origin.

5. Wind is modeled as a constant (vector) force $\mathbf{w}$, which can be thought of as a force acting in a given direction. Suppose the wind speed is W and that it acts along the direction vector $\langle \cos \theta, 0, \sin \theta \rangle$ (remember that here we are modeling y as being "up", so that the horizontal plane is the xz plane). Then the acceleration vector due to the wind is $\langle W \cos \theta, 0, W \sin \theta \rangle$, so that the total acceleration on the object is

$$\mathbf{a}(t) = \langle 0, -g, 0 \rangle + \langle W \cos \theta, 0, W \sin \theta \rangle.$$

The motion then becomes a three-dimensional motion, and the equations are slightly more complicated to solve.

7. The displacement of the object from $t = a$ to $t = b$ is the vector from $\mathbf{r}(a)$ to $\mathbf{r}(b)$ - that is, the straight line between the two points, not the distance along the curve.

9. The tangential component of acceleration is the component of acceleration in the direction of motion (i.e., in the direction of the tangent). The normal component of acceleration is the component of acceleration normal to the direction of motion. Centripetal acceleration is another name for the normal component of acceleration. The tangential and normal components of acceleration are given by

$$a_{\mathbf{T}} = \frac{\mathbf{v} \cdot \mathbf{a}}{\|\mathbf{v}\|}, \qquad a_{\mathbf{N}} = \frac{\|\mathbf{v} \times \mathbf{a}\|}{\|\mathbf{v}\|}.$$

See Theorem 11.27.

11. The tangential component of acceleration always points in the direction of motion, tangent to the curve defined by your path. The normal component of acceleration is orthogonal to your direction of motion, and points "into" the curve, so if you are curving to the left, it points to your left. If you are curving to the right, it points to your right. If you are ascending or descending in a straight line, it is zero. If you are ascending at a decreasing rate, it points down, and so forth.

13. (a) If we write $x = f(t)$, then $\mathbf{r}(t) = \langle x, x^2 \rangle$, so that every point (x, y) on $\mathbf{r}(t)$ satisfies $y = x^2$, so lies on a parabola.

(b) Differentiating gives

$$\mathbf{r}'(t) = \langle f'(t), 2f(t)f'(t) \rangle, \qquad \mathbf{a}(t) = \mathbf{r}''(t) = \left\langle f''(t), 2f'(t)^2 + 2f(t)f''(t) \right\rangle.$$

Then

$$\begin{aligned}
\mathbf{r}(t) \cdot \mathbf{a}(t) &= \left\langle f(t), f(t)^2 \right\rangle \cdot \left\langle f''(t), 2f'(t)^2 + 2f(t)f''(t) \right\rangle \\
&= f(t)f''(t) + 2f(t)^2 f'(t)^2 + 2f(t)^3 f''(t) \\
&= f(t) \left(f''(t) + 2f(t)f'(t)^2 + 2f(t)^2 f''(t) \right).
\end{aligned}$$

Assuming $f(t) \neq 0$, f must satisfy the equation $f''(t) + 2f(t)f'(t)^2 + 2f(t)^2 f''(t) = 0$.

Skills

15. (a) $\mathbf{r}(\pi) - \mathbf{r}(0) = \langle \pi \sin \pi, \pi \cos \pi \rangle - \langle 0, 0 \rangle = \langle 0, -\pi \rangle - \langle 0, 0 \rangle = \langle 0, -\pi \rangle.$

(b) $\|\mathbf{r}(\pi) - \mathbf{r}(0)\| = \sqrt{0^2 + \pi^2} = \pi.$

(c) Since $\mathbf{r}'(t) = \langle \sin t + t \cos t, \ \cos t - t \sin t \rangle$, we have

$$\|\mathbf{r}'(t)\| = \sqrt{(\sin t + t \cos t)^2 + (\cos t - t \sin t)^2} = \sqrt{1 + t^2}.$$

Thus the distance travelled by the particle is

$$\int_0^\pi \sqrt{1 + t^2}\, dt.$$

To integrate, use the trigonometric substitution $t = \tan u$, so that $dt = \sec^2 u\, du$. Then

$$\begin{aligned}
\int_0^\pi \sqrt{1 + t^2}\, dt &= \int_{t=0}^{t=\pi} \sqrt{1 + \tan^2 u} \cdot \sec^2 u\, du \\
&= \int_{t=0}^{t=\pi} \sec^3 u\, du \\
&= \frac{1}{2}\left[\sec u \tan u + \ln|\sec u + \tan u|\right]_{t=0}^{t=\pi} \\
&= \frac{1}{2}\left[t\sqrt{1 + t^2} + \ln\left|t + \sqrt{1 + t^2}\right|\right]_0^\pi \\
&= \frac{1}{2}\left(\pi\sqrt{\pi^2 + 1} + \ln\left(\pi + \sqrt{\pi^2 + 1}\right)\right).
\end{aligned}$$

17. (a) $\mathbf{r}(1) - \mathbf{r}(0) = \langle \alpha \sin \beta, \alpha \cos \beta, \gamma \rangle - \langle \alpha \sin 0, \alpha \cos 0, 0 \rangle = \langle \alpha \sin \beta, -\alpha + \alpha \cos \beta, \gamma \rangle$.

   (b) $\|\mathbf{r}(1) - \mathbf{r}(0)\| = \sqrt{\alpha^2 \sin \beta^2 + (-\alpha + \alpha \cos \beta)^2 + \gamma^2} = \sqrt{2\alpha^2 + \gamma^2 - 2\alpha^2 \cos \beta}$.

   (c) Since

$$\mathbf{r}'(t) = \langle \alpha\beta \cos \beta t, \ -\alpha\beta \sin \beta t, \ \gamma \rangle,$$

   we have

$$\|\mathbf{r}'(t)\| = \sqrt{\alpha^2\beta^2 \cos^2 \beta t + \alpha^2\beta^2 \sin^2 \beta t + \gamma^2} = \sqrt{\alpha^2\beta^2 + \gamma^2},$$

   so that the distance travelled by the particle is

$$\int_0^1 \sqrt{\alpha^2\beta^2 + \gamma^2}\, dt = \sqrt{\alpha^2\beta^2 + \gamma^2}.$$

19. Thinking of $\mathbf{r}(t)$ as lying in the xy plane, we get

$$\mathbf{v}(t) = \mathbf{r}'(t) = \langle 3 \cos 3t, -4 \sin 4t, 0 \rangle, \qquad \mathbf{a}(t) = \mathbf{v}'(t) = \langle -9 \sin 3t, -16 \cos 4t, 0 \rangle.$$

Then by Theorem 11.27,

$$\begin{aligned}
a_{\mathbf{T}} &= \frac{\mathbf{v} \cdot \mathbf{a}}{\|\mathbf{v}\|} = \frac{\langle 3 \cos 3t, -4 \sin 4t, 0 \rangle \cdot \langle -9 \sin 3t, -16 \cos 4t, 0 \rangle}{\sqrt{9 \cos^2 3t + 16 \sin^2 4t}} = \frac{-27 \sin 3t \cos 3t + 64 \sin 4t \cos 4t}{\sqrt{9 \cos^2 3t + 16 \sin^2 4t}} \\
a_{\mathbf{N}} &= \frac{\|\mathbf{v} \times \mathbf{a}\|}{\|\mathbf{v}\|} = \frac{\|\langle 3 \cos 3t, -4 \sin 4t, 0 \rangle \times \langle -9 \sin 3t, -16 \cos 4t, 0 \rangle\|}{\sqrt{9 \cos^2 3t + 16 \sin^2 4t}} \\
&= \frac{\|\langle 0, 0, -48 \cos 3t \cos 4t - 36 \sin 3t \sin 4t \rangle\|}{\sqrt{9 \cos^2 3t + 16 \sin^2 4t}} \\
&= \frac{|-36 \sin 3t \sin 4t - 48 \cos 3t \cos 4t|}{\sqrt{9 \cos^2 3t + 16 \sin^2 4t}} \\
&= \frac{|36 \sin 3t \sin 4t + 48 \cos 3t \cos 4t|}{\sqrt{9 \cos^2 3t + 16 \sin^2 4t}}.
\end{aligned}$$

21. Differentiating gives

$$\mathbf{v}(t) = \mathbf{r}'(t) = \left\langle e^t \sin t + e^t \cos t, e^t \cos t - e^t \sin t, e^t \right\rangle = \left\langle e^t (\sin t + \cos t), e^t (\cos t - \sin t), e^t \right\rangle$$
$$\mathbf{a}(t) = \mathbf{v}'(t) = \left\langle e^t (\sin t + \cos t) + e^t (\cos t - \sin t), e^t (\cos t - \sin t) + e^t (-\sin t - \cos t), e^t \right\rangle$$
$$= \left\langle 2e^t \cos t, -2e^t \sin t, e^t \right\rangle.$$

Note that

$$\|\mathbf{v}(t)\| = e^t \sqrt{(\sin t + \cos t)^2 + (\cos t - \sin t)^2 + 1} = \sqrt{3}\, e^t.$$

Then by Theorem 11.27,

$$
\begin{aligned}
a_{\mathbf{T}} = \frac{\mathbf{v} \cdot \mathbf{a}}{\|\mathbf{v}\|} &= \frac{\left\langle e^t (\sin t + \cos t), e^t (\cos t - \sin t), e^t \right\rangle \cdot \left\langle 2e^t \cos t, -2e^t \sin t, e^t \right\rangle}{\sqrt{3}e^t} \\
&= \frac{e^{2t} (2 \sin t \cos t + 2 \cos^2 t - 2 \sin t \cos t + 2 \sin^2 t + 1)}{\sqrt{3}e^t} \\
&= \frac{3e^{2t}}{\sqrt{3}e^t} = \sqrt{3}e^t
\end{aligned}
$$

$$
\begin{aligned}
a_{\mathbf{N}} = \frac{\|\mathbf{v} \times \mathbf{a}\|}{\|\mathbf{v}\|} &= \frac{\left\| \left\langle e^t (\sin t + \cos t), e^t (\cos t - \sin t), e^t \right\rangle \times \left\langle 2e^t \cos t, -2e^t \sin t, e^t \right\rangle \right\|}{\sqrt{3}e^t} \\
&= \frac{\left\| \left\langle e^{2t} (\sin t + \cos t), e^{2t} (\cos t - \sin t), -2e^{2t} \right\rangle \right\|}{\sqrt{3}e^t} \\
&= \frac{e^{2t} \sqrt{(\sin t + \cos t)^2 + (\cos t - \sin t)^2 + 4}}{\sqrt{3}e^t} \\
&= \frac{e^t \sqrt{6}}{\sqrt{3}} = \sqrt{2}\, e^t.
\end{aligned}
$$

23. Differentiating gives

$$\mathbf{v}(t) = \mathbf{r}'(t) = \left\langle -\sin t, \cos t, 1 \right\rangle, \qquad \mathbf{a}(t) = \mathbf{v}'(t) = \left\langle -\cos t, -\sin t, 0 \right\rangle.$$

Then by Theorem 11.27,

$$
\begin{aligned}
a_{\mathbf{T}} = \frac{\mathbf{v} \cdot \mathbf{a}}{\|\mathbf{v}\|} &= \frac{\left\langle -\sin t, \cos t, 1 \right\rangle \cdot \left\langle -\cos t, -\sin t, 0 \right\rangle}{\sqrt{\sin^2 t + \cos^2 t + 1}} = \frac{\sin t \cos t - \sin t \cos t}{\sqrt{2}} = 0 \\
a_{\mathbf{N}} = \frac{\|\mathbf{v} \times \mathbf{a}\|}{\|\mathbf{v}\|} &= \frac{\left\| \left\langle -\sin t, \cos t, 1 \right\rangle \times \left\langle -\cos t, -\sin t, 0 \right\rangle \right\|}{\sqrt{\sin^2 t + \cos^2 t + 1}} \\
&= \frac{\left\| \left\langle \sin t, -\cos t, \sin^2 t + \cos^2 t \right\rangle \right\|}{\sqrt{2}} \\
&= \frac{\left\| \left\langle \sin t, -\cos t, 1 \right\rangle \right\|}{\sqrt{2}} = \frac{\sqrt{\sin^2 t + \cos^2 t + 1}}{\sqrt{2}} = 1.
\end{aligned}
$$

Applications

25. From the first equation, $\cos \theta = \frac{40}{44t} = \frac{10}{11t}$. From the second equation, $\sin \theta = \frac{16t^2 - 4}{44t} = \frac{4t^2 - 1}{11t}$. Then

$$1 = \sin^2 \theta + \cos^2 \theta = \frac{100 + (4t^2 - 1)^2}{121t^2} = \frac{16t^4 - 8t^2 + 101}{121t^2},$$

so that $16t^4 - 8t^2 + 101 = 121t^2$ and thus $16t^4 - 129t^2 + 101 = 0$. Solving this equation numerically gives $t \approx \pm 0.9374$ and $t \approx \pm 2.6803$. Since $t \geq 0$, we reject the negative roots. When $t \approx 0.9374$, then $\cos\theta \approx \frac{10}{11\cdot 0.9374}$ and then $\theta \approx 14.12°$. When $t \approx 2.6803$, then $\cos\theta \approx \frac{10}{11\cdot 2.6803}$ and then $\theta \approx 70.17°$.

27. (a) We have

$$\mathbf{r}(t) = \left\langle (\|\mathbf{v_0}\|\cos\theta)t, -\frac{1}{2}gt^2 + (\|\mathbf{v_0}\|\sin\theta)t + h \right\rangle = \left\langle \frac{\sqrt{3}}{2}\|\mathbf{v_0}\|t, -4.9t^2 + \frac{1}{2}\|\mathbf{v_0}\|t \right\rangle.$$

He hit the ground when the y coordinate was zero, i.e. when $-4.9t^2 + \frac{1}{2}\|\mathbf{v_0}\|t = 0$. This occurs at $t = 0$, when he leaves the ground, and at $t = \frac{1}{9.8}\|\mathbf{v_0}\| \approx 0.1\|\mathbf{v_0}\|$, when he lands. The horizontal distance he travels is

$$8.9 = \frac{\sqrt{3}}{2}\|\mathbf{v_0}\|\,(0.1\|\mathbf{v_0}\|) \approx 0.087\|\mathbf{v_0}\|^2,$$

so that $\|\mathbf{v_0}\| \approx 10.11$ meters per second.

(b) On the Moon, the equation of motion would have been (with $\|\mathbf{v_0}\| = 10.11$)

$$\mathbf{r}(t) = \left\langle (\|\mathbf{v_0}\|\cos\theta)t, -\frac{1}{2}gt^2 + (\|\mathbf{v_0}\|\sin\theta)t + h \right\rangle$$

$$= \left\langle \frac{\sqrt{3}}{2}\cdot 10.11t, -0.8t^2 + \frac{1}{2}\cdot 10.11t \right\rangle$$

$$= \left\langle 8.76t, -0.8t^2 + 5.06t \right\rangle.$$

Thus he would hit the ground when $t = \frac{5.06}{0.8} \approx 6.33$; at that time he would have traveled $6.33 \cdot 8.76 \approx 55.45$ meters.

29. (a) Since

$$100.9 \text{ miles/hour} = 100.9 \cdot 5280 \text{ feet/mile} \cdot \frac{1}{3600} \text{ hours/second} \approx 150 \text{ feet/second},$$

the equation of motion is

$$\mathbf{r}(t) = \langle x(t), y(t) \rangle = \left\langle 150t\cos\theta, -16t^2 + 150t\sin\theta + 6 \right\rangle.$$

The ball is 60 feet away when $150t\cos\theta = 60$, so when $t = \frac{2}{5\cos\theta}$. Since $y(t) = 3$ at that time, we must have

$$3 = -16\left(\frac{2}{5\cos\theta}\right)^2 + 150\left(\frac{2}{5\cos\theta}\right)\sin\theta + 6.$$

Solving numerically gives an infinite number of solutions; the four nearest zero are in degrees $\theta \approx 179.58$, $\theta \approx 87.56$, $\theta \approx -0.42$, and $\theta \approx -92.44$. Of these, the only one that makes sense is $\theta \approx -0.42°$. So the angle of elevation was 0.42° below horizontal at the time of release.

(b) The equation of motion from part (a), with $\cos\theta = \sin\theta = \frac{\sqrt{2}}{2}$, is

$$\mathbf{r}(t) = \langle x(t), y(t) \rangle = \left\langle 75t\sqrt{2}, -16t^2 + 75t\sqrt{2} + 6 \right\rangle.$$

Then the ball hits the ground when $-16t^2 + 75t\sqrt{2} + 6 = 0$, or when $t \approx -0.06$ or $t \approx 6.69$ seconds. Rejecting the negative root, the ball hits the ground after 6.69 seconds, at which time it has travelled $x(6.69) = 75 \cdot 6.69\sqrt{2} \approx 709.6$ feet.

31. The equation of motion of the projectile is (from part (b) of Exercise 30)

$$\mathbf{r}(t) = \langle x(t), y(t) \rangle = \left\langle 1500t \cos\theta, -4.9t^2 + 1500t \sin\theta + 5 \right\rangle.$$

We wish to find the maximum height, i.e., to maximize $y(t)$. The projectile is at its maximum height when $0 = y'(t) = -9.8t + 1500 \sin\theta$, which is when $t = \frac{1500}{9.8} \sin\theta \approx 153.06 \sin\theta$. The maximum height of the projectile is thus

$$y(153.06 \sin\theta) = -4.9(153.06 \sin\theta)^2 + 1500(153.06 \sin\theta) \sin\theta + 5 \approx 114976 \sin^2\theta + 5.$$

Since $\sin\theta$ is increasing on $[0°, 60°]$, it reaches its maximum at $\theta = 60°$, so that the maximum height of the projectile is

$$114976 \sin^2 60° + 5 = 114976 \left(\frac{\sqrt{3}}{2}\right)^2 + 5 = 86102 \text{ m} \approx 86 \text{ km}.$$

Proofs

33. Since the initial and final heights are equal, we may as well assume that they are both zero. We also assume the initial velocity is fixed, at $\mathbf{v}_0$. Then the equation of motion is

$$\mathbf{r}(t) = \left\langle (\|\mathbf{v}_0\| \cos\theta)t, -\frac{1}{2}gt^2 + (\|\mathbf{v}_0\| \sin\theta)t \right\rangle.$$

The ball hits the ground when $y(t) = -\frac{1}{2}gt^2 + (\|\mathbf{v}_0\| \sin\theta)t = 0$. Solving gives

$$t = 0, \qquad t = \frac{2\|\mathbf{v}_0\| \sin\theta}{g}.$$

So when the ball hits the ground, its horizontal distance will be

$$x\left(\frac{2\|\mathbf{v}_0\| \sin\theta}{g}\right) = \|\mathbf{v}_0\| \cos\theta \left(\frac{2\|\mathbf{v}_0\| \sin\theta}{g}\right) = \frac{2\|\mathbf{v}_0\|^2}{g} \sin\theta \cos\theta.$$

For simplicity, write $K = \frac{2\|\mathbf{v}_0\|^2}{g}$, which is a positive constant. Then we want to maximize $x(\theta) = K \sin\theta \cos\theta$ with respect to θ. We have $x'(\theta) = K \cos^2\theta - K \sin^2\theta = K \cos 2\theta$, which is zero for $\theta = \frac{\pi}{4} = 45°$. (The angle of elevation is between $0°$ and $90°$, so this is the only valid solution). Also, since $x''(\theta) = -2K \sin 2\theta$, we have $x''(45°) = -2K < 0$, so that $45°$ is a maximum. Hence the maximum horizontal position is achieved with a launch angle of $45°$.

35. Since $\mathbf{r}''(t) = \langle 0, 0 \rangle$, integrating gives $\mathbf{r}'(t) = \langle C_1, C_2 \rangle$. Integrate again to get

$$\mathbf{r}(t) = \langle C_1 t + D_1, C_2 t + D_2 \rangle = \langle D_1, D_2 \rangle + t \langle C_1, C_2 \rangle,$$

which is the vector equation for a straight line.

37. If $y = f(x)$, we can write that in vector form as $\mathbf{r}(t) = \langle t, f(t) \rangle$. Then $\mathbf{v}(t) = \mathbf{r}'(t) = \langle 1, f'(t) \rangle$ and $\mathbf{a}(t) = \mathbf{v}'(t) = \langle 0, f''(t) \rangle$. Thinking of the graph as a space curve lying in the xy plane, we get for the normal component of acceleration

$$a_{\mathbf{N}} = \frac{\|\mathbf{v}(t) \times \mathbf{a}(t)\|}{\|\mathbf{v}(t)\|} = \frac{\|\langle 1, f'(t), 0 \rangle \times \langle 0, f''(t), 0 \rangle\|}{\|\mathbf{v}(t)\|}.$$

At an inflection point, $f''(t) = 0$, so this becomes

$$a_{\mathbf{N}} = \frac{\|\langle 0, f'(t), 0 \rangle \times \mathbf{0}\|}{\|\mathbf{v}(t)\|} = 0.$$

Thinking Forward

The graph of a function $f : \mathbb{R}^2 \to \mathbb{R}$

For each pair (x, y), the function assigns a third number, $f(x, y)$. If we think of $f(x, y)$ as the z coordinate in $\mathbb{R}^3$, we see that the graph of such a function is a surface, in which exactly one point of the surface lies above each point (x, y) in the domain of f.

The graph of a function $f : \mathbb{R}^3 \to \mathbb{R}$

For each triple (x, y, z), the function assigns a fourth number, $f(x, y, z)$. If we think of $f(x, y, z)$ as the w coordinate in $\mathbb{R}^4$, we see that the graph of such a function is a hypersurface, in which exactly one point of the hypersurface lies above each point (x, y, z) in the domain of f.

Chapter Review and Self-Test

1. A plot of $\mathbf{r}(t) = \langle t, t^3 \rangle$ is

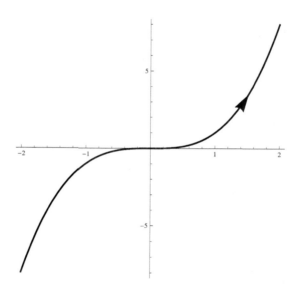

3. A plot of $\mathbf{r}(t) = \langle t, \sin 2t, \cos 2t \rangle$ is

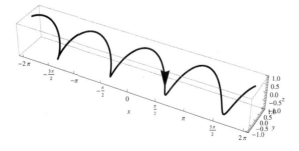

5. We have

$$\lim_{t\to 0}\left\langle \frac{\frac{1}{3+t}-\frac{1}{3}}{t}, \frac{(3+t)^2-9}{t} \right\rangle = \left\langle \lim_{t\to 0}\frac{\frac{1}{3+t}-\frac{1}{3}}{t}, \lim_{t\to 0}\frac{(3+t)^2-9}{t} \right\rangle$$

$$= \left\langle \lim_{t\to 0}\frac{3-(3+t)}{3t(t+3)}, \lim_{t\to 0}\frac{9+6t+t^2-9}{t} \right\rangle$$

$$= \left\langle \lim_{t\to 0}\frac{-t}{3t(t+3)}, \lim_{t\to 0}\frac{6t+t^2}{t} \right\rangle$$

$$= \left\langle \lim_{t\to 0}\frac{-1}{3(t+3)}, \lim_{t\to 0}(t+6) \right\rangle$$

$$= \left\langle -\frac{1}{9}, 6 \right\rangle.$$

7. We have

$$\lim_{t\to 0}\left\langle \frac{\sin t}{t}, \frac{1-\cos t}{t}, (1+t)^{1/t} \right\rangle = \left\langle \lim_{t\to 0}\frac{\sin t}{t}, \lim_{t\to 0}\frac{1-\cos t}{t}, \lim_{t\to 0}(1+t)^{1/t} \right\rangle = \langle 1,0,1 \rangle.$$

9. Differentiating gives

$$\mathbf{v}(t) = \mathbf{r}'(t) = \langle 1,2,3 \rangle, \qquad \mathbf{a}(t) = \mathbf{v}'(t) = \langle 0,0,0 \rangle.$$

11. Differentiating gives

$$\mathbf{v}(t) = \mathbf{r}'(t) = \langle -2\sin 2t, 3\cos 3t \rangle, \qquad \mathbf{a}(t) = \mathbf{v}'(t) = \langle -4\cos 2t, -9\sin 3t \rangle.$$

13. Since $\mathbf{r}'(t) = \langle 1,3t^2 \rangle$, we get $r'(2) = \langle 1,12 \rangle$, so that

$$\mathbf{T}(2) = \frac{\mathbf{r}'(2)}{\|\mathbf{r}'(2)\|} = \frac{\langle 1,12 \rangle}{\sqrt{1^2+12^2}} = \left\langle \frac{1}{\sqrt{145}}, \frac{12}{\sqrt{145}} \right\rangle.$$

15. Since $\mathbf{r}'(t) = \langle \alpha\beta\cos\beta t, -\alpha\beta\sin\beta t \rangle$, we get $\mathbf{r}'(0) = \langle \alpha\beta,0 \rangle$, so that

$$\mathbf{T}(0) = \frac{\mathbf{r}'(0)}{\|\mathbf{r}'(0)\|} = \frac{\langle \alpha\beta,0 \rangle}{\sqrt{\alpha^2\beta^2+0}} = \left\langle \frac{\alpha\beta}{|\alpha\beta|}, 0 \right\rangle.$$

17. Since $\mathbf{r}'(t) = \langle 1, 2\sin t + 2t\cos t, 2\cos t - 2t\sin t \rangle$, we get

$$\mathbf{r}'(\pi) = \langle 1, 2\sin\pi + 2\pi\cos\pi, 2\cos\pi - 2\pi\sin\pi \rangle = \langle 1, -2\pi, -2 \rangle,$$

so that

$$\mathbf{T}(\pi) = \frac{\mathbf{r}'(\pi)}{\|\mathbf{r}'(\pi)\|} = \frac{\langle 1,-2\pi,-2 \rangle}{\sqrt{1^2+(2\pi)^2+2^2}} = \left\langle \frac{1}{\sqrt{5+4\pi^2}}, -\frac{2\pi}{\sqrt{5+4\pi^2}}, -\frac{2}{\sqrt{5+4\pi^2}} \right\rangle.$$

19. From Exercise 13, $\mathbf{r}'(t) = \langle 1,3t^2 \rangle$, so that

$$\mathbf{T}(t) = \frac{\mathbf{r}'(t)}{\|\mathbf{r}'(t)\|} = \frac{\langle 1,3t^2 \rangle}{\sqrt{1+9t^4}} = \left\langle \frac{1}{\sqrt{1+9t^4}}, \frac{3t^2}{\sqrt{1+9t^4}} \right\rangle$$

$$\mathbf{T}'(t) = \left\langle -\frac{36t^3}{2(1+9t^4)^{3/2}}, \frac{(1+9t^4)^{1/2}(6t)-3t^2\left(\frac{1}{2}(1+9t^4)^{-1/2}(36t^3)\right)}{1+9t^4} \right\rangle$$

$$= \left\langle -\frac{18t^3}{(1+9t^4)^{3/2}}, \frac{(1+9t^4)(6t)-54t^5}{(1+9t^4)^{3/2}} \right\rangle$$

$$= \left\langle -\frac{18t^3}{(1+9t^4)^{3/2}}, \frac{6t}{(1+9t^4)^{3/2}} \right\rangle.$$

Then

$$\mathbf{T}'(2) = \left\langle -\frac{144}{145\sqrt{145}}, \frac{12}{145\sqrt{145}} \right\rangle,$$

$$\|\mathbf{T}'(2)\| = \sqrt{\frac{144^2}{145^3} + \frac{12^2}{145^3}} = \frac{12}{145}\sqrt{\frac{145}{145}} = \frac{12}{145}.$$

Thus

$$\mathbf{N}(2) = \frac{\mathbf{T}'(2)}{\|\mathbf{T}'(2)\|} = \frac{145}{12}\left\langle -\frac{144}{145\sqrt{145}}, \frac{12}{145\sqrt{145}} \right\rangle = \left\langle -\frac{12}{\sqrt{145}}, \frac{1}{\sqrt{145}} \right\rangle.$$

21. From Exercise 15, $\mathbf{r}'(t) = \langle \alpha\beta\cos\beta t, -\alpha\beta\sin\beta t \rangle$, so that (assuming α and β are positive)

$$\mathbf{T}(t) = \frac{\mathbf{r}'(t)}{\|\mathbf{r}'(t)\|} = \frac{\langle \alpha\beta\cos\beta t, -\alpha\beta\sin\beta t \rangle}{\sqrt{\alpha^2\beta^2\cos^2 t + \alpha^2\beta^2\sin^2 t}} = \langle \cos\beta t, -\sin\beta t \rangle$$

$$\mathbf{T}'(t) = \langle -\beta\sin\beta t, -\beta\cos\beta t \rangle$$

$$\|\mathbf{T}'(t)\| = \sqrt{\beta^2\sin^2\beta + \beta^2\cos^2\beta} = \beta.$$

Thus

$$\mathbf{N}(0) = \frac{\mathbf{T}'(0)}{\|\mathbf{T}'(0)\|} = \frac{\langle 0, -\beta \rangle}{\sqrt{0^2 + \beta^2}} = \langle 0, -1 \rangle.$$

23. From Exercise 17, $\mathbf{r}'(t) = \langle 1, 2\sin t + 2t\cos t, 2\cos t - 2t\sin t \rangle$, so that

$$\mathbf{T}(t) = \frac{\mathbf{r}'(t)}{\|\mathbf{r}'(t)\|}$$
$$= \frac{\langle 1, 2\sin t + 2t\cos t, 2\cos t - 2t\sin t \rangle}{\sqrt{1 + 4 + 4t^2}}$$
$$= \left\langle \frac{1}{\sqrt{5+4t^2}}, \frac{2\sin t + 2t\cos t}{\sqrt{5+4t^2}}, \frac{2\cos t - 2t\sin t}{\sqrt{5+4t^2}} \right\rangle$$

$$\mathbf{T}'(t) = \left\langle -\frac{4t}{(5+4t^2)^{3/2}}, \right.$$
$$\frac{(5+4t^2)^{1/2}(4\cos t - 2t\sin t) - (2\sin t + 2t\cos t)\left(\frac{1}{2}(5+4t^2)^{-1/2}(8t)\right)}{5+4t^2},$$
$$\left. \frac{(5+4t^2)^{1/2}(-4\sin t - 2t\cos t) - (2\cos t - 2t\sin t)\left(\frac{1}{2}(5+4t^2)^{-1/2}(8t)\right)}{5+4t^2} \right\rangle$$

$$= \left\langle -\frac{4t}{(5+4t^2)^{3/2}}, \right.$$
$$\frac{(5+4t^2)(4\cos t - 2t\sin t) - 4t(2\sin t + 2t\cos t)}{(5+4t^2)^{3/2}},$$
$$\left. \frac{(5+4t^2)(-4\sin t - 2t\cos t) - 4t(2\cos t - 2t\sin t)}{(5+4t^2)^{3/2}} \right\rangle$$
$$= \left\langle -\frac{4t}{(5+4t^2)^{3/2}}, \frac{4(5+2t^2)\cos t - 2t(9+4t^2)\sin t}{(5+4t^2)^{3/2}}, \frac{-2t(9+4t^2)\cos t - 4(5+2t^2)\sin t}{(5+4t^2)^{3/2}} \right\rangle$$

Then

$$\mathbf{T}'(\pi) = \left\langle -\frac{4\pi}{(5+4\pi^2)^{3/2}}, -\frac{4(5+2\pi^2)}{(5+4\pi^2)^{3/2}}, \frac{2\pi(9+4\pi^2)}{(5+4\pi^2)^{3/2}} \right\rangle$$

$$\|\mathbf{T}'(\pi)\| = \frac{\sqrt{16\pi^2 + 16(5+2\pi^2)^2 + 4\pi^2(9+4\pi^2)^2}}{(5+4\pi^2)^{3/2}}$$

$$= \frac{\sqrt{64\pi^6 + 352\pi^4 + 660\pi^2 + 400}}{(5+4\pi^2)^{3/2}}$$

$$= \frac{2\sqrt{(5+4\pi^2)(4\pi^4 + 17\pi^2 + 20)}}{(5+4\pi^2)^{3/2}}$$

$$= \frac{2\sqrt{4\pi^4 + 17\pi^2 + 20}}{5+4\pi^2}.$$

Thus

$$\mathbf{N}(\pi) = \frac{\mathbf{T}'(\pi)}{\|\mathbf{T}'(\pi)\|}$$

$$= \frac{5+4\pi^2}{2\sqrt{4\pi^4 + 17\pi^2 + 20}} \left\langle -\frac{4\pi}{(5+4\pi^2)^{3/2}}, -\frac{4(5+2\pi^2)}{(5+4\pi^2)^{3/2}}, \frac{2\pi(9+4\pi^2)}{(5+4\pi^2)^{3/2}} \right\rangle$$

$$= \frac{1}{\sqrt{4\pi^4 + 17\pi^2 + 20}} \left\langle -\frac{2\pi}{\sqrt{5+4\pi^2}}, -\frac{2(5+2\pi^2)}{\sqrt{5+4\pi^2}}, \frac{\pi(9+4\pi^2)}{\sqrt{5+4\pi^2}} \right\rangle.$$

25. From Exercise 19,

$$\mathbf{T}(2) = \left\langle \frac{1}{\sqrt{145}}, \frac{12}{\sqrt{145}} \right\rangle, \qquad \mathbf{N}(2) = \left\langle -\frac{12}{\sqrt{145}}, \frac{1}{\sqrt{145}} \right\rangle.$$

Then (regarding the curve as a space curve lying in the xy plane)

$$\mathbf{B}(2) = \mathbf{T}(2) \times \mathbf{N}(2)$$

$$= \frac{1}{145}\langle 1, 12, 0 \rangle \times \langle -12, 1, 0 \rangle$$

$$= \frac{1}{145}\langle 0, 0, 145 \rangle$$

$$= \langle 0, 0, 1 \rangle.$$

Since $\mathbf{r}(2) = \langle 2, 2^3, 0 \rangle$, we get for the equation of the osculating plane

$$\mathbf{B}(2) \cdot \langle x - 2, y - 8, z - 0 \rangle = 0, \quad \text{or} \quad z - 0 = 0, \quad \text{or} \quad z = 0.$$

(Since the curve is planar and lies in the xy plane, it is obvious that this is its osculating plane).

27. From Exercise 21,

$$\mathbf{T}(0) = \langle \cos 0, -\sin 0 \rangle = \langle 1, 0 \rangle, \qquad \mathbf{N}(0) = \langle 0, -1 \rangle,$$

so that (regarding the curve as a space curve lying in the xy plane)

$$\mathbf{B}(0) = \langle 1, 0, 0 \rangle \times \langle 0, -1, 0 \rangle = \langle 0, 0, -1 \rangle.$$

Since $\mathbf{r}(0) = \langle 0, \alpha \rangle$, the equation of the osculating plane is

$$\mathbf{B}(0) \cdot \langle x - 0, y - \alpha, z - 0 \rangle = \langle 0, 0, -1 \rangle \cdot \langle x, y - \alpha, z \rangle = -z = 0,$$

or $z = 0$. (Since the curve is planar and lies in the xy plane, it is obvious that this is its osculating plane).

29. From Exercise 23,

$$\mathbf{T}(\pi) = \left\langle \frac{1}{\sqrt{5+4\pi^2}}, -\frac{2\pi}{\sqrt{5+4\pi^2}}, -\frac{2}{\sqrt{5+4\pi^2}} \right\rangle$$

$$\mathbf{N}(\pi) = \frac{1}{\sqrt{4\pi^4+17\pi^2+20}} \left\langle -\frac{2\pi}{\sqrt{5+4\pi^2}}, -\frac{2(5+2\pi^2)}{\sqrt{5+4\pi^2}}, \frac{\pi(9+4\pi^2)}{\sqrt{5+4\pi^2}} \right\rangle.$$

Then

$$\mathbf{B}(\pi) = \mathbf{T}(\pi) \times \mathbf{N}(\pi)$$

$$= \frac{1}{(5+4\pi^2)\sqrt{4\pi^4+17\pi^2+20}} \langle 1, -2\pi, -2 \rangle \times \left\langle -2\pi, -10-4\pi^2, 9\pi+4\pi^3 \right\rangle$$

$$= \frac{1}{(5+4\pi^2)\sqrt{4\pi^4+17\pi^2+20}} \left\langle -8\pi^4 - 26\pi^2 - 20, -4\pi^3 - 5\pi, -8\pi^2 - 10 \right\rangle.$$

The equation of the osculating plane is

$$\mathbf{B}(\pi) \cdot (\langle x, y, z \rangle - \mathbf{r}(\pi)) = 0,$$

so that the constant factor in front of $\mathbf{B}(\pi)$ is irrelevant. Since $\mathbf{r}(\pi) = \langle \pi, 0, -2\pi \rangle$, the equation for the osculating plane is

$$\left\langle -8\pi^4 - 26\pi^2 - 20, -4\pi^3 - 5\pi, -8\pi^2 - 10 \right\rangle \cdot \langle x - \pi, y, z + 2\pi \rangle = 0, \text{ or}$$

$$(8\pi^4 + 26\pi^2 + 20)(x - \pi) + (4\pi^3 + 5\pi)y + (8\pi^2 + 10)(z + 2\pi) = 0.$$

31. The curvature of $\mathbf{r}(t)$ at $t = 2$, from Theorem 11.23 (and using Exercise 19) is

$$\kappa = \frac{\|\mathbf{T}'(2)\|}{\|\mathbf{r}'(2)\|} = \frac{\frac{12}{145}}{\|\langle 1, 12 \rangle\|} = \frac{12}{145\sqrt{145}}.$$

Then by Definition 11.25, the radius of the osculating circle is $\rho = \frac{1}{\kappa} = \frac{145\sqrt{145}}{12}$ and its center is at

$$\mathbf{r}(2) + \rho\mathbf{N}(2) = \langle 2, 8 \rangle + \frac{145\sqrt{145}}{12} \left\langle -\frac{12}{\sqrt{145}}, \frac{1}{\sqrt{145}} \right\rangle = \langle 2, 8 \rangle + \left\langle -145, \frac{145}{12} \right\rangle = \left\langle -143, \frac{241}{12} \right\rangle.$$

Thus the equation of the osculating circle is

$$(x + 143)^2 + \left(y - \frac{241}{12} \right)^2 = \frac{145^3}{144}.$$

33. The curvature of $\mathbf{r}(t)$ at $t = 0$, from Theorem 11.23 (and using Exercise 21) is

$$\kappa = \frac{\|\mathbf{T}'(0)\|}{\|\mathbf{r}'(0)\|} = \frac{\beta}{\|\langle \alpha\beta, 0 \rangle\|} = \frac{\beta}{\alpha\beta} = \frac{1}{\alpha}.$$

Then by Definition 11.25, the radius of the osculating circle is $\rho = \frac{1}{\kappa} = \alpha$ and its center is at

$$\mathbf{r}(0) + \rho\mathbf{N}(0) = \langle 0, \alpha \rangle + \alpha \langle 0, -1 \rangle = \langle 0, 0 \rangle.$$

Thus the equation of the osculating circle is $x^2 + y^2 = \alpha^2$.

35. The curvature of $\mathbf{r}(t)$ at $t = \pi$, from Theorem 11.23 (and using Exercise 23) is

$$\kappa = \frac{\|\mathbf{T}'(\pi)\|}{\|\mathbf{r}'(\pi)\|} = \frac{\frac{2\sqrt{4\pi^4 + 17\pi^2 + 20}}{5 + 4\pi^2}}{\sqrt{5 + 4\pi^2}} = \frac{2\sqrt{4\pi^4 + 17\pi^2 + 20}}{(5 + 4\pi^2)^{3/2}}.$$

Then by Definition 11.25, the radius of the osculating circle is

$$\rho = \frac{1}{\kappa} = \frac{(5 + 4\pi^2)^{3/2}}{2\sqrt{4\pi^4 + 17\pi^2 + 20}}$$

and its center is at

$$\mathbf{r}(\pi) + \rho \mathbf{N}(\pi) = \langle \pi, 0, -2\pi \rangle + \frac{(5 + 4\pi^2)^{3/2}}{2\sqrt{4\pi^4 + 17\pi^2 + 20}}$$
$$\cdot \frac{1}{\sqrt{4\pi^4 + 17\pi^2 + 20}\sqrt{5 + 4\pi^2}} \left\langle -2\pi, -2(5 + 2\pi^2), \pi(9 + \pi^2) \right\rangle$$
$$= \langle \pi, 0, -2\pi \rangle + \frac{5 + 4\pi}{2(4\pi^4 + 17\pi^2 + 20)} \left\langle -2\pi, -2(5 + 2\pi^2), \pi(9 + \pi^2) \right\rangle.$$

37. Using Theorem 11.24, since $f'(x) = 2x$ and $f''(x) = 2$, the curvature is

$$\kappa = \frac{|f''(x)|}{(1 + f'(x)^2)^{3/2}} = \frac{2}{(1 + 4x^2)^{3/2}},$$

so that at $x = 0$ it is $\kappa = 2$. A tangent vector to the curve is $\langle 1, f'(x) \rangle$, which at $x = 0$ is the unit vector $\mathbf{T}(0) = \langle 1, 0 \rangle$, so that the two unit normals are $\langle 0, \pm 1 \rangle$. Since the curve is concave up, we choose the normal that points upwards, which is $\mathbf{N}(0) = \langle 0, 1 \rangle$. Then the radius of the osculating circle is $\rho = \frac{1}{\kappa} = \frac{1}{2}$, and its center is at

$$\mathbf{r}(0) + \rho \mathbf{N}(0) = \langle 0, 0 \rangle + \frac{1}{2} \langle 0, 1 \rangle = \left\langle 0, \frac{1}{2} \right\rangle.$$

Thus the equation for the osculating circle is $x^2 + \left(y - \frac{1}{2} \right)^2 = \frac{1}{2}^2 = \frac{1}{4}$.

39. Using Theorem 11.24, since $f'(x) = f''(x) = e^x$, the curvature is

$$\kappa = \frac{|e^x|}{(1 + e^{2x})^{3/2}},$$

so that at $x = 0$ it is $\kappa = \frac{1}{2^{3/2}} = \frac{1}{2\sqrt{2}}$. A tangent vector to the curve is $\langle 1, f'(x) \rangle$, which at $x = 0$ is $\langle 1, e^0 \rangle = \langle 1, 1 \rangle$. The norm of this vector is $\sqrt{2}$, so that the unit tangent vector is $\left\langle \frac{1}{\sqrt{2}}, \frac{1}{\sqrt{2}} \right\rangle$. Thus the two unit normals at 0 are $\left\langle -\frac{1}{\sqrt{2}}, \frac{1}{\sqrt{2}} \right\rangle$ and $\left\langle \frac{1}{\sqrt{2}}, -\frac{1}{\sqrt{2}} \right\rangle$. Since e^x is concave up everywhere, we choose $\mathbf{N}(0) = \left\langle -\frac{1}{\sqrt{2}}, \frac{1}{\sqrt{2}} \right\rangle$. Then the radius of the osculating circle is $\frac{1}{\kappa} = 2\sqrt{2}$, and its center is at

$$\mathbf{r}(0) + \rho \mathbf{N}(0) = \langle 0, 1 \rangle + 2\sqrt{2} \left\langle -\frac{1}{\sqrt{2}}, \frac{1}{\sqrt{2}} \right\rangle = \langle -2, 3 \rangle.$$

Thus the equation of the osculating circle is $(x + 2)^2 + (y - 3)^2 = 8$.

Part V

Multivariable Calculus

Chapter 12

Multivariable Functions

12.1 Functions of Two and Three Variables

Thinking Back

The definition of a function

Definition 0.1 simply says that a function f assigns to each element x of the *domain* set exactly one element $f(x)$ of the target set. In the event that f is a function of two (or three) variables, the domain consists of ordered pairs (or ordered triples) of coordinates. Nothing in the definition of a function implies that the domain must consist of a set of real numbers.

The domain and range of a function of two variables

The definitions of domain and range in Definition 0.2 explicitly restrict those concepts to apply to functions between subsets of $\mathbb{R}$, the real numbers. Thus if a function is a function of two or three variables, those definitions will not apply.

Concepts

1. (a) True. The two variables are the x coordinate and the y coordinate. See Definition 12.1.

    (b) False. The range is typically a subset of $\mathbb{R}$ (see Definition 12.1), although one may define functions from $\mathbb{R}^2$ to (say) $\mathbb{R}^2$.

    (c) True. Since the function has domain a subset of $\mathbb{R}^2$ and range $\mathbb{R}$, the graph may be thought of as the set of points in $\mathbb{R}^3$ of the form $(x, y, f(x, y))$.

    (d) False. The domain is a subset of $\mathbb{R}^3$ (see Definition 12.3).

    (e) True. The range is the set of possible values, which is a set of real numbers (see Definition 12.3).

    (f) True. Since the function has domain a subset of $\mathbb{R}^3$ and range $\mathbb{R}$, the graph may be thought of as the set of points in $\mathbb{R}^4$ of the form $(x, y, z, f(x, y, z))$.

    (g) True. Such a function is of the form $z = ax + by + c$, which is the same as $ax + by - z = c$, which is the equation of a plane. See also Exercise 69.

    (h) True. See the discussion following Definition 12.2. A point $(a, f(a))$ in the xz plane on the graph of f becomes under rotation the circle $x^2 + y^2 = a^2$ at the level $z = f(a)$.

3. To each element $x \in \mathbb{R}$ in the domain of f, we get a single point $f(x) \in \mathbb{R}$. Thus the set of points on the curve, which is its graph, is $\{(x, f(x)) \mid x \text{ is in the domain of } f\} \subseteq \mathbb{R}^2$.

5. To each element $(x, y, z) \in \mathbb{R}^3$ in the domain of f, we get a single point $f(x, y, z) \in \mathbb{R}$. Thus the set of points on the curve, which is its graph, is $\{(x, y, z, f(x, y, z)) \mid (x, y, z) \text{ is in the domain of } f\} \subseteq \mathbb{R}^4$.

7. (a) The graph of $f(x) = \sqrt{1 - x^2}$ is the graph of the upper half of the circle $x^2 + y^2 = 1$:

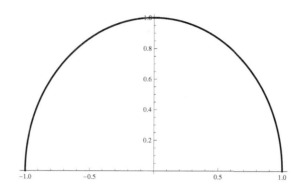

(b) The graph of $f(x, y) = \sqrt{1 - x^2 - y^2}$ is graph of the upper half of the sphere $x^2 + y^2 + z^2 = 1$:

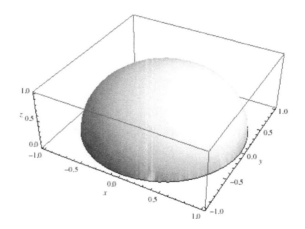

(c) Squaring both sides and simplifying gives $x^2 + y^2 + z^2 + f(x, y, z)^2 = 1$, which is the set of points in $\mathbb{R}^4$ at distance 1 from the origin. This is analogous to a sphere in $\mathbb{R}^3$, hence the term hypersphere. In the form given, $f(x, y, z) = \sqrt{1 - x^2 - y^2 - z^2}$, we are restricted to the positive square root, so we get only the graph of the upper half of the sphere.

(d) The graph of $f(x_1, x_2, \ldots, x_n) = \sqrt{1 - x_1^2 - x_2^2 - \cdots - x_n^2}$ is half of a hypersphere of dimension n in $\mathbb{R}^{n+1}$.

9. There is a point above (x, y) on the graph of f exactly when $f(x, y)$ is defined, which is to say, exactly when (x, y) is in the domain of f. But there is a point above (x, y) on the graph of f if and only if there is some z such that (x, y, z) is on the graph of f.

11. There is a point above (x, y, z) on the graph of f exactly when $f(x, y, z)$ is defined, which is to say, exactly when (x, y, z) is in the domain of f. But there is a point above (x, y, z) on the graph of f if and only if there is some w such that (x, y, z, w) is on the graph of f.

13. For example, $z = x^2 + y^2$. The level curve at $z = 0$ is the single point $(0,0,0)$:

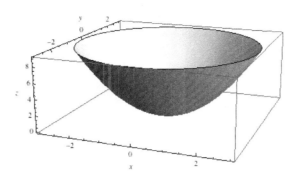

15. For example, in the surface below the level curve at $z = 4$ consists of the circle $x^2 + y^2 = 16$ together with the point $(0,0,4)$.

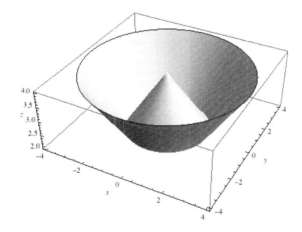

17. For example, the surface in Exercise 13, $z = x^2 + y^2$. The level curve at $z = 0$ is the single point $(0,0,0)$, while the level curve at any $z_0 > 0$ is the circle of radius $\sqrt{z_0}$:

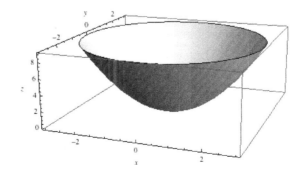

19. For example, the surface in Exercise 15. The level curves for $0 < z < 4$ consist of two concentric circles:

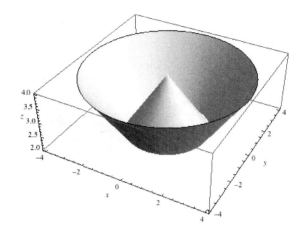

21. For example, the surface below:

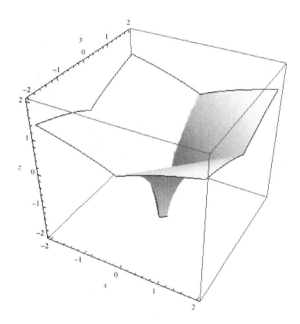

Skills

23. We have

$$f(1,5) = 1^2 - 5^2 = -24$$
$$f(-3,-2) = (-3)^2 - (-2)^2 = 9 - 4 = 5.$$

This function is defined for all x and y, so its domain is $\mathbb{R}^2$. Its range is all of $\mathbb{R}$: letting $y = 0$ we can vary x to get any positive real; letting $x = 0$ we can vary y to get any negative real, and finally, $f(a, a) = 0$.

25. We have

$$g(-e, -1) = \frac{\ln((-e)(-1))}{\sqrt{1-(-e)}} = \frac{\ln e}{\sqrt{e+1}} = \frac{1}{\sqrt{e+1}}$$

$$g\left(\frac{1}{2}, 2\right) = \frac{\ln\left(\frac{1}{2} \cdot 2\right)}{\sqrt{1-\frac{1}{2}}} = \frac{\ln 1}{\sqrt{\frac{1}{2}}} = 0.$$

g is defined only for $x < 1$, since if $x \geq 1$ the denominator is either zero or is undefined. In addition, however, we must also have $xy > 0$, since otherwise the numerator is not defined. Thus the domain of g is

$$\{(x, y) \mid xy > 0, x < 1\}$$

This consists of the third quadrant (minus the two axes) together with a vertical strip in the first quadrant between $x = 0$ and $x = 1$. g has range $\mathbb{R}$. To see this, let $x = \frac{1}{2}$, say; then for any given y, the value of the function is $\frac{\ln\left(\frac{y}{2}\right)}{\sqrt{\frac{1}{2}}}$, and it is clear that by choosing y appropriately we can get this expression to equal any given real number.

27. We have

$$f(1, 0, -5) = 1^2 + 0^2 + (-5)^2 = 26$$

$$f\left(\frac{1}{2}, -1, \frac{1}{3}\right) = \left(\frac{1}{2}\right)^2 + (-1)^2 + \left(\frac{1}{3}\right)^2 = \frac{49}{36}.$$

The domain of f is all of $\mathbb{R}^3$. Since x^2, y^2, and z^2 are always nonnegative, the range is all nonnegative real numbers. (If $a \geq 0$, then $f(\sqrt{a}, 0, 0) = a$).

29. $f_1(g_1(t), g_2(t)) = f_1(\sin t, \cos t) = \sin^2 t + \cos^2 t = 1.$

31. $g_1(g_2(t)) = g_1(\cos t) = \sin(\cos t).$

33. $f_3(g_2(t), g_1(t), g_3(t)) = f_3(\cos t, \sin t, 1 - t) = \frac{\cos t + \sin t}{1 - t + \sin t}.$

35. Since $g_3(t)$ outputs a real number, and $\mathbf{r}_2$ takes as its input a single real number, this composition is defined, and

$$\mathbf{r}_2(g_3(t)) = \mathbf{r}_2(1 - t) = \left\langle 1 - t, (1 - t)^2, (1 - t)^3 \right\rangle.$$

37. Following the procedure in the text, substitute $\sqrt{x^2 + y^2}$ for x to get $z = \sqrt{x^2 + y^2}$. Since x varies from 0 to 3, the surface lies over the disk $x^2 + y^2 \leq 3^2 = 9$ in the xy plane:

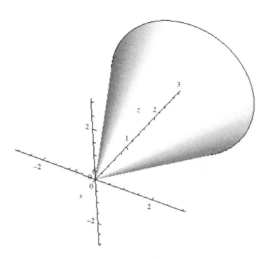

39. Following the procedure in the text, substitute $\sqrt{x^2 + y^2}$ for x to get $z = x^2 + y^2$. Since x varies from 0 to 2, the surface lies over the disk $x^2 + y^2 \leq 2^2 = 4$ in the xy plane:

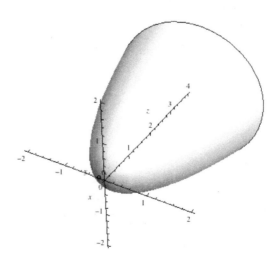

41. Following the procedure in the text, substitute $\sqrt{x^2 + y^2}$ for x to get $z = \sin \sqrt{x^2 + y^2}$. Since x varies from 0 to $\frac{\pi}{2}$, the surface lies over the disk $x^2 + y^2 \leq \left(\frac{\pi}{2}\right)^2 = \frac{\pi^2}{4}$ in the xy plane:

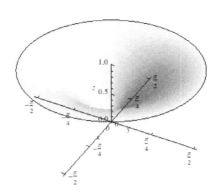

43. The level curve at c is the function $f(x,y) = c$, or $x + y = c$. This is a line with slope -1 and y-intercept c:

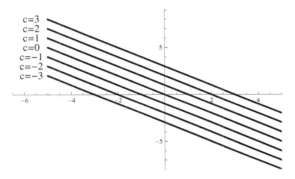

45. The level curve at c is the function $f(x,y) = c$, or $cy^2 = 3x$. For $c = 0$, this is the y axis ($3x = 0$), while for $c \neq 0$ it is the parabola $x = \frac{c}{3}y^2$:

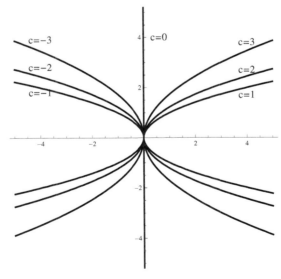

47. The level curve at c is the curve $c = 4 - (x^2 + y^2)$, or, simplifying, $x^2 + y^2 = 4 - c$. Since $4 - c > 0$ for the given values of c, all these level curves exist and are circles of radius $\sqrt{4 - c}$ centered at the origin:

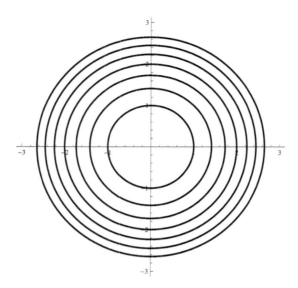

There is not enough room on the graph to label these, but the innermost circle is the level curve for $c = 3$, and increasing larger circles correspond to increasing smaller values of c, so that the outermost circle corresponds to $c = -3$.

49. Since $-1 \leq \sin(x + y) \leq 1$, the level curves for $c = -3$, $c = -2$, $c = 2$, and $c = 3$ do not exist. For $c = -1$, the level curve is $\sin(x + y) = -1$, which consists of the lines $x + y = \frac{3\pi}{2} + 2n\pi$ where n is an integer. For $c = 0$, the level curve is $\sin(x + y) = 0$, which consists of the lines $x + y = n\pi$ for n an integer. Finally, for $c = 1$, the level curve is $\sin(x + y) = 1$, which consists of the lines $x + y = \frac{\pi}{2} + 2n\pi$. A graph of these level curves is below, with $c = -1$ in black, $c = 0$ in dark gray, and $c = 1$ in light gray:

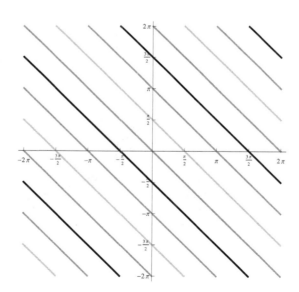

51. The level curve $c = y \csc x$ is $y = \frac{c}{\csc x} = c \sin x$ wherever $f(x, y) = y \csc x$ is defined, which is for $x \neq n\pi$ for n an integer. Note that for $c = 0$, this is the curve $y = 0$, which is the x axis.

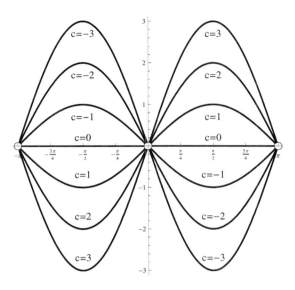

53. The level surfaces are the planes $x + 2y + 3z = c$ for c in the specified range. These are parallel but distinct planes with normal $\langle 1, 2, 3 \rangle$.

55. $c = \frac{x}{y-z}$ is the same as $x = cy - cz$, or $x - cy + cz = 0$, as long as $y \neq z$. Thus the level surfaces are planes with normal vector $\langle 1, -c, c \rangle$, except that each level surface is missing all points on the plane where $y = z$.

57. Since $x^2 + y^2 + z^2 \geq 0$, the level surfaces for $c = -3$, $c = -2$, and $c = -1$ do not exist. The level surface for $c = 0$ consists of solutions to $x^2 + y^2 + z^2 = 0$, which is just the origin. For $c > 0$, the level surface $x^2 + y^2 + z^2 = c$ is the sphere of radius $\sqrt{c}$ centered at the origin.

59. Along the x axis, where $y = 0$, this curve is the upward-opening parabola $z = x^2$. Along the y axis, where $x = 0$, it is the downward-opening parabola $z = -y^2$. The only surface having this shape along the two orthogonal axes is IV, so this must be $f(x, y) = x^2 - y^2$. Its level curves must be A, since each level curve of surface I is a hyperbola except for the level curve through the origin, which is a pair of crossing lines. This corresponds to A.

61. The maximum of this function clearly occurs at $(0, 0)$, where both terms are $e^0 = 1$. Along either axis, the function approaches 1, since $\lim_{y \to \infty} \left(e^{-0^2} + e^{-y^2} \right) = e^0 = 1$ and similarly along the x axis. This matches graph III. The level curves are the B set of curves, since this set of curves has a maximum at the origin and large flat spaces away from both axes.

63. Since $V(r, h)$ is a function of the radius and the height, the domain is $\{(r, h) \mid r > 0, h > 0\}$, since both the radius and the height must be positive. The surface area is the area of the two ends plus the lateral surface area. Each end is a disk of radius r, so it has area πr^2. The lateral surface has area $2\pi r h$, since it is composed of circles of radius r through a height h. Thus the total surface area is

$$S(r, h) = 2\pi r^2 + 2\pi r h = 2\pi r(r + h).$$

65. The volume is $V(x, y, z) = xyz$. The domain is $\{(x, y, z) \mid x > 0, y > 0, z > 0\}$, since all three dimensions must be positive. The surface area is the sum of the area of each of the six faces, so is $S(x, y, z) = 2xy + 2xz + 2yz$. The domain of S is the same as the domain of V.

Applications

67. (a) A contour plot of the caribou distribution is

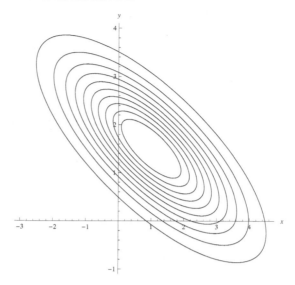

The caribou are most likely to be found at the highest level point on this surface, which looks to be at about $(1, 1.5)$.

(b) The direction of the stream follows a diagonal straight line from northwest to southeast; eyeballing the line shows that it looks like the exponent of the first exponential, $y + 0.5x - 2 = 0$. above, so it has a direction vector $\langle 1, -0.5 \rangle$, or $\langle 2, -1 \rangle$.

Proofs

69. Clearly $(0, 0, c)$ is on the graph of the curve, since $a \cdot 0 + b \cdot 0 + c = c$. Rewrite $f(x, y) = ax + by + c$ as $ax + by - z = -c$. We recognize this function as a plane with normal vector $\langle a, b, -1 \rangle$ from Section 10.6.

71. If the level curves intersect, then by Exercise 70, we have $c_1 = c_2$. Then the two level curves are both $f(x, y) = c$, so they are identical.

73. If the level curves intersect, then by Exercise 72, we have $c_1 = c_2$. Then the two level curves are both $f(x, y, z) = c$, so they are identical.

Thinking Forward

A kind of derivative for a function of two variables

If we think of y as a constant, we can think of $\frac{3x}{y^4}$ as Kx where $K = \frac{3}{y^4}$ is a constant. The derivative of this function is then K, which equals $\frac{3}{y^4}$. Similarly, if we think of x as a constant, then we can think of $\frac{3x}{y^4}$ as $\frac{C}{y^4}$ where $C = 3x$ is a constant. The derivative of this function is $-4\frac{C}{y^5}$, which equals $-4\frac{3x}{y^5} = -\frac{12x}{y^5}$. If we think of both x and y as constants, then the entire expression is a constant, so its derivative is zero.

A kind of derivative for a function of three variables

If we think of y and z as constants, then $\sin y$ and e^{-4z} are also constants, so that the function is Ax where $A = e^{-4z} \sin y$ is a constant. So its derivative is $A = e^{-4z} \sin y$. If we think of x and z as constants, then the function is $B \sin y$ where $B = xe^{-4z}$ is a constant. So its derivative is $B \cos y = xe^{-4z} \cos y$. Finally, thinking of x and y as constants, the function becomes Ce^{-4z} where $C = x \sin y$ is a constant. So its derivative is $-4Ce^{-4z} = -4xe^{-4z} \sin y$. If all three variables are thought of as constant, then the entire expression is a constant, so its derivative is zero.

12.2 Open Sets, Closed Sets, Limits and Continuity

Thinking Back

Intervals

In $\mathbb{R}$, an open interval is an interval of the form (a, b) for $a < b$, which is the set $\{x \mid a < x < b\}$. Note that it does not contain the endpoints. (The interval $(a, \infty) = \{x \mid a < x\}$ is also an open interval). A closed interval is written $[a, b]$ for $a < b$, and is the set $\{x \mid a \leq x \leq b\}$. Note that in a closed interval, the endpoints are part of the interval. The interval $[a, b) = \{x \mid a \leq x < b\}$, for example, is neither open nor closed.

Limits

Intuitively, the interpretation is that if as x gets very close to a, the values of $f(x)$ get close to some number L, then $L = \lim_{x \to a} f(x)$. More formally, the definition is: $L = \lim_{x \to a} f(x)$ if for every $\epsilon > 0$ there exists some $\delta > 0$ such that whenever $0 < |x - a| < \delta$, then $|f(x) - L| < \epsilon$.

Concepts

1. (a) True. This is the definition of a closed set in $\mathbb{R}^2$; see Definition 12.11.

   (b) False. For example, the set $S = \{(x, y) \mid x^2 + y^2 \leq 1, x \neq 1\}$ is neither open nor closed. It is not open since the point $(-1, 0) \in S$ yet any open ball around $(-1, 0)$ contains some point with $x < -1$, so it does not lie in S. Its complement is not open either, since the point $(1, 0) \in S^c$, yet any open ball around $(1, 0)$ contains some point $(x, 0)$ with $x < 1$ but close to 1, so it lies in S. Since S^c is not open, it follows that S is not closed.

   (c) True. $\mathbb{R}^2$ and $\varnothing$ are two sets that are both open and closed. To see this, note that $\mathbb{R}^2$ is open since any point has an open ball around it contained wholly in $\mathbb{R}^2$ (any open ball will work), and $\varnothing$ is open since the condition for being open is satisfied vacuously. Since $\mathbb{R}^2$ and $\varnothing$ are complements of each other in $\mathbb{R}^2$, they are both closed in $\mathbb{R}^2$ as well.

   (d) True. If S^c were closed, then $(S^c)^c = S$ would be open (see Theorems 12.10 and 12.12), while if S^c were open, then S would be closed by Definition 12.11.

   (e) False. As in the case of functions of a single variable, the value of f at the point in question does not matter. For example, let $f(x, y) = 1$ for $(x, y) \neq (0, 0)$, and $f(0, 0) = 0$. Then $\lim_{(x,y) \to (0,0)} f(x, y) = 1$, since $f(x, y) = 1$ whenever $(x, y) \neq (0, 0)$, but $f(0, 0) \neq 1$.

   (f) True. By Definition 12.17, $\lim_{(x,y) \to (a,b)} f(x) = L$ if $\lim_{\substack{(x,y) \to (a,b) \\ C}} f(x(t), y(t)) = L$ for every curve

   $C = \langle x(t), y(t) \rangle$ such that $\lim_{t \to t_0} f(x(t), y(t)) = (a, b)$.

   (g) False. For example, let $f(x, y) = 1$ for $(x, y) \neq (0, 0)$ but $f(0, 0) = 0$. Then clearly the limit along any path is 1, yet f is discontinuous at $(0, 0)$.

   (h) True. By Definition 12.19, f is continuous at $(0, 0)$ if $\lim_{(x,y) \to (0,0)} f(x, y) = f(0, 0)$.

3. All you can say is that if the limit exists, it must equal 5. If the limit along all paths C through $(1, -3)$ is 5, then Theorem 12.18 tells us that $\lim\limits_{(x,y)\to(1,-3)} f(x,y) = 5$. However, there could be a third curve, C_3, for which $\lim\limits_{\substack{(x,y)\to(1,-3) \\ C_3}} f(x,y) = 12$, in which case by Theorem 12.18, $\lim\limits_{(x,y)\to(1,-3)} f(x,y)$ does not exist.

5. By Theorem 12.18, since the limit along any such curve is equal to 5, we get $\lim\limits_{(x,y)\to(3,-7)} f(x,y) = 5$. However, we know nothing about $f(3, -7)$, since the limit has nothing to do with the value of the function at the given point in the absence of an assumption about continuity.

7. The definition of a limit says that $0 < |x - a| < \delta$ implies that $|f(x) - L| < \epsilon$. So if $0 < |x - a| < \delta$, then $f(x)$ must at least be defined. But $0 < |x - a| < \delta$ is equivalent to $x \in (a - \delta, a) \cup (a, a + \delta)$, which is the union of two open intervals.

9. Definition 12.15 says that the limit of f at $\mathbf{a}$ is L if for every $\epsilon > 0$ there is some $\delta > 0$ such that $|f(\mathbf{x}) - L| < \epsilon$ whenever $0 < \|\mathbf{x} - \mathbf{a}\| < \delta$. Since $|f(\mathbf{x}) - L| < \epsilon$ in this case, it follows in particular that $f(\mathbf{x})$ is *defined* whenever $0 < \|\mathbf{x} - \mathbf{a}\| < \delta$. But this is a punctured ball around $\mathbf{a}$, consisting of an open ball around $\mathbf{a}$ minus the point $\mathbf{a}$ itself. This is an open set, since for any $\mathbf{x}$ in that ball, $\mathbf{x}$ is a positive distance d_1 from the sphere of radius δ around $\mathbf{a}$, and is a positive distance d_2 from $\mathbf{a}$ itself. Let $d = \frac{1}{2}\min(d_1, d_2)$. Then the open ball of radius d around $\mathbf{x}$ must be wholly within this punctured ball, so the punctured ball is open.

11. The definition of continuity of $f(x)$ at $x = a$ says that $0 < |x - a| < \delta$ implies that $|f(x) - f(a)| < \epsilon$. So if $0 < |x - a| < \delta$, then $f(x)$ must at least be defined. But $0 < |x - a| < \delta$ is equivalent to $x \in (a - \delta, a) \cup (a, a + \delta)$, which is the union of two open intervals.

13. If C is the x axis, then C is the path $(t, 0)$, so that

$$\lim_{\substack{(x,y)\to(0,0) \\ C}} \frac{x^2 y}{x^4 + y^2} = \lim_{t\to 0} \frac{t^2 \cdot 0}{t^4 + 0^2} = \lim_{t\to 0} \frac{0}{t^4} = 0.$$

If C is instead the y axis, then C is the path $(0, t)$, so that

$$\lim_{\substack{(x,y)\to(0,0) \\ C}} \frac{x^2 y}{x^4 + y^2} = \lim_{t\to 0} \frac{0 \cdot t}{0^4 + t^2} = \lim_{t\to 0} \frac{0}{t^2} = 0.$$

15. The limit of $f(x, y)$ at (a, b) is ∞ if for every $M > 0$ there is a $\delta > 0$ such that $f(x, y) > M$ whenever $0 < \|\mathbf{x} - \mathbf{a}\| < \delta$, where $\mathbf{x} = \langle x, y \rangle$ and $\mathbf{a} = \langle a, b \rangle$. In this case we write $\lim\limits_{(x,y)\to(a,b)} f(x,y) = \infty$.

17. Let

$$f(x, y) = \begin{cases} 1, & x \le 0, \\ 0, & x > 0 \end{cases}$$

$$g(x, y) = \begin{cases} 0, & x \le 0, \\ 1, & x > 0 \end{cases}.$$

Let $C_1(t) = \left\langle -\frac{1}{t}, 0 \right\rangle$ for $t > 0$ and $C_2(t) = \left\langle \frac{1}{t}, 0 \right\rangle$ for $t > 0$. Then

$$\lim_{\substack{(x,y)\to(0,0) \\ C_1}} f(x,y) = \lim_{t\to\infty} 1 = 1, \qquad \lim_{\substack{(x,y)\to(0,0) \\ C_2}} f(x,y) = \lim_{t\to\infty} 0 = 0$$

$$\lim_{\substack{(x,y)\to(0,0) \\ C_1}} g(x,y) = \lim_{t\to\infty} 0 = 0, \qquad \lim_{\substack{(x,y)\to(0,0) \\ C_2}} g(x,y) = \lim_{t\to\infty} 1 = 1,$$

so that neither $\lim\limits_{(x,y)\to(0,0)} f(x,y)$ nor $\lim\limits_{(x,y)\to(0,0)} g(x,y)$ exists. However, $f(x,y) + g(x,y) = 1$ for all x,y, so that $\lim\limits_{(x,y)\to(0,0)} (f(x,y) + g(x,y)) = 1$. This does not contradict Theorem 12.16 since the hypothesis of that theorem is that both $\lim\limits_{(x,y)\to(0,0)} f(x,y)$ and $\lim\limits_{(x,y)\to(0,0)} g(x,y)$ exist.

19. (a) If $x = y = a$, then $\frac{f(a,a)}{a-a}$ is undefined. Thus $\frac{f(x,y)}{x-y}$ is not continuous when $x = y$. When $x \neq y$, however, say $x = a$ and $y = b$ where $a \neq b$. Then f is continuous at (a,b) and so is $x - y$. Also, $\lim\limits_{(x,y)\to(a,b)} (x - y) = a - b \neq 0$. So it follows from Theorem 12.16, Quotient Rule, that

$$\lim\limits_{(x,y)\to(a,b)} \frac{f(x,y)}{x - y} = \frac{\lim\limits_{(x,y)\to(a,b)} f(x,y)}{\lim\limits_{(x,y)\to(a,b)} (x - y)} = \frac{f(a,b)}{a - b},$$

so that $\frac{f(x,y)}{x-y}$ is continuous at (a,b).

(b) Choose any $\epsilon > 0$. Then for any δ, consider the open disk of radius δ around (a,a). That disk contains the point $\left(a + \frac{\delta}{2}, a + \frac{\delta}{2}\right)$, at which $\frac{f(x,y)}{x-y}$ is not defined. Thus the condition of Definition 12.15, which is that $|f(\mathbf{x}) - L| < \epsilon$ when $0 < \|\mathbf{x} - \mathbf{a}\| < \delta$, does not hold, so that $\lim\limits_{(x,y)\to(a,b)} \frac{f(x,y)}{x-y}$ does not exist at $(x,y) = (a,a)$.

Skills

21. (a) This set, $S = \{(x,y) \mid x > 0, y > 0\}$ is open. To see this, choose (a,b) such that $a > 0$ and $b > 0$, and let $\delta = \min\left(\frac{a}{2}, \frac{b}{2}\right)$. Then clearly the open disk of radius δ around (a,b) is wholly contained in S.

(b) $S^c = \{(x,y) \mid x \leq 0 \text{ or } y \leq 0\}$. It is not open, for consider any point $(a,0)$ on the positive x axis. A neighborhood of that point contains points in S (points with y coordinate greater than zero). Thus S is not closed.

(c) The boundary is $\partial S = \{(x,0) \mid x \geq 0\} \cup \{(0,y) \mid y \geq 0\}$. For any open disk around $(x,0)$ contains some point (x,a) for $a > 0$ and some point (x,b) with $b < 0$, so it contains points of both S and S^c. Similarly, any open disk around $(0,y)$ contains some point (a,y) for $a > 0$ and some point (b,y) with $b < 0$, so it contains points of both S and S^c. We know that no point of S is in ∂S since S is open. To see that no other point of S^c is in ∂S, choose $(a,b) \in S^c$ not lying on either the positive x axis or the positive y axis. Then the point has a positive distance from each of those axes. So choose an open disk of radius smaller than that positive distance. That disk lies wholly within S^c, so that the (a,b) is not in ∂S.

23. (a) This set S is the set of all points inside, or on the boundary of, a diamond-shaped region bounded by the four lines $x + y = 1$, $x + y = -1$, $x - y = 1$, and $x - y = -1$. S is closed. To see this, we need only prove that its complement is open. Its complement is $S^c = \{(x,y) \mid |x| + |y| > 1\}$. Choose $(a,b) \in S^c$, and suppose that $|a| + |b| = c > 1$. Then an open disk of radius $\frac{c-1}{2}$ around (a,b) will consist wholly of points (x_0, y_0) with $|x_0| + |y_0| \geq \frac{c+1}{2} > 1$, so that the disk is contained in S^c. Thus S^c is open, so that S is closed. However, S is not open. For example, the point $(1,0)$ is in S, but any open disk around $(1,0)$ will contain some point $(a,0)$ for $a > 1$, so that point is not in S. Thus S is not open.

(b) From part (a), $S^c = \{(x,y) \mid |x| + |y| > 1\}$.

(c) $\partial S = \{(x,y) \mid |x| + |y| = 1\}$. To see this, note that an open disk around any point in that set must contain points (a,b) with $|a| + |b| > 1$ (any point in the disk on the other side of ∂S

from the origin) and points (a, b) with $|a| + |b| < 1$ (any point in the disk on the same side of ∂S as the origin). Thus any such point is in ∂S. No point of S^c is in ∂S since S^c is open. And no point of $S - \{(x, y) \mid |x| + |y| = 1\}$ is in ∂S, since any such point (a, b) is such that $|a| + |b| = c < 1$, so that an open disk of radius $\frac{1-c}{2}$ around (a, b) will consist wholly of points (x_0, y_0) with $|x_0| + |y_0| \leq \frac{1-c}{2} < 1$, so that the disk is contained in S.

25. (a) This set is both open and closed. It is open since it vacuously satisfies the condition that every open disk around any $x \in \emptyset$ is contained in $\emptyset$. Its complement, $\mathbb{R}^2$, is also open, since if $(a, b) \in \mathbb{R}^2$, any open disk around (a, b) lies in $\mathbb{R}^2$. Since $\mathbb{R}^2$ is open, $\emptyset$ is also closed.

    (b) $\emptyset^c = \mathbb{R}^2$.

    (c) The boundary is $\partial S = \emptyset$. Since $S^c = \mathbb{R}^2$ is open, any point in S^c has an open disk around it wholly contained in S^c, so that no point of $S^c = \mathbb{R}^2$ is in ∂S. Thus ∂S is empty.

27. (a) This set S is open but not closed. It is open since if (a, b, c) has $a > 0$, $b < 0$, and $c < 0$, consider the open ball around (a, b, c) whose radius is $\frac{1}{2} \min(a, b, c)$. All points in that ball lie in S. Thus S is open. To see that S is not closed, we must show that its complement is not open. See part (b).

    (b) The complement S^c is the set $\{(x, y, z) \mid x \leq 0, y \geq 0, \text{ or } z \geq 0\}$. This set is not open (and thus S is not closed). To see this, choose $(0, 0, 0) \in S^c$. The open ball of radius a around $(0, 0, 0)$ will contain the point $\left(\frac{a}{2}, -\frac{a}{2}, \frac{a}{2}\right) \in S$. Thus S^c is not closed.

    (c) $\partial S = \{(x, y, z) \mid x = 0, y \leq 0, z \leq 0\} \cup \{(x, y, z) \mid x \geq 0, y = 0, z \leq 0\} \cup \{(x, y, z) \mid x \geq 0, y \leq 0, z = 0\}$. Clearly for any point in this region, any open ball around that point will contain points both in S and in S^c (for example, if the point is of the form $(0, y, z)$ for $y, z < 0$, then it will contain both $(-\epsilon, y, z)$ and (ϵ, y, z) for ϵ smaller than the radius of the ball). Clearly no point in S is in ∂S. Similarly, any point in S^c minus the set above is not in ∂S either, since such a point has no coordinates equal to zero, so a small enough open ball around the point will not touch any of the coordinate axes, so it will not intersect ∂S.

29. (a) This is $S = \{(x, y, 0)\}$. S is closed but not open. It is not open since, for example, any open ball around $(1, 1, 0)$ will contain a point $(1, 1, \epsilon)$ whenever ϵ is smaller than the radius of the ball. To see it is closed, we prove that its complement is open; see part (b).

    (b) $S^c = \{(x, y, z) \mid z \neq 0\}$. To see that S^c is open, choose $(a, b, c) \in S^c$ and let $\epsilon = \frac{1}{2}|c|$. Then an open ball of radius ϵ around (a, b, c) will contain no points whose z coordinate is zero. Thus S^c is open, so S is closed.

    (c) $\partial S = S$. First, note that from part (a), any open ball around a point of S contains a point of S^c as well, so that all points of S are in the boundary. But since S^c is open, no point of S^c is in ∂S, since we can find an open ball around any point of S^c containing only points of S^c.

31. (a) This set is both open and closed. It is open since it vacuously satisfies the condition that every open disk around any $x \in \emptyset$ is contained in $\emptyset$. Its complement, $\mathbb{R}^3$, is also open, since if $(a, b, c) \in \mathbb{R}^3$, any open disk around (a, b, c) lies in $\mathbb{R}^3$. Since $\mathbb{R}^3$ is open, $\emptyset$ is also closed.

    (b) $\emptyset^c = \mathbb{R}^3$.

    (c) The boundary is $\partial S = \emptyset$. Since $S^c = \mathbb{R}^3$ is open, any point in S^c has an open disk around it wholly contained in S^c, so that no point of $S^c = \mathbb{R}^3$ is in ∂S. Thus ∂S is empty.

33. By the product rule,

$$\lim_{(x,y) \to (-2, \pi)} x^2 y^3 \sin y = \lim_{(x,y) \to (-2, \pi)} x^2 \cdot \lim_{(x,y) \to (-2, \pi)} y^3 \cdot \lim_{(x,y) \to (-2, \pi)} \sin y = (-2)^2 \cdot \pi^3 \cdot \sin \pi = 0.$$

35. Since $\lim\limits_{(x,y)\to(1,2)} (x^2 - y^2) = -3 \neq 0$, the quotient rule applies, so using the quotient and sum rules we get

$$\lim_{(x,y)\to(1,2)} \frac{x^2 + y^2}{x^2 - y^2} = \frac{\lim\limits_{(x,y)\to(1,2)} x^2 + \lim\limits_{(x,y)\to(1,2)} y^2}{\lim\limits_{(x,y)\to(1,2)} x^2 - \lim\limits_{(x,y)\to(1,2)} y^2} = \frac{1+4}{1-4} = -\frac{5}{3}.$$

37. Since the path is the path $x = 3$, we can substitute $x = 3$ in the expression to get

$$\lim_{\substack{(x,y)\to(3,3) \\ x=3}} \frac{x^3 - y^3}{x^2 - y^2} = \lim_{\substack{(x,y)\to(3,3) \\ x=3}} \frac{27 - y^3}{9 - y^2} = \lim_{\substack{(x,y)\to(3,3) \\ x=3}} \frac{(3-y)(9 + 3y + y^2)}{(3-y)(3+y)}$$

$$= \lim_{\substack{(x,y)\to(3,3) \\ x=3}} \frac{9 + 3y + y^2}{3 + y} = \frac{9}{2}.$$

39. This limit does not exist, since the given function is not defined on the line $y = x$ (its denominator is zero).

41. Substituting, we get

$$\lim_{(x,y)\to(0,0)} \frac{x^2}{x^2 + y^2} = \lim_{r\to0} \frac{r^2 \cos^2 \theta}{r^2} = \lim_{r\to0} \cos^2 \theta = \cos^2 \theta.$$

Since the limit depends on the angle θ, this limit does not exist, since approaching the origin along different lines will result in different values of θ.

43. Substituting, we get

$$\lim_{(x,y)\to(0,0)} \frac{x^2 y^2}{x^2 + y^2} = \lim_{r\to0} \frac{r^2 \cos^2 \theta \cdot r^2 \sin^2 \theta}{r^2} \lim_{r\to0} r^2 \cos^2 \theta \sin^2 \theta = 0.$$

Since the limit is zero as $r \to 0$, it is zero for every value of θ, so that the limit is independent of the path through the origin. Thus this limit is 0.

45. Substituting, we get

$$\lim_{(x,y)\to(0,0)} \frac{xy}{\sqrt{x^2 + y^2}} = \lim_{r\to0} \frac{r \cos \theta \cdot r \sin \theta}{r} = \lim_{r\to0} r \cos \theta \sin \theta = 0.$$

Since the limit is zero as $r \to 0$, it is zero for every value of θ, so that the limit is independent of the path through the origin. Thus this limit is 0.

47. $f(x,y)$ is defined unless $x^2 - y^2 = 0$; this happens when $y = x$ or $y = -x$. So the domain of the function is $\{(x,y) \mid x \neq y \text{ and } x \neq -y\}$. On its domain, $f(x,y)$ is a rational function, so it is continuous everywhere on its domain.

49. $f(x,y)$ is defined as long as $x^2 + y \geq 0$, since otherwise the argument of the square root is negative. On its domain, $f(x,y)$ is the composition of the square root function, which is continuous on the nonnegative real numbers, and the function $x^2 + y$, which is polynomial and thus continuous. However, f is not continuous on the boundary, which is $\{(x,y) \mid y = -x^2\}$, since the limit from the other side does not exist. So $f(x,y)$ is continuous on $\{(x,y) \mid y > -x^2\}$.

51. $f(x,y)$ is defined when $x + y + z > 0$, since if $x + y + z = 0$ then the denominator is zero, while if $x + y + z < 0$, then $\sqrt{x + y + z}$ is undefined. Now, $\sin(x + y + z)$ is continuous everywhere, and $\sqrt{x + y + z}$ is continuous where it is defined since its is the composition of the continuous square root function with the continuous polynomial function $x + y + z$. Thus the quotient, $f(x,y)$, is continuous on its domain.

53. For $(x, y) \neq (0, 0)$, numerator and denominator are both defined, and the denominator is nonzero. Both numerator and denominator are continuous away from $(0, 0)$, so that $f(x, y)$ is continuous for $(x, y) \neq (0, 0)$. To see what happens at $(0, 0)$, evaluate the limit:

$$\lim_{(x,y) \to (0,0)} \frac{xy}{\sqrt{x^2 + y^2}} = \lim_{r \to 0} \frac{r \cos \theta \cdot r \sin \theta}{r} = \lim_{r \to 0} r \cos \theta \sin \theta = 0.$$

Since the limit is zero as $r \to 0$, it is zero for every value of θ, so that the limit is independent of the path through the origin. Thus this limit is 0. Since $\lim_{(x,y) \to (0,0)} f(x, y) = f(0, 0)$, we see that $f(x, y)$ is also continuous at $(0, 0)$.

55. For $(x, y) \neq (0, 0)$, numerator and denominator are both defined, and the denominator is nonzero. Both numerator and denominator are continuous away from $(0, 0)$, so that $f(x, y)$ is continuous for $(x, y) \neq (0, 0)$. To see what happens at $(0, 0)$, evaluate the limit:

$$\lim_{(x,y) \to (0,0)} \frac{sin(x^2 + y^2)}{x^2 + y^2} = \lim_{r \to 0} \frac{\sin(r^2)}{r^2}.$$

Now, $r^2 \to 0$ as $r \to 0$, so this is the same as

$$\lim_{r^2 \to 0} \frac{\sin(r^2)}{r^2} = 1.$$

Since the limit is 1 as $r \to 0$, it is 1 for every value of θ, so that the limit is independent of the path through the origin. Thus this limit is 1. Since $\lim_{(x,y) \to (0,0)} f(x, y) = f(0, 0) = 1$, we have that $f(x, y)$ is also continuous at $(0, 0)$.

Applications

57. Treating n, R, and V as constants, we have

$$\lim_{T \to 0} P = \lim_{T \to 0} \frac{nRT}{V} = \frac{nR \cdot 0}{V} = 0.$$

This makes physical sense, since as the temperature cools, the particle motion decreases, which decreases the gas pressure. As the temperature approaches absolute zero, the particle motion approaches a state of no motion.

59. Treating n, R, and T as constants, we have

$$\lim_{V \to 0^+} P = \lim_{V \to 0^+} \frac{nRT}{V} = \infty.$$

This makes physical sense, since as the volume is decreased towards zero, the same number of particles must occupy a smaller space, so that the pressure will increase.

61. (a) The two planes are given by

$$\frac{1}{350}x - \frac{1}{725}y + z = -40, \quad \text{with normal vector } \left\langle \frac{1}{350}, -\frac{1}{725}, 1 \right\rangle,$$

$$\frac{1}{150}x - \frac{1}{300}y + z = -39.5, \quad \text{with normal vector } \left\langle \frac{1}{150}, -\frac{1}{300}, 1 \right\rangle.$$

The line of intersection of the planes thus has direction vector equal to

$$\left\langle \frac{1}{350}, -\frac{1}{725}, 1 \right\rangle \times \left\langle \frac{1}{150}, -\frac{1}{300}, 1 \right\rangle = \left\langle \frac{17}{8700}, \frac{2}{525}, -\frac{1}{3045000} \right\rangle =$$

$$\frac{1}{3045000} \langle 5950, 11600, -1 \rangle .$$

To find a point on the line, we want a point that lies on both planes. Set $z = 0$ and clear fractions to get the equations $29x - 14y = -406000$ and $2x - y = -11850$. Solving gives $x = -240100$ and $y = -468350$, so that $(-240100, -468350, 0)$ is a point on the line. Thus the parametric equation of intersection of the two planes is

$$\langle -240100 + 5950t, -468350 + 11600t, -t \rangle .$$

Moving this line to the surface of the earth gives the parametric equation

$$\langle -240100 + 5950t, -468350 + 11600t \rangle .$$

This line has slope $\frac{11600}{5950} = \frac{232}{119} \approx 1.95$. The y-intercept occurs when the x coordinate is zero, so for $t = \frac{240100}{5950} \approx 40.35$; at that value of t, the y coordinate is $-468250 + 11600 \cdot 40.35 \approx -255.88$. Thus the equation of the line is $y = 1.95x - 255.88$.

(b) When $x = 150$ on the line of intersection, $t = \frac{240100 + 150}{5950} \approx 40.38$, so the z coordinate at that point is $-t \approx -40.38$ feet.

(c) Since the basalt is not at the predicted point, clearly one or both of her determinations of the basalt planes b_1 and b_2 is incorrect.

Proofs

63. Suppose $x \in S$. Then $x \notin S^c$, since $S^c = \{z \mid z \notin S\}$. But $x \notin S^c$ means that $x \in (S^c)^c$, since $(S^c)^c = \{z \mid z \notin S^c\}$. Thus $S \subseteq (S^c)^c$. Conversely, suppose $x \in (S^c)^c$. Then $x \notin S^c$, by the definition of $(S^c)^c$ just given. Finally, $x \notin S^c$ means that $x \in S$, by the definition of S^c given above. Thus $(S^c)^c \subseteq S$, so that $S = (S^c)^c$.

65. Suppose S is open. Then any point $x \in S$ has a open disk (or ball) around it all of whose points are in S. Since that open disk/ball exists, we conclude that $x \notin \partial S$, since any open disk/ball around a point on ∂S will by definition contain both points in S and points in ∂S. Thus $S \cap \partial S = \emptyset$. Conversely, suppose $\partial S \cap S = \emptyset$. Then every point of S has an open disk/ball around it that contains no points of S^c (since otherwise the point would be in ∂S). But this is exactly the definition of S being open.

67. $x \in \partial S$ if and only if (by Definition 12.13) every open disk/ball containing x contains points of both S and S^c. But this happens if and only if every open disk/ball containing x contains points of $(S^c)^c = S$ and S^c. Then again by definition 12.13, this happens if and only if $x \in \partial S^c$. Thus $\partial S = \partial S^c$.

69. Since ∂S is closed by Exercise 68, it follows from Exercise 66 that $\partial(\partial S) \subseteq \partial S$.

71. $\mathbb{R}^2$ is open since if $(a, b) \in \mathbb{R}^2$, any open disk around (a, b) lies in $\mathbb{R}^2$. Its complement, which is $\emptyset$, is also open since it vacuously satisfies the condition that every open disk around any $x \in \emptyset$ is contained in $\emptyset$. Since $\mathbb{R}^{2^c} = \emptyset$ is open, $\mathbb{R}^2$ is also closed.

Thinking Forward

The derivative along a cut edge

When cut by the plane with equation $x = 2$, we are considering points on $f(x, y)$ with $x = 2$, so the function becomes $g(y) = f(2, y) = 4y^3$. The rate of change of f on that plane is then $g'(y) = 12y^2$. Similarly, when cut by the plane with equation $y = 1$, we are considering points on $f(x, y)$ with $y = 1$, so the function becomes $h(x) = f(x, 1) = x^2$. The rate of change of f on that plane is thus $h'(x) = 2x$.

The derivative when two variables are held fixed

If x and y are thought of as constants, then the rate of change in the z direction is the derivative where z is the variable: $x^2 y^3 \cdot \frac{1}{2} z^{-1/2} = \frac{x^2 y^3}{2\sqrt{z}}$. When x and z are thought of as constants, the rate of change in the y direction is the derivative where y is the variable: $x^2 \cdot 3y^2 \cdot \sqrt{z} = 3x^2 y^2 \sqrt{z}$. Finally, when y and z are thought of as constants, the rate of change in the x direction is the derivative where x is the variable: $2xy^3 \sqrt{z}$.

12.3   Partial Derivatives

Thinking Back

Finding a direction vector for a tangent line

Parametrize $f(x) = x^3$ as $\mathbf{r}(t) = \langle t, t^3 \rangle$. Then $\mathbf{r}'(t) = \langle 1, 3t^2 \rangle$, so that $\mathbf{r}'(2) = \langle 1, 12 \rangle$. A direction vector in the direction of the tangent line is $\langle 1, 12 \rangle$.

Finding the equation of the plane containing two intersecting lines

To determine the point of intersection, parametrize the second curve by u and equate components:

$$3t - 4 = -u + 2, \quad -4t + 1 = 2u - 9, \quad t = -2u + 7.$$

Substitute $-2u + 7$ for t in the first equation to get $3(-2u + 7) - 4 = -6u + 17 = -u + 2$, so that $u = 3$ and then $t = 1$. Thus the point of intersection is

$$\langle 3 \cdot 1 - 4, -4 \cdot 1 + 1, 1 \rangle = \langle -1, -3, 1 \rangle.$$

Since both lines pass through that point, and they have direction vectors $\langle 3, -4, 1 \rangle$ and $\langle -1, 2, -2 \rangle$, they determine a plane whose normal vector is the cross product of these two vectors:

$$\langle 3, -4, 1 \rangle \times \langle -1, 2, -2 \rangle = \langle -4 \cdot (-2) - 1 \cdot 2, 1 \cdot (-1) - 3 \cdot (-2), 3 \cdot 2 - (-4) \cdot (-1) \rangle = \langle 6, 5, 2 \rangle.$$

Thus the equation of the plane containing the lines is

$$\langle 6, 5, 2 \rangle \cdot \langle x + 1, y + 3, z - 1 \rangle = 0, \text{ or } 6(x + 1) + 5(y + 3) + 2(z - 1) = 0.$$

Simplifying gives $6x + 5y + 2z = -19$.

Concepts

1. (a) True. See Definition 12.20(a) with x and y in place of x_0 and y_0.

   (b) True. See Definition 12.20(b) with r and s in place of x_0 and y_0, and g in place of f.

   (c) False. The second order partial derivative with respect to x and y is the derivative with respect to y of the derivative with respect to x of $f(x, y)$, so it is $\lim\limits_{h \to 0} \frac{f_x(x, y+h) - f_x(x, y)}{h}$.

(d) False. The partials with respect to x and y are generally quite different. What is true is that if f has continuous second partials, then $f_{xy} = f_{yx}$.

(e) True. By Clairaut's Theorem (12.24), if the second-order partials are continuous, then mixed partials are equal.

(f) False. These two expressions are identical. When writing derivatives using the fractional ∂ notation, they are read from left to right, so the right-hand side means the partial with respect to y of the partial with respect to x, which is what f_{xy} means as well.

(g) False. $\lim\limits_{h \to 0} \frac{f(0+h,0)-f(0,0)}{h} = \lim\limits_{h \to 0} \frac{|h| \cdot 0 - |0| \cdot 0}{h} = 0$, so that the partial with respect to x is zero at $(0,0)$.

(h) False. $\lim\limits_{h \to 0} \frac{f(0,0+h)-f(0,0)}{h} = \lim\limits_{h \to 0} \frac{|0| \cdot (0+h) - |0| \cdot 0}{h} = 0$, so that the partial with respect to y is zero at $(0,0)$.

3. If you rotate the piece of paper on the tabletop, the pitch of the line in three dimensions remains zero, since it remains on the tabletop.

5. The partial derivative with respect to x at (x_0, y_0) is

$$f_x(x_0, y_0) = \lim_{h \to 0} \frac{f(x_0 + h, y_0) - f(x_0, y_0)}{h}.$$

Define $g(x) = f(x, y_0)$. Then

$$g'(x_0) = \lim_{h \to 0} \frac{g(x_0 + h) - g(x_0)}{h} = \lim_{h \to 0} \frac{f(x_0 + h, y_0) - f(x_0, y_0)}{h} = f_x(x_0, y_0).$$

But $g(x)$ is just the equation of f along a path where $y = y_0$, so that $g'(x_0)$, which is the slope of the line tangent to $g(x)$ at x_0, is the slope of the line tangent to $f(x, y)$ at (x_0, y_0) in the direction of constant y (i.e., parallel to the x axis).

Similarly, the partial derivative with respect to y at (x_0, y_0) is

$$f_y(x_0, y_0) = \lim_{h \to 0} \frac{f(x_0, y_0 + h) - f(x_0, y_0)}{h}.$$

Define $h(y) = f(x_0, y)$. Then

$$h'(y_0) = \lim_{h \to 0} \frac{h(y_0 + h) - h(y_0)}{h} = \lim_{h \to 0} \frac{f(x_0, y_0 + h) - f(x_0, y_0)}{h} = f_y(x_0, y_0).$$

But $h(y)$ is just the equation of f along a path where $x = x_0$, so that $h'(y_0)$, which is the slope of the line tangent to $h(y)$ at y_0, is the slope of the line tangent to $f(x, y)$ at (x_0, y_0) in the direction of constant x (i.e., parallel to the y axis).

7. (a) The graph of $g(x, y) = f(y)$ is, for each value of x, a copy of the graph of $z = f(y)$. That is, if you plot $z = f(y)$ and then stretch that graph out over the x axis, you will have the graph of $z = g(x, y)$.

   (b) The first order partials of g exist for every x and y since f is differentiable:

$$\frac{\partial g}{\partial y} = \lim_{h \to 0} \frac{g(x, y+h) - g(x, y)}{h} = \lim_{h \to 0} \frac{f(y+h) - f(y)}{h} = f'(y)$$

$$\frac{\partial g}{\partial x} = \lim_{h \to 0} \frac{g(x+h, y) - g(x, y)}{h} = \lim_{h \to 0} \frac{f(y) - f(y)}{h} = 0.$$

(c) The values of the two partials are given in part (b). The partial derivative with respect to y makes sense, since if x is held constant, we are taking the derivative on a graph that is a copy of the graph of f, so that the derivative should be $f'(y)$. The partial derivative with respect to x also makes sense, since if y is held constant, then $z = g(x, y_0) = f(y_0)$ is a constant, so that its partial derivative with respect to x should be zero.

9. We have

$$g_x(0,0) = \lim_{h \to 0} \frac{(0+h)^2 - 0^2 - (0^2 - 0^2)}{h} = \lim_{h \to 0} \frac{h^2}{h} = \lim_{h \to 0} h = 0$$

$$g_y(0,0) = \lim_{h \to 0} \frac{0^2 - (0+h)^2 - (0^2 - 0^2)}{h} = -\lim_{h \to 0} \frac{h^2}{h} = -\lim_{h \to 0} h = 0.$$

Since the slopes are zero, the direction vectors of the tangent line in the x direction is $\mathbf{i} + 0\mathbf{k} = \mathbf{i}$, and in the y direction is $\mathbf{j} + 0\mathbf{k} = \mathbf{j}$. Thus the equations of these tangent lines are $\langle 0,0,0 \rangle + t \langle 1,0,0 \rangle = \langle t,0,0 \rangle$ and $\langle 0,0,0 \rangle + t \langle 0,1,0 \rangle = \langle 0,t,0 \rangle$. The plane containing these two lines has normal vector $\mathbf{i} \times \mathbf{j} = \mathbf{k}$, so that the equation of the plane containing the lines is $z = c$ for some c. Since $(0,0,0)$ lies on the plane, we have $c = 0$ so that the equation of the plane is $z = 0$, or the xy plane.

11. We have

$$f_x(0,0) = \lim_{h \to 0} \frac{f(0+h,0) - f(0,0)}{h} = \lim_{h \to 0} \frac{0-0}{h} = 0$$

$$f_y(0,0) = \lim_{h \to 0} \frac{f(0,0+h) - f(0,0)}{h} = \lim_{h \to 0} \frac{0-0}{h} = 0.$$

However, $f(x,y)$ is not continuous at $(0,0)$, since for example the limit along the path $x = y$ is

$$\lim_{\substack{(x,y) \to (0,0) \\ x=y}} f(x,y) = \lim_{h \to 0} f(0+h, 0+h) = \lim_{h \to 0} 1 = 1 \neq f(0,0),$$

so that f is not continuous at $(0,0)$.

13. We have

$$f_x(0,0,0) = \lim_{h \to 0} \frac{f(0+h,0,0) - f(0,0,0)}{h} = \lim_{h \to 0} \frac{0-0}{h} = 0$$

$$f_y(0,0,0) = \lim_{h \to 0} \frac{f(0,0+h,0) - f(0,0,0)}{h} = \lim_{h \to 0} \frac{0-0}{h} = 0$$

$$f_z(0,0,0) = \lim_{h \to 0} \frac{f(0,0,0+h) - f(0,0,0)}{h} = \lim_{h \to 0} \frac{0-0}{h} = 0.$$

However, $f(x,y,z)$ is not continuous at $(0,0)$, since for example the limit along the path $x = y = z$ is

$$\lim_{\substack{(x,y,z) \to (0,0,0) \\ x=y=z}} f(x,y,z) = \lim_{h \to 0} f(0+h, 0+h, 0+h) = \lim_{h \to 0} 1 = 1 \neq f(0,0,0),$$

so that f is not continuous at $(0,0,0)$.

15. (a) There are four. There are two choices for which variable to differentiate first, and two choices as to which to differentiate second. The four derivatives are f_{xx}, f_{xy}, f_{yx}, and f_{yy}.

(b) There are eight. There are two choices for which variable to differentiate each time. The eight derivatives are f_{xxx}, f_{xxy}, f_{xyx}, f_{xyy}, f_{yxx}, f_{yxy}, f_{yyx}, and f_{yyy}.

(c) There are 2^n. At each stage, there are two choices for the variable to differentiate, and n stages.

17. (a) There are three: $f_{xx}, f_{xy} = f_{yx},$ and f_{yy}.

    (b) There are four: $f_{xxx}, f_{xxy}, f_{xyy},$ and f_{yyy}.

    (c) There are $n + 1$. Since the orders of the partials derivatives are not important, the identity of the derivative depends only on how many x's there are in the subscript, and there can be anywhere from 0, to n x's.

19. Since $f_x(x, y) = g_x(x, y)$ for all x, y, it follows that $f_x - g_x$ is the zero function. But $f_x - g_x = (f - g)_x$, so that $f(x, y) - g(x, y)$ must have a zero partial with respect to x, so that it must be a function of y alone: $f(x, y) - g(x, y) = h(y)$.

21. From Exercise 19, since the partials with respect to x are equal, $f(x, y) - g(x, y)$ does not depend on x. But from Exercise 20, since the partials with respect to y are equal, $f(x, y) - g(x, y)$ does not depend on y. Thus $f(x, y) - g(x, y)$ does not depend on either x or y, so it is a constant.

Skills

23. We have
$$f_y(2, 1) = \lim_{h \to 0} \frac{(9 - 2^2 - (1 + h)^2) - (9 - 2^2 - 1^2)}{h} = \lim_{h \to 0} \frac{-2h - h^2}{h} = \lim_{h \to 0} (-2 - h) = -2.$$

25. We have
$$f_x(x, y, z) = \lim_{h \to 0} \frac{\frac{(x+h)y^2}{z} - \frac{xy^2}{z}}{h} = \lim_{h \to 0} \frac{xy^2 + hy^2 - xy^2}{zh} = \lim_{h \to 0} \frac{hy^2}{zh} = \lim_{h \to 0} \frac{y^2}{z} = \frac{y^2}{z}$$
$$f_y(x, y, z) = \lim_{h \to 0} \frac{\frac{x(y+h)^2}{z} - \frac{xy^2}{z}}{h} = \lim_{h \to 0} \frac{xy^2 + 2xyh + xh^2 - xy^2}{zh} = \lim_{h \to 0} \frac{2xyh + xh^2}{zh}$$
$$= \lim_{h \to 0} \frac{2xy + xh}{z} = \frac{2xy}{z}$$
$$f_z(x, y, z) = \lim_{h \to 0} \frac{\frac{xy^2}{z+h} - \frac{xy^2}{z}}{h} = \lim_{h \to 0} \frac{xy^2 z - xy^2(z + h)}{hz(z + h)} = \lim_{h \to 0} \frac{-xy^2 h}{hz(z + h)} = - \lim_{h \to 0} \frac{xy^2}{z(z + h)} = -\frac{xy^2}{z^2}.$$

27. Using Theorem 12.22 and the product rule and chain rule,
$$f_x(x, y) = e^x \sin(xy) + e^x \cos(xy) \cdot y = e^x(\sin(xy) + y \cos(xy))$$
$$f_y(x, y) = e^x \cos(xy) \cdot x = xe^x \cos(xy).$$

29. Using Theorem 12.22,
$$f_x(x, y) = yx^{y-1}$$
$$f_y(x, y) = x^y \ln x.$$

31. Using Theorem 12.22,
$$f_x(x, y) = \sin y$$
$$f_y(x, y) = x \cos y.$$

33. Using Theorem 12.22,
$$f_r(r, \theta) = \sin \theta$$
$$f_\theta(r, \theta) = r \cos \theta.$$

35. Using Theorem 12.22 and the quotient rule,

$$f_x(x,y,z) = \frac{(x+z)(y^2) - xy^2(1)}{(x+z)^2} = \frac{y^2z}{(x+z)^2}$$

$$f_y(x,y,z) = \frac{2xy}{x+z}$$

$$f_z(x,y,z) = -\frac{xy^2}{(x+z)^2} \cdot 1 = -\frac{xy^2}{(x+z)^2}.$$

37. (a) In the x direction, the slope is the value of $f_x(x,y)$ at the given point $\left(0, \frac{\pi}{2}\right)$, which is

$$e^0\left(\sin\left(0\cdot\frac{\pi}{2}\right) + \frac{\pi}{2}\cos\left(0\cdot\frac{\pi}{2}\right)\right) = \frac{\pi}{2}.$$

Thus the tangent line in the x direction has direction vector $\left\langle 1, 0, \frac{\pi}{2}\right\rangle$. Also note that $f\left(0, \frac{\pi}{2}\right) = e^0\sin(0) = 0$, so that an equation for the tangent line in the x direction is

$$\left\langle 0, \frac{\pi}{2}, 0\right\rangle + t\left\langle 1, 0, \frac{\pi}{2}\right\rangle.$$

(b) In the y direction, the slope is the value of $f_y(x,y)$ at the given point $\left(0, \frac{\pi}{2}\right)$, which is

$$0\cdot e^0\cdot\cos\left(0\cdot\frac{\pi}{2}\right) = 0.$$

Thus the tangent line in the y direction has direction vector $\langle 0,1,0\rangle$. Since $f\left(0, \frac{\pi}{2}\right) = 0$ from part (a), an equation for the tangent line in the y direction is

$$\left\langle 0, \frac{\pi}{2}, 0\right\rangle + t\langle 0,1,0\rangle.$$

(c) The plane containing those two lines has normal vector

$$\left\langle 1, 0, \frac{\pi}{2}\right\rangle \times \langle 0,1,0\rangle = \left\langle 0\cdot 0 - \frac{\pi}{2}\cdot 1, \frac{\pi}{2}\cdot 0 - 1\cdot 0, 1\cdot 1 - 0\cdot 0\right\rangle = \left\langle -\frac{\pi}{2}, 0, 1\right\rangle.$$

Thus the equation of the plane is $-\frac{\pi}{2}(x - 0) + (z - 0) = 0$, or $\frac{\pi}{2}x - z = 0$.

39. (a) In the x direction, the slope is the value of $f_x(x,y)$ at the given point $(e,3)$, which is $3e^2$. Thus the tangent line in the x direction has direction vector $\langle 1, 0, 3e^2\rangle$. Also note that $f(e,3) = e^3$, so that an equation for the tangent line in the x direction is

$$\left\langle e, 3, e^3\right\rangle + t\left\langle 1, 0, 3e^2\right\rangle.$$

(b) In the y direction, the slope is the value of $f_y(x,y)$ at the given point $(e,3)$, which is

$$e^3\ln e = e^3.$$

Thus the tangent line in the y direction has direction vector $\langle 0, 1, e^3\rangle$. Since $f(e,3) = e^3$ from part (a), an equation for the tangent line in the y direction is

$$\left\langle e, 3, e^3\right\rangle + t\left\langle 0, 1, e^3\right\rangle.$$

(c) The plane containing those two lines has normal vector

$$\left\langle 1, 0, 3e^2\right\rangle \times \left\langle 0, 1, e^3\right\rangle = \left\langle 0\cdot e^3 - 3e^2\cdot 1, 3e^2\cdot 0 - 1\cdot e^3, 1\cdot 1 - 0\cdot 0\right\rangle = \left\langle -3e^2, -e^3, 1\right\rangle.$$

Thus the equation of the plane is $-3e^2(x - e) - e^3(y - 3) + 1(z - e^3) = 0$, or $3e^2x + e^3y - z = 5e^3$.

41. (a) In the x direction, the slope is the value of $f_x(x, y)$ at the given point $(2, \frac{\pi}{3})$, which is $\sin \frac{\pi}{3} = \frac{\sqrt{3}}{2}$. Thus the tangent line in the x direction has direction vector $\left\langle 1, 0, \frac{\sqrt{3}}{2} \right\rangle$. Also note that $f\left(2, \frac{\pi}{3}\right) = 2 \sin \frac{\pi}{3} = \sqrt{3}$, so that an equation for the tangent line in the x direction is

$$\left\langle 2, \frac{\pi}{3}, \sqrt{3} \right\rangle + t \left\langle 1, 0, \frac{\sqrt{3}}{2} \right\rangle.$$

(b) In the y direction, the slope is the value of $f_y(x, y)$ at the given point $(2, \frac{\pi}{3})$, which is $2 \cos \frac{\pi}{3} = 1$. Thus the tangent line in the y direction has direction vector $\langle 0, 1, 1 \rangle$. Since $f\left(2, \frac{\pi}{3}\right) = \sqrt{3}$ from part (a), an equation for the tangent line in the y direction is

$$\left\langle 2, \frac{\pi}{3}, \sqrt{3} \right\rangle + t \langle 0, 1, 1 \rangle.$$

(c) The plane containing those two lines has normal vector

$$\left\langle 1, 0, \frac{\sqrt{3}}{2} \right\rangle \times \langle 0, 1, 1 \rangle = \left\langle 0 \cdot 1 - \frac{\sqrt{3}}{2} \cdot 1, \frac{\sqrt{3}}{2} \cdot 0 - 1 \cdot 1, 1 \cdot 1 - 0 \cdot 0 \right\rangle = \left\langle -\frac{\sqrt{3}}{2}, -1, 1 \right\rangle.$$

Thus the equation of the plane is $-\frac{\sqrt{3}}{2}(x - 2) - 1 \left(y - \frac{\pi}{3}\right) + 1(z - \sqrt{3}) = 0$, or $\frac{\sqrt{3}}{2} x + y - z = \frac{\pi}{3}$.

43. Using the results of Exercise 27, we get

$$\begin{aligned}
f_{xx}(x, y) &= \frac{\partial}{\partial x}(f_x(x, y)) = e^x(\sin(xy) + y \cos(xy)) + e^x(y \cos(xy) - y^2 \sin(xy)) \\
&= e^x((1 - y^2) \sin(xy) + 2y \cos(xy)) \\
f_{xy}(x, y) &= \frac{\partial}{\partial y}(f_x(x, y)) = e^x(x \cos(xy) + \cos(xy) - xy \sin(xy)) \\
&= e^x(-xy \sin(xy) + (1 + x) \cos(xy)) \\
f_{yy}(x, y) &= \frac{\partial}{\partial y}(f_y(x, y)) = xe^x(-\sin(xy)) \cdot x = -x^2 e^x \sin(xy) \\
f_{yx}(x, y) &= \frac{\partial}{\partial x}(f_y(x, y)) = e^x \cos(xy) + xe^x \cos(xy) - xe^x \sin(xy) \cdot y \\
&= e^x(-xy \sin(xy) + (1 + x) \cos(xy)).
\end{aligned}$$

The mixed partials are equal.

45. Using the results of Exercise 29, we get

$$\begin{aligned}
f_{xx}(x, y) &= \frac{\partial}{\partial x}(f_x(x, y)) = y(y - 1)x^{y-2} \\
f_{xy}(x, y) &= \frac{\partial}{\partial y}(f_x(x, y)) = x^{y-1} + yx^{y-1} \ln x = x^{y-1}(1 + y \ln x) \\
f_{yy}(x, y) &= \frac{\partial}{\partial y}(f_y(x, y)) = x^y (\ln x)^2 \\
f_{yx}(x, y) &= \frac{\partial}{\partial x}(f_y(x, y)) = yx^{y-1} \ln x + x^y \cdot \frac{1}{x} = yx^{y-1} \ln x + x^{y-1} = x^{y-1}(1 + y \ln x).
\end{aligned}$$

The mixed partials are equal.

47. Using the results of Exercise 31, we get

$$f_{xx}(x,y) = \frac{\partial}{\partial x}(f_x(x,y)) = 0$$

$$f_{xy}(x,y) = \frac{\partial}{\partial y}(f_x(x,y)) = \cos y$$

$$f_{yy}(x,y) = \frac{\partial}{\partial y}(f_y(x,y)) = -x\sin y$$

$$f_{yx}(x,y) = \frac{\partial}{\partial x}(f_y(x,y)) = \cos y.$$

The mixed partials are equal.

49. Using the results of Exercise 33, we get

$$f_{rr}(r,\theta) = \frac{\partial}{\partial r}(f_r(r,\theta)) = 0$$

$$f_{r\theta}(r,\theta) = \frac{\partial}{\partial \theta}(f_r(r,\theta)) = \cos\theta$$

$$f_{\theta\theta}(r,\theta) = \frac{\partial}{\partial \theta}(f_\theta(r,\theta)) = -r\sin\theta$$

$$f_{\theta r}(r,\theta) = \frac{\partial}{\partial r}(f_\theta(r,\theta)) = \cos\theta.$$

The mixed partials are equal.

51. Since $\frac{\partial f}{\partial x} = 0$, $f(x,y)$ must not depend on x; that is, it must be a function of y, so that $f(x,y) = g(y)$.

53. Integrating $\frac{\partial^2 f}{\partial x^2} = 0$ gives $\frac{\partial f}{\partial x} = g(y)$, since $g(y)$ is the most general function that does not depend on x. Integrating again gives $f(x,y) = xg(y) + h(y)$. So $f(x,y)$ must be a linear function in x whose coefficients are functions of y.

55. Integrating with respect to x gives $f(x,y,z) = g(y,z)$, where g is any function of y and z.

57. Integrating with respect to x gives $\frac{\partial f}{\partial x} = g(y,z)$. Then integrate again with respect to x to get $f(x,y,z) = xg(y,z) + h(y,z)$. So $f(x,y,z)$ must be a linear function in x whose coefficients are functions of y and z.

59. Since $g_y(x,y) = -e^x \sin y = h_x(x,y)$, such a function $F(x,y)$ exists, by Theorem 12.25. To find the function, first integrate $g(x,y)$ with respect to x:

$$F(x,y) = \int g(x,y)\,dx = e^x \cos y + f(y).$$

If we differentiate with respect to y, we must get $h(x,y)$:

$$F_y(x,y) = -e^x \sin y + f'(y) = h(x,y) = -e^x \sin y + 2y.$$

Thus $f'(y) = 2y$, so that $f(y) = y^2 + C$. Hence $F(x,y) = e^x \cos y + y^2 + C$.

61. We have

$$g_y(x,y) = \frac{(1 + x^2 y^2)(1) - y(2x^2 y)}{(1 + x^2 y^2)^2} = \frac{1 - x^2 y^2}{(1 + x^2 y^2)^2}$$

$$h_x(x,y) = \frac{(1 + x^2 y^2)(1) - x(2xy^2)}{(1 + x^2 y^2)^2} = \frac{1 - x^2 y^2}{(1 + x^2 y^2)^2}.$$

Since the two are equal, such a function $F(x, y)$ exists, by Theorem 12.25. To find it, first integrate $g(x, y)$ with respect to x, using the substitution $u = xy$ so that $du = y\,dx$:

$$F(x, y) = \int g(x, y)\,dx = \int \frac{y}{1 + x^2 y^2}\,dx = \int \frac{1}{1 + u^2}\,du = \tan^{-1} u + f(y) = \tan^{-1}(xy) + f(y).$$

If we differentiate this with respect to y, we must get $h(x, y)$:

$$F_y(x, y) = \frac{x}{1 + x^2 y^2} + f'(y) = h(x, y) = \frac{x}{1 + x^2 y^2}.$$

Thus $f'(y) = 0$, so that $f(y) = C$, and thus $F(x, y) = \tan^{-1}(xy) + C$.

63. Let $g(x, y) = e^y$ and $h(x, y) = xe^y - 7$. Then $g_y(x, y) = e^y = h_x(x, y)$, so the differential equation is exact. To solve it, we must find $F(x, y)$ with $F_x = g$ and $F_y = h$. First integrate $g(x, y)$ with respect to x:

$$F(x, y) = \int g(x, y)\,dx = xe^y + f(y).$$

If we differentiate this with respect to y, we must get $h(x, y)$:

$$F_y(x, y) = xe^y + f'(y) = h(x, y) = xe^y - 7.$$

Thus $f'(y) = -7$, so that $f(y) = -7y + C$. It follows that $F(x, y) = xe^y - 7y + C$.

65. Let $g(x, y) = e^x \ln y + x^3$ and $h(x, y) = \frac{e^x}{y}$. Then

$$g_y(x, y) = \frac{e^x}{y} = h_x(x, y),$$

so the differential equation is exact. To solve it, we must find $F(x, y)$ with $F_x = g$ and $F_y = h$. First integrate $g(x, y)$ with respect to x:

$$F(x, y) = \int g(x, y)\,dx = \int (e^x \ln y + x^3)\,dx = e^x \ln y + \frac{1}{4}x^4 + f(y).$$

If we differentiate this with respect to y, we must get $h(x, y)$:

$$F_y(x, y) = \frac{e^x}{y} + f'(y) = h(x, y) = \frac{e^x}{y}.$$

Thus $f'(y) = 0$, so that $f(y) = C$, and then $F(x, y) = e^x \ln y + \frac{1}{4}x^4 + C$.

Applications

67. Treating h as a constant, we have $\frac{\partial V}{\partial r} = 2\pi rh$; treating r as a constant, we have $\frac{\partial V}{\partial h} = \pi r^2$. The first of these means that if the height of a cylinder is held constant, then as the radius changes, the volume will change by $2\pi rh$. The second means that if the radius is held constant, then as the height changes, the volume will change by πr^2.

69. (a) We have

$$\frac{\partial}{\partial x}(\alpha e^{nx} \sin ny) = \alpha \sin ny \frac{\partial}{\partial x}e^{nx} = n\alpha e^{nx} \sin ny$$

$$\frac{\partial^2}{\partial x^2}(\alpha e^{nx} \sin ny) = \frac{\partial}{\partial x}(n\alpha e^{nx} \sin ny) = n\alpha \sin ny \frac{\partial}{\partial x}e^{nx} = n^2 \alpha e^{nx} \sin ny$$

$$\frac{\partial}{\partial y}(\alpha e^{nx} \sin ny) = \alpha e^{nx}\frac{\partial}{\partial y}\sin ny = n\alpha e^{nx} \cos ny$$

$$\frac{\partial^2}{\partial y^2}(\alpha e^{nx} \sin ny) = \frac{\partial}{\partial y}(n\alpha e^{nx} \cos ny) = n\alpha e^{nx}\frac{\partial}{\partial y}(\cos ny) = -n^2 \alpha e^{nx} \sin ny.$$

Clearly

$$\frac{\partial^2}{\partial x^2}\left(\alpha e^{nx}\sin ny\right) + \frac{\partial^2}{\partial y^2}\left(\alpha e^{nx}\sin ny\right) = 0.$$

Similarly,

$$\frac{\partial}{\partial x}\left(\beta e^{ny}\sin nx\right) = \beta e^{ny}\frac{\partial}{\partial x}\sin nx = n\beta e^{ny}\cos nx$$

$$\frac{\partial^2}{\partial x^2}\left(\beta e^{ny}\sin nx\right) = \frac{\partial}{\partial x}\left(n\beta e^{ny}\cos nx\right) = n\beta e^{ny}\frac{\partial}{\partial x}(\cos nx) = -n^2\beta e^{ny}\sin nx$$

$$\frac{\partial}{\partial y}\left(\beta e^{ny}\sin nx\right) = \beta\sin nx\frac{\partial}{\partial y}e^{ny} = n\beta e^{ny}\sin nx$$

$$\frac{\partial^2}{\partial y^2}\left(\beta e^{ny}\sin nx\right) = \frac{\partial}{\partial y}\left(n\beta e^{ny}\sin nx\right) = n\beta\sin nx\frac{\partial}{\partial y}e^{ny} = n^2\beta e^{ny}\sin nx.$$

Clearly

$$\frac{\partial^2}{\partial x^2}\left(\beta e^{ny}\sin nx\right) + \frac{\partial^2}{\partial y^2}\left(\beta e^{ny}\sin nx\right) = 0.$$

(b) Since $f(x,0)$ contains a sin term but no exponential term, $f(x,y)$ must equal $\beta e^{ny}\sin nx$. Then $f(x,0) = \beta\sin nx$, so that $\beta = 0.04$ and $n = \frac{\pi}{2}$ is one reasonable possibility. The solution $f(x,y) = 0.04e^{\pi y/2}\sin\frac{\pi x}{2}$ also satisfies the other three conditions, so this is in fact the answer. A plot of the solution is

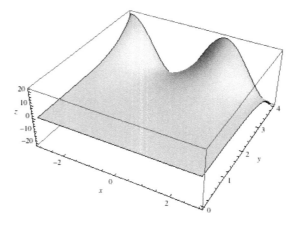

Proofs

71. By the chain rule, regarding y as a constant in the first line and x as a constant in the second,

$$\frac{\partial z}{\partial x} = n(f(x,y))^{n-1}\frac{\partial}{\partial x}f(x,y) = n(f(x,y))^{n-1}\frac{\partial f}{\partial x}$$

$$\frac{\partial z}{\partial y} = n(f(x,y))^{n-1}\frac{\partial}{\partial y}f(x,y) = n(f(x,y))^{n-1}\frac{\partial f}{\partial y}.$$

73. To differentiate with respect to x, think of y as a constant, so that both $f(x,y)$ and $g(x,y)$ are thought of as functions of x. Then using the quotient rule for single-variable functions,

$$\frac{\partial z}{\partial x} = \frac{g(x,y)\frac{\partial}{\partial x}f(x,y) - f(x,y)\frac{\partial}{\partial x}g(x,y)}{(g(x,y))^2} = \frac{f_x(x,y)g(x,y) - f(x,y)g_x(x,y)}{(g(x,y))^2}.$$

Similarly, to differentiate with respect to y, think of x as a constant, so that both $f(x, y)$ and $g(x, y)$ are thought of as functions of y. Then using the quotient rule for single-variable functions,

$$\frac{\partial z}{\partial y} = \frac{g(x,y)\frac{\partial}{\partial y}f(x,y) - f(x,y)\frac{\partial}{\partial y}g(x,y)}{(g(x,y))^2} = \frac{f_y(x,y)g(x,y) - f(x,y)g_y(x,y)}{(g(x,y))^2}.$$

75. We have

$$\frac{\partial^2 h}{\partial x\,\partial y} = \frac{\partial}{\partial x}\left(\frac{\partial h}{\partial y}\right) = \frac{\partial}{\partial x}(0 + g'(y)) = \frac{\partial}{\partial x}g'(y) = 0$$

$$\frac{\partial^2 h}{\partial y\,\partial x} = \frac{\partial}{\partial y}\left(\frac{\partial h}{\partial x}\right) = \frac{\partial}{\partial y}(f'(x) + 0) = \frac{\partial}{\partial y}f'(x) = 0,$$

so that the two are equal. The last equality in each line requires some explanation. In the first line, note that we have assumed only that g is once differentiable, so that there is no reason to assume that $g'(y)$ is continuous or differentiable. However, since we are taking its derivative with respect to x, this does not matter - for this purpose, $g'(y)$ is a constant, so its derivative is zero. An identical argument applies to $f'(x)$ in the second line, with the roles of the variables reversed.

Thinking Forward

A chain rule for functions of two variables

▷ $\dfrac{\partial f}{\partial x} = 2xy^3$ and $\dfrac{\partial f}{\partial y} = 3x^2y^2$.

▷ $\dfrac{\partial x}{\partial s} = \cos t$ and $\dfrac{\partial x}{\partial t} = -s\sin t$.

▷ $\dfrac{\partial y}{\partial s} = \sin t$ and $\dfrac{\partial y}{\partial t} = s\cos t$.

▷ Substituting into $f(x, y)$ gives

$$f(s, t) = x^2y^3 = (s\cos t)^2(s\sin t)^3 = s^5\cos^2 t\sin^3 t.$$

▷ Differentiating, we get

$$\frac{\partial f}{\partial s} = 5s^4\cos^2 t\sin^3 t$$

$$\frac{\partial f}{\partial t} = s^5(\cos^2 t \cdot 3\sin^2 t\cos t + (2\cos t(-\sin t))\sin^3 t) = s^5(3\sin^2 t\cos^3 t - 2\sin^4 t\cos t).$$

▷ It appears that

$$\frac{\partial f}{\partial x}\cdot\frac{dx}{ds} + \frac{\partial f}{\partial y}\cdot\frac{dy}{ds} = 2xy^3\cos t + 3x^2y^2\sin t$$

$$= 2(s\cos t)(s\sin t)^3\cos t + 3(s\cos t)^2(s\sin t)^2\sin t$$

$$= 5s^4\cos^2 t\sin^3 t = \frac{\partial f}{\partial s}.$$

▷ It appears that

$$\frac{\partial f}{\partial x}\cdot\frac{dx}{dt} + \frac{\partial f}{\partial y}\cdot\frac{dy}{dt} = 2xy^3(-s\sin t) + 3x^2y^2(s\cos t)$$

$$= 2(s\cos t)(s\sin t)^3(-s\sin t) + 3(s\cos t)^2(s\sin t)^2(s\cos t)$$

$$= -2s^5\cos t\sin^4 t + 3s^5\cos^3 t\sin^2 t = \frac{\partial f}{\partial t}.$$

12.4 Directional Derivatives and Differentiability

Thinking Back

Collinear tangent lines

The left- and right-hand derivatives of $f(x) = x^3$ at $x = 2$ are:

$$f'_-(2) = \lim_{h \to 0^-} \frac{f(2+h) - f(2)}{h} = \lim_{h \to 0^-} \frac{8 + 12h + 6h^2 + h^3 - 8}{h} = \lim_{h \to 0^-} (12 + 6h + h^2) = 12$$

$$f'_+(2) = \lim_{h \to 0^+} \frac{f(2+h) - f(2)}{h} = \lim_{h \to 0^+} \frac{8 + 12h + 6h^2 + h^3 - 8}{h} = \lim_{h \to 0^+} (12 + 6h + h^2) = 12,$$

so that both derivatives exist. Since the values of the left and right derivatives are equal, the lines with those slopes passing through $(2, 8)$ are the same line.

Noncollinear tangent lines

The left- and right-hand derivatives of $f(x) = |x| + 3$ at $x = 0$ are:

$$f'_-(0) = \lim_{h \to 0^-} \frac{(|0 + h| + 3) - (|0| + 3)}{h} = \lim_{h \to 0^-} \frac{-h}{h} = -1$$

$$f'_+(0) = \lim_{h \to 0^+} \frac{(|0 + h| + 3) - (|0| + 3)}{h} = \lim_{h \to 0^+} \frac{h}{h} = 1,$$

so that both derivatives exist. However, the two values are unequal, so that the lines with those slopes passing through $(0, f(3)) = (0, 3)$ are not the same line (so not collinear).

Concepts

1. (a) True. Since $\mathbf{j} = \langle 0, 1 \rangle$, substituting $\alpha = 0$ and $\beta = 1$ in Definition 12.26 gives

$$D_{\mathbf{j}} f(a, b) = \lim_{h \to 0} \frac{f(a + 0h, b + 1h) - f(a, b)}{h} = \lim_{h \to 0} \frac{f(a, b + h) - f(a, b)}{h} = f_y(a, b).$$

   (b) True. If $\mathbf{u} = \langle \alpha, \beta \rangle$, then

$$\begin{aligned} D_{\mathbf{u}} f(a, b) &= \lim_{h \to 0} \frac{f(a + \alpha h, b + \beta h) - f(a, b)}{h} \\ &= \lim_{h \to 0} \frac{f(a + (-\alpha)(-h), b + (-\beta)(-h)) - f(a, b)}{h} \\ &= -\lim_{h \to 0} \frac{f(a + (-\alpha)(-h), b + (-\beta)(-h)) - f(a, b)}{-h} \\ &= -\lim_{k \to 0} \frac{f(a + (-\alpha)k, b + (-\beta)k) - f(a, b)}{k} \\ &= -D_{-\mathbf{u}} f(a, b). \end{aligned}$$

   (c) False. For example, let $f(x, y) = 1$ if $y \neq 0$ and $f(x, y) = 0$ if $y = 0$. Then

$$D_{\mathbf{i}} f(0, 0) = f_x(0, 0) = \lim_{h \to 0} \frac{f(0 + h, 0) - f(0, 0)}{h} = \lim_{h \to 0} \frac{0 - 0}{h} = 0,$$

   but

$$D_{\mathbf{j}} f(0, 0) = f_y(0, 0) = \lim_{h \to 0} \frac{f(0, h) - f(0, 0)}{h} = \lim_{h \to 0} \frac{1 - 0}{h} = \infty.$$

   Thus $D_{\mathbf{j}} f(0, 0)$ does not exist.

(d) False. $D_{\mathbf{v}}f(a,b)$ is defined only if $\mathbf{v}$ is a unit vector.

(e) False. For example, let $f(x,y) = \frac{xy}{x^2+y^2}$. Then

$$f_x(0,0) = \lim_{h\to 0} \frac{f(0+h,0) - f(0,0)}{h} = \lim_{h\to 0} \frac{f(h,0) - 0}{h} = \lim_{h\to 0} \frac{0-0}{h} = 0$$

$$f_y(0,0) = \lim_{h\to 0} \frac{f(0,0+h) - f(0,0)}{h} = \lim_{h\to 0} \frac{f(0,h) - 0}{h} = \lim_{h\to 0} \frac{0-0}{h} = 0.$$

However, if f is differentiable at $(0,0)$, then by Definition 12.28 there must exist ϵ_1 and ϵ_2, function of Δx and Δy, such that

$$f(0+\Delta x, 0+\Delta y) - f(0,0) = f_x(0,0)\Delta x + f_y(0,0)\Delta y + \epsilon_1 \Delta x + \epsilon_2 \Delta y, \text{ or}$$

$$\frac{\Delta x \Delta y}{\Delta x^2 + \Delta y^2} = \epsilon_1 \Delta x + \epsilon_2 \Delta y.$$

In addition, we must have $\epsilon_1 \to 0$ and $\epsilon_2 \to 0$ as $(\Delta x, \Delta y) \to (0,0)$. Assume these exist, and let $(\Delta x, \Delta y) \to (0,0)$ along the path $\Delta y = \Delta x$. Then the above equation becomes

$$\frac{\Delta x^2}{\Delta x^2 + \Delta x^2} = (\epsilon_1 + \epsilon_2)\Delta x, \text{ or}$$

$$\frac{1}{2} = \Delta x(\epsilon_1 + \epsilon_2).$$

But this is impossible, since the right side goes to zero as $\Delta x \to 0$. Thus $f(x,y)$ is not differentiable at $(0,0)$.

(f) True. This is Theorem 12.29. If f_s and f_t are continuous in an open set containing (a,b), then by Theorem 12.29, f is differentiable at (a,b).

(g) False. For example, let $f(x,y) = x$ and $(a,b) = (0,0)$. Then

$$D_{\mathbf{i}}f(0,0) = f_x(0,0) = 1, \quad \text{but} \quad D_{\mathbf{j}}f(0,0) = f_y(0,0) = 0.$$

(h) True. Suppose instead that $D_{\mathbf{i}}f(a,b) = D_{-\mathbf{i}}f(a,b) = 1$. Then

$$1 = D_{\mathbf{i}}f(a,b) = \lim_{h\to 0} \frac{f(a+h,b) - f(a,b)}{h}$$

$$1 = D_{-\mathbf{i}}f(a,b) = \lim_{h\to 0} \frac{f(a-h,b) - f(a,b)}{h} = -\lim_{h\to 0} \frac{f(a-h,b) - f(a,b)}{-h}$$

$$= -\lim_{k\to 0} \frac{f(a+k,b) - f(a,b)}{k}.$$

But this is impossible, since the final expression above is $-D_{\mathbf{i}}f(a,b) = -1$. Thus f is not differentiable.

3. In $\mathbb{R}^1$, unit vectors are numbers x with $\|x\| = \sqrt{x^2} = |x| = 1$, so the only unit vectors are ± 1. In $\mathbb{R}^n$, unit vectors are all points $(x_1, \ldots, x_n)$ such that $\sqrt{x_1^2 + \cdots + x_n^2} = 1$, so all points such that $x_1^2 + \cdots + x_n^2 = 1$. These points are the surface of the hypersphere of radius 1 around the origin, so there are an infinite number of them.

5. If $\mathbf{u} = \langle \alpha, \beta \rangle$ is a unit vector, then the directional derivative of f in the direction $\mathbf{u}$ is

$$D_{\mathbf{u}}f(x,y) = \lim_{h\to 0} \frac{f(x+\alpha h, y+\beta h) - f(x,y)}{h}$$

provided the limit exists. See Definition 12.26 for details.

7. If $\mathbf{u} = \langle \alpha_1, \ldots, \alpha_n \rangle$ is a unit vector, then the directional derivative of f in the direction $\mathbf{u}$ at $\mathbf{v} = \langle x_1, x_2, \ldots, x_n \rangle$ is

$$D_\mathbf{u} f(x_1, x_2, \ldots, x_n) = \lim_{h \to 0} \frac{f(x_1 + \alpha_1 h, x_2 + \alpha_2 h, \ldots, x_n + \alpha_n h) - f(x_1, x_2, \ldots, x_n)}{h}$$

provided the limit exists.

9. (a) The directional derivative is

$$\begin{aligned}
D_\mathbf{u} f(-1, 3) &= \lim_{h \to 0} \frac{f(-1 + \alpha h, 3 + \beta h) - f(-1, 3)}{h} \\
&= \lim_{h \to 0} \frac{(-1 + \alpha h)(3 + \beta h) - (-1)(3)}{h} = \lim_{h \to 0} \frac{(3\alpha - \beta + \alpha\beta h)h}{h} \\
&= \lim_{h \to 0} (3\alpha - \beta + \alpha\beta h) = 3\alpha - \beta.
\end{aligned}$$

(b) We get

$$\begin{aligned}
\lim_{h \to 0} \frac{(-1 + (k\alpha)h)(3 + (k\beta)h) - (-1)(3)}{h} &= \lim_{h \to 0} \frac{(3k\alpha - k\beta + k^2\alpha\beta h)}{h} \\
&= \lim_{h \to 0} (3k\alpha - k\beta + k^2\alpha\beta h) = k(3\alpha - \beta).
\end{aligned}$$

(c) The result in part (b) is what would happen if you applied the formula for a directional derivative to the vector $k\mathbf{u}$. While $\mathbf{u}$ and $k\mathbf{u}$ give the same direction vectors (at least if $k > 0$), they do not give the same directional derivative if the formula is simply applied blindly. So it is necessary to specify a particular value of k. But in the one-dimensional case, $k = 1$ is required in order to get the usual derivative, so we use $k = 1$ here as well.

11. From Exercise 3, the only unit vectors are $\mathbf{u} = \langle 1 \rangle$ and $\mathbf{u} = \langle -1 \rangle$. If $\mathbf{u} = \langle \alpha \rangle$ is either of these, then the directional derivative of f in the direction of $\mathbf{u}$ is, mimicking Definition 12.26,

$$D_\mathbf{u} f(c) = \lim_{h \to 0} \frac{f(c + \alpha h) - f(c)}{h}$$

if the limit exists.

13. See Definition 12.28. $z = f(x, y)$ is differentiable at (a, b) if it is defined on an open set containing (a, b), if the partial derivatives f_x and f_y both exist at (a, b), and if there are functions ϵ_1 and ϵ_2 depending on Δx and Δy such that

$$f(a + \Delta x, b + \Delta y) - f(a, b) = f_x(a, b)\Delta x + f_y(a, b)\Delta y + \epsilon_1 \Delta x + \epsilon_2 \Delta y$$

and if in addition ϵ_1 and ϵ_2 go to zero as $(\Delta x, \Delta y) \to (0, 0)$.

15. If f is differentiable at (a, b), the tangent lines to the graph of f in the $\mathbf{u}_1$ and $\mathbf{u}_2$ directions obviously have direction vectors $\mathbf{u}_1$ and $\mathbf{u}_2$. Since these are nonparallel direction vectors, the tangent lines are nonparallel. But two nonparallel vectors through (a, b) determine a unique plane, which is the tangent plane.

17. If f is differentiable at (a, b, c), then the tangent lines in the x, y, and z directions have direction vectors $\mathbf{i}$, $\mathbf{j}$, and $\mathbf{k}$, so they determine non-coplanar vectors that all pass through (a, b, c). Thus they determine a unique hyperplane, which is the tangent hyperplane.

19. A definition might be: Let $f(x_1, \ldots, x_n)$ be a function from $\mathbb{R}^n$ to $\mathbb{R}$ defined on an open set containing $(a_1, \ldots, a_n)$. f is said to be differentiable at $(a_1, \ldots, a_n)$ if the partial derivatives $f_{x_i}(a_1, \ldots, a_n)$ all exist and if

$$f(a_1 + \Delta x_1, a_2 + \Delta x_2, \ldots, a_n + \Delta x_n) - f(a_1, \ldots, a_n) = \sum_{i=1}^{n} f_{x_i}(a_1, \ldots, a_n)\Delta x_i + \sum_{i=1}^{n} \epsilon_i \Delta x_i,$$

where the ϵ_i are functions depending on the Δx_k, and they all go to zero as $(\Delta x_1, \ldots, \Delta x_k) \to (0, 0, \ldots, 0)$.

Skills

21. Computing gives

$$\begin{aligned}
D_{\mathbf{u}}f(2,3) &= \lim_{h \to 0} \frac{\left(2 + \frac{\sqrt{2}}{2}h\right)^2 - \left(3 + \frac{\sqrt{2}}{2}h\right)^2 - (2^2 - 3^2)}{h} \\
&= \lim_{h \to 0} \frac{2^2 + 2h\sqrt{2} + \frac{1}{2}h^2 - 3^2 - 3h\sqrt{2} - \frac{1}{2}h^2 - (2^2 - 3^2)}{h} \\
&= \lim_{h \to 0} \frac{-h\sqrt{2}}{h} = -\sqrt{2}.
\end{aligned}$$

23. Computing gives

$$\begin{aligned}
D_{\mathbf{u}}f(-2,1) &= \lim_{h \to 0} \frac{\frac{-2 + \frac{\sqrt{10}}{10}h}{\left(1 - \frac{3\sqrt{10}}{10}h\right)^2} - \frac{-2}{1^2}}{h} \\
&= \lim_{h \to 0} \frac{\frac{-200 + 10h\sqrt{10}}{\left(10 - 3h\sqrt{10}\right)^2} + 2}{h} \\
&= \lim_{h \to 0} \frac{-200 + 10h\sqrt{10} + 2\left(10 - 3h\sqrt{10}\right)^2}{h\left(10 - 3h\sqrt{10}\right)^2} \\
&= \lim_{h \to 0} \frac{-200 + 10h\sqrt{10} + 200 - 120h\sqrt{10} + 180h^2}{h\left(10 - 3h\sqrt{10}\right)^2} \\
&= \lim_{h \to 0} \frac{-110h\sqrt{10} + 180h^2}{h\left(10 - 3h\sqrt{10}\right)^2} \\
&= \lim_{h \to 0} \frac{-110\sqrt{10} + 180h}{\left(10 - 3h\sqrt{10}\right)^2} \\
&= -\frac{110\sqrt{10}}{100} = -\frac{11}{10}\sqrt{10}.
\end{aligned}$$

25. Computing gives

$$D_{\mathbf{u}}f(4,9) = \lim_{h \to 0} \frac{\sqrt{\frac{9 - \frac{4\sqrt{17}}{17}h}{4 - \frac{\sqrt{17}}{17}h}} - \sqrt{\frac{9}{4}}}{h}$$

$$= \lim_{h \to 0} \frac{\sqrt{\frac{153 - 4h\sqrt{17}}{68 - h\sqrt{17}}} - \frac{3}{2}}{h}$$

$$= \lim_{h \to 0} \frac{2\sqrt{153 - 4h\sqrt{17}} - 3\sqrt{68 - h\sqrt{17}}}{2h\sqrt{68 - h\sqrt{17}}}$$

$$= \lim_{h \to 0} \frac{\left(2\sqrt{153 - 4h\sqrt{17}} - 3\sqrt{68 - h\sqrt{17}}\right)\left(2\sqrt{153 - 4h\sqrt{17}} + 3\sqrt{68 - h\sqrt{17}}\right)}{2h\sqrt{68 - h\sqrt{17}}\left(2\sqrt{153 - 4h\sqrt{17}} + 3\sqrt{68 - h\sqrt{17}}\right)}$$

$$= \lim_{h \to 0} \frac{4(153 - 4h\sqrt{17}) - 9(68 - h\sqrt{17})}{2h\sqrt{68 - h\sqrt{17}}\left(2\sqrt{153 - 4h\sqrt{17}} + 3\sqrt{68 - h\sqrt{17}}\right)}$$

$$= \lim_{h \to 0} \frac{-7h\sqrt{17}}{2h\sqrt{68 - h\sqrt{17}}\left(2\sqrt{153 - 4h\sqrt{17}} + 3\sqrt{68 - h\sqrt{17}}\right)}$$

$$= \lim_{h \to 0} \frac{-7\sqrt{17}}{2\sqrt{68 - h\sqrt{17}}\left(2\sqrt{153 - 4h\sqrt{17}} + 3\sqrt{68 - h\sqrt{17}}\right)}$$

$$= -\frac{7\sqrt{17}}{2\sqrt{68}(2\sqrt{153} + 3\sqrt{68})} = -\frac{7}{2\sqrt{68}(2 \cdot 3 + 3 \cdot 2)} = -\frac{7}{48\sqrt{17}}.$$

27. Computing gives

$$D_{\mathbf{u}}f(2,-2,2) = \lim_{h \to 0} \frac{\left(2 + \frac{3}{5}h\right)^2 + (-2 + 0h)^2 - \left(2 + \frac{4}{5}h\right)^3 - (2^2 + (-2)^2 - 2^3)}{h}$$

$$= \lim_{h \to 0} \frac{4 + \frac{12}{5}h + \frac{9}{25}h^2 + 4 - 8 - \frac{48}{5}h - \frac{96}{25}h^2 - \frac{64}{125}h^3 - 4 - 4 + 8}{h}$$

$$= \lim_{h \to 0} \frac{-\frac{36}{5}h - \frac{87}{25}h^2 - \frac{64}{125}h^3}{h}$$

$$= \lim_{h \to 0} \left(-\frac{36}{5} - \frac{87}{25}h - \frac{64}{125}h^2\right)$$

$$= -\frac{36}{5}.$$

29. The equation of the line tangent to $f(x,y)$ at $P = (2,3)$ in the direction of $\mathbf{u} = \left\langle \frac{\sqrt{2}}{2}, \frac{\sqrt{2}}{2} \right\rangle$ is a line through $(2,3)$ with direction vector (using Exercise 21)

$$\left\langle \frac{\sqrt{2}}{2}, \frac{\sqrt{2}}{2}, D_{\mathbf{u}}f(2,3) \right\rangle = \left\langle \frac{\sqrt{2}}{2}, \frac{\sqrt{2}}{2}, -\sqrt{2} \right\rangle.$$

The line is then parametrized by

$$x = 2 + \frac{\sqrt{2}}{2}t, \quad y = 3 + \frac{\sqrt{2}}{2}t, \quad z = f(2,3) - \sqrt{2}t = -5 - \sqrt{2}t.$$

31. The equation of the line tangent to $f(x,y)$ at $P = (-2,1)$ in the direction of $\mathbf{u} = \left\langle \frac{\sqrt{10}}{10}, -\frac{3\sqrt{10}}{10} \right\rangle$ is a line through $(-2,1)$ with direction vector (using Exercise 23)

$$\left\langle \frac{\sqrt{10}}{10}, -\frac{3\sqrt{10}}{10}, -\frac{11\sqrt{10}}{10} \right\rangle.$$

We can multiply through by the scalar $\sqrt{10}$ to simplify and get the equivalent direction vector $\langle 1, -3, -11 \rangle$. The tangent line is then parametrized by

$$x = -2 + t, \quad y = 1 - 3t, \quad z = f(-2,1) - 11t = -2 - 11t.$$

33. The equation of the line tangent to $f(x,y)$ at $P = (4,9)$ in the direction of $\mathbf{u} = \left\langle -\frac{\sqrt{17}}{17}, -\frac{4\sqrt{17}}{17} \right\rangle$ is a line through $(4,9)$ with direction vector (using Exercise 25)

$$\left\langle -\frac{\sqrt{17}}{17}, -\frac{4\sqrt{17}}{17}, -\frac{7}{48\sqrt{17}} \right\rangle.$$

We can multiply through by the scalar $\sqrt{17}$ to simplify and get the equivalent direction vector $\left\langle -1, -4, -\frac{7}{48} \right\rangle$. The tangent line is then parametrized by

$$x = 4 - t, \quad y = 9 - 4t, \quad z = f(4,9) - \frac{7}{48}t = \frac{3}{2} - \frac{7}{48}t.$$

35. First normalize $\mathbf{v}$. Since $\|\mathbf{v}\| = \sqrt{1^2 + 5^2} = \sqrt{26}$, a unit vector in the direction of $\mathbf{v}$ is $\mathbf{u} = \left\langle -\frac{1}{\sqrt{26}}, \frac{5}{\sqrt{26}} \right\rangle$. Then computing gives

$$
\begin{aligned}
D_{\mathbf{u}}f(3,3) &= \lim_{h \to 0} \frac{\left(3 - \frac{1}{\sqrt{26}}h\right)^2 - \left(3 + \frac{5}{\sqrt{26}}h\right)^2 - (3^2 - 3^2)}{h} \\
&= \lim_{h \to 0} \frac{9 - \frac{6}{\sqrt{26}}h + \frac{1}{26}h^2 - 9 - \frac{30}{\sqrt{26}}h - \frac{25}{26}h^2 - 9 + 9}{h} \\
&= \lim_{h \to 0} \frac{-\frac{36}{\sqrt{26}}h - \frac{12}{13}h^2}{h} = \lim_{h \to 0} \left(-\frac{36}{\sqrt{26}} - \frac{12}{13}h \right) \\
&= -\frac{36}{\sqrt{26}} = -\frac{18}{13}\sqrt{26}.
\end{aligned}
$$

37. First normalize $\mathbf{v}$. Since $\|\mathbf{v}\| = \sqrt{2^2 + 1^2} = \sqrt{5}$, a unit vector in the direction of $\mathbf{v}$ is $\mathbf{u} =$

$\left\langle \frac{2}{\sqrt{5}}, -\frac{1}{\sqrt{5}} \right\rangle$. Then computing gives

$$D_{\mathbf{u}}f(1,16) = \lim_{h\to 0} \frac{\sqrt{\frac{16-\frac{1}{\sqrt{5}}h}{1+\frac{2}{\sqrt{5}}h}} - 4}{h} = \lim_{h\to 0} \frac{\sqrt{\frac{16\sqrt{5}-h}{\sqrt{5}+2h}} - 4}{h}$$

$$= \lim_{h\to 0} \frac{\sqrt{16\sqrt{5}-h} - 4\sqrt{\sqrt{5}+2h}}{h\sqrt{\sqrt{5}+2h}}$$

$$= \lim_{h\to 0} \frac{\left(\sqrt{16\sqrt{5}-h} - 4\sqrt{\sqrt{5}+2h}\right)\left(\sqrt{16\sqrt{5}-h} + 4\sqrt{\sqrt{5}+2h}\right)}{h\sqrt{\sqrt{5}+2h}\left(\sqrt{16\sqrt{5}-h} + 4\sqrt{\sqrt{5}+2h}\right)}$$

$$= \lim_{h\to 0} \frac{16\sqrt{5}-h - 16(\sqrt{5}+2h)}{h\sqrt{\sqrt{5}+2h}\left(\sqrt{16\sqrt{5}-h} + 4\sqrt{\sqrt{5}+2h}\right)}$$

$$= \lim_{h\to 0} \frac{-33h}{h\sqrt{\sqrt{5}+2h}\left(\sqrt{16\sqrt{5}-h} + 4\sqrt{\sqrt{5}+2h}\right)}$$

$$= \lim_{h\to 0} \frac{-33}{\sqrt{\sqrt{5}+2h}\left(\sqrt{16\sqrt{5}-h} + 4\sqrt{\sqrt{5}+2h}\right)}$$

$$= -\frac{33}{5^{1/4}(4\cdot 5^{1/4} + 4\cdot 5^{1/4})} = -\frac{33}{8\sqrt{5}}.$$

39. Let $\mathbf{u} = \langle a, b \rangle$ be a unit vector. Then

$$D_{\mathbf{u}}f(1,-2) = \lim_{h\to 0} \frac{(1+ah)(-2+bh) + 2(1+ah) - (-2+bh) - ((1)(-2) + 2(1) - (-2))}{h}$$

$$= \lim_{h\to 0} \frac{-2+bh - 2ah + abh^2 + 2 + 2ah + 2 - bh - 2}{h}$$

$$= \lim_{h\to 0} \frac{abh^2}{h}$$

$$= \lim_{h\to 0} abh = 0.$$

41. Let $\mathbf{u} = \langle a, b \rangle$ be a unit vector. Then

$$D_{\mathbf{u}}f(-1,0) = \lim_{h\to 0} \frac{((-1+ah)+1)(0+bh)^2 - (-1+1)\cdot 0^2}{h} = \lim_{h\to 0} \frac{ab^2h^3}{h} = \lim_{h\to 0} ab^2h^2 = 0.$$

43. We have

$$f_x(x,y) = 2x, \qquad f_y(x,y) = -2y.$$

Both partial derivatives are continuous everywhere, so by Theorem 12.29, f is differentiable everywhere.

45. We have

$$f_x(x,y) = \frac{(x^2+y^2-1)(1) - x(2x)}{(x^2+y^2-1)^2} = \frac{y^2-x^2-1}{(x^2+y^2-1)^2}, \qquad f_y(x,y) = -\frac{2y}{(x^2+y^2-1)^2}.$$

Then both partial derivatives are defined except where the denominator vanishes, which is on the unit circle. Thus they are continuous in some open set around any point (a,b) not on the unit circle. So by Theorem 12.29, f is differentiable everywhere on $\mathbb{R}^2$ except for the unit circle.

47. We have
$$f_x(x,y) = y\cos(xy), \qquad f_y(x,y) = x\cos(xy).$$

Both partial derivatives are continuous everywhere, so by Theorem 12.29, f is differentiable everywhere.

49. We have
$$f_x(x,y) = y\sec^2(xy), \qquad f_y(xy) = x\sec^2(xy).$$

Both partial derivatives are continuous except where $\sec^2(xy)$ does not exist, which is when xy is $\frac{\pi}{2} + n\pi$ where n is an integer. For any other point, the partial derivatives are continuous in some open set around that point. Thus by Theorem 12.29, f is differentiable everywhere except on the hyperbolas $xy = \frac{\pi}{2} + n\pi$ where n is an integer.

51. We have
$$f_x(x,y) = -\frac{1}{2} \cdot \frac{y^{1/2}}{x^{3/2}} = -\frac{\sqrt{y}}{2x\sqrt{x}}, \qquad f_y(x,y) = \frac{1}{2} \cdot \frac{y^{-1/2}}{x^{1/2}} = \frac{1}{2\sqrt{xy}}.$$

Then $f_x(x,y)$ is continuous where it is defined, which is for $x > 0$ and $y \geq 0$, and $f_y(x,y)$ is continuous where it is defined, which is for $xy > 0$. Putting these together gives $x > 0$ and $xy > 0$, so that $y > 0$. The region where both partials are continuous is the first quadrant (not including the axes), so by Theorem 12.29, f is differentiable in that region.

53. We have
$$f_x(x,y,z) = 2x, \qquad f_y(x,y,z) = 2y, \qquad f_z(x,y,z) = -3z^2.$$

Since all three partials are continuous everywhere, Theorem 12.32 tells us that f is differentiable everywhere.

55. By Theorem 12.30, and using Exercise 43, the equation of the tangent plane to the graph of f at $(1,-3)$ is
$$f_x(1,-3)(x-1) + f_y(1,-3)(y+3) = z - f(1,-3), \text{ or } 2(x-1) + 6(y+3) = z - (-8).$$

Simplify to get $2x + 6y - z = -8$.

57. By Theorem 12.30, and using Exercise 46, the equation of the tangent plane to the graph of f at $(-3,0)$ is
$$f_x(-3,0)(x+3) + f_y(-3,0)y = z - f(-3,0), \text{ or}$$
$$\frac{0^2 - (-3)^2}{((-3)^2 + 0^2)^2}(x+3) + 0y = z - \frac{-3}{(-3)^2 + 0^2}, \text{ or}$$
$$-\frac{1}{9}(x+3) = z + \frac{1}{3}.$$

Multiply through by 9 and simplify to get $x + 9z = -6$.

59. By Theorem 12.30, and using Exercise 47, the equation of the tangent plane to the graph of f at $\left(2, \frac{\pi}{2}\right)$ is
$$f_x\left(2, \frac{\pi}{2}\right)(x-2) + f_y\left(2, \frac{\pi}{2}\right)\left(y - \frac{\pi}{2}\right) = z - \sin\left(2 \cdot \frac{\pi}{2}\right), \text{ or}$$
$$\frac{\pi}{2}(\cos\pi)(x-2) + 2(\cos\pi)\left(y - \frac{\pi}{2}\right) = z, \text{ or}$$
$$-\frac{\pi}{2}(x-2) - 2\left(y - \frac{\pi}{2}\right) = z.$$

Multiply through by 2 and simplify to get $\pi x + 4y + 2z = 4\pi$.

61. By Theorem 12.30, and using Exercise 49, the equation of the tangent plane to the graph of f at $\left(1, -\frac{\pi}{4}\right)$ is

$$f_x\left(1, -\frac{\pi}{4}\right)(x-1) + f_y\left(1, -\frac{\pi}{4}\right)\left(y + \frac{\pi}{4}\right) = z - \tan\left(-\frac{\pi}{4}\right), \text{ or}$$

$$-\frac{\pi}{4}\sec^2\left(-\frac{\pi}{4}\right)(x-1) + \sec^2\left(-\frac{\pi}{4}\right)\left(y + \frac{\pi}{4}\right) = z + 1, \text{ or}$$

$$-\frac{\pi}{2}(x-1) + 2\left(y + \frac{\pi}{4}\right) = z + 1.$$

Multiply through by 2 and simplify to get $\pi x - 4y + 2z = 2\pi - 2$.

63. By Theorem 12.30, and using Exercise 51, the equation of the tangent plane to the graph of f at $(1, 9)$ is

$$f_x(1,9)(x-1) + f_y(1,9)(y-9) = z - \sqrt{\frac{9}{1}}, \text{ or}$$

$$-\frac{\sqrt{9}}{2}(x-1) + \frac{1}{2\sqrt{9}}(y-9) = z - 3, \text{ or}$$

$$-\frac{3}{2}(x-1) + \frac{1}{6}(y-9) = z - 3.$$

Multiply through by 6 to clear fractions and simplify to get $9x - y + 6z = 18$.

65. By the discussion following Theorem 12.32, and using Exercise 53, the equation of the tangent plane to the graph of f at $(1, -5, 3)$ is

$$f_x(1,-5,3)(x-1) + f_y(1,-5,3)(y+5) + f_z(1,-5,3)(z-3) = w - f(1,-5,3), \text{ or}$$

$$2(x-1) - 10(y+5) - 27(z-3) = w + 1.$$

Expand and simplify to get $w = 2x - 10y - 27z + 28$.

Applications

67. (a) If $\mathbf{u} = \langle a, b \rangle$ is a unit vector, then

$$D_{\mathbf{u}}f(0,0) = \lim_{h \to 0} \frac{f(ah, bh) - f(0,0)}{h}$$

$$= \lim_{h \to 0} \frac{1.2 - 0.2(ah)^2 - 0.3(bh)^2 + 0.1(ah)(bh) - 0.25ah - 1.2}{h}$$

$$= \lim_{h \to 0} \frac{-0.2a^2h^2 - 0.3b^2h^2 + 0.1abh^2 - 0.25ah}{h}$$

$$= \lim_{h \to 0}\left(-0.2a^2h - 0.3b^2h + 0.1abh - 0.25a\right) = -0.25a.$$

Northeast has the direction vector $\langle 1, 1 \rangle$, and a unit vector in that direction is $\left\langle \frac{\sqrt{2}}{2}, \frac{\sqrt{2}}{2} \right\rangle$. Thus

$$D_{\left\langle \frac{\sqrt{2}}{2}, \frac{\sqrt{2}}{2} \right\rangle}f(0,0) = -0.25\frac{\sqrt{2}}{2} = -0.125\sqrt{2} \approx -0.177.$$

This is about a $10°$ slope.

(b) At $(0,0)$, the directional derivative is, from part (a), $-0.25a$, so for Ian to be walking along a level curve, we must have $a = 0$, so that the direction vector of travel is $\langle 0, 1 \rangle$, or due north.

Proofs

69. If $\mathbf{u} = \langle \alpha, \beta \rangle$ is a unit vector, then

$$
\begin{aligned}
D_{\mathbf{u}} f(a, b) &= \lim_{h \to 0} \frac{f(a + \alpha h, b + \beta h) - f(a, b)}{h} \\
&= \lim_{h \to 0} \frac{f(a + (-\alpha)(-h), b + (-\beta)(-h)) - f(a, b)}{h} \\
&= -\lim_{h \to 0} \frac{f(a + (-\alpha)(-h), b + (-\beta)(-h)) - f(a, b)}{-h} \\
&= -\lim_{k \to 0} \frac{f(a + (-\alpha)k, b + (-\beta)k) - f(a, b)}{k} \\
&= -D_{-\mathbf{u}} f(a, b).
\end{aligned}
$$

71. If $D_{\mathbf{u}} f(a, b) = 1$ for all unit vectors $\mathbf{u}$, then by Exercise 70, f cannot be differentiable at (a, b). This is a contradiction to the assumption that f is differentiable there, so there is some unit vector $\mathbf{u}$ with $D_{\mathbf{u}} f(a, b) \neq 1$.

Thinking Forward

Paraboloid

A graph of the surface is

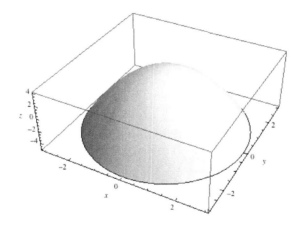

The surface apparently attains its maximum at $(0, 0)$, where $f(x, y) = 4$. In fact, this can be seen analytically, since $f(x, y) = 4 - (x^2 + y^2)$, and $x^2 + y^2 \geq 0$, so that f is at its maximum when $x^2 + y^2 = 0$, or $(x, y) = (0, 0)$. At $(0, 0)$, if $\mathbf{u} = \langle a, b \rangle$ is a unit vector, then

$$
\begin{aligned}
D_{\mathbf{u}} f(0, 0) &= \lim_{h \to 0} \frac{(4 - (0 + ah)^2 - (0 + bh)^2) - (4 - 0^2 - 0^2)}{h} \\
&= \lim_{h \to 0} \frac{4 - a^2 h^2 - b^2 h^2 - 4}{h} = \lim_{h \to 0} \frac{h^2(-a^2 - b^2)}{h} = \lim_{h \to 0} h(-a^2 - b^2) = 0.
\end{aligned}
$$

Hyperboloid

A graphh of the surface is

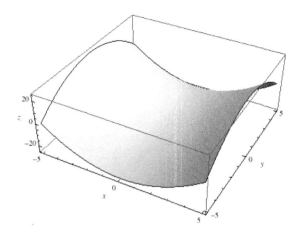

It does not appear that g has any extreme values. Nonetheless, if $\mathbf{u} = \langle a, b \rangle$ is a unit vector, then

$$D_{\mathbf{u}}f(0,0) = \lim_{h \to 0} \frac{(0 + ah)^2 - (0 + bh)^2 - (0^2 - 0^2)}{h} = \lim_{h \to 0} \frac{a^2 h^2 - b^2 h^2}{h} = \lim_{h \to 0} (a^2 h - b^2 h) = 0.$$

Even though g does not have a local extremum at $(0,0)$, all its directional derivatives are defined and vanish there.

12.5   The Chain Rule and The Gradient

Thinking Back

Chain Rule

The chain rule states that $\frac{df}{dt} = \frac{df}{dx} \cdot \frac{dx}{dt}$.

Critical points

A point $x = c$ in the domain of $f(x)$ is a critical point if either $f'(c) = 0$ or if f is not differentiable at c. See Definition 3.2. All local extrema of f on an interval $[a, b]$ must occur either at a critical point or at one of the endpoints of the interval.

Concepts

1. (a) False. By Theorem 12.34, $\frac{\partial z}{\partial s} = \frac{\partial z}{\partial x} \frac{\partial x}{\partial s} + \frac{\partial z}{\partial y} \frac{\partial y}{\partial s}$.

   (b) False. By Theorem 12.33, $\frac{dz}{dt} = \frac{\partial z}{\partial x} \frac{dx}{dt} + \frac{\partial z}{\partial y} \frac{dy}{dt}$.

   (c) False. By Definition 12.36, $\nabla f = \frac{\partial f}{\partial x}\mathbf{i} + \frac{\partial f}{\partial y}\mathbf{j}$. ∇f is a vector, not a scalar.

   (d) False. Just because the partial derivatives of a function exist at a point, that does not mean that the function is differentiable there.

   (e) False. f must be differentiable, according to Theorem 12.37, in order for this to be true.

   (f) True. This is the content of Theorem 12.37.

   (g) True. Since $\nabla f(a, b, c)$ points in the direction in which f is increasing most rapidly at (a, b, c), the opposite vector, which is $-\nabla f(a, b, c)$, must point in the direction in which f is decreasing most rapidly at (a, b, c).

(h) True. By Theorem 12.39, since f is differentiable at $(-1,0)$, then $\nabla f(-1,0)$ is orthogonal to the level curve $f(x,y) = 4$ at $(-1,0)$.

3. (a) By Theorem 12.33,

$$\frac{dz}{dt} = \frac{\partial z}{\partial x}\frac{dx}{dt} + \frac{\partial z}{\partial y}\frac{dy}{dt}$$

$$= \left(-e^{-x}(3xy - 4x + y^2) + e^{-x}(3y - 4)\right)\cos t + \left(e^{-x}(3x + 2y)\right)(-\sin t)$$

$$= e^{-x}(3y - 4 - 3xy + 4x - y^2)\cos t - e^{-x}(3x + 2y)\sin t$$

$$= e^{-\sin t}\left(\left(3\cos t - 4 - 3\sin t\cos t + 4\sin t - \cos^2 t\right)\cos t - (3\sin t + 2\cos t)\sin t\right)$$

$$= e^{-\sin t}\left(3\cos^2 t - 4\cos t - 3\sin t\cos^2 t + 4\sin t\cos t - \cos^3 t - 3\sin^2 t - 2\sin t\cos t\right)$$

$$= e^{-\sin t}\left(3\cos 2t - 4\cos t - 3\sin t\cos^2 t + 2\sin t\cos t - \cos^3 t\right)$$

$$= e^{-\sin t}\left(3\cos 2t - 4\cos t - 3\sin t\cos^2 t + \sin 2t - \cos^3 t\right).$$

(b) Substituting gives

$$f(\sin t, \cos t) = e^{-\sin t}(3\sin t\cos t - 4\sin t + \cos^2 t),$$

and differentiating then gives

$$f'(t) = -e^{-\sin t}\cos t(3\sin t\cos t - 4\sin t + \cos^2 t)$$
$$+ e^{-\sin t}(3\cos^2 t - 3\sin^2 t - 4\cos t - 2\sin t\cos t)$$
$$= e^{-\sin t}(-3\sin t\cos^2 t + 4\sin t\cos t - \cos^3 t + 3\cos^2 t - 3\sin^2 t - 4\cos t - 2\sin t\cos t)$$
$$= e^{-\sin t}(-3\sin t\cos^2 t + 2\sin t\cos t - \cos^3 t + 3\cos 2t - 4\cos t).$$

(c) The two answers are the same except for the order of the terms. The second method was clearly easier.

5. If $n = m = 1$, then $\frac{\partial z}{\partial x_i}$ makes sense only for $i = 1$, and z depends only on x_1, so $\frac{\partial z}{\partial x_1} = \frac{dz}{dx_1}$. Similarly, $\frac{\partial x_i}{\partial t_j}$ makes sense only for $i = j = 1$, in which case x_1 depends only on t_1, so that $\frac{\partial x_i}{\partial t_j} = \frac{dx_1}{dt_1}$. Finally, since z is a function of x_1, which is a function of t_1, it follows that z is a function of t_1 alone, so that $\frac{\partial z}{\partial t_1} = \frac{dz}{dt_1}$. Then Theorem 12.35 becomes

$$\frac{dz}{dt_1} = \frac{dz}{dx_1}\frac{dx_1}{dt_1},$$

which is the same as the single-variable version of the chain rule.

7. Rewriting Theorem 12.35 with $n = m = 2$ gives

$$\frac{\partial z}{\partial t_j} = \frac{\partial z}{\partial x_1}\frac{\partial x_1}{\partial t_j} + \frac{\partial z}{\partial x_2}\frac{\partial x_2}{\partial t_j}, \text{ for } 1 \leq j \leq 2.$$

Letting $s = t_1, t = t_2, x = x_1$, and $y = x_2$ gives

$$\frac{\partial z}{\partial s} = \frac{\partial z}{\partial x}\frac{\partial x}{\partial s} + \frac{\partial z}{\partial y}\frac{\partial y}{\partial s} \quad \text{and} \quad \frac{\partial z}{\partial t} = \frac{\partial z}{\partial x}\frac{\partial x}{\partial t} + \frac{\partial z}{\partial y}\frac{\partial y}{\partial t},$$

which is Theorem 12.34.

9. (a) The level curve of f at $z = c$ is the line $2x + 3y = c$, or $3y = c - 2x$. Thus a direction vector for this line is $\langle -3, 2 \rangle$.

    (b) The gradient $\nabla f(x, y) = \langle 2, 3 \rangle$, and $\langle 2, 3 \rangle \cdot \langle -3, 2 \rangle = 0$. Thus the gradient is orthogonal to each of the level curves.

11. (a) The level curve of f at $z = c$ is the line $ax + by = c$, or $by = c - ax$. Thus a direction vector for this line is $\langle -b, a \rangle$.

    (b) The gradient $\nabla f(x, y) = \langle a, b \rangle$, and $\langle a, b \rangle \cdot \langle -b, a \rangle = 0$. Thus the gradient is orthogonal to each of the level curves.

13. By the discussion following Theorem 12.32, the equation of the hyperplane tangent to $f(x, y, z)$ at (a, b, c) is

$$
\begin{aligned}
w - f(a, b, c) &= f_x(a, b, c)(x - a) + f_y(a, b, c)(y - b) + f_z(a, b, c)(z - c) \\
&= \langle f_x(a, b, c), f_y(a, b, c), f_z(a, b, c) \rangle \cdot \langle x - a, y - b, z - c \rangle \\
&= \nabla f \cdot \langle x - a, y - b, z - c \rangle .
\end{aligned}
$$

15. The level curves and gradient vectors are:

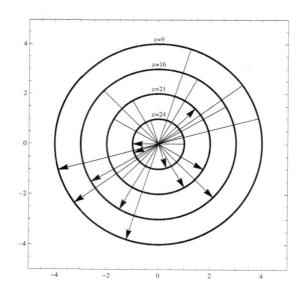

All the level curves are circles, which makes sense, since their equations are $25 - x^2 - y^2 = c$, or $x^2 + y^2 = 25 - c$. Second, the gradient vectors all point inwards, and are orthogonal to the level curves. Finally, the size of the gradient vector, which reflects the size of the change in f along the gradient, decreases as z increases. This implies that the surface is less steep for larger values of z. However, all the gradients on a particular level curve have the same length, so that the steepness of the surface depends only on z, not on the values of x and y.

17. The level curves and gradient vectors are:

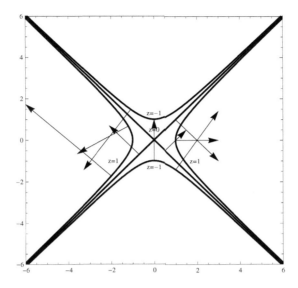

The level curves for $z = -1$ and 1 are the hyperbolas $x^2 - y^2 = \pm 1$, while the level curve for $z = 0$ is the curve $x^2 - y^2 = 0$, which is $(x - y)(x + y) = 0$, so it is the union of the two lines $y = x$ and $y = -x$. The gradient vectors along $z = 1$ point outwards, while those along $z = -1$ point inwards. The gradient vectors along $z = 0$ at a point (a, b) seems to point in the direction $(a, -b)$. Finally, the size of the gradient appears to increase the further from the origin the point is. So the curvature and steepness of this surface seem pretty complex.

19. If your path remains on a single contour line, then it is level, since a contour line has a fixed value of z, which is your altitude. If instead you hike perpendicular to contour lines, then you are hiking parallel to the gradient, so you are either hiking upwards as steeply as possible if you are moving in the direction of the gradient, or downwards as steeply as possible if you are moving in the opposite direction.

Skills

21. Using Theorem 12.33 gives

$$\frac{dz}{dt} = \frac{\partial z}{\partial x}\frac{dx}{dt} + \frac{\partial z}{\partial y}\frac{dy}{dt} = \cos x \cos y \cdot e^t - \sin x \sin y \cdot 3t^2 = e^t \cos e^t \cos t^3 - 3t^2 \sin e^t \sin t^3.$$

23. Using Theorem 12.33 gives

$$\frac{dx}{dt} = \frac{\partial x}{\partial r}\frac{dr}{dt} + \frac{\partial x}{\partial \theta}\frac{d\theta}{dt} = \cos \theta \cdot (2t) - r \sin \theta \cdot (3t^2)$$
$$= 2t \cos \theta - 3t^2 r \sin \theta = 2t \cos(t^3 + 1) - 3t^2(t^2 - 5) \sin(t^3 + 1).$$

25. Using Theorem 12.33 gives

$$
\begin{aligned}
\frac{dr}{dt} &= \frac{\partial r}{\partial x}\frac{dx}{dt} + \frac{\partial r}{\partial y}\frac{dy}{dt} \\
&= \frac{1}{2}(x^2 + y^2)^{-1/2} \cdot 2x \cdot \left(\frac{1}{2\sqrt{t}}\right) + \frac{1}{2}(x^2 + y^2)^{-1/2} \cdot 2y \cdot (2t) \\
&= \frac{x}{2\sqrt{t}}(x^2 + y^2)^{-1/2} + 2ty(x^2 + y^2)^{-1/2} \\
&= (x^2 + y^2)^{-1/2}\left(\frac{x}{2\sqrt{t}} + 2ty\right) \\
&= (t + t^4)^{-1/2}\left(\frac{1}{2} + 2t^3\right).
\end{aligned}
$$

27. Using Theorem 12.34 gives

$$
\frac{\partial z}{\partial s} = \frac{\partial z}{\partial x}\frac{\partial x}{\partial s} + \frac{\partial z}{\partial y}\frac{\partial y}{\partial s} = 2xy^3 t\cos s + 3x^2 y^2 \cdot \cos t = 2t^2 s^3 \sin s \cos s \cos^3 t + 3s^2 t^2 \sin^2 s \cos^3 t.
$$

29. Using Theorem 12.34 gives

$$
\begin{aligned}
\frac{\partial z}{\partial r} &= \frac{\partial z}{\partial x}\frac{\partial x}{\partial r} + \frac{\partial z}{\partial y}\frac{\partial y}{\partial r} \\
&= (2x + y)e^y \cdot \cos\theta + \left(xe^y + (x^2 + xy)e^y\right) \cdot \sin\theta \\
&= e^y\left((2x + y)\cos\theta + (x^2 + xy + x)\sin\theta\right) \\
&= e^{r\sin\theta}\left((2r\cos\theta + r\sin\theta)\cos\theta + (r^2\cos^2\theta + r^2\sin\theta\cos\theta + r\cos\theta)\sin\theta\right) \\
&= re^{r\sin\theta}\left(2\cos^2\theta + \sin\theta\cos\theta + r\cos^2\theta\sin\theta + r\sin^2\theta\cos\theta + \sin\theta\cos\theta\right) \\
&= re^{r\sin\theta}\left(2\cos^2\theta + 2\sin\theta\cos\theta + r\cos^2\theta\sin\theta + r\sin^2\theta\cos\theta\right).
\end{aligned}
$$

31. Using Theorem 12.35 gives

$$
\begin{aligned}
\frac{dx}{dt} &= \frac{\partial x}{\partial\rho}\frac{d\rho}{dt} + \frac{\partial x}{\partial\phi}\frac{d\phi}{dt} + \frac{\partial x}{\partial\theta}\frac{d\theta}{dt} \\
&= \sin\phi\cos\theta \cdot 2t + \rho\cos\phi\cos\theta \cdot 3t^2 - \rho\sin\phi\sin\theta \cdot 4t^3 \\
&= 2t\sin t^3\cos t^4 + 3t^4\cos t^3\cos t^4 - 4t^5\sin t^3\sin t^4.
\end{aligned}
$$

33. Using Theorem 12.35 gives

$$
\begin{aligned}
\frac{\partial w}{\partial\rho} &= \frac{\partial w}{\partial x}\frac{\partial x}{\partial\rho} + \frac{\partial w}{\partial y}\frac{\partial y}{\partial\rho} + \frac{\partial w}{\partial z}\frac{\partial z}{\partial\rho} \\
&= 2xe^y \cdot \sin\phi\cos\theta + (x^2 + z)e^y \cdot \sin\phi\sin\theta + e^y \cdot \cos\phi \\
&= 2(\rho\sin\phi\cos\theta)e^{\rho\sin\phi\sin\theta}\sin\phi\cos\theta + (\rho^2\sin^2\phi\cos^2\theta + \rho\cos\phi)e^{\rho\sin\phi\sin\theta}\sin\phi\sin\theta \\
&\quad + e^{\rho\sin\phi\sin\theta}\cos\phi \\
&= e^{\rho\sin\phi\sin\theta}(2\rho\sin^2\phi\cos^2\theta + \rho^2\sin^3\phi\cos^2\theta\sin\theta + \rho\sin\phi\cos\phi\sin\theta + \cos\phi).
\end{aligned}
$$

35. Using Theorem 12.34 gives

$$\frac{d\rho}{dt} = \frac{\partial \rho}{\partial x} \cdot \frac{dx}{dt} + \frac{\partial \rho}{\partial y} \cdot \frac{dy}{dt} + \frac{\partial \rho}{\partial z} \cdot \frac{dz}{dt}$$

$$= \frac{1}{2}(x^2 + y^2 + z^2)^{-1/2} \cdot 2x \cdot \frac{1}{2}t^{-1/2} + \frac{1}{2}(x^2 + y^2 + z^2)^{-1/2} \cdot 2y \cdot 2t + \frac{1}{2}(x^2 + y^2 + z^2)^{-1/2} \cdot 2z \cdot 3t^2$$

$$= \frac{x}{2\sqrt{x^2 + y^2 + z^2}}t^{-1/2} + \frac{2y}{\sqrt{x^2 + y^2 + z^2}}t + \frac{3z}{\sqrt{x^2 + y^2 + z^2}}t^2$$

$$= \frac{t^{1/2}}{2\sqrt{t + t^4 + t^6}}t^{-1/2} + \frac{2t^2}{\sqrt{t + t^4 + t^6}}t + \frac{3t^3}{\sqrt{t + t^4 + t^6}}t^2$$

$$= \frac{1}{\sqrt{t + t^4 + t^6}}\left(\frac{1}{2} + 2t^3 + 3t^5\right).$$

37. Computing using Definition 12.36 gives

$$\nabla z = \langle z_x, z_y \rangle = \left\langle 2x \sin y + y \cos x, \; x^2 \cos y + \sin x \right\rangle.$$

39. Computing using Definition 12.36 gives

$$\nabla f(x, y) = \langle f_x(x, y), f_y(x, y) \rangle$$

$$= \left\langle \frac{1}{2\sqrt{x^2 + y^2}} \cdot 2x, \; \frac{1}{2\sqrt{x^2 + y^2}} \cdot 2y \right\rangle = \left\langle \frac{x}{\sqrt{x^2 + y^2}}, \; \frac{y}{\sqrt{x^2 + y^2}} \right\rangle.$$

41. Computing using Definition 12.36 gives

$$\nabla f(x, y, z) = \langle f_x(x, y, z), f_y(x, y, z), f_z(x, y, z) \rangle$$

$$= \left\langle \frac{1}{2\sqrt{x^2 + y^2 + z^2}} \cdot 2x, \; \frac{1}{2\sqrt{x^2 + y^2 + z^2}} \cdot 2y, \; \frac{1}{2\sqrt{x^2 + y^2 + z^2}} \cdot 2z \right\rangle$$

$$= \left\langle \frac{x}{\sqrt{x^2 + y^2 + z^2}}, \; \frac{y}{\sqrt{x^2 + y^2 + z^2}}, \; \frac{z}{\sqrt{x^2 + y^2 + z^2}} \right\rangle.$$

43. (a) Evaluating the gradient from Exercise 37 at $\left(\pi, \frac{\pi}{2}\right)$ gives

$$\nabla z \left(\pi, \frac{\pi}{2}\right) = \left\langle 2\pi \sin \frac{\pi}{2} + \frac{\pi}{2} \cos \pi, \; \pi^2 \cos \frac{\pi}{2} + \sin \pi \right\rangle = \left\langle \frac{3\pi}{2}, 0 \right\rangle.$$

This is the direction in which z is changing most rapidly; dividing through by the constant $\frac{3\pi}{2}$ gives the unit direction vector $\langle 1, 0 \rangle$.

(b) The rate of change in this direction is

$$D_{\langle 1,0 \rangle} z \left(\pi, \frac{\pi}{2}\right) = \nabla z \left(\pi, \frac{\pi}{2}\right) \cdot u = \left\langle \frac{3\pi}{2}, 0 \right\rangle \cdot \langle 1, 0 \rangle = \frac{3\pi}{2}.$$

(c) Since $\langle 1, 0 \rangle$ is the direction greatest increase, its negative, $\langle -1, 0 \rangle$, is the direction of greatest decrease.

45. (a) Evaluating the gradient from Exercise 39 at $(2, -3)$ gives

$$\nabla f(2, -3) = \left\langle \frac{2}{\sqrt{2^2 + 3^2}}, \; \frac{-3}{\sqrt{2^2 + 3^2}} \right\rangle = \left\langle \frac{2}{\sqrt{13}}, \; -\frac{3}{\sqrt{13}} \right\rangle.$$

Note that this is a unit vector, which we call **u**.

(b) The rate of change in this direction is

$$D_{\mathbf{u}}f(2,-3) = \nabla f(2,-3) \cdot \mathbf{u} = \left\langle \frac{2}{\sqrt{13}}, -\frac{3}{\sqrt{13}} \right\rangle \cdot \left\langle \frac{2}{\sqrt{13}}, -\frac{3}{\sqrt{13}} \right\rangle = \frac{4}{13} + \frac{9}{13} = 1.$$

(c) The direction of most rapid decrease is $-\mathbf{u} = \left\langle -\frac{2}{\sqrt{13}}, \frac{3}{\sqrt{13}} \right\rangle$.

47. (a) Evaluating the gradient from Exercise 41 at $(2,-1,-2)$ gives

$$\nabla f(2,-1,-2) = \left\langle \frac{2}{\sqrt{2^2+1^2+2^2}}, \frac{-1}{\sqrt{2^2+1^2+2^2}}, \frac{-2}{\sqrt{2^2+1^2+2^2}} \right\rangle = \left\langle \frac{2}{3}, -\frac{1}{3}, -\frac{2}{3} \right\rangle.$$

Note that this is a unit vector, which we call $\mathbf{u}$.

(b) The rate of change in this direction is

$$D_{\mathbf{u}}f(2,-1,-2) = \nabla f(2,-1,-2) \cdot \mathbf{u} = \left\langle \frac{2}{3}, -\frac{1}{3}, -\frac{2}{3} \right\rangle \cdot \left\langle \frac{2}{3}, -\frac{1}{3}, -\frac{2}{3} \right\rangle = \frac{4}{9} + \frac{1}{9} + \frac{4}{9} = 1.$$

(c) The direction of most rapid decrease is $-\mathbf{u} = \left\langle -\frac{2}{3}, \frac{1}{3}, \frac{2}{3} \right\rangle$.

49. Note that $e^x \tan y$ is differentiable in an open set around P, so that Theorem 12.37 applies. Since $\|\mathbf{v}\| = \sqrt{3^2+1^2} = \sqrt{10}$, a unit vector in the direction of $\mathbf{v}$ is $\mathbf{u} = \left\langle \frac{3}{\sqrt{10}}, -\frac{1}{\sqrt{10}} \right\rangle$, so that

$$\begin{aligned}
D_{\mathbf{u}}z\left(0, \frac{\pi}{4}\right) &= \nabla z\left(0, \frac{\pi}{4}\right) \cdot \mathbf{u} \\
&= \left\langle z_x\left(0, \frac{\pi}{4}\right), z_y\left(0, \frac{\pi}{4}\right) \right\rangle \cdot \mathbf{u} \\
&= \left\langle e^0 \tan \frac{\pi}{4}, e^0 \sec^2 \frac{\pi}{4} \right\rangle \cdot \left\langle \frac{3}{\sqrt{10}}, -\frac{1}{\sqrt{10}} \right\rangle \\
&= \langle 1, 2 \rangle \cdot \left\langle \frac{3}{\sqrt{10}}, -\frac{1}{\sqrt{10}} \right\rangle \\
&= \frac{3}{\sqrt{10}} - \frac{2}{\sqrt{10}} = \frac{1}{\sqrt{10}}.
\end{aligned}$$

51. Note that z is differentiable in an open set around $(7,1)$, so that Theorem 12.37 applies. Since $\|\mathbf{v}\| = \sqrt{1^2+4^2} = \sqrt{17}$, a unit vector in the direction of $\mathbf{v}$ is $\mathbf{u} = \left\langle -\frac{1}{\sqrt{17}}, -\frac{4}{\sqrt{17}} \right\rangle$, so that

$$\begin{aligned}
D_{\mathbf{u}}z(7,1) &= \nabla z(7,1) \cdot \mathbf{u} \\
&= \langle z_x(7,1), z_y(7,1) \rangle \cdot \mathbf{u} \\
&= \left\langle \left(\frac{y^2}{x} \cdot \frac{1}{y^2}\right) \Big|_{(7,1)}, \left(\frac{y^2}{x} \cdot \left(-\frac{2x}{y^3}\right)\right) \Big|_{(7,1)} \right\rangle \cdot \mathbf{u} \\
&= \left\langle \frac{1}{7}, -2 \right\rangle \cdot \left\langle -\frac{1}{\sqrt{17}}, -\frac{4}{\sqrt{17}} \right\rangle \\
&= -\frac{1}{7\sqrt{17}} + \frac{8}{\sqrt{17}} = \frac{55}{7\sqrt{17}}.
\end{aligned}$$

53. Note that w is differentiable everywhere, so that Theorem 12.37 applies. Since

$$\|\mathbf{v}\| = \sqrt{1^2 + 2^2 + 3^2} = \sqrt{14},$$

a unit vector in the direction of $\mathbf{v}$ is $\mathbf{u} = \left\langle \frac{1}{\sqrt{14}}, -\frac{2}{\sqrt{14}}, \frac{3}{\sqrt{14}} \right\rangle$, so that

$$
\begin{aligned}
D_{\mathbf{u}}w\left(3, \frac{\pi}{4}, -\frac{\pi}{2}\right) &= \nabla w\left(3, \frac{\pi}{4}, -\frac{\pi}{2}\right) \cdot \mathbf{u} \\
&= \left\langle w_x\left(3, \frac{\pi}{4}, -\frac{\pi}{2}\right), w_y\left(3, \frac{\pi}{4}, -\frac{\pi}{2}\right), w_z\left(3, \frac{\pi}{4}, -\frac{\pi}{2}\right) \right\rangle \cdot \mathbf{u} \\
&= \left\langle \sin\frac{\pi}{4}\cos\left(-\frac{\pi}{2}\right), 3\cos\frac{\pi}{4}\cos\left(-\frac{\pi}{2}\right), -3\sin\frac{\pi}{4}\sin\left(-\frac{\pi}{2}\right) \right\rangle \cdot \mathbf{u} \\
&= \left\langle 0, 0, \frac{3\sqrt{2}}{2} \right\rangle \cdot \left\langle \frac{1}{\sqrt{14}}, -\frac{2}{\sqrt{14}}, \frac{3}{\sqrt{14}} \right\rangle \\
&= \frac{9\sqrt{2}}{2\sqrt{14}} = \frac{9}{2\sqrt{7}}.
\end{aligned}
$$

55. Since $f_x(x,y) = -\frac{3}{x}$, integrating with respect to x gives $f(x,y) = -3\ln|x| + g(y)$. Then

$$
f_y(x,y) = g'(y) = \frac{1}{y},
$$

so that integrating $g'(y)$ with respect to y gives $g(y) = \ln|y| + C$. Thus $f(x,y) = -3\ln|x| + \ln|y| + C = \ln\left|\frac{y}{x^3}\right| + C$.

57. Since $f_x(x,y) = -\frac{y}{x^2+y^2}$, integrate with respect to x:

$$
-\int \frac{y}{x^2+y^2}\,dx = -\int \frac{1/y}{\left(\frac{x}{y}\right)^2 + 1}\,dx = -\int \frac{1}{u^2+1}\,du = -\tan^{-1}u + g(y) = -\tan^{-1}\left(\frac{x}{y}\right) + g(y).
$$

Differentiating with respect to y must give the y component of the gradient, so that

$$
f_y(x,y) = -\frac{1}{1+\left(\frac{x}{y}\right)^2} \cdot \left(-\frac{x}{y^2}\right) + g'(y) = \frac{x}{x^2+y^2} + g'(y) = \frac{x}{x^2+y^2}.
$$

Thus $g'(y) = 0$, so that g is a constant and then $f(x,y) = -\tan^{-1}\left(\frac{x}{y}\right) + C$.

59. Since $f_x(x,y) = \frac{y}{(x+y)^2}$, integrate with respect to x:

$$
f(x,y) = \int \frac{y}{(x+y)^2}\,dx = y\int \frac{1}{(x+y)^2}\,dx = -\frac{y}{x+y} + g(y).
$$

Differentiating with respect to y must give the y component of the gradient, so that

$$
f_y(x,y) = -\frac{(x+y)(1) - y(1)}{(x+y)^2} + g'(y) = -\frac{x}{(x+y)^2} + g'(y) = -\frac{x}{(x+y)^2}.
$$

Thus $g'(y) = 0$, so that g is a constant and then $f(x,y) = -\frac{y}{x+y} + C$.

Applications

61. (a) The slope is greatest upwards in the direction of the gradient, so that Ian wants to go in the negative direction from the gradient. With $f(x,y) = 1.2 - 0.2x^2 - 0.3y^2 + 0.1xy - 0.25x$ the gradient is

$$
\nabla f(x,y) = \left\langle \frac{\partial f}{\partial x}, \frac{\partial f}{\partial y} \right\rangle = \langle -0.4x + 0.1y - 0.25, -0.6y + 0.1x \rangle.
$$

At $(0.5, -0.5)$, this is $\langle -0.5, 0.35 \rangle$, so that Ian should walk in the direction $\langle 0.5, -0.35 \rangle$, or somewhat east of southeast.

(b) The contour is perpendicular to the gradient; since the gradient at $(0.5, -0.5)$ is $\langle -0.5, 0.35 \rangle$, the contour is in the direction $\langle 0.35, 0.5 \rangle$ (and $\langle -0.35, -0.5 \rangle$, in the opposite direction).

Proofs

63. Computing using Definition 12.36 gives

$$\nabla f(x, y, z) = \langle f_x(x, y, z), f_y(x, y, z), f_z(x, y, z) \rangle$$

$$= \left\langle \frac{1}{2\sqrt{x^2 + y^2 + z^2}} \cdot 2x, \ \frac{1}{2\sqrt{x^2 + y^2 + z^2}} \cdot 2y, \ \frac{1}{2\sqrt{x^2 + y^2 + z^2}} \cdot 2z \right\rangle$$

$$= \left\langle \frac{x}{\sqrt{x^2 + y^2 + z^2}}, \ \frac{y}{\sqrt{x^2 + y^2 + z^2}}, \ \frac{z}{\sqrt{x^2 + y^2 + z^2}} \right\rangle.$$

Then if $(x, y, z) \neq (0, 0, 0)$, then $\nabla f(x, y, z)$ is defined, and

$$\|\nabla f(x, y, z)\| = \frac{1}{\sqrt{x^2 + y^2 + z^2}} \sqrt{x^2 + y^2 + z^2} = 1.$$

If $(x, y, z) = (0, 0, 0)$, then f is not differentiable at $(0, 0, 0)$, since for example $f_x(x, y, z)$ does not exist at $(0, 0, 0)$. To see this, note that

$$f_x(0, 0, 0) = \lim_{h \to 0} \frac{f(0 + h, 0, 0) - f(0, 0, 0)}{h} = \lim_{h \to 0} \frac{\sqrt{h^2}}{h} = \lim_{h \to 0} |h|.$$

Thus the left and right limits along this path are unequal, so the partial with respect to x does not exist.

65. Let (s_0, t_0) be a point in the domains of both u and v where both u and v are differentiable, and such that f is differentiable at $(x_0, y_0) = (u(s_0, t_0), v(s_0, t_0))$. Then by Definition 12.28,

$$\Delta z = \frac{\partial f}{\partial x} \Delta x + \frac{\partial f}{\partial y} \Delta y + \epsilon_1 \Delta x + \epsilon_2 \Delta y,$$

where $\epsilon_1, \epsilon_2 \to 0$ as $(\Delta x, \Delta y) \to (0, 0)$.

First, divide both sides of this equation by Δs to get

$$\frac{\Delta z}{\Delta s} = \frac{\partial f}{\partial x} \frac{\Delta x}{\Delta s} + \frac{\partial f}{\partial y} \frac{\Delta y}{\Delta s} + \epsilon_1 \frac{\Delta x}{\Delta s} + \epsilon_2 \frac{\Delta y}{\Delta s}.$$

Now let $(\Delta x, \Delta y) = (u(s_0 + \Delta s, t_0), v(s_0 + \Delta s, t_0))$. Then $\Delta x \to 0$ and $\Delta y \to 0$ as $\Delta s \to 0$ since u and v are both continuous at (s_0, t_0). It follows that $\epsilon_1, \epsilon_2 \to 0$ as $\Delta s \to 0$. Taking limits of these quotients, then, gives

$$\frac{\partial z}{\partial s} = \lim_{\Delta s \to 0} \frac{\Delta z}{\Delta s} = \lim_{\Delta s \to 0} \left(\frac{\partial f}{\partial x} \frac{\Delta x}{\Delta s} + \frac{\partial f}{\partial y} \frac{\Delta y}{\Delta s} + \epsilon_1 \frac{\Delta x}{\Delta s} + \epsilon_2 \frac{\Delta y}{\Delta s} \right) = \frac{\partial f}{\partial x} \frac{\partial x}{\partial s} + \frac{\partial f}{\partial y} \frac{\partial y}{\partial s}.$$

For the other equation, the computation is very similar, using t instead of s: divide both sides of the equation by Δt to get

$$\frac{\Delta z}{\Delta t} = \frac{\partial f}{\partial x} \frac{\Delta x}{\Delta t} + \frac{\partial f}{\partial y} \frac{\Delta y}{\Delta t} + \epsilon_1 \frac{\Delta x}{\Delta t} + \epsilon_2 \frac{\Delta y}{\Delta t}.$$

Now let $(\Delta x, \Delta y) = (u(s_0, t_0 + \Delta t), v(s_0, t_0 + \Delta t))$. Then $\Delta x \to 0$ and $\Delta y \to 0$ as $\Delta t \to 0$ since u and v are both continuous at (s_0, t_0). It follows that $\epsilon_1, \epsilon_2 \to 0$ as $\Delta t \to 0$. Taking limits of these quotients, then, gives

$$\frac{\partial z}{\partial t} = \lim_{\Delta t \to 0} \frac{\Delta z}{\Delta t} = \lim_{\Delta t \to 0} \left(\frac{\partial f}{\partial x} \frac{\Delta x}{\Delta t} + \frac{\partial f}{\partial y} \frac{\Delta y}{\Delta t} + \epsilon_1 \frac{\Delta x}{\Delta t} + \epsilon_2 \frac{\Delta y}{\Delta t} \right) = \frac{\partial f}{\partial x} \frac{\partial x}{\partial t} + \frac{\partial f}{\partial y} \frac{\partial y}{\partial t}.$$

67. If $f(x,y) = c$, then $\frac{\partial f}{\partial x} = \frac{\partial f}{\partial y} = 0$, so that $\nabla f(x,y) = \left\langle \frac{\partial f}{\partial x}, \frac{\partial f}{\partial y} \right\rangle = \mathbf{0}$.

69. We have

$$\nabla(f(x,y) + g(x,y)) = \left\langle \frac{\partial}{\partial x}(f(x,y) + g(x,y)), \frac{\partial}{\partial y}(f(x,y) + g(x,y)) \right\rangle$$

$$= \left\langle \frac{\partial f}{\partial x} + \frac{\partial g}{\partial x}, \frac{\partial f}{\partial y} + \frac{\partial g}{\partial y} \right\rangle$$

$$= \left\langle \frac{\partial f}{\partial x}, \frac{\partial f}{\partial y} \right\rangle + \left\langle \frac{\partial g}{\partial x}, \frac{\partial g}{\partial y} \right\rangle$$

$$= \nabla f(x,y) + \nabla g(x,y).$$

71. We have

$$\nabla(f(x,y)g(x,y)) = \left\langle \frac{\partial}{\partial x}(f(x,y)g(x,y)), \frac{\partial}{\partial y}(f(x,y)g(x,y)) \right\rangle$$

$$= \left\langle f(x,y)\frac{\partial g}{\partial x} + g(x,y)\frac{\partial f}{\partial x}, f(x,y)\frac{\partial g}{\partial y} + g(x,y)\frac{\partial f}{\partial y} \right\rangle$$

$$= \left\langle f(x,y)\frac{\partial g}{\partial x}, f(x,y)\frac{\partial g}{\partial y} \right\rangle + \left\langle g(x,y)\frac{\partial f}{\partial x}, g(x,y)\frac{\partial f}{\partial y} \right\rangle$$

$$= f(x,y)\left\langle \frac{\partial g}{\partial x}, \frac{\partial g}{\partial y} \right\rangle + g(x,y)\left\langle \frac{\partial f}{\partial x}, \frac{\partial f}{\partial y} \right\rangle$$

$$= f(x,y)\nabla g(x,y) + g(x,y)\nabla f(x,y).$$

73. These proofs would all work more or less as is in the case of three variables. You would of course have to add a third variable, and a third component to each of the vectors involved. Other than that, however, the proofs are identical.

Thinking Forward

The gradient at a maximum

Since the function has a maximum at (x_0, y_0), it follows that the single variable function $f(x, y_0)$ has a maximum at $x = x_0$, so that $\frac{d}{dx}f(x, y_0) = \frac{\partial f}{\partial x}(x_0, y_0) = 0$. Similarly, the single variable function $f(x_0, y)$ has a maximum at $y = y_0$, so that $\frac{d}{dy}f(x_0, y) = \frac{\partial f}{\partial y}(x_0, y_0) = 0$. But this means that

$$\nabla f(x_0, y_0) = \left\langle \frac{\partial f}{\partial x}(x_0, y_0), \frac{\partial f}{\partial y}(x_0, y_0) \right\rangle = \langle 0, 0 \rangle.$$

Thus the gradient is zero.

The gradient at a minimum

Since the function has a minimum at (x_0, y_0, z_0), it follows that the single variable function $f(x, y_0, z_0)$ has a minimum at $x = x_0$, so that $\frac{d}{dx}f(x, y_0, z_0) = \frac{\partial f}{\partial x}(x_0, y_0, z_0) = 0$. Similarly, the single variable function $f(x_0, y, z_0)$ has a minimum at $y = y_0$, so that $\frac{d}{dy}f(x_0, y, z_0) = \frac{\partial f}{\partial y}(x_0, y_0, z_0) = 0$. A similar argument applies to the function $f(x_0, y_0, z)$. But this means that

$$\nabla f(x_0, y_0, z_0) = \left\langle \frac{\partial f}{\partial x}(x_0, y_0, z_0), \frac{\partial f}{\partial y}(x_0, y_0, z_0), \frac{\partial f}{\partial z}(x_0, y_0, z_0) \right\rangle = \langle 0, 0, 0 \rangle.$$

Thus the gradient is zero.

12.6   Extreme Values

Thinking Back

First-Derivative Test

See Theorem 3.8. If $x = c$ is a critical point of f, and (a, b) is an open interval containing c on which f is continuous, and differentiable except possibly at $x = c$, and which contains no other critical points of f, then:

- If $f'(x) > 0$ on (a, c) and < 0 on (c, b), then f has a local maximum at $x = c$. Conversely, if $f'(x) < 0$ on (a, c) and > 0 on (c, b), then f has a local minimum at $x = c$.

- If $f'(x) > 0$ on $(a, c) \cup (c, b)$, or if $f'(x) < 0$ on $(a, c) \cup (c, b)$, then f does not have a local extremum at $x = c$.

The test works as follows. First, if $f'(x)$ changes sign at $x = c$, then either the curve is decreasing on (a, c) and increasing on (c, b), in which case it must have a local minimum at $x = c$, or it is increasing on (a, c) and decreasing on (c, b), in which case it must have a local maximum at $x = c$. If it does not change sign at $x = c$, then it is increasing both to the left and to the right of $x = c$, or decreasing both to the left and to the right of $x = c$, so that the curve just flattens out at $x = c$ momentarily; it does not change direction.

Second-Derivative Test

See Theorem 3.11. If $x = c$ is a critical point of f, with $f'(c) = 0$, and both f and f' are differentiable, and f'' is continuous, in some open interval around $x = c$, then:

- If $f''(c)$ is positive, then f has a local minimum at $x = c$.

- If $f''(c)$ is negative, then f has a local maximum at $x = c$.

- If $f''(c) = 0$, we cannot conclude anything about local extrema at $x = c$.

The test works as follows. If $f''(c) > 0$, then the first derivative is increasing. Since it is zero at c, it must be negative to the left of c and positive to the right, so that it follows from the first derivative test that f has a local minimum at $x = c$. Similarly, if $f''(c) < 0$, then the first derivative is decreasing. Since it is zero at c, it must be positive to the left of c and negative to the right, so that it follows from the first derivative test that f has a local maximum at $x = c$. Finally, if $f''(c) = 0$, we cannot draw any conclusion about the behavior of f' near $x = c$, so that the first derivative test gives us no conclusions.

Concepts

1. (a) False. In order to be able to conclude that f has a stationary point at (a, b), we must also know that f is differentiable there, by Definition 12.41.

    (b) True. If (a, b) is a stationary point, then by Definition 12.42, (a, b) is a critical point of f.

    (c) False. By Definition 12.42, if (a, b) is a critical point of f, then *either* (a, b) is a stationary point of f or else f is not differentiable at (a, b).

    (d) True. By Definition 12.44, a saddle point can only occur at a critical point.

    (e) False. By Definition 12.40(a), all that is required is that $f(a, b) \geq f(x, y)$ for every point in some open disc containing (a, b) (not $f(a, b) > f(x, y)$).

    (f) True. If $(a, b) \in \mathbb{R}^2$, then $f(a, b) = \pi \leq f(x, y) = \pi$ for every $(x, y) \in \mathbb{R}^2$. By Theorem 12.40(e), this means that f has an absolute minimum at (a, b).

(g) True. For example, the function in part (f), $f(x,y) = \pi$, has an absolute minimum at every point of $\mathbb{R}^2$. But also if $(a,b) \in \mathbb{R}^2$, then $f(a,b) = \pi \geq f(x,y) = \pi$ for any $(x,y) \in \mathbb{R}^2$, so that by Theorem 12.40(d), f has an absolute maximum at (a,b) as well.

(h) True. If the graph of $f(x,y)$ is a plane, then f has the form $f(x,y) = ax + by + c$. This function is clearly differentiable everywhere, and $\nabla f(x,y) = \langle a,b \rangle$. Thus if $\mathbf{u}$ is a unit vector, $D_{\mathbf{u}}f(x_0,y_0) = \langle a,b \rangle \cdot \mathbf{u}$. If $D_{\mathbf{u}}f(x_0,y_0) \neq 0$, it follows that $\langle a,b \rangle \neq 0$, so that $\nabla f(x,y)$ is never zero. Then no point (x,y) is a stationary point (Definition 12.41), so that no point is a critical point (Definition 12.42), so that f has no extrema (Theorem 12.43).

3. By definition 12.42, the critical points of f are the stationary points of f together with the points where f is not differentiable. So to find the critical points, locate the points at which f is not differentiable. For the remaining points, a point (a,b) is stationary if $\nabla f(a,b) = 0$, by Definition 12.41.

   The critical points of f are important since, by Theorem 12.43, local extrema of f can occur only at critical points.

5. A saddle point of $f(x,y)$ is a point (a,b) which is a stationary point of f but at which f does not have a local extremum. See Definition 12.44.

7. The first-derivative test is discussed in the first Thinking Back exercise for this section. This test might be difficult to conceptualize for a function of more than one variable since it depends on the sign of f', and if f is a function of more than one variable, then the sign of f' may well depend on the direction in which you take the derivative. This difficulty is clearly closely related to the existence of things such as saddle points, which look like minima when the derivative is taken in one direction, and look like maxima when it is taken in the other direction.

9. If $AC - B^2 > 0$, then adding B^2 to both sides gives $AC > B^2$. Since $B^2 \geq 0$ for all B, we get $AC > 0$. If A and C had opposite signs, then $AC < 0$, while if A and C have the same sign, then $AC > 0$. Thus $AC > 0$ if and only if A and C have the same sign.

11. Note that $g(x,y)$ is differentiable everywhere since it is a polynomial function. Now, $g(x,y)$ has a stationary point at the origin since

$$\nabla g(0,0) = \left\langle -4x^3 \big|_{(0,0)}, \ -4y^3 \big|_{(0,0)} \right\rangle = \langle 0,0 \rangle.$$

Further, we have

$$g_x(x,y) = -4x^3, \qquad g_y(x,y) = -4y^3,$$
$$g_{xx}(x,y) = -12x^2, \qquad g_{yy}(x,y) = -12y^2, \qquad g_{xy}(x,y) = g_{yx}(x,y) = 0.$$

Thus the Hessian has determinant

$$\det\left(\begin{bmatrix} g_{xx}(x,y) & g_{xy}(x,y) \\ g_{yx}(x,y) & g_{yy}(x,y) \end{bmatrix} \right) = \det\left(\begin{bmatrix} -12x^2 & 0 \\ 0 & -12y^2 \end{bmatrix} \right) = 144x^2y^2.$$

So at $(0,0)$, $\det(H_g(0,0)) = 144 \cdot 0^2 \cdot 0^2 = 0$. For this function, however, since $x^4 + y^4 \geq 0$ everywhere, it follows that $g(x,y) \leq 0$ everywhere. Since $g(0,0) = 0$, we see that $(0,0)$ is an absolute maximum.

13. Collecting terms in each equation gives

$$\left(2 + \frac{2a^2}{c^2} \right) x + \frac{2ab}{c^2} y = 2x_0 + \frac{2ad}{c^2} - \frac{2az_0}{c}$$
$$\frac{2ab}{c^2} x + \left(2 + \frac{2b^2}{c^2} \right) y = 2y_0 + \frac{2bd}{c^2} - \frac{2bz_0}{c}.$$

Multiply through by c^2 to clear fractions and divide through by 2 to get

$$(c^2 + a^2)x + aby = c^2x_0 + ad - acz_0 \qquad (*)$$
$$abx + (c^2 + b^2)y = c^2y_0 + bd - bcz_0. \qquad (**)$$

Now multiply $(**)$ through by $\frac{c^2+a^2}{ab}$ to get

$$(c^2 + a^2)x + \frac{(c^2 + b^2)(c^2 + a^2)}{ab}y = \frac{(c^2 + a^2)(c^2y_0 + bd - bcz_0)}{ab}.$$

Subtract from $(*)$, giving

$$\left(ab - \frac{(c^2 + b^2)(c^2 + a^2)}{ab}\right)y = c^2x_0 + ad - acz_0 - \frac{(c^2 + a^2)(c^2y_0 + bd - bcz_0)}{ab}, \text{ or}$$
$$(a^2b^2 - (c^2 + b^2)(c^2 + a^2))y = (c^2x_0 + ad - acz_0)(ab) - (c^2 + a^2)(c^2y_0 + bd - bcz_0).$$

Expand and simplify, giving

$$-c^2(a^2 + b^2 + c^2)y = abc^2x_0 + a^2bd - a^2bcz_0 - c^4y_0 - bc^2d + bc^3z_0 - a^2c^2y_0 - a^2bd + a^2bcz_0$$
$$= abc^2x_0 - c^4y_0 - bc^2d + bc^3z_0 - a^2c^2y_0.$$

Finally, divide through by $-c^2(a^2 + b^2 + c^2)$ and cancel a c^2 from numerator and denominator to get

$$y = \frac{bd - abx_0 - bcz_0 + a^2y_0 + c^2y_0}{a^2 + b^2 + c^2}.$$

Substitute this value for y into $(**)$:

$$abx + (c^2 + b^2)\left(\frac{bd - abx_0 - bcz_0 + a^2y_0 + c^2y_0}{a^2 + b^2 + c^2}\right) = c^2y_0 + bd - bcz_0.$$

Multiply through by $a^2 + b^2 + c^2$ to clear fractions, and get all of the constant terms onto the right-hand side:

$$ab(a^2 + b^2 + c^2)x = (c^2y_0 + bd - bcz_0)(a^2 + b^2 + c^2) - (b^2 + c^2)(bd - abx_0 - bcz_0 + a^2y_0 + c^2y_0).$$

Now expand the right-hand side:

$$ab(a^2 + b^2 + c^2)x = a^2c^2y_0 + b^2c^2y_0 + c^4y_0 + a^2bd + b^3d + bc^2d - a^2bcz_0 - b^3cz_0 - bc^3z_0$$
$$- b^3d + ab^3x_0 + b^3cz_0 - a^2b^2y_0 - b^2c^2y_0 - bc^2d + abc^2x_0 + bc^3z_0 - a^2c^2y_0 - c^4y_0.$$

Simplify:

$$ab(a^2 + b^2 + c^2)x = a^2bd - a^2bcz_0 + ab^3x_0 - a^2b^2y_0 + abc^2x_0.$$

Finally, divide both sides by $ab(a^2 + b^2 + c^2)$ and cancel an ab from numerator and denominator to get

$$x = \frac{ad - acz_0 + b^2x_0 - aby_0 + c^2x_0}{a^2 + b^2 + c^2}.$$

15. We have

$$D\left(\frac{ad - acz_0 + b^2x_0 - aby_0 + c^2x_0}{a^2 + b^2 + c^2}, \frac{bd - abx_0 - bcz_0 + a^2y_0 + c^2y_0}{a^2 + b^2 + c^2}\right)$$

$$= \left(\frac{ad - acz_0 + b^2x_0 - aby_0 + c^2x_0}{a^2 + b^2 + c^2} - x_0\right)^2 +$$

$$\left(\frac{bd - abx_0 - bcz_0 + a^2y_0 + c^2y_0}{a^2 + b^2 + c^2} - y_0\right)^2 +$$

$$\left(\frac{d - a\left(\frac{ad-acz_0+b^2x_0-aby_0+c^2x_0}{a^2+b^2+c^2}\right) - b\left(\frac{bd-abx_0-bcz_0+a^2y_0+c^2y_0}{a^2+b^2+c^2}\right)}{c} - z_0\right)^2$$

$$= \frac{1}{(a^2 + b^2 + c^2)^2}\left(\left(ad - acz_0 + b^2x_0 - aby_0 + c^2x_0 - x_0(a^2 + b^2 + c^2)\right)^2\right.$$

$$+ \left(bd - abx_0 - bcz_0 + a^2y_0 + c^2y_0 - y_0(a^2 + b^2 + c^2)\right)^2$$

$$\left.+ \left(\frac{d(a^2+b^2+c^2)-a(ad-acz_0+b^2x_0-aby_0+c^2x_0)-b(bd-abx_0-bcz_0+a^2y_0+c^2y_0)}{c} - z_0(a^2 + b^2 + c^2)\right)^2\right)$$

$$= \frac{1}{(a^2 + b^2 + c^2)^2}\left((ad - acz_0 - aby_0 - a^2x_0)^2 + (bd - abx_0 - bcz_0 - b^2y_0)^2\right.$$

$$\left.+ \left(\frac{a^2d+b^2d+c^2d-a^2d+a^2cz_0-ab^2x_0+a^2by_0-ac^2x_0-b^2d+ab^2x_0+b^2cz_0-a^2by_0-bc^2y_0-a^2cz_0-b^2cz_0-c^3z_0}{c}\right)^2\right)$$

$$= \frac{1}{(a^2 + b^2 + c^2)^2}\left((ad - acz_0 - aby_0 - a^2x_0)^2 + (bd - abx_0 - bcz_0 - b^2y_0)^2\right.$$

$$\left.+ \left(\frac{c^2d - ac^2x_0 - bc^2y_0 - c^3z_0}{c}\right)^2\right)$$

$$= \frac{a^2(d - ax_0 - by_0 - cz_0)^2 + b^2(d - ax_0 - by_0 - cz_0)^2 + c^2(d - ax_0 - by_0 - cz_0)^2}{(a^2 + b^2 + c^2)^2}$$

$$= \frac{(d - ax_0 - by_0 - cz_0)^2}{a^2 + b^2 + c^2}.$$

17. Let $D(t)$ be the distance from P to the point $\mathbf{r}(t) = P_0 + \mathbf{d}t$. Then

$$S(t) = D(t)^2 = (x_1 - (x_0 + at))^2 + (y_1 - (y_0 + bt))^2 + (z_1 - (z_0 + ct))^2$$
$$= (-at + x_1 - x_0)^2 + (-bt + y_1 - y_0)^2 + (-ct + z_1 - z_0)^2.$$

Thus the square $S(t)$ of the distance $D(t)$ is minimized when $S'(t) = 0$. But

$$S'(t) = 2(-at + x_1 - x_0)(-a) + 2(-bt + y_1 - y_0)(-b) + 2(-ct + z_1 - z_0)(-c)$$
$$= 2(a^2 + b^2 + c^2)t - 2(a(x_1 - x_0) + b(y_1 - y_0) + c(z_1 - z_0)).$$

Thus $S'(t) = 0$ for

$$t = \frac{a(x_1 - x_0) + b(y_1 - y_0) + c(z_1 - z_0)}{a^2 + b^2 + c^2}.$$

(Note that $S''(t) = 2(a^2 + b^2 + c^2) > 0$, so that this value of t is indeed a minimum). Note that we may write t as

$$t = \frac{\mathbf{d} \cdot \overrightarrow{P_0P}}{\|\mathbf{d}\|^2}.$$

Then for this value of t,

$$S(t) = \left(x_1 - x_0 - a \cdot \frac{\mathbf{d} \cdot \overrightarrow{P_0 P}}{\|\mathbf{d}\|^2}\right)^2 + \left(y_1 - y_0 - b \cdot \frac{\mathbf{d} \cdot \overrightarrow{P_0 P}}{\|\mathbf{d}\|^2}\right)^2 + \left(z_1 - z_0 - c \cdot \frac{\mathbf{d} \cdot \overrightarrow{P_0 P}}{\|\mathbf{d}\|^2}\right)^2$$

$$= (x_1 - x_0)^2 + (y_1 - y_0)^2 + (z_1 - z_0)^2 + (a^2 + b^2 + c^2)\left(\frac{\mathbf{d} \cdot \overrightarrow{P_0 P}}{\|\mathbf{d}\|^2}\right)^2$$

$$- 2\frac{\mathbf{d} \cdot \overrightarrow{P_0 P}}{\|\mathbf{d}\|^2}\langle x_1 - x_0, y_1 - y_0, z_1 - z_0\rangle \cdot \langle a, b, c\rangle$$

$$= \left\|\overrightarrow{P_0 P}\right\|^2 + \|\mathbf{d}\|^2\left(\frac{\mathbf{d} \cdot \overrightarrow{P_0 P}}{\|\mathbf{d}\|^2}\right)^2 - 2\frac{\mathbf{d} \cdot \overrightarrow{P_0 P}}{\|\mathbf{d}\|^2}\overrightarrow{P_0 P} \cdot \mathbf{d}$$

$$= \left\|\overrightarrow{P_0 P}\right\|^2 + \frac{(\mathbf{d} \cdot \overrightarrow{P_0 P})^2}{\|\mathbf{d}\|^2} - 2\frac{(\mathbf{d} \cdot \overrightarrow{P_0 P})^2}{\|\mathbf{d}\|^2}$$

$$= \left\|\overrightarrow{P_0 P}\right\|^2 - \frac{(\mathbf{d} \cdot \overrightarrow{P_0 P})^2}{\|\mathbf{d}\|^2}.$$

Now, recall that $\mathbf{u} \cdot \mathbf{v} = \|\mathbf{u}\|\,\|\mathbf{v}\|\cos\theta$, so that

$$(\mathbf{u} \cdot \mathbf{v})^2 = \|\mathbf{u}\|^2\,\|\mathbf{v}\|^2\cos^2\theta = \|\mathbf{u}\|^2\,\|\mathbf{v}\|^2 - \|\mathbf{u}\|^2\,\|\mathbf{v}\|^2\sin\theta = \|\mathbf{u}\|^2\,\|\mathbf{v}\|^2 - \|\mathbf{u} \times \mathbf{v}\|^2,$$

so that, substituting in the above, we get

$$S(t) = \left\|\overrightarrow{P_0 P}\right\|^2 - \frac{(\mathbf{d} \cdot \overrightarrow{P_0 P})^2}{\|\mathbf{d}\|^2}$$

$$= \left\|\overrightarrow{P_0 P}\right\|^2 - \frac{\|\mathbf{d}\|^2\left\|\overrightarrow{P_0 P}\right\|^2 - \left\|\mathbf{d} \times \overrightarrow{P_0 P}\right\|^2}{\|\mathbf{d}\|^2}$$

$$= \frac{\left\|\overrightarrow{P_0 P}\right\|^2\|\mathbf{d}\|^2 - \|\mathbf{d}\|^2\left\|\overrightarrow{P_0 P}\right\|^2 + \left\|\mathbf{d} \times \overrightarrow{P_0 P}\right\|^2}{\|\mathbf{d}\|^2}$$

$$= \frac{\left\|\mathbf{d} \times \overrightarrow{P_0 P}\right\|^2}{\|\mathbf{d}\|^2}.$$

Taking square roots gives

$$D(t) = \frac{\left\|\mathbf{d} \times \overrightarrow{P_o P}\right\|}{\|\mathbf{d}\|}$$

as desired.

19. $f(x)$ will have a minimum at $x = 0$ when $f'(x) = nx^{n-1}$ changes sign from negative to positive at the origin (note that for $n > 2$ that $f''(0) = 0$, so that the second derivative test will be inconclusive). For $x < 0$ and close to zero, $f'(x) < 0$ if $n - 1$ is odd and $f'(x) > 0$ if $n - 1$ is even. Further, $f'(x) > 0$ whenever $x > 0$. Thus $f'(x)$ changes sign from negative to positive at the origin if and

only if $n - 1$ is odd, i.e., if and only if n is even. Hence $f(x)$ has a minimum at $x = 0$ if and only if n is even.

$f(x)$ has an inflection point at $x = 0$ when $f''(x) = n(n-1)x^{n-2}$ changes sign at the origin. For $x < 0$ and close to zero, $f''(x) < 0$ if $n - 2$ is odd and $f''(x) > 0$ if $n - 2$ is even. Further, $f''(x) > 0$ whenever $x > 0$. Thus $f''(x)$ changes sign from negative to positive at the origin if and only if $n - 2$ is odd, i.e., if and only if n is odd. Hence $f(x)$ has an inflection point at $x = 0$ if and only if n is odd.

Since n must be either even or odd, there are no other possibilities: either f has a minimum at $x = 0$ or it has an inflection point there.

Skills

21. Taking derivatives gives

$$f_x(x,y) = 2e^{2x} \cos y, \qquad f_y(x,y) = -e^{2x} \sin y$$

$$f_{xx}(x,y) = 4e^{2x} \cos y, \qquad f_{yy}(x,y) = -e^{2x} \cos y, \qquad f_{xy}(x,y) = f_{yx}(x,y) = -2e^{2x} \sin y.$$

Then the discriminant of f is

$$\det(H_f(x,y)) = \det \begin{bmatrix} 4e^{2x} \cos y & -2e^{2x} \sin y \\ -2e^{2x} \sin y & -e^{2x} \cos y \end{bmatrix}$$

$$= (4e^{2x} \cos y)(-e^{2x} \cos y) - (-2e^{2x} \sin y)^2$$

$$= -4e^{4x} \cos^2 y - 4e^{4x} \sin^2 y$$

$$= -4e^{4x}.$$

23. Taking derivatives gives

$$f_s(s,t) = 3s^2 e^{t/2}, \qquad f_t(s,t) = \frac{1}{2}s^3 e^{t/2}$$

$$f_{ss}(s,t) = 6s e^{t/2}, \qquad f_{tt}(s,t) = \frac{1}{4}s^3 e^{t/2}, \qquad f_{st}(s,t) = f_{ts}(s,t) = \frac{3}{2}s^2 e^{t/2}.$$

Then the discriminant of f is

$$\det(H_f(s,t)) = \det \begin{bmatrix} 6s e^{t/2} & \frac{3}{2}s^2 e^{t/2} \\ \frac{3}{2}s^2 e^{t/2} & \frac{1}{4}s^3 e^{t/2} \end{bmatrix}$$

$$= \left(6s e^{t/2}\right)\left(\frac{1}{4}s^3 e^{t/2}\right) - \left(\frac{3}{2}s^2 e^{t/2}\right)^2$$

$$= \frac{3}{2}s^4 e^t - \frac{9}{4}s^4 e^t$$

$$= -\frac{3}{4}s^4 e^t.$$

25. Taking derivatives gives

$$f_\theta(\theta,\phi) = -\sin\theta \sin\phi, \qquad f_\phi(\theta,\phi) = \cos\theta \cos\phi$$

$$f_{\theta\theta}(\theta,\phi) = -\cos\theta \sin\phi, \qquad f_{\phi\phi}(\theta,\phi) = -\cos\theta \sin\phi, \qquad f_{\theta\phi}(\theta,\phi) = -\sin\theta \cos\phi.$$

Then the discriminant of f is

$$\det(H_f(\theta,\phi)) = \det \begin{bmatrix} -\cos\theta \sin\phi & -\sin\theta \cos\phi \\ -\sin\theta \cos\phi & -\cos\theta \sin\phi \end{bmatrix}$$

$$= (-\cos\theta \sin\phi)^2 - (\sin\theta \cos\phi)^2$$

$$= \cos^2\theta \sin^2\phi - \sin^2\theta \cos^2\phi.$$

27. From Example 4, the distance is

$$\frac{|1 \cdot 3 - 3 \cdot 5 + 4 \cdot (-2) - 8|}{\sqrt{1^2 + 3^2 + 4^2}} = \frac{28}{\sqrt{26}}.$$

29. From Example 4, the distance is

$$\frac{|0 \cdot 4 + 12 \cdot 0 - 5 \cdot (-3) - 7|}{\sqrt{0^2 + 12^2 + 5^2}} = \frac{8}{13}.$$

31. Note that f is differentiable everywhere since it is a polynomial. Taking derivatives gives

$$f_x(x,y) = 6x + 3, \qquad f_y(x,y) = 12y,$$
$$f_{xx}(x,y) = 6, \qquad f_{yy}(x,y) = 12, \qquad f_{xy}(x,y) = f_{yx}(x,y) = 0.$$

The gradient is

$$\nabla f(x,y) = \langle f_x(x,y), f_y(x,y) \rangle = \langle 6x + 3, 12y \rangle.$$

Thus the only stationary point is $\left(-\frac{1}{2}, 0\right)$.

Now, $\det H_f\left(-\frac{1}{2}, 0\right) = 6 \cdot 12 - 0^2 = 72$, so that at the stationary point, both the discriminant and f_{xx} are positive. Thus this point is a local minimum, and is the only local extremum. Since $\lim_{x \to \pm\infty} (3x^2 + 3x) = \infty$ and $\lim_{y \to \pm\infty} 6y^2 = \infty$, it follows that f increases without bound as (x, y) gets far away from the origin. Thus $\left(-\frac{1}{2}, 0\right)$ is a global minimum.

33. Note that f is differentiable everywhere since it is a polynomial. Taking derivatives gives

$$f_x(x,y) = 3x^2 - 12, \qquad f_y(x,y) = 3y^2 - 3$$
$$f_{xx}(x,y) = 6x, \qquad f_{yy}(x,y) = 6y, \qquad f_{xy}(x,y) = 0.$$

The gradient is

$$\nabla f(x,y) = \langle f_x(x,y), f_y(x,y) \rangle = \left\langle 3x^2 - 12, 3y^2 - 3 \right\rangle.$$

Then the gradient is zero when $3x^2 = 12$ and $3y^2 = 3$, so it is zero at the points $(\pm 2, \pm 1)$, which are the stationary points.

Now, $\det H_f(x,y) = 6x \cdot 6y - 0^2 = 36xy$, so that at $(2, -1)$ and $(-2, 1)$, we have a negative discriminant, so that these are saddle points. At $(2, 1)$, the discriminant is positive and $f_{xx}(2, 1) = 12 > 0$, so $(2, 1)$ is a relative minimum. At $(-2, -1)$, the discriminant is positive ant $f_{xx}(-2, -1) = -12 < 0$ so that this is a relative maximum.

Finally, since

$$\lim_{x \to \infty} f(x, 0) = \lim_{x \to \infty} (x^3 - 12x + 15) = \infty$$
$$\lim_{x \to -\infty} f(x, 0) = \lim_{x \to -\infty} (x^3 - 12x + 15) = -\infty,$$

f has neither a global minimum nor a global maximum.

35. Note that f is differentiable everywhere since it is a polynomial. Taking derivatives gives

$$f_x(x,y) = 3x^2 - 12y, \qquad f_y(x,y) = -12x + 3y^2,$$
$$f_{xx}(x,y) = 6x, \qquad f_{yy}(x,y) = 6y, \qquad f_{xy}(x,y) = -12.$$

The gradient is

$$\nabla f(x,y) = \langle f_x(x,y), f_y(x,y) \rangle = \left\langle 3x^2 - 12y, 3y^2 - 12x \right\rangle.$$

The gradient is zero when both components are zero, so when $x^2 = 4y$ and $y^2 = 4x$. Substituting $\frac{y^2}{4}$ for x in the first equation gives $\frac{y^4}{16} = 4y$, so that $y^4 = 64y$. This holds when $y = 0$ or $y = 4$. If $y = 0$, then $x = 0$, while if $y = 4$, then $x = 4$. Thus the two stationary points are $(0,0)$ and $(4,4)$.

Now, $\det H_f(x,y) = 36xy - 144$. At $(0,0)$, the discriminant is $-144 < 0$, so that $(0,0)$ is a saddle point. At $(4,4)$, the discriminant is $432 > 0$, and $f_{xx}(4,4) = 24 > 0$ so that $(4,4)$ is a relative minimum.

Finally, since

$$\lim_{x \to \infty} f(x,0) = \lim_{x \to \infty} x^3 = \infty,$$
$$\lim_{x \to -\infty} f(x,0) = \lim_{x \to -\infty} x^3 = -\infty,$$

f has neither a global minimum nor a global maximum.

37. Note that f is differentiable everywhere since it is a polynomial. Taking derivatives gives

$$f_x(x,y) = 4x - 2y + 3, \qquad f_y(x,y) = -2x + 8y + 9,$$
$$f_{xx}(x,y) = 4, \qquad f_{yy}(x,y) = 8, \qquad f_{xy}(x,y) = -2.$$

The gradient is

$$\nabla f(x,y) = \langle f_x(x,y), f_y(x,y) \rangle = \langle 4x - 2y + 3, -2x + 8y + 9 \rangle.$$

The gradient is zero when both of the components are zero. Setting them to zero and solving the pair of linear equations gives $(x,y) = \left(-\frac{3}{2}, -\frac{3}{2}\right)$ as the only stationary point.

Now, $\det H_f(x,y) = 4 \cdot 8 - (-2)^2 = 28 > 0$, and $f_{xx}(x,y) = 4 > 0$ for all (x,y), so that $\left(-\frac{3}{2}, -\frac{3}{2}\right)$ is a relative minimum.

Finally, since
$$f(x,y) = (x - y)^2 + x^2 + 3y^2 + 3x + 9y - 5,$$

we see that as either x or $y \to \pm\infty$, the function increases without bound, so that f has no global maximum. However, it does have a global minimum, which must be at $(x,y) = \left(-\frac{3}{2}, -\frac{3}{2}\right)$.

39. f is differentiable everywhere except on the x and y axes, since the function is not defined there. On its domain, we have

$$f_x(x,y) = 8y - \frac{1}{x^2}, \qquad f_y(x,y) = 8x - \frac{1}{y^2},$$
$$f_{xx}(x,y) = \frac{2}{x^3}, \qquad f_{yy}(x,y) = \frac{2}{y^3}, \qquad f_{xy}(x,y) = 8.$$

Then the gradient is

$$\nabla f(x,y) = \langle f_x(x,y), f_y(x,y) \rangle = \left\langle 8y - \frac{1}{x^2}, 8x - \frac{1}{y^2} \right\rangle.$$

Stationary points occur when both components are zero, so when $8y = \frac{1}{x^2}$ and $8x = \frac{1}{y^2}$. Thus $y = \frac{1}{8x^2}$; substituting into the second equation gives $8x = \frac{1}{1/64x^4}$ so that $8x^4 = x$. Now, $x = 0$ is not a valid solution since $f(x,y)$ is not defined for $x = 0$, so we must have $8x^3 = 1$, or $x = \frac{1}{2}$. Then $8y = 4$ so that $y = \frac{1}{2}$ as well. Thus the only stationary point is $\left(\frac{1}{2}, \frac{1}{2}\right)$.

The discriminant is $\det H_f(x, y) = \frac{2}{x^3} \cdot \frac{2}{y^3} - 64$, so at the stationary point we get

$$\det H_f\left(\frac{1}{2}, \frac{1}{2}\right) = 16 \cdot 16 - 64 > 0.$$

Also, $f_{xx}\left(\frac{1}{2}, \frac{1}{2}\right) = 16 > 0$, so that $\left(\frac{1}{2}, \frac{1}{2}\right)$ is a relative minimum.

Finally, since

$$\lim_{x \to 0^+} f(x, 1) = \lim_{x \to 0^+} \left(8x + \frac{1}{x} + 1\right) = \infty,$$

$$\lim_{x \to 0^-} f(x, 1) = \lim_{x \to 0^-} \left(8x + \frac{1}{x} + 1\right) = -\infty,$$

we see that f has no global minimum or maximum.

41. f is differentiable everywhere since both x and $\sin y$ are. Differentiating gives

$$f_x(x, y) = \sin y, \qquad f_y(x, y) = x \cos y,$$
$$f_{xx}(x, y) = 0, \qquad f_{yy}(x, y) = -x \sin y, \qquad f_{xy}(x, y) = \cos y.$$

Then the gradient is

$$\nabla f(x, y) = \langle f_x(x, y), f_y(x, y) \rangle = \langle \sin y, x \cos y \rangle.$$

The gradient is zero when $\sin y = x \cos y = 0$. If $x = 0$, then $\sin y = 0$, so that $y = n\pi$ for n an integer. If $x \neq 0$, then $\sin y = x \cos y = 0$ means that both $\sin y$ and $\cos y$ are zero, which is impossible. So the only stationary points are $(0, n\pi)$ for n an integer.

The discriminant is $\det H_f(x, y) = 0 \cdot (-x \sin y) - \cos^2 y = -\cos^2 y$, so that $\det H_f(0, n\pi) = -1 < 0$ and thus $(0, n\pi)$ is a saddle point for all integral n.

Finally, since

$$\lim_{x \to \infty} f\left(x, \frac{\pi}{2}\right) = \lim_{x \to \infty} x = \infty,$$

$$\lim_{x \to \infty} f\left(x, \frac{3\pi}{2}\right) = \lim_{x \to \infty} (-x) = -\infty,$$

f has no global maximum or minimum.

43. f is differentiable everywhere since both e^{x^2} and $\cos y$ are. Differentiating gives

$$f_x(x, y) = 2xe^{x^2} \cos y, \qquad f_y(x, y) = -e^{x^2} \sin y,$$
$$f_{xx}(x, y) = 2e^{x^2} \cos y + 4x^2 e^{x^2} \cos y = (4x^2 + 2)e^{x^2} \cos y,$$
$$f_{yy}(x, y) = -e^{x^2} \cos y, \qquad f_{xy}(x, y) = -2xe^{x^2} \sin y.$$

The gradient is

$$\nabla f(x, y) = \left\langle 2xe^{x^2} \cos y, -e^{x^2} \sin y \right\rangle.$$

In order for the gradient to be zero, its second component, $-e^{x^2} \sin y$ must be zero, which happens only when $y = n\pi$ for n an integer. At those points, the first component is $\pm 2xe^{x^2}$, so it is zero if and only if $x = 0$. Thus the stationary points are $(0, n\pi)$ for n an integer.

The discriminant is

$$\det H_f(x,y) = \left((4x^2+2)e^{x^2}\cos y\right)\left(-e^{x^2}\cos y\right) - \left(-2xe^{x^2}\sin y\right)^2$$
$$= -4x^2e^{2x^2}\cos^2 y - 2e^{2x^2}\cos^2 y - 4x^2e^{2x^2}\sin^2 y$$
$$= -2e^{2x^2}\cos^2 y - 4x^2e^{2x^2}.$$

This is negative everywhere, so that all the stationary points are saddle points.

Finally, since

$$\lim_{x\to\infty} f(x,0) = \lim_{x\to\infty} e^{x^2} = \infty,$$
$$\lim_{x\to\infty} f(x,\pi) = \lim_{x\to\infty} (-e^{x^2}) = -\infty,$$

f has neither a global minimum nor a global maximum.

45. f is differentiable everywhere since both xy and e^z are. Differentiating gives

$$f_x(x,y) = ye^{xy}, \qquad f_y(x,y) = xe^{xy},$$
$$f_{xx}(x,y) = y^2e^{xy}, \qquad f_{yy}(x,y) = x^2e^{xy}, \qquad f_{xy}(x,y) = e^{xy} + xye^{xy}.$$

The gradient is

$$\nabla f(x,y) = \langle f_x(x,y), f_y(x,y)\rangle = \langle ye^{xy}, xe^{xy}\rangle.$$

Since e^{xy} is never zero, the gradient is zero if and only if $(x,y) = (0,0)$, so that this is the only stationary point.

The discriminant is

$$\det H_f(x,y) = (y^2e^{xy})(x^2e^{xy}) - (e^{xy} + xye^{xy})^2$$
$$= x^2y^2e^{2xy} - e^{2xy} - 2xye^{2xy} - x^2y^2e^{2xy} = -(2xy+1)e^{2xy}.$$

Thus $\det H_f(0,0) = -1$, so that $(0,0)$ is a saddle point.

Finally, since $\lim_{x\to\infty} f(x,1) = \lim_{x\to\infty} e^x = \infty$, we see that f has no global maximum. Also, clearly $f(x,y) > 0$ for all (x,y), but $\lim_{x\to-\infty} f(x,1) = \lim_{x\to-\infty} e^{-x} = 0$, so f asymptotically approaches 0 as $x \to -\infty$. Thus f has no global minimum either.

47. f is differentiable everywhere except where $x^2 + y^2 = 1$, which is the unit circle, since except at those points, f is the quotient of a differentiable function (1) by a nonzero differentiable function

$(x^2 + y^2 - 1)$. Differentiating gives

$$f_x(x,y) = -\frac{2x}{(x^2+y^2-1)^2},$$

$$f_y(x,y) = -\frac{2y}{(x^2+y^2-1)^2},$$

$$f_{xx}(x,y) = -\frac{(x^2+y^2-1)^2(2) - 2x\left(2(x^2+y^2-1)(2x)\right)}{(x^2+y^2-1)^4} = -\frac{2x^2+2y^2-2-8x^2}{(x^2+y^2-1)^3}$$

$$= \frac{6x^2-2y^2+2}{(x^2+y^2-1)^3},$$

$$f_{yy}(x,y) = -\frac{(x^2+y^2-1)^2(2) - 2y\left(2(x^2+y^2-1)(2y)\right)}{(x^2+y^2-1)^4} = -\frac{2x^2+2y^2-2-8y^2}{(x^2+y^2-1)^3}$$

$$= \frac{6y^2-2x^2+2}{(x^2+y^2-1)^3},$$

$$f_{xy}(x,y) = -(-2)\frac{2x}{(x^2+y^2-1)^3}(2y) = \frac{8xy}{(x^2+y^2-1)^3}.$$

The gradient is

$$\nabla f(x,y) = \langle f_x(x,y), f_y(x,y)\rangle = \left\langle -\frac{2x}{(x^2+y^2-1)^2}, \ -\frac{2y}{(x^2+y^2-1)^2}\right\rangle.$$

The gradient is zero when both numerators vanish, which is only at the point $(0,0)$. Since the denominators are nonzero there, this is in fact a stationary point.

The discriminant at $(0,0)$ is

$$\det H_f(0,0) = \left(\frac{6x^2-2y^2+2}{(x^2+y^2-1)^3}\cdot\frac{6y^2-2x^2+2}{(x^2+y^2-1)^3} - \left(\frac{8xy}{(x^2+y^2-1)^3}\right)^2\right)\Bigg|_{(0,0)}$$

$$= \frac{2}{-1}\cdot\frac{2}{-1} - 0 = 4 > 0.$$

Since $f_{xx}(0,0) = \frac{2}{-1} = -2 < 0$, we see that $(0,0)$ is a local maximum.

Finally, since

$$\lim_{x\to1^-} f(x,0) = \lim_{x\to1^-}\frac{1}{x^2-1} = -\infty,$$

$$\lim_{x\to1^+} f(x,0) = \lim_{x\to1^+}\frac{1}{x^2-1} = \infty,$$

f has neither a global minimum nor a global maximum.

49. f is differentiable everywhere since it is a polynomial. Differentiating gives

$$f_x(x,y) = 2xy^2, \qquad f_y(x,y) = 2x^2y,$$

$$f_{xx}(x,y) = 2y^2, \qquad f_{yy}(x,y) = 2x^2, \qquad f_{xy}(x,y) = 4xy.$$

The gradient is

$$\nabla f(x,y) = \langle f_x(x,y), f_y(x,y)\rangle = \left\langle 2xy^2, 2x^2y\right\rangle.$$

Thus the gradient is zero if (x,y) is either on the x axis or the y axis, so these are all stationary points.

The discriminant is $\det H_f(x,y) = (2y^2)(2x^2) - (4xy)^2 = 4x^2y^2 - 16x^2y^2 = -12x^2y^2$. So at all these stationary points, since either x or y is zero, the discriminant is zero, so we get no information about the behavior of f at the stationary points. However, since $f(x,y) \geq 0$ everywhere, and $f(x,y) = 0$ at all of the stationary points, we see that they are all local minima.

Finally, since $\lim\limits_{x\to\infty} f(x,1) = \lim\limits_{x\to\infty} x^2 = \infty$, we see that f has no global maximum. Since $f(x,y) \geq 0$ for all (x,y), and $f(0,0) = 0$, it follows that f has a global minimum at $(0,0)$ (as well as at all other points on the x and y axes).

51. f is differentiable everywhere since it is a polynomial. Differentiating gives

$$f_x(x,y) = 3x^2y^2, \qquad f_y(x,y) = 2x^3y,$$
$$f_{xx}(x,y) = 6xy^2, \qquad f_{yy}(x,y) = 2x^3, \qquad f_{xy}(x,y) = 6x^2y.$$

The gradient is

$$\nabla f(x,y) = \langle f_x(x,y), f_y(x,y) \rangle = \left\langle 3x^2y^2, 2x^3y \right\rangle.$$

The gradient is zero exactly when either $x = 0$ or $y = 0$, so that all points on either axis are stationary points.

The discriminant is $\det H_f(x,y) = 6xy^2 \cdot 2x^3 - (6x^2y)^2 = -24x^4y^2$, so that at the stationary points, the discriminant is zero and we get no information about the behavior of f. For stationary points of the form $(0,y)$, there are points (x,y') arbitrarily close to $(0,y)$ with $x < 0$ and $y' \neq 0$, so that $f(x,y') = x(y')^2 < 0$, and there are points (x,y') arbitrarily close to $(0,y)$ with $x > 0$ and $y' \neq 0$, so that $f(x,y') = x(y')^2 > 0$. Thus $(0,y)$ is neither a local maximum nor a local minimum, so all of those points are saddle points. Next look at stationary points of the form $(x,0)$ with $x > 0$. We have $f(x,y) = 0$, and every point in a small disc around $(x,0)$ is of the form (x',y) with $x' > 0$, so that $f(x',y) = x'y^2 \geq 0$. Thus the points $(x,0)$ with $x > 0$ are all local minima. Finally, look at stationary points of the form $(x,0)$ for $x < 0$. We have $f(x,y) = 0$, and every point in a small disc around $(x,0)$ is of the form (x',y) with $x' < 0$, so that $f(x',y) = x'y^2 \leq 0$. Thus the points $(x,0)$ with $x < 0$ are all local maxima.

Finally, since

$$\lim_{x\to\infty} f(x,1) = \lim_{x\to\infty} x^3 = \infty,$$
$$\lim_{x\to-\infty} f(x,1) = \lim_{x\to-\infty} x^3 = -\infty,$$

f has neither a global minimum nor a global maximum.

Applications

53. Let the dimensions of the base be $x \times y$, and the height be z. Then the volume of the box is $xyz = 20$, and the cost of building the box is

$$C(x,y) = 2xy + 2(5xz) + 2(5yz) + 7xy = 9xy + 10(x+y)z = 9xy + 10(x+y)\frac{20}{xy}$$
$$= 9xy + \frac{200(x+y)}{xy} = 9xy + \frac{200}{y} + \frac{200}{x}.$$

We wish to minimize $C(x,y)$. The various derivatives are:

$$C_x = 9y - \frac{200}{x^2}, \qquad C_y = 9x - \frac{200}{y^2}, \qquad C_{xx} = \frac{400}{x^3}, \qquad C_{yy} = \frac{400}{y^3}, \qquad C_{xy} = 9.$$

The stationary points occur where $C_x = C_y = 0$. Setting $C_x = 0$ gives $9x^2y = 200$, so that $y = \frac{200}{9x^2}$. Substitute into $C_y = 0$ to get

$$9x - \frac{200}{200^2/(9x^2)^2} = 9x - \frac{81}{200}x^4 = 0.$$

Thus $x = 0$, which is impossible, or $x = 2\sqrt[3]{\frac{5^2}{3^2}}$. Substituting this value into $C_x = 0$ gives

$$C_x = 9y - \frac{200}{4\sqrt[3]{\frac{5^4}{3^4}}} = 9y - \frac{200 \cdot 3^2}{100\sqrt[3]{\frac{3^2}{5^2}}} = 9y - 18\sqrt[3]{\frac{5^2}{3^2}} = 0,$$

so that $y = 2\left(\frac{5^2}{3^2}\right)^{1/3} = x$. Now, at this point, the discriminant is

$$C_{xx}\left(2\left(\frac{5^2}{3^2}\right)^{1/3}\right) C_{yy}\left(2\left(\frac{5^2}{3^2}\right)^{1/3}\right) - C_{xy}\left(2\left(\frac{5^2}{3^2}\right)^{1/3}\right)^2$$

$$= \frac{400}{8 \cdot \frac{25}{9}} \cdot \frac{400}{8 \cdot \frac{25}{9}} - 81$$

$$= \frac{400 \cdot 400 \cdot 81}{200 \cdot 200} - 81$$

$$= 324 - 81 = 243 > 0.$$

Since $C_{xx}\left(2\left(\frac{5^2}{3^2}\right)^{1/3}\right) > 0$ as well, Theorem 12.46 tells us that this point is a minimum. So the cost is minimized when the base is square, with side length $2\left(\frac{5^2}{3^2}\right)^{1/3}$, and the height is

$$z = \frac{20}{4\left(\frac{5^2}{3^2}\right)^{2/3}} = 5 \cdot \frac{3}{5}\left(\frac{3}{5}\right)^{1/3} = 3\left(\frac{3}{5}\right)^{1/3}.$$

Proofs

55. Since f has continuous second partials, certainly f, f_x and f_y are all differentiable. Further, we know that the mixed partials are equal, so that $f_{xy}(x,y) = f_{yx}(x,y)$. Finally, we know that

$$D_{\mathbf{u}}f(x,y) = \nabla f(x,y) \cdot \mathbf{u}, \quad D_{\mathbf{u}}f_x(x,y) = \nabla f_x(x,y) \cdot \mathbf{u}, \quad D_{\mathbf{u}}f_y(x,y) = \nabla f_y(x,y) \cdot \mathbf{u}.$$

Thus

$$D_{\mathbf{u}}f(x,y) = \langle f_x(x,y), f_y(x,y)\rangle \cdot \langle a, b\rangle = af_x(x,y) + bf_y(x,y).$$

Then

$$D_{\mathbf{u}}^2 f(x,y) = D_{\mathbf{u}}\left(D_{\mathbf{u}}f(x,y)\right)$$
$$= D_{\mathbf{u}}\left(af_x(x,y) + bf_y(x,y)\right)$$
$$= \nabla(af_x(x,y) + bf_y(x,y)) \cdot \mathbf{u}$$
$$= \langle(af_x(x,y) + bf_y(x,y))_x, (af_x(x,y) + bf_y(x,y))_y\rangle \cdot \langle a, b\rangle$$
$$= \langle af_{xx}(x,y) + bf_{yx}(x,y), af_{xy}(x,y) + bf_{yy}(x,y)\rangle \cdot \langle a, b\rangle$$
$$= a^2 f_{xx}(x,y) + abf_{yx}(x,y) + abf_{xy}(x,y) + b^2 f_{yy}(x,y)$$
$$= a^2 f_{xx}(x,y) + 2abf_{xy}(x,y) + b^2 f_{yy}(x,y).$$

57. Let $d(x, y)$ be the distance from (x_0, y_0) to the point (x, y) on the line $\alpha x + \beta y = \gamma$. Note that $d(x, y) \geq 0$ for all x and y. Let $s(x, y) = d(x, y)^2$. Suppose that the point (x_1, y_1) minimizes $s(x, y)$. Then $s(x_1, y_1) = d(x_1, y_1)^2$. If there were another point (x_2, y_2) with $0 \leq d(x_2, y_2) < d(x_1, y_1)$. Then $s(x_2, y_2) = d(x_2, y_2)^2 < d(x_1, y_1)^2 = s(x_1, y_1)$, which contradicts the assumption that (x_1, y_1) minimizes $s(x, y)$. Thus (x_1, y_1) minimizes $d(x, y)$ as well.

Thinking Forward

A function without extrema

This is a polynomial function, so it is defined on all of $\mathbb{R}^2$. However, it has no local extrema; for consider a point (a, b). Then for any $\epsilon > 0$, we have

$$f(a + \epsilon, b) = 3(a + \epsilon) - 4b = f(a, b) + 3\epsilon > f(a, b), \quad f(a - \epsilon, b) = 3(a - \epsilon) - 4b = f(a, b) - 3\epsilon < f(a, b).$$

Since any open disk around (a, b) contains $(a + \epsilon, b)$ and $(a - \epsilon, b)$ for some $\epsilon > 0$, it follows that (a, b) can be neither a local minimum nor a local maximum.

Extrema on a closed and bounded set

If $a^2 + b^2 < 1$, then there is an open set around (a, b) that is contained in the disc $D = \{(x, y) \mid x^2 + y^2 \leq 1\}$, so by the argument in the previous part, (a, b) cannot be an extremum. So to find the extremum on D we need only look on the $\partial D = \{(x, y) \mid x^2 + y^2 = 1\}$. Then $3x - 4y = 3x \pm 4\sqrt{1 - x^2}$, so that finding the local extrema of $3x - 4y$ on ∂D amounts to finding the local extrema of $3x \pm 4\sqrt{1 - x^2}$ on $[-1, 1]$. But the derivative is

$$3 \pm 4 \cdot \frac{1}{2} \frac{1}{\sqrt{1 - x^2}} \cdot (-2x) = 3 \mp \frac{4x}{\sqrt{1 - x^2}}.$$

Setting this equal to zero and solving gives $3\sqrt{1 - x^2} = \pm 4x$, so that $9 - 9x^2 = 16x^2$, or $x = \pm\frac{3}{5}$ and then $y = \pm\frac{4}{5}$. Testing all four possibilities gives

$$f\left(-\frac{3}{5}, -\frac{4}{5}\right) = 3\left(-\frac{3}{5}\right) - 4\left(-\frac{4}{5}\right) = \frac{7}{5}$$

$$f\left(-\frac{3}{5}, \frac{4}{5}\right) = 3\left(-\frac{3}{5}\right) - 4\left(\frac{4}{5}\right) = -5$$

$$f\left(\frac{3}{5}, -\frac{4}{5}\right) = 3\left(\frac{3}{5}\right) - 4\left(-\frac{4}{5}\right) = 5$$

$$f\left(\frac{3}{5}, \frac{4}{5}\right) = 3\left(\frac{3}{5}\right) - 4\left(\frac{4}{5}\right) = -\frac{7}{5}.$$

Thus the maximum value is 5 at $\left(\frac{3}{5}, -\frac{4}{5}\right)$, and the minimum value is -5 at $\left(-\frac{3}{5}, \frac{4}{5}\right)$.

12.7 Lagrange Multipliers

Thinking Back

Optimizing a function of two variables subject to a constraint

Since $x^2 = 16 - 4y^2$, we can substitute for x to get $g(y) = y\sqrt{16 - 4y^2}$ or $h(y) = -y\sqrt{16 - 4y^2}$. Taking $g(y)$ first, it is defined on $[-2, 2]$, and $g(2) = g(-2) = 0$. On $(-2, 2)$, we can determine its extrema by looking at

$$g'(y) = \sqrt{16 - 4y^2} + \frac{y}{2\sqrt{16 - 4y^2}} \cdot (-8y) = \frac{16 - 4y^2 - 4y^2}{\sqrt{16 - 4y^2}} = \frac{16 - 8y^2}{\sqrt{16 - 4y^2}}.$$

Then $g'(y) = 0$ for $y = \pm\sqrt{2}$, so that $x^2 = 16 - 4y^2 = 8$ and $x = \pm 2\sqrt{2}$. Finally,

$$f(2\sqrt{2}, -\sqrt{2}) = f(-2\sqrt{2}, \sqrt{2}) = -4, \qquad f(2\sqrt{2}, \sqrt{2}) = f(-2\sqrt{2}, -\sqrt{2}) = 4,$$

so that f is maximized at either of the second pair of points and minimized at either of the first pair. Considering $h(y)$, it has the same domain, and $h(-2) = h(2) = 0$ as well. Also, $h'(y) = -g'(y)$, so we get the same solutions for y upon setting $h'(y) = 0$, so the same solutions for x.

Optimizing a function of three variables subject to a constraint

Since $x^2 = 16 - 4y^2 - 9z^2$, we can substitute for x to get $g(y, z) = yz\sqrt{16 - 4y^2 - 9z^2}$. (As in the previous problem, we can also use the negative square root, but we will again get the same set of solutions since $g(y, z)$ and $-g(y, z)$ have the same stationary points). Note that the domain of $g(y, z)$ is $\{(y, z) \mid 4y^2 + 9z^2 \le 16\}$, and that on the boundary of that region, $g(y, z) = 0$. Inside the region, we determine the extrema of g by finding the stationary points. We have

$$g_y(y, z) = z\sqrt{16 - 4y^2 - 9z^2} + \frac{yz}{2\sqrt{16 - 4y^2 - 9z^2}} \cdot (-8y) = \frac{z(16 - 4y^2 - 9z^2) - 4y^2 z}{\sqrt{16 - 4y^2 - 9z^2}}$$

$$= \frac{16z - 8y^2 z - 9z^3}{\sqrt{16 - 4y^2 - 9z^2}}$$

$$g_z(y, z) = y\sqrt{16 - 4y^2 - 9z^2} + \frac{yz}{2\sqrt{16 - 4y^2 - 9z^2}} \cdot (-18z) = \frac{y(16 - 4y^2 - 9z^2) - 9yz^2}{\sqrt{16 - 4y^2 - 9z^2}}$$

$$= \frac{16y - 4y^3 - 18yz^2}{\sqrt{16 - 4y^2 - 9z^2}}.$$

Then $g_y(y, z) = 0$ when $z(16 - 8y^2 - 9z^2) = 0$, and $g_z(y, z) = 0$ when $y(16 - 4y^2 - 18z^2) = 0$. If $z = 0$ so that $g_y(y, z) = 0$, then we must have $y(16 - 4y^2) = 0$, so that $y = 0$ or $y = \pm 2$. But $y = \pm 2$ is impossible, since $g_y(y, z)$ is undefined at $(\pm 2, 0)$. Thus $(0, 0)$ is a stationary point. Otherwise, neither y nor z is zero, so we must have

$$16 - 8y^2 - 9z^2 = 16 - 4y^2 - 18z^2 = 0,$$

so that $4y^2 = 9z^2$. Substitute $4y^2$ for $9z^2$ in the first equation to get $16 - 12y^2 = 0$, so that $y = \pm\frac{2}{\sqrt{3}}$. Then $9z^2 = 4\frac{4}{3}$, so that $z = \pm\frac{4}{3\sqrt{3}}$. So we get four more stationary points,

$$\left(\pm\frac{2}{\sqrt{3}}, \pm\frac{4}{3\sqrt{3}} \right).$$

Then at $(y, z) = (0, 0)$, we have $x = \pm\sqrt{16 - 4y^2 - 9z^2} = \pm 4$, so that we need to check $(4, 0, 0)$ and $(-4, 0, 0)$. At

$$(y, z) = \left(\pm\frac{2}{\sqrt{3}}, \pm\frac{4}{3\sqrt{3}} \right), \text{ we get } x = \pm\sqrt{16 - 4 \cdot \frac{4}{3} - 9 \cdot \frac{16}{27}} = \pm\sqrt{16 - \frac{16}{3} - \frac{16}{3}} = \pm\frac{4}{\sqrt{3}},$$

so that we have eight more points

$$(x, y, z) = \left(\pm\frac{2}{\sqrt{3}}, \pm\frac{4}{3\sqrt{3}}, \pm\frac{4}{\sqrt{3}} \right).$$

Considering the value of $f(x, y, z) = xyz$ at these points gives

$$f(\pm 4, 0, 0) = 0,$$

$$f\left(\pm\frac{2}{\sqrt{3}}, \pm\frac{4}{3\sqrt{3}}, \pm\frac{4}{\sqrt{3}} \right) = \pm\frac{2}{\sqrt{3}} \frac{4}{3\sqrt{3}} \frac{4}{\sqrt{3}} = \pm\frac{32}{9\sqrt{3}}.$$

Thus f achieves a maximum of $\frac{32}{9\sqrt{3}}$ at the four points where either zero or two of the signs are negative, and a minimum of $-\frac{32}{9\sqrt{3}}$ at the four points where either one or three signs are negative.

Concepts

1. (a) False. There are many examples; one simple one is $f(x,y) = x^2 + y^2$ subject to the constraint $x = 0$. Since $x = 0$, we have $f(x,y) = f(0,y) = y^2$, which has a minimum of 0 but no maximum.

   (b) False. Since the given region includes the point $\left(0, -\frac{1}{2}, \frac{1}{2}\right)$, where f is undefined, limits as (x,y,z) approach that point may approach either ∞ or $-\infty$ depending on the direction. So f has neither a minimum nor a maximum in that region. Note that this does not violate Theorem 12.49 since $f(x,y,z)$ is not continuous on the given region.

   (c) True. Since $f(x,y,z)$ is continuous on the given region (in fact, it is continuous everywhere), the Extreme Value Theorem 12.49 shows that $f(x,y,z)$ achieves both a minimum and a maximum.

   (d) False. First, even though $f(x,y)$ is continuous everywhere, Theorem 12.49 does not apply since the region is not bounded. Since

   $$\lim_{y \to \infty} f(0,y) = \lim_{y \to \infty} 2y = \infty,$$

   $$\lim_{x \to \infty} f(x,1) = \lim_{x \to \infty} 2 - 4x = -\infty,$$

   we see that f has neither a minimum nor a maximum on the given region.

   (e) False. It does have a minimum, at $x = 0$, clearly. However, since $(-2)^2 = 2^2 = 4$, we have $f(x) \le 4$ for all $x \in (-2, 2)$, and $\lim_{x \to \pm 2} x^2 = 4$. Thus $f(x)$ approaches 4 asymptotically as $x \to \pm 2$, so that f has no maximum on the open interval. Note that this does not violate the Extreme Value Theorem since the interval is not closed.

   (f) False. For example, let $f(x,y) = x^2 + y^2$. On the region $x^2 + y^2 \le 1$, the maximum value of $f(x,y)$ is 1. But clearly $f(x,y)$ has no maximum value on $\mathbb{R}^2$.

   (g) False. f may not achieve that absolute maximum on a given region. For example, let $f(x,y) = e^{-(x^2+y^2)}$, which has a maximum value of 1. On $1 \le x^2 + y^2 \le 2$, however, its maximum value is $e^{-1} < 1$.

   (h) True. Since m is the minimum of f anywhere on its domain, surely it is no greater than the minimum on any region in the domain. Similarly, since M is the maximum of f anywhere on its domain, surely it is no less than the maximum on any region in the domain.

3. The constraint function $g(x,y) = 0$ represents a relationship that the variables x and y in any solution must have to each other. The objective function $z = f(x,y)$ represents the quantity z that we wish to maximize or minimize.

5. The constraint gives a relationship between x and y. If it is possible to solve that equation for either x or y, then substituting the result into $f(x,y)$ gives a function of one variable, either x or y, which we can optimize using methods from earlier chapters.

   Another way of approaching the problem is using Lagrange multipliers. Write

   $$\nabla f(x,y) = \lambda \nabla g(x,y)$$

   and solve the resulting pair of equations, together with the equation $g(x,y) = 0$, to find the set of maximum and minimum points. See the discussion in the text preceding Theorem 12.47.

7. (a) For example, $\mathbb{R}^2 - \{(0,0)\} = \{(x,y) \in \mathbb{R}^2 \mid x \ne 0 \text{ or } y \ne 0\}$ is not closed, since its complement is the origin, and the origin is not open. To see that the origin is not open, note that there is no open disc around the origin that is wholly contained within the set containing only the origin.

(b) For example, $S = \{(x,y) \mid |y| \leq 1\}$. This set is not bounded since it contains points with arbitrarily large x coordinates, so it is not contained in any disc around the origin. However, it is closed, since its complement is $S^c = \{(x,y) \mid |y| > 1\}$, and S^c is open: choose $(a,b) \in S^c$ with $b > 1$ and let $\epsilon = |b-1|$. Then the open ball of radius $\epsilon/2$ around (a,b) is wholly contained in S^c. Similarly, if $b < -1$, let $\epsilon = |-1-b|$; then again the open ball of radius $\epsilon/2$ around (a,b) is wholly contained in S^c.

(c) For example, the disc $S = \{(x,y) \mid x^2 + y^2 < 1\}$. It is bounded since it is contained in the disc $\{(x,y) \mid x^2 + y^2 \leq 1\}$, but it is not closed, since for example the point $(1,0) \in \partial S$ but $(1,0) \notin S$.

(d) For example, the disk $S = \{(x,y) \mid x^2 + y^2 \leq 1\}$. It is bounded since it is contained in the disc $\{(x,y) \mid x^2 + y^2 \leq 1\}$. It is closed since its complement $S^c = \{(x,y) \mid x^2 + y^2 > 1\}$ is open: choose $(a,b) \in S^c$ with $a^2 + b^2 = 1 + \epsilon$ for $\epsilon > 0$; then the open ball of radius $\epsilon/2$ around (a,b) is wholly contained in S^c.

9. (a) For example,

$$\mathbb{R}^3 - \{(0,0,0)\} = \{(x,y,z) \in \mathbb{R}^3 \mid x \neq 0 \text{ or } y \neq 0 \text{ or } z \neq 0\}$$

is not closed, since its complement is the origin, and the origin is not open. To see that the origin is not open, note that there is no open disc around the origin that is wholly contained within the set containing only the origin.

(b) For example, $S = \{(x,y,z) \mid |y| \leq 1\}$. This set is not bounded since it contains points with arbitrarily large x coordinates, so it is not contained in any disc around the origin. However, it is closed, since its complement is $S^c = \{(x,y,z) \mid |y| > 1\}$, and S^c is open: choose $(a,b,c) \in S^c$ with $b > 1$ and let $\epsilon = |b-1|$. Then the open ball of radius $\epsilon/2$ around (a,b,c) is wholly contained in S^c. Similarly, if $b < -1$, let $\epsilon = |-1-b|$; then again the open ball of radius $\epsilon/2$ around (a,b,c) is wholly contained in S^c.

(c) For example, the ball $S = \{(x,y,z) \mid x^2 + y^2 + z^2 < 1\}$. It is bounded since it is contained in the ball $\{(x,y,z) \mid x^2 + y^2 + z^2 \leq 1\}$, but it is not closed, since for example the point $(1,0,0) \in \partial S$ but $(1,0,0) \notin S$.

(d) For example, the ball $S = \{(x,y,z) \mid x^2 + y^2 + z^2 \leq 1\}$. It is bounded since it is contained in the ball $\{(x,y,z) \mid x^2 + y^2 + z^2 \leq 1\}$. It is closed since its complement $S^c = \{(x,y,z) \mid x^2 + y^2 + z^2 > 1\}$ is open: choose $(a,b,c) \in S^c$ with $a^2 + b^2 + c^2 = 1 + \epsilon$ for $\epsilon > 0$; then the open ball of radius $\epsilon/2$ around (a,b,c) is wholly contained in S^c.

11. There would be four equations: one for each of the components of the equation

$$\nabla f(x,y,z) = \lambda \nabla g(x,y,z),$$

and one from the equation $g(x,y,z) = 0$.

13. (a) Differentiating gives

$$f_x(x,y) = 3x^2 + 18x, \qquad f_y(x,y) = 12y,$$
$$f_{xx}(x,y) = 6x + 18, \qquad f_{yy}(x,y) = 12, \qquad f_{xy}(x,y) = 0.$$

Thus the discriminant is $\det H_f(x,y) = 12(6x + 18)$, so that $H_f(-6,0) = -216 < 0$ and $(-6,0)$ is a saddle point.

(b) From part (a), $\det H_f(0,0) = 216 > 0$, and $f_{yy}(0,0) = 12 > 0$, so that $(0,0)$ is a local minimum.

(c) Note that $f(0,0) = 0$. Since $f(x,y) = x^3 + 9x^2 + 6y^2 = x^2(x+9) + 6y^2$, we see that $f(x,y) \geq 0$ for $|x| \leq 9$, so certainly in an open disc around the origin. Thus $(0,0)$ is a local minimum. However, $f(-10,0) = -1000 + 900 + 0 = -100$, so that $(0,0)$ is not a global minimum.

15. $(0,0)$ is not an extremum, since for any $\epsilon > 0$,

$$f(-\epsilon, \epsilon) = (-\epsilon)^2 \epsilon = \epsilon^3 > 0,$$
$$f(\epsilon, -\epsilon) = \epsilon^2(-\epsilon) = -\epsilon^3 < 0,$$

and both points satisfy the constraint $x + y = 0$. Thus there are points arbitrarily close to $(0,0)$ satisfying the constraint for which f takes values both larger and smaller than $f(0,0) = 0$. Thus $(0,0)$ is not in fact a local extremum of $f(x,y)$ subject to the constraint $x + y = 0$.

17. If $ax + by = 0$ for a and b nonzero, then $y = -\frac{a}{b}x$, so that on the constraint curve, $f(x,y) = h(y) = x^2\left(-\frac{a}{b}x\right) = -\frac{a}{b}x^3$. This has no extrema, but has only an inflection point at $x = 0$. The method of Lagrange multipliers will therefore fail, producing $(0,0)$: we have $\nabla f(x,y) = \langle 2xy, x^2 \rangle$ and $\nabla g(x,y) = \langle a, b \rangle$, so that we get

$$2xy = \lambda a, \qquad x^2 = \lambda b, \qquad x + y = 0.$$

If $x = 0$, then by the third equation, $y = 0$, and $(0,0)$ satisfies all three equations, so is a solution. If $x \neq 0$, then $x = -y$, so that the first two equations become

$$\lambda = -\frac{2}{a}x^2 = \frac{1}{b}x^2.$$

This is impossible unless $-\frac{2}{a} = \frac{1}{b}$, so that $a = -2b$. But then x is arbitrary, so we do not get any reasonable solution (not surprisingly, since there is no optimum point).

19. First determine the stationary points of f in the interior of the rectangle, by determining where $\nabla f = \mathbf{0}$ on the interior and examining the discriminant at those points. Next, use four separate applications of the Lagrange multiplier method, one to optimize f subject to $x = a$, one subject to $x = b$, one subject to $y = c$, and one subject to $y = d$. Each time, however, we must be careful to allow only solutions in the appropriate range given by the boundary rectangle. The union of the extrema from these two processes gives the set of extrema on or within the given rectangle.

21. First determine the stationary points of f in the interior of the triangle, by determining where $\nabla f = \mathbf{0}$ on the interior and examining the discriminant at those points. Next, use three separate applications of the Lagrange multiplier method, one along each edge of the triangle. Determine the equation of the line marking each edge of the triangle, and optimize f using Lagrange multipliers with that line, set to 0, as a constraint. Each time, however, we must be careful to allow only solutions in the appropriate range given by the boundary triangle. The union of the extrema from these two processes gives the set of extrema on or within the given triangle.

23. If the maximum or minimum occurred in the interior of the region, then it would occur at a critical point. Since f is differentiable, that critical point would be a stationary point. But $\nabla f(x,y) \neq \mathbf{0}$, so there are no stationary points. So the maximum or minimum, if it occurs, must occur on the boundary of the region. Theorem 12.49 assures us that such a maximum and minimum do in fact occur.

Skills

25. Let $g(x,y) = x^2 + 4y^2 - 16$. Note that a maximum and a minimum must exist by Theorem 12.49 since the region defined by the constraint, $x^2 + 4y^2 = 16$, is an ellipse, which is closed and bounded. Then we have

$$\nabla f(x,y) = \lambda \nabla g(x,y), \text{ or } \langle 1, 1 \rangle = \lambda \langle 2x, 8y \rangle.$$

Together with the constraint equation, that gives

$$2x\lambda = 1, \qquad 8y\lambda = 1, \qquad x^2 + 4y^2 = 16.$$

The first two equations give $x = 4y$, and then the third equation gives $(4y)^2 + 4y^2 = 16$, or $20y^2 = 16$, so that $y = \pm\frac{2}{\sqrt{5}}$ and $x = \pm\frac{8}{\sqrt{5}}$. Since

$$f\left(\frac{8}{\sqrt{5}}, \frac{2}{\sqrt{5}}\right) = \frac{10}{\sqrt{5}} = 2\sqrt{5}, \qquad f\left(-\frac{8}{\sqrt{5}}, -\frac{2}{\sqrt{5}}\right) = -\frac{10}{\sqrt{5}} = -2\sqrt{5},$$

the first point is the maximum and the second is the minimum.

27. Let $g(x,y) = x^2 + 4y^2 - 16$. Note that a maximum and a minimum must exist by Theorem 12.49 since the region defined by the constraint, $x^2 + 4y^2 = 16$, is an ellipse, which is closed and bounded. Then we have

$$\nabla f(x,y) = \lambda \nabla g(x,y), \text{ or } \langle y, x \rangle = \lambda \langle 2x, 8y \rangle.$$

Together with the constraint equation, that gives

$$y = 2x\lambda, \qquad x = 8y\lambda, \qquad x^2 + 4y^2 = 16.$$

From the first two equations we get $\frac{y}{2x} = \frac{x}{8y}$, so that $8y^2 = 2x^2$ and thus $x = \pm 2y$. Plugging this into the third equation gives $(\pm 2y)^2 + 4y^2 = 8y^2 = 16$ so that $y = \pm\sqrt{2}$. We get the four points $(\pm 2\sqrt{2}, \pm\sqrt{2})$. Evaluating f gives

$$f(2\sqrt{2}, \sqrt{2}) = f(-2\sqrt{2}, -\sqrt{2}) = 4, \qquad f(2\sqrt{2}, -\sqrt{2}) = f(-2\sqrt{2}, \sqrt{2}) = -4,$$

so that the maximum of 4 occurs at either of the first two points, and the minimum of -4 at either of the second two.

29. Let $g(x,y,z) = x^2 + 4y^2 + 16z^2 - 64$. Note that a maximum and a minimum must exist by Theorem 12.49 since the region defined by the constraint, $x^2 + 4y^2 + 16z^2 = 64$, is an ellipsoid, which is closed and bounded. Then we have

$$\nabla f(x,y,z) = \lambda \nabla g(x,y,z), \text{ or } \langle 1,1,1 \rangle = \lambda \langle 2x, 8y, 32z \rangle.$$

Together with the constraint equation, that gives

$$2x\lambda = 1, \qquad 8y\lambda = 1, \qquad 32z\lambda = 1, \qquad x^2 + 4y^2 + 16z^2 = 64.$$

From the first three equations we get $2x = 8y = 32z$, so that $x = 16z$ and $y = 4z$. Substituting into the third equation gives

$$(16z)^2 + 4(4z)^2 + 16z^2 = 336z^2 = 64,$$

so that $z = \pm\frac{2}{\sqrt{21}}$. So we get two points:

$$\pm\left(\frac{32}{\sqrt{21}}, \frac{8}{\sqrt{21}}, \frac{2}{\sqrt{21}}\right).$$

Evaluating f gives

$$f\left(\frac{32}{\sqrt{21}}, \frac{8}{\sqrt{21}}, \frac{2}{\sqrt{21}}\right) = \frac{42}{\sqrt{21}} = 2\sqrt{21}, \qquad f\left(-\frac{32}{\sqrt{21}}, -\frac{8}{\sqrt{21}}, -\frac{2}{\sqrt{21}}\right) = -\frac{42}{\sqrt{21}} = -2\sqrt{21},$$

so that the maximum of $2\sqrt{21}$ occurs at the first point, and the minimum of $-2\sqrt{21}$ at the second.

31. Let $g(x,y,z) = x^2 + 4y^2 + 16z^2 - 64$. Note that a maximum and a minimum must exist by Theorem 12.49 since the region defined by the constraint, $x^2 + 4y^2 + 16z^2 = 64$, is an ellipsoid, which is closed and bounded. Then we have

$$\nabla f(x,y,z) = \lambda \nabla g(x,y,z), \text{ or } \langle yz, xz, xy \rangle = \lambda \langle 2x, 8y, 32z \rangle.$$

Together with the constraint equation, that gives

$$2x\lambda = yz, \qquad 8y\lambda = xz, \qquad 32z\lambda = xy, \qquad x^2 + 4y^2 + 16z^2 = 64.$$

Note that none of x, y, or z can be zero since the first three equations then force the other two to be zero as well, and $(0,0,0)$ does not lie on the constraint curve. Then from the first three equations we get

$$\frac{yz}{2x} = \frac{xz}{8y} = \frac{xy}{32z},$$

so that $8y^2z = 2x^2z$ and $32xz^2 = 8xy^2$. Simplifying these equations gives $4y^2 = x^2$ and $4z^2 = y^2$. Thus $x^2 = 4y^2 = 4(4z^2) = 16z^2$. Substituting into the third equation gives

$$16z^2 + 4(4z^2) + 16z^2 = 48z^2 = 64,$$

so that $z = \pm\frac{2}{\sqrt{3}}$. So we get eight points:

$$\left(\pm\frac{8}{\sqrt{3}}, \pm\frac{4}{\sqrt{3}}, \pm\frac{2}{\sqrt{3}} \right).$$

Now, f evaluates to $\frac{64}{3\sqrt{3}}$ if either one or three plus signs are chosen, and to $-\frac{64}{3\sqrt{3}}$ if either zero or two plus signs are chosen. Thus four of these points give the maximum of $\frac{64}{3\sqrt{3}}$ and the other four give the minimum of $-\frac{64}{3\sqrt{3}}$.

33. We wish to minimize $(x - 0)^2 + (y - 0)^2 = x^2 + y^2$ subject to the constraint $x^3 + y^3 = 1$. So let $f(x,y) = x^2 + y^2$ and $g(x,y) = x^3 + y^3 - 1$. Then we have

$$\nabla f(x,y) = \lambda \nabla g(x,y), \text{ or } \langle 2x, 2y \rangle = \lambda \left\langle 3x^2, 3y^2 \right\rangle.$$

Together with the constraint equation, this gives

$$2x = 3x^2\lambda, \qquad 2y = 3y^2\lambda, \qquad x^3 + y^3 = 1.$$

$x = 0$ satisfies the first equation; the third equation then gives $y = 1$. Similarly, $y = 0$ satisfies the second equation; the third equation then gives $x = 1$. If neither x nor y is zero, then the first and second equations together give $\frac{3x^2}{2x} = \frac{3y^2}{2y}$ so that $x = y$; then from the third equation we get $2x^3 = 1$ so that $x = y = \frac{1}{\sqrt[3]{2}}$. So we get a total of three points:

$$(1,0), \qquad (0,1), \qquad \left(\frac{1}{\sqrt[3]{2}}, \frac{1}{\sqrt[3]{2}} \right).$$

Evaluating f gives

$$f(1,0) = 1, \qquad f(0,1) = 1,$$

$$f\left(\frac{1}{\sqrt[3]{2}}, \frac{1}{\sqrt[3]{2}} \right) = \frac{1}{\sqrt[3]{4}} + \frac{1}{\sqrt[3]{4}} = \frac{2}{\sqrt[3]{4}} = \frac{\sqrt[3]{8}}{\sqrt[3]{4}} = \sqrt[3]{2} \approx 1.26.$$

So the minimum distance (squared) from the origin to the curve $x^3 + y^3 = 1$ is 1, achieved at the points $(1,0)$ and $(0,1)$, and hence the minimum distance is also $1 = \sqrt{1}$.

35. We wish to minimize $(x - 0)^2 + (y - 0)^2 = x^2 + y^2$ subject to the constraint $\sqrt{x} + \sqrt{y} = 1$. So let $f(x, y) = x^2 + x^2$ and $g(x, y) = \sqrt{x} + \sqrt{y} - 1$. Then we have

$$\nabla f(x, y) = \lambda \nabla g(x, y), \text{ or } \langle 2x, 2y \rangle = \lambda \left\langle \frac{1}{2\sqrt{x}}, \frac{1}{2\sqrt{y}} \right\rangle.$$

Together with the constraint equation, this gives

$$2x = \frac{1}{2\sqrt{x}}\lambda, \qquad 2y = \frac{1}{2\sqrt{y}}\lambda, \qquad \sqrt{x} + \sqrt{y} = 1.$$

Then $4x^{3/2} = 4y^{3/2}$, so that $x = y$. Substituting into the third equation gives $2\sqrt{x} = 1$, so that $x = \frac{1}{4} = y$. So these are the coordinates of the point giving the minimum squared distance from the origin, and it is

$$f\left(\frac{1}{4}, \frac{1}{4}\right) = \frac{1}{16} + \frac{1}{16} = \frac{1}{8}.$$

Thus the minimum distance from the origin is $\frac{1}{2\sqrt{2}}$.

37. We wish to minimize $(x - 0)^2 + (y - 0)^2 = x^2 + y^2$ subject to the constraint $x^n + y^n = 1$, where n is a positive odd integer. So let $f(x, y) = x^2 + x^2$ and $g(x, y) = x^n + y^n - 1$. Then we have

$$\nabla f(x, y) = \lambda \nabla g(x, y), \text{ or } \langle 2x, 2y \rangle = \lambda \left\langle nx^{n-1}, ny^{n-1} \right\rangle.$$

Together with the constraint equation, this gives

$$2x = nx^{n-1}\lambda, \qquad 2y = ny^{n-1}\lambda, \qquad x^n + y^n = 1.$$

This has the two solutions $(0, 1)$ and $(1, 0)$ if either x or y is zero (recall that n is odd). Otherwise, neither is zero, and $\frac{nx^{n-1}}{2x} = \frac{ny^{n-1}}{2y}$, so that $x^{n-2} = y^{n-2}$. Since n is odd, $n - 2$ is odd, so that $x = y$. Substituting into the third equation gives $2x^n = 1$, so that $x = y = \frac{1}{\sqrt[n]{2}}$. Then the squared distances of these three points from the origin are

$$f(0, 1) = 0^2 + 1^2 = 1, \qquad f(1, 0) = 1^2 + 0^2 = 1,$$

$$f\left(\frac{1}{\sqrt[n]{2}}, \frac{1}{\sqrt[n]{2}}\right) = \frac{1}{\sqrt[n]{4}} + \frac{1}{\sqrt[n]{4}} = \frac{2}{\sqrt[n]{4}}.$$

The last distance computed above is equal to $\frac{1}{2}$ if $n = 1$, so the minimum squared distance from the origin is $\frac{1}{2}$, achieved at $\left(\frac{1}{2}, \frac{1}{2}\right)$. If $n > 1$, then $\frac{2}{\sqrt[n]{4}} > 1$, so that the minimum distance from the origin is 1, achieved at the points $(1, 0)$ and $(0, 1)$.

39. We wish to minimize $f(x, y, z) = (x - 0)^2 + (y - 0)^2 + (z - 0)^2 = x^2 + y^2 + z^2$ subject to the constraint $g(x, y, z) = xyz - 1 = 0$. Then

$$\nabla f(x, y, z) = \lambda \nabla g(x, y, z), \text{ or } \langle 2x, 2y, 2z \rangle = \lambda \langle yz, xz, xy \rangle.$$

Together with the constraint equation, this gives

$$2x = \lambda yz, \qquad 2y = \lambda xz, \qquad 2z = \lambda xy, \qquad xyz = 1.$$

Note that none of x, y, or z can be zero, since then the point cannot lie on the constraint curve $xyz = 1$. So from the first three equations, we get

$$\frac{yz}{x} = \frac{xz}{y} = \frac{xy}{z},$$

so that $y^2 z = x^2 z$ and $xz^2 = xy^2$. It follows that $x^2 = y^2 = z^2$. So x, y, and z are either equal or are negatives of one another in some combination, and their product is 1. So they are all ± 1, and we get the points

$$(1,1,1), \quad (1,-1,-1), \quad (-1,1,-1), \quad (-1,-1,1).$$

All of these points have distance

$$\sqrt{1^2 + 1^2 + 1^2} = \sqrt{3}$$

from the origin, so this is the minimum distance.

41. We wish to minimize $f(x,y,z) = (x-0)^2 + (y-0)^2 + (z-0)^2 = x^2 + y^2 + z^2$ subject to the constraint $g(x,y,z) = ax + by + cz - d = 0$. Then

$$\nabla f(x,y,z) = \lambda \nabla g(x,y,z), \text{ or } \langle 2x, 2y, 2z \rangle = \lambda \langle a, b, c \rangle.$$

Together with the constraint equation, this gives

$$2x = a\lambda, \quad 2y = b\lambda, \quad 2z = c\lambda, \quad ax + by + cz = d.$$

Then $x = \frac{a}{c}z$ and $y = \frac{b}{c}z$. Plugging these into the constraint equation gives

$$a \cdot \frac{a}{c}z + b \cdot \frac{b}{c}z + cz = d, \text{ or } a^2 z + b^2 z + c^2 z = cd.$$

This gives $z = \frac{cd}{a^2+b^2+c^2}$, so that $x = \frac{ad}{a^2+b^2+c^2}$ and $y = \frac{bd}{a^2+b^2+c^2}$. Thus the minimum distance from this plane to the origin is

$$\frac{1}{a^2+b^2+c^2}\sqrt{c^2 d^2 + a^2 d^2 + b^2 d^2} = \frac{\sqrt{d^2(a^2+b^2+c^2)}}{a^2+b^2+c^2} = \frac{|d|}{\sqrt{a^2+b^2+c^2}}.$$

Note that this is consistent with the formula in Exercise 16, Section 12.6.

43. Using Exercise 42, the point is

$$\left(\alpha + \frac{a(d - a\alpha - b\beta - c\gamma)}{a^2+b^2+c^2}, \ \beta + \frac{b(d - a\alpha - b\beta - c\gamma)}{a^2+b^2+c^2}, \ \gamma + \frac{c(d - a\alpha - b\beta - c\gamma)}{a^2+b^2+c^2} \right).$$

Now,

$$d - a\alpha - b\beta - c\gamma = 4 - 1 \cdot (-1) - 2 \cdot 5 - (-3) \cdot 3 = 4$$
$$a^2 + b^2 + c^2 = 1^2 + 2^2 + 3^2 = 14,$$

so that the point is

$$\left(-1 + \frac{1 \cdot 4}{14}, \ 5 + \frac{2 \cdot 4}{14}, \ 3 + \frac{-3 \cdot 4}{14} \right) = \left(-\frac{5}{7}, \frac{39}{7}, \frac{15}{7} \right).$$

45. On the interior of the square, we find the maxima and minima by using the gradient. We have

$$\nabla f(x,y) = \langle 6x, 10y \rangle.$$

Thus the gradient is zero only at $(0,0)$. At that point, the discriminant is

$$\det H_f(0,0) = f_{xx}(0,0) f_{yy}(0,0) - f_{xy}(0,0)^2 = 60 > 0,$$

and $f_{xx}(0,0) = 6 > 0$, so that $(0,0)$ is a local minimum, with $f(0,0) = 0$.

Along the boundary, consider first $x = -1$. It is simpler to substitute than it is to use Lagrange optimization. We get $f(-1, y) = 3 + 5y^2$, and this function has a local extremum where $f'(-1, y) = 10y = 0$, so at $y = 0$. So $(-1, 0)$ is a local extremum with $f(-1, 0) = 3$. (We will deal with the corners later). On the boundary $x = 1$, we get $f(1, y) = 3 + 5y^2$, so as before $y = 0$ is a local extremum, and $f(1, 0) = 3$. On the boundary $y = -1$, we get $f(x, -1) = 3x^2 + 5$, and $f'(x, -1) = 6x$, so that $x = 0$ is a local extremum, with $f(0, -1) = 5$. Similarly, on the boundary $y = 1$, we get $f(x, 1) = 3x^2 + 5$, and again $y = 0$ is a local extremum with $f(0, 1) = 5$.

Finally, at the corners, we have $f(\pm 1, \pm 1) = 3 + 5 = 8$.

Summarizing, we have the following data:

$$f(0, 0) = 0, \qquad f(\pm 1, 0) = 3, \qquad f(0, \pm 1) = 5, \qquad f(\pm 1, \pm 1) = 8.$$

So f has a maximum value of 8 at $(\pm 1, \pm 1)$ and a minimum value of 0 at $(0, 0)$.

47. On the interior of the square, we find the maxima and minima by using the gradient. We have

$$\nabla f(x, y) = \langle 2x, 1 \rangle,$$

so that clearly the gradient is never zero, so there are no local extrema on the interior.

Along the boundary, it is easier to use substitution than to use Lagrange optimization.

- On the edge between $(1, 0)$ and $(0, 1)$, we have $x + y = 1$ so that $y = 1 - x$. Then $f(x, 1 - x) = x^2 - x + 1$. Since $f'(x) = 2x - 1$, this has a local extremum at $x = \frac{1}{2}$, so that $y = \frac{1}{2}$ as well, so we get the point $\left(\frac{1}{2}, \frac{1}{2} \right)$.

- On the edge between $(0, 1)$ and $(-1, 0)$, we have $y - x = 1$, so that $y = x + 1$. Then $f(x, x + 1) = x^2 + x + 1$. Since $f'(x) = 2x + 1$, this has a local extremum at $x = -\frac{1}{2}$, so that $y = \frac{1}{2}$, and we get the point $\left(-\frac{1}{2}, \frac{1}{2} \right)$.

- On the edge between $(-1, 0)$ and $(0, -1)$, we have $x + y = -1$, so that $y = -1 - x$. Then $f(x, -1 - x) = x^2 - x - 1$. Since $f'(x) = 2x - 1$, this has a local extremum at $x = \frac{1}{2}$, which cannot be on the line segment between $(-1, 0)$ and $(0, -1)$. So we get no point on this edge.

- Finally, on the edge between $(0, -1)$ and $(1, 0)$, we have $x - y = 1$, so that $y = x - 1$. Then $f(x, x - 1) = x^2 + x - 1$. Since $f'(x) = 2x + 1$, this has a local extremum at $x = -\frac{1}{2}$, which cannot be on the line segment between $(0, -1)$ and $(1, 0)$. So we get no point on this edge.

Summarizing, together with the four corners as possible extrema, we have

$$f\left(\frac{1}{2}, \frac{1}{2} \right) = \frac{3}{4}, \quad f\left(-\frac{1}{2}, \frac{1}{2} \right) = \frac{3}{4}, \quad f(1, 0) = 1, \quad f(0, 1) = 1, \quad f(-1, 0) = 1, \quad f(0, -1) = -1.$$

Thus f achieves a maximum value of 1 at the three points $(\pm 1, 0)$ and $(0, 1)$, and a minimum value of -1 at $(0, -1)$.

49. On the interior of the region, we find the maxima and minima by using the gradient. Since the region does not contain points where $y = 0$, the function is continuous and differentiable on the entire region. We have

$$\nabla f(x, y) = \left\langle \frac{1}{y}, -\frac{x}{y^2} \right\rangle.$$

Clearly the gradient is never zero (its first component is not zero), so there are no local extrema on the interior.

Along the boundary, it is easier to use substitution rather than the Lagrange method.

- On the boundary $y = 1$, we get $f(x,1) = x$, so that $f'(x,1) = 1$ and f has no extrema.

- On the boundary $y = 4$, we get $f(x,4) = \frac{x}{4}$, so that $f'(x,4) = \frac{1}{4}$ and again f has no extrema.

- On the boundary $x = -1$, we get $f(-1,y) = -\frac{1}{y}$, so that $f'(-1,y) = \frac{1}{y^2}$, which has no zeros, so that f has no extrema on this edge either.

- Finally, on the boundary $x = 1$, we get $f(1,y) = \frac{1}{y}$, so that $f'(1,y) = -\frac{1}{y^2}$, which has no zeros, so that f has no extrema on this edge either.

Since f has no extrema on the interior or the edges of the region, all that is left are the corner points. Evaluating f here gives

$$f(1,4) = \frac{1}{4}, \qquad f(-1,4) = -\frac{1}{4}, \qquad f(-1,1) = -1, \qquad f(1,1) = 1.$$

Thus f achieves a maximum value of 1 at $(1,1)$, and a minimum value of -1 at $(-1,1)$.

51. If the vertex diagonally opposite the origin is (x,y,z), then the volume of the box is xyz. So we wish to maximize $f(x,y,z) = xyz$ subject to the constraint $g(x,y,z) = ax + by + cz - d = 0$ (and with x, y, z all positive). Using Lagrange optimization, we have

$$\nabla f(x,y,z) = \lambda \nabla g(x,y,z), \text{ or } \langle yz, xz, xy \rangle = \lambda \langle a,b,c \rangle.$$

Together with the constraint equation, this gives

$$yz = a\lambda, \qquad xz = b\lambda, \qquad xy = c\lambda, \qquad ax + by + cz = d.$$

Note that none of x, y, or z can be zero since then the box has zero volume. From the first three equations we get

$$\frac{yz}{a} = \frac{xz}{b} = \frac{xy}{c}.$$

Clearing fractions between the first two and canceling a z gives $by = ax$. Clearing fractions between the second two and canceling an x gives $by = cz$. Thus $ax = by = cz$ and $ax + by + cz = d$, so each of the three terms in the sum must be $\frac{d}{3}$, so that

$$x = \frac{d}{3a}, \qquad y = \frac{d}{3b}, \qquad z = \frac{d}{3c}.$$

These are the sides of the box, which then has volume $\frac{d}{3a} \cdot \frac{d}{3b} \cdot \frac{d}{3c} = \frac{d^3}{27abc}$.

53. We wish to minimize the squared distance function $f(x,y,z) = x^2 + y^2 + z^2$ subject to the pair of constraints $g(x,y,z) = x^2 + y^2 - z^2 = 0$ and $h(x,y,z) = x + 2y - 6 = 0$. Using Lagrange optimization, we have

$$\nabla f(x,y,z) = \lambda \nabla g(x,y,z) + \mu \nabla h(x,y,z), \text{ or } \langle 2x, 2y, 2z \rangle = \lambda \langle 2x, 2y, -2z \rangle + \mu \langle 1,2,0 \rangle.$$

Together with the constraint equations, this gives

$$2x = 2x\lambda + \mu, \quad 2y = 2y\lambda + 2\mu, \quad 2z = -2z\lambda, \quad x^2 + y^2 = z^2, \quad x + 2y = 6.$$

Note that z cannot be zero, as the first constraint equation then forces $x = y = 0$, and $(0,0,0)$ does not lie on the second constraint curve. Thus the third equation forces $\lambda = -1$, so that

$$2x = -2x + \mu, \qquad 2y = -2y + 2\mu, \qquad x^2 + y^2 = z^2, \qquad x + 2y = 6.$$

Thus from the first two equations, $4x = 2y$ so that $y = 2x$. Then $x + 2y = 6$ becomes $5x = 6$, so that $x = \frac{6}{5}$ and then $y = \frac{12}{5}$. Hence $z^2 = \frac{180}{25}$ from the first constraint equation so that $z = \pm\frac{6}{\sqrt{5}}$. Evaluating at these two points gives

$$f\left(\frac{6}{5}, \frac{12}{5}, \frac{6}{\sqrt{5}}\right) = \frac{36}{25} + \frac{144}{25} + \frac{36}{5} = \frac{360}{25} = \frac{72}{5}, f\left(\frac{6}{5}, \frac{12}{5}, -\frac{6}{\sqrt{5}}\right) = \frac{36}{25} + \frac{144}{25} + \frac{36}{5} = \frac{360}{25} = \frac{72}{5}.$$

so that the minimum squared distance is $\frac{72}{5}$ and the minimum distance is $\frac{6\sqrt{10}}{5}$. This is achieved twice, at each of the points above.

Applications

55. Let the dimensions of the base be $x \times y$, and the height be z. The cost of the box is $c(x,y,z) = 2xy + 2(5xz + 5yz) = 2xy + 10xz + 10yz$. The volume of the box is $v(x,y,z) = xyz$. Thus we want to minimize $c(x,y,z)$ subject to the constraint $v(x,y,z) = 10$. Using Lagrange multipliers, we must have

$$\nabla c(x,y,z) = \lambda \nabla v(x,y,z).$$

This translates into the four equations

$$2y + 10z = \lambda yz,$$
$$2x + 10z = \lambda xz,$$
$$10x + 10y = \lambda xy,$$
$$xyz = 10.$$

The first three equations become, solving for λ,

$$\frac{2}{z} + \frac{10}{y} = \frac{2}{z} + \frac{10}{x} = \frac{10}{y} + \frac{10}{x} = \lambda.$$

From the first equation, we have $\frac{10}{y} = \frac{10}{x}$, so that $x = y$. From the second pair we get $\frac{2}{z} = \frac{10}{y}$, so that $y = 5z$. Thus $xyz = 10$ becomes $25z^3 = 10$, so that $z = \left(\frac{2}{5}\right)^{1/3}$ and $x = y = 5\left(\frac{2}{5}\right)^{1/3} = 50^{1/3}$. The box should have a square base of side length $50^{1/3}$ feet and a height of $\left(\frac{2}{5}\right)^{1/3}$ feet.

Proofs

57. Suppose the rectangle has sides of length x and y and has perimeter P. We wish to maximize the area, $f(x,y) = xy$, subject to the constraint that the perimeter, $2x + 2y$ is P, i.e. subject to the constraint $g(x,y) = 2x + 2y - P = 0$. Then using Lagrange optimization we get

$$\nabla f(x,y) = \lambda \nabla g(x,y), \text{ or } \langle y, x \rangle = \lambda \langle 2, 2 \rangle.$$

Together with the constraint equation, this gives

$$y = 2\lambda, \qquad x = 2\lambda, \qquad 2x + 2y = P.$$

The first two equations give $y = x$; substituting into the constraint equation gives $4x = P$ so that $x = \frac{P}{4} = y$ and the rectangle is a square.

59. Let $S(x,y,z)$ be the square of the distance from (x,y,z) to the origin, and $D(x,y,z)$ the distance from (x,y,z) to the origin. Then clearly $S(x,y,z) = D(x,y,z)^2$. Thus if $S(x,y,z)$ is minimized, so is $D(x,y,z)^2$. But $D(x,y,z) \geq 0$, so that if $D(x,y,z)^2$ is minimized, so is $D(x,y,z)$.

61. Translate the diagram by $(-\alpha, -\beta)$. The equation of the circle becomes

$$((x+\alpha) - a)^2 + ((y+\beta) - b)^2 = r^2,$$

and the point (α, β) becomes $(0,0)$. So by Exercise 60, the point closest to the origin on this circle lies on the line between $(0,0)$ and $(a - \alpha, b - \beta)$, which in the original set of coordinates is the line between (α, β) and (a, b).

63. Translate the diagram by $(-\alpha, -\beta, -\gamma)$. The equation of the circle becomes

$$((x+\alpha) - a)^2 + ((y+\beta) - b)^2 + ((z+\gamma) - c)^2 = r^2,$$

and the point (α, β, γ) becomes $(0,0,0)$. So by Exercise 62, the point closest to the origin on this circle lies on the line between $(0,0,0)$ and $(a - \alpha, b - \beta, c - \gamma)$, which in the original set of coordinates is the line between (α, β, γ) and (a, b, c).

65. Choose $A > 0$. Suppose that we showed that $f(x,y,z) = \sqrt[3]{xyz}$ has its maximum value when $x = y = z = A$ when maximized subject to the constraint $\frac{1}{3}(x+y+z) = A$. Then $\sqrt[3]{xyz} = \sqrt[3]{A^3} = A = \frac{1}{3}(x+y+z)$, so that $\sqrt[3]{xyz} \le \frac{1}{3}(x+y+z)$ for all x, y, and z for which it is defined, so in particular for positive x, y, and z. Since this holds for all A, it holds for all positive x, y, and z. So we want to maximize $f(x,y,z) = \sqrt[3]{xyz}$ subject to the constraint $g(x,y,z) = \frac{1}{3}(x+y+z) - A = 0$. We have

$$\nabla f(x,y,z) = \lambda \nabla g(x,y,z), \text{ or } \left\langle \frac{yz}{3(xyz)^{2/3}}, \frac{xz}{3(xyz)^{2/3}}, \frac{xy}{3(xyz)^{2/3}} \right\rangle = \lambda \left\langle \frac{1}{3}, \frac{1}{3}, \frac{1}{3} \right\rangle.$$

Together with the constraint equation (and canceling the 3's from the first three equations), we get the equations

$$\frac{yz}{(xyz)^{2/3}} = \lambda, \qquad \frac{xz}{(xyz)^{2/3}} = \lambda, \qquad \frac{xy}{(xyz)^{2/3}} = \lambda, \qquad \frac{1}{3}(x+y+z) = A.$$

The first three equations give

$$\frac{yz}{(xyz)^{2/3}} = \frac{xz}{(xyz)^{2/3}} = \frac{xy}{(xyz)^{2/3}}, \text{ so that } yz = xz = xy.$$

Since we are assuming that x, y, z are positive, none of them is zero, so that $yz = xz$ gives $x = y$ and $xz = xy$ gives $y = z$. Thus all three are equal, and the constraint equation forces their values to be $x = y = z = A$. Note that this must be a maximum, not a minimum, since for example $x = 0$, $y = 0$, $z = 3A$ gives $f(x,y,z) = 0$.

Thinking Forward

A double summation

Consider the inner sum $\sum_{j=1}^{10} ij^2$ for a particular value of i. For this sum, i is constant, so that the sum is the same as $i\sum_{j=1}^{10} j^2 = i \cdot \frac{10(11)(21)}{6} = 385i$. Then the entire sum becomes

$$\sum_{i=1}^{15}\sum_{j=1}^{10} ij^2 = \sum_{i=1}^{15}\left(i\sum_{j=1}^{10} j^2 \right) = \sum_{i=1}^{15}(385i) = 385\sum_{i=1}^{15} i = 385 \cdot \frac{15 \cdot 16}{2} = 46200.$$

Reordering a double summation

Since there are a finite number of terms, write them out, conceptually, on a piece of paper. There are mn terms. Now reorder the terms so that all the terms with a particular value of j appear together, in order of increasing i. Put all those terms together to get $\sum_{i=1}^{m} f(i,j)$. Now order those sums in order of increasing j, and put them together to get $\sum_{j=1}^{n}\sum_{i=1}^{m} f(i,j)$.

Chapter Review and Self-Test

1. The level curve at $f(x,y) = c$ is the curve $-\frac{2y}{x} = c$, or $-2y = cx$, so that $y = -\frac{c}{2}x$, except that the origin is excluded since $x = 0$ is not in the domain of f. The level curves are:

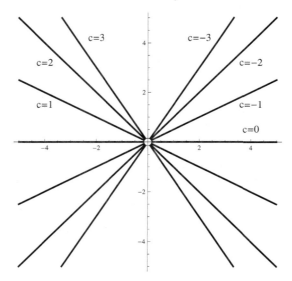

3. The level curves are $x - \sin^{-1} y = c$:

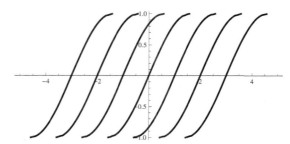

   Moving left to right, these curves are the level curves for $c = -3$, $c = -2$, $c = -1$, $c = 0$, $c = 1$, $c = 2$, and $c = 3$.

5. We can evaluate this limit by simple substitution, since $\frac{x^2+y^2}{x^2-y^2}$ is continuous at $(1,2)$:

$$\lim_{(x,y)\to(1,2)} \frac{x^2 + y^2}{x^2 - y^2} = \frac{1^2 + 2^2}{1^2 - 2^2} = -\frac{5}{3}.$$

7. Use polar coordinates:

$$\lim_{(x,y)\to(0,0)} \frac{x^3 + y^3}{x^2 + y^2} = \lim_{r\to 0} \frac{r^3(\cos^3 \theta + \sin^3 \theta)}{r^2} = \lim_{r\to 0}(r(\cos^3 \theta + \sin^3 \theta)) = 0.$$

   Since the limit is zero independent of θ, it exists and its value is 0.

9. Since numerator and denominator are both continuous everywhere, and since the denominator is nonzero at $(1,0,-1)$, we can simply substitute:

$$\lim_{(x,y,z)\to(1,0,-1)} \frac{\sin(xy)}{x^2 - y^2 + z^2} = \frac{\sin 0}{1^2 - 0^2 + 1^2} = 0.$$

11. When $x = 0$, this becomes

$$\lim_{\substack{(x,y)\to(0,0) \\ x=0}} \frac{x^2 + y^2}{x^2 - y^2} = \lim_{y\to 0} \frac{y^2}{-y^2} = \lim_{y\to 0}(-1) = -1.$$

13. Since the denominator vanishes on the line $x = y$, and since both numerator and denominator are continuous, $f(x, y)$ is continuous for $x \neq y$. Since $f(x, y)$ is not defined for $x = y$, we see that $f(x, y)$ is continuous everywhere on its domain, which is $\{(x, y) \mid x \neq y\}$.

15. $\ln xy$ is defined when $xy > 0$, so the domain of this function is $\{(x, y) \mid x > 0, y > 0\} \cup \{(x, y) \mid x < 0, y < 0\}$, or the first and third quadrants (except for the axes). On that domain, $\ln xy$ is continuous since log is continuous on its domain, so $f(x, y)$ is continuous everywhere on its domain.

17. $f(x, y, z)$ is defined except if (x, y, z) is on the plane $x - 2y + 3z = 0$. Thus this is the domain of f. Further, $f(x, y, z)$ is continuous for all (x, y, z) in its domain, since then both numerator and denominator are continuous functions and the denominator is nonzero.

19. We have

$$f_x(x, y) = \frac{(x^2 + y^2)(1) - (x + y)(2x)}{(x^2 + y^2)^2} = \frac{y^2 - 2xy - x^2}{(x^2 + y^2)^2}$$

$$f_y(x, y) = \frac{(x^2 + y^2)(1) - (x + y)(2y)}{(x^2 + y^2)^2} = \frac{x^2 - 2xy - y^2}{(x^2 + y^2)^2}$$

$$f_{xx}(x, y) = \frac{(x^2 + y^2)^2(-2y - 2x) - (y^2 - 2xy - x^2)\left(2(x^2 + y^2)(2x)\right)}{(x^2 + y^2)^4}$$

$$= \frac{(x^2 + y^2)(-2x - 2y) - (y^2 - 2xy - x^2)(4x)}{(x^2 + y^2)^3}$$

$$= \frac{2x^3 - 6xy^2 + 6x^2y - 2y^3}{(x^2 + y^2)^3}$$

$$f_{yy}(x, y) = \frac{(x^2 + y^2)^2(-2x - 2y) - (x^2 - 2xy - y^2)\left(2(x^2 + y^2)(2y)\right)}{(x^2 + y^2)^4}$$

$$= \frac{(x^2 + y^2)(-2x - 2y) - (x^2 - 2xy - y^2)(4y)}{(x^2 + y^2)^3}$$

$$= \frac{2y^3 - 6x^2y + 6xy^2 - 2x^3}{(x^2 + y^2)^3}$$

$$f_{xy}(x, y) = \frac{(x^2 + y^2)^2(2y - 2x) - (y^2 - 2xy - x^2)\left(2(x^2 + y^2)(2y)\right)}{(x^2 + y^2)^4}$$

$$= \frac{(x^2 + y^2)(2y - 2x) - (y^2 - 2xy - x^2)(4y)}{(x^2 + y^2)^3}$$

$$= \frac{-2y^3 + 6xy^2 + 6x^2y - 2x^3}{(x^2 + y^2)^3}$$

$$f_{yx}(x, y) = \frac{(x^2 + y^2)^2(2x - 2y) - (x^2 - 2xy - y^2)\left(2(x^2 + y^2)(2x)\right)}{(x^2 + y^2)^4}$$

$$= \frac{(x^2 + y^2)(2x - 2y) - (x^2 - 2xy - y^2)(4x)}{(x^2 + y^2)^3}$$

$$= \frac{-2y^3 + 6xy^2 + 6x^2y - 2x^3}{(x^2 + y^2)^3}.$$

21. We have

$$f_x(x,y) = ye^{x^2} + xye^{x^2}(2x) = (2x^2y + y)e^{x^2}, \qquad f_y(x,y) = xe^{x^2}$$
$$f_{xx}(x,y) = 4xye^{x^2} + (2x^2y + y)e^{x^2}(2x) = (4x^3y + 6xy)e^{x^2}, \qquad f_{yy}(x,y) = 0$$
$$f_{xy}(x,y) = (2x^2 + 1)e^{x^2}, \qquad f_{yx}(x,y) = e^{x^2} + xe^{x^2}(2x) = (2x^2 + 1)e^{2x}.$$

23. We have

$$f_x(x,y,z) = z^2e^y, \qquad f_y(x,y,z) = xz^2e^y, \qquad f_z(x,y,z) = 2xze^y,$$
$$f_{xx}(x,y,z) = 0, \qquad f_{yy}(x,y,z) = xz^2e^y, \qquad f_{zz}(x,y,z) = 2xe^y,$$
$$f_{xy}(x,y,z) = z^2e^y, \qquad f_{yx}(x,y,z) = z^2e^y,$$
$$f_{yz}(x,y,z) = 2xze^y, \qquad f_{zy}(x,y,z) = 2xze^y,$$
$$f_{xz}(x,y,z) = 2ze^y, \qquad f_{zx}(x,y,z) = 2ze^y.$$

25. Since $\|\mathbf{v}\| = \sqrt{2^2 + 3^2} = \sqrt{13}$, a unit vector in the direction of $\mathbf{v}$ is $\mathbf{u} = \left\langle \frac{2}{\sqrt{13}}, -\frac{3}{\sqrt{13}} \right\rangle$. Since $f(x,y)$ is differentiable at $(4,3)$ (it is the quotient of a polynomial by a nonzero polynomial), the directional derivative is

$$D_\mathbf{u}f(4,3) = \langle f_x(4,3), f_y(4,3) \rangle \cdot \mathbf{u} = \left\langle -\frac{3}{4^2}, \frac{1}{4} \right\rangle \cdot \left\langle \frac{2}{\sqrt{13}}, -\frac{3}{\sqrt{13}} \right\rangle = -\frac{3}{8\sqrt{13}} - \frac{3}{4\sqrt{13}} = -\frac{9}{8\sqrt{13}}.$$

27. Since $\|\mathbf{v}\| = \sqrt{3^2 + 2^2} = \sqrt{13}$, a unit vector in the direction of $\mathbf{v}$ is $\mathbf{u} = \left\langle \frac{3}{\sqrt{13}}, -\frac{2}{\sqrt{13}} \right\rangle$. Since $f(x,y)$ is differentiable at $(1,2)$ (it is the quotient of a polynomial by a nonzero polynomial), differentiating gives

$$f_x(x,y) = \frac{(x^2 + y^2)(1) - (x + y)(2x)}{(x^2 + y^2)^2} = \frac{y^2 - 2xy - x^2}{(x^2 + y^2)^2}$$
$$f_y(x,y) = \frac{(x^2 + y^2)(1) - (x + y)(2y)}{(x^2 + y^2)^2} = \frac{x^2 - 2xy - y^2}{(x^2 + y^2)^2}.$$

Then the directional derivative is

$$D_\mathbf{u}f(1,2) = \langle f_x(1,2), f_y(1,2) \rangle \cdot \mathbf{u}$$
$$= \left\langle \frac{2^2 - 2 \cdot 1 \cdot 2 - 1^2}{(1^2 + 2^2)^2}, \frac{1^2 - 2 \cdot 1 \cdot 2 - 2^2}{(1^2 + 2^2)^2} \right\rangle \cdot \mathbf{u}$$
$$= \left\langle -\frac{1}{25}, -\frac{7}{25} \right\rangle \cdot \left\langle \frac{3}{\sqrt{13}}, -\frac{2}{\sqrt{13}} \right\rangle$$
$$= -\frac{3}{25\sqrt{13}} + \frac{14}{25\sqrt{13}} = \frac{11}{25\sqrt{13}}.$$

29. Since $\|\mathbf{v}\| = \sqrt{1^2 + 2^2 + 1^1} = \sqrt{6}$, a unit vector in the direction of $\mathbf{v}$ is $\mathbf{u} = \left\langle \frac{1}{\sqrt{6}}, -\frac{2}{\sqrt{6}}, -\frac{1}{\sqrt{6}} \right\rangle$. Now, $f(x,y)$ is differentiable everywhere since it is a polynomial, and differentiating gives

$$f_x(x,y,z) = y^2z^3, \qquad f_y(x,y,z) = 2xyz^3, \qquad f_z(x,y,z) = 3xy^2z^2.$$

Then the directional derivative is

$$D_\mathbf{u}f(0,0,0) = \langle f_x(0,0,0), f_y(0,0,0), f_z(0,0,0) \rangle \cdot \mathbf{u} = \langle 0,0,0 \rangle \cdot \mathbf{u} = 0.$$

31. The gradient is

$$\nabla f(x,y) = \langle f_x(x,y), f_y(x,y) \rangle = \left\langle -\frac{y}{x^2}, \frac{1}{x} \right\rangle.$$

At $(4,3)$, this evaluates to

$$\left\langle -\frac{3}{4^2}, \frac{1}{4} \right\rangle = \left\langle -\frac{3}{16}, \frac{1}{4} \right\rangle.$$

$f(x,y)$ is increasing most rapidly in the direction $\left\langle -\frac{3}{16}, \frac{1}{4} \right\rangle$ at $(4,3)$. We can simplify the direction vector by multiplying through by 16 to get $\langle -3,4 \rangle$.

33. The gradient is

$$\nabla f(x,y) = \langle f_x(x,y), f_y(x,y) \rangle = \langle \sin y, x \cos y \rangle.$$

At $(3, \frac{\pi}{2})$, this evaluates to

$$\left\langle \sin \frac{\pi}{2}, 3 \cos \frac{\pi}{2} \right\rangle = \langle 1,0 \rangle.$$

$f(x,y)$ is increasing most rapidly in the direction $\langle 1,0 \rangle$ at $(3, \frac{\pi}{2})$.

35. The gradient is

$$\nabla f(x,y,z) = \langle f_x(x,y,z), f_y(x,y,z), f_z(x,y,z) \rangle$$
$$= \left\langle \frac{2xy}{z}, \frac{x^2}{z}, -\frac{x^2 y}{z^2} \right\rangle.$$

At $(0,3,-1)$, this evaluates to $\langle 0,0,0 \rangle$, so that $(0,3,-1)$ is a stationary point of f.

37. f is differentiable everywhere, since it is a polynomial. Differentiating gives

$$f_x(x,y) = 4x, \qquad f_y(x,y) = 2y + 1,$$
$$f_{xx}(x,y) = 4, \qquad f_{yy}(x,y) = 2, \qquad f_{xy}(x,y) = 0.$$

The gradient is

$$\nabla f(x,y) = \langle f_x(x,y), f_y(x,y) \rangle = \langle 4x, 2y+1 \rangle.$$

The gradient is zero when both components vanish, which is only at the point $\left(0, -\frac{1}{2}\right)$, so this is the only stationary point.

The discriminant is $H_f(x,y) = 4 \cdot 2 - 0^2 = 8 > 0$, so it is greater than zero everywhere. Since $f_{xx}(x,y) = 4 > 0$ everywhere as well, we see that $\left(0, -\frac{1}{2}\right)$ is a local minimum.

39. f is differentiable everywhere, since it is a polynomial. Differentiating gives

$$f_x(x,y) = 3x^2 - 12x, \qquad f_y(x,y) = 3y^2 + 6y,$$
$$f_{xx}(x,y) = 6x - 12, \qquad f_{yy}(x,y) = 6y + 6, \qquad f_{xy}(x,y) = 0.$$

The gradient is

$$\nabla f(x,y) = \langle f_x(x,y), f_y(x,y) \rangle = \left\langle 3x^2 - 12x, 3y^2 + 6y \right\rangle.$$

The gradient is zero when both components vanish. $3x^2 - 12x = 0$ for $x = 0$ and $x = 4$, while $3y^2 + 6y = 0$ for $y = 0$ and $y = -2$. Thus there are four stationary points. The discriminant is

$H_f(x,y) = (6x - 12)(6y + 6)$. Evaluating H_f and the second derivatives at each of the four points gives the following:

(x_0, y_0)	$\det(H_f(x_0, y_0))$	$f_{xx}(x_0, y_0)$	$f_{yy}(x_0, y_0)$	conclusion
$(0,0)$	-72	-12	6	saddle point
$(4,0)$	72	12	6	local minimum
$(0,-2)$	72	-12	-6	local maximum
$(4,-2)$	-72	12	-6	saddle point

Chapter 13

Double and Triple Integrals

13.1 Double Integrals over Rectangular Regions

Thinking Back

Using summation formulas

For $n = 4$ we have

$$\sum_1^4 k = 1 + 2 + 3 + 4 = 10 = \frac{4(4+1)}{2}$$

$$\sum_1^4 k^2 = 1 + 4 + 9 + 16 = 30 = frac4(4+1)(2 \cdot 4 + 1)6 = \frac{180}{6}.$$

For $n = 7$ we have

$$\sum_1^7 k = 1 + 2 + 3 + 4 + 5 + 6 + 7 = 28 = \frac{7(7+1)}{2} = \frac{56}{2}$$

$$\sum_1^7 k^2 = 1 + 4 + 9 + 16 + 25 + 36 + 49 = 140 = \frac{7(7+1)(2 \cdot 7 + 1)}{6} = \frac{840}{6}.$$

Using the definition to compute a definite integral

With n subintervals, the right edges of the subintervals are $x_k = 1 + \frac{3}{n}k$ for $k = 1, 2, \ldots, n$, and $\Delta x = \frac{3}{n}$. So the value of the integral is

$$
\begin{aligned}
\int_1^4 x^2\, dx &= \lim_{n\to\infty} \sum_{k=1}^n f(x_k)\Delta x \\
&= \lim_{n\to\infty} \frac{3}{n} \sum_{k=1}^n \left(1 + \frac{3}{n}k\right)^2 \\
&= \lim_{n\to\infty} \frac{3}{n} \left(\sum_{k=1}^n 1 + \frac{6}{n}\sum_{k=1}^n k + \frac{9}{n^2}\sum_{k=1}^n k^2 \right) \\
&= \lim_{n\to\infty} \left(\frac{3}{n}\cdot n + \frac{18}{n^2}\cdot\frac{n(n+1)}{2} + \frac{27}{n^3}\cdot\frac{n(n+1)(2n+1)}{6} \right) \\
&= \lim_{n\to\infty} \left(3 + \frac{9n^2 + 9n}{n^2} + \frac{18n^3 + 27n^2 + 9n}{2n^3} \right) \\
&= \lim_{n\to\infty} \left(3 + \frac{9 + 9/n}{1} + \frac{18 + 27/n + 9/n^2}{2} \right) \\
&= 21.
\end{aligned}
$$

Concepts

1. (a) True. Looking carefully at the two sums, in the sum on the right, j plays the role of k and k plays the role of j, so the two sums are the same.

   (b) True. First,

   $$
   \sum_{j=1}^m \sum_{k=1}^n j^2 k^3 = \sum_{j=1}^m \left(j^2 \sum_{k=1}^n k^3 \right),
   $$

   since for a given inner sum, the value of j is constant so that j^2 may be factored out of the sum. Next, note that each term of the outer sum has the constant factor $\sum_{k=1}^n k^3$, so that can be factored out of the outer sum, giving

   $$
   \left(\sum_{k=1}^n k^3 \right)\left(\sum_{j=1}^m j^2 \right) = \left(\sum_{j=1}^m j^2 \right)\left(\sum_{k=1}^n k^3 \right).
   $$

   (c) False. The right-hand side, by an argument identical to that in part (b), is equal to

   $$
   \sum_{j=1}^m \sum_{k=1}^n \left(e^j e^{k^2} \right) = \sum_{j=1}^m \sum_{k=1}^n e^{j+k^2}.
   $$

   (d) True. By Definition 13.4, this integral, defined as the limit of a double Riemann sum, is defined if f is integrable, and continuous functions are integrable.

   (e) False. The integral on the right integrates x from c to d (since dx corresponds to the inner integral) and then y from a to b, which is not the definition of the rectangle. The correct integral is given in part (f).

   (f) True. This is Fubini's Theorem, Theorem 13.7, which holds if f is continuous.

(g) True. We have

$$\int_b^a \int_d^c f(x,y)\,dy\,dx = \int_b^a \left(-\int_c^d f(x,y)\,dy\right) dx$$

$$= -\int_a^b \left(-\int_c^d f(x,y)\,dy\right) dx = \int_a^b \int_c^d f(x,y)\,dy\,dx.$$

(h) True. By Fubini's Theorem, Theorem 13.7, we may interchange the integrals without affecting the answer.

3. We can use an index j going from 3 to 4 for the coefficients, and an index k going from 2 to 4 for the exponents k^2:

$$3e^4 + 3e^9 + 3e^{16} + 4e^4 + 4e^9 + 4e^{16} = \sum_{j=3}^{4} \sum_{k=2}^{4} je^{k^2}.$$

5. For each of the $13 - 3 + 1 = 11$ outer indices (j's), there are $20 - 5 + 1 = 16$ values of k. So there are $11 \cdot 16 = 176$ summands.

7. There are $15 - 2 + 1 = 14$ values of i, and for each of those, there are $17 - 3 + 1 = 15$ values of j; for each of those pairs of values there are $19 - 4 + 1 = 16$ values of k. So there are a total of $14 \cdot 15 \cdot 16 = 3360$ summands.

9. A definite integral is the limit of a Riemann sum for a function $f(x)$ of a single variable over a closed interval $[a, b]$ as the number of subintervals goes to ∞. A double integral is the limit of a double Riemann sum for a function $f(x, y)$ of two variables over a rectangular region that has been subdivided into smaller subrectangles in both the x and y directions as the number of divisions in each direction goes to ∞. These are both limiting processes involving finer subdivisions of the appropriate region.

11. In Exercise 10, the point (x_j^*, y_k^*) was selected arbitrarily in the subrectangle $[x_{j-1}, x_j] \times [y_{k-1}, y_k]$. For a midpoint sum, simply select the point in the center of each subrectangle, which is

$$\left(\frac{x_{j-1} + x_j}{2}, \frac{y_{k-1} + y_k}{2}\right).$$

13. Fubini's Theorem, Theorem 13.7, states that if f is continuous on a rectangle $\mathcal{R} = [a, b] \times [c, d]$, then the iterated integrals are the same regardless of the order of evaluation, and both are equal to the double integral:

$$\iint_{\mathcal{R}} f(x,y)\,dA = \int_a^b \int_c^d f(x,y)\,dy\,dx = \int_c^d \int_a^b f(x,y)\,dx\,dy.$$

15. When evaluating any integral of a single-variable function, we use the Fundamental Theorem of Calculus as follows: if we are integrating $f(x)$ on $[a, b]$, and if F is any antiderivative of f, then the Fundamental Theorem tells us that $\int_a^b f(x)\,dx = [F(x)]_a^b = F(b) - F(a)$. When evaluating an iterated integral, we use that process twice, once for the inner integral and once for the outer integral. So to evaluate $\int_a^b \int_c^d f(x,y)\,dy\,dx$, we find an antiderivative of $f(x,y)$ thought of as a function of y, say $F(x, y)$ (here F depends on x, but for the purposes of the integration we have thought of x as being a constant). Then the value of the inner integral is $F(x, d) - F(x, c)$, by the Fundamental Theorem. Next let $G(x)$ be an antiderivative of $F(x, d) - F(x, c)$ as a function of x; then the Fundamental Theorem again gives us $G(b) - G(a)$ for the value of the outer, and thus of the iterated, integral.

17. We have

$$
\begin{aligned}
\iint_{\mathcal{R}} x^2 y \, dA &= \int_1^3 \int_2^5 x^2 y \, dy \, dx \\
&= \int_1^3 \left(\int_2^5 x^2 y \, dy \right) dx \\
&= \int_1^3 \left[\frac{1}{2} x^2 y^2 \right]_{y=2}^{y=5} dx \\
&= \int_1^3 \left(\frac{25}{2} x^2 - \frac{4}{2} x^3 \right) dx = \int_1^3 \frac{21}{2} x^2 \, dx \\
&= \left[\frac{7}{2} x^3 \right]_{x=1}^{x=3} \\
&= \frac{189}{2} - \frac{7}{6} = 91.
\end{aligned}
$$

19. Suppose we chose to evaluate the integral with respect to x first. Then

$$
\iint_{\mathcal{R}} \cos(xy) \, dA = \int_{\pi/2}^{\pi} \int_{\pi/4}^{\pi/2} \cos(xy) \, dx \, dy
$$

$$
= \int_{\pi/2}^{\pi} \left[\frac{1}{y} \sin(xy) \right]_{x=\frac{\pi}{4}}^{x=\frac{\pi}{2}} = \int_{\pi/2}^{\pi} \left(\frac{\sin\left(\frac{\pi}{2} y\right)}{y} - \frac{\sin\left(\frac{\pi}{4} y\right)}{y} \right) dy.
$$

We do not know how to integrate either term in this integrand (and in fact they do not have antiderivatives that are expressible in terms of elementary functions). The same problem occurs if we choose to integrate first with respect to y; we will get the almost identical integral

$$
\int_{\pi/4}^{\pi/2} \left(\frac{\sin(\pi x)}{x} - \frac{\sin\left(\frac{\pi}{2} x\right)}{x} \right) dx.
$$

21. To evaluate the inner integral

$$
\int_0^1 \frac{x - y}{(x + y)^3} \, dx,
$$

use integration by parts with $u = x - y$ and $dv = \frac{1}{(x+y)^3}$. Then $du = dx$ and $v = -\frac{1}{2(x+y)^2}$, so that

$$
\begin{aligned}
\int_0^1 \frac{x - y}{(x + y)^3} \, dx &= \left[-\frac{x - y}{2(x + y)^2} \right]_{x=0}^{x=1} + \frac{1}{2} \int_0^1 \frac{1}{(x + y)^2} \, dx \\
&= \frac{y - 1}{2(y + 1)^2} - \frac{y}{2y^2} - \frac{1}{2} \left[\frac{1}{x + y} \right]_{x=0}^{x=1} \\
&= \frac{y - 1}{2(y + 1)^2} - \frac{1}{2y} - \frac{1}{2(y + 1)} + \frac{1}{2y} \\
&= \frac{(y - 1) - (y + 1)}{2(y + 1)^2} = -\frac{1}{(y + 1)^2}.
\end{aligned}
$$

Then integrating again gives

$$
\int_0^1 \int_0^1 \frac{x - y}{(x + y)^3} \, dx \, dy = \int_0^1 \left(-\frac{1}{(y + 1)^2} \right) dy = \left[\frac{1}{y + 1} \right]_0^1 = -\frac{1}{2}.
$$

To integrate in the other direction, we want to evaluate for the inner integral

$$\int_0^1 \frac{x-y}{(x+y)^3}\, dy = -\int_0^1 \frac{y-x}{(y+x)^3}\, dy.$$

This is exactly the same as the inner integral above, with the roles of x and y reversed, and with a minus sign. Thus the value of this integral is

$$\frac{1}{(x+1)^2}.$$

So the value of the iterated integral is

$$\int_0^1 \int_0^1 \frac{x-y}{(x+y)^3}\, dy\, dx = \int_0^1 \frac{1}{(x+1)^2}\, dy = \left[-\frac{1}{x+1}\right]_0^1 = \frac{1}{2}.$$

The two results are obviously not the same. This does not violate Fubini's Theorem since $f(x,y)$ is not continuous at $(0,0)$, and it cannot even be made continuous by defining it at $(0,0)$, since

$$\lim_{(h,0)\to(0,0)} f(x,y) = \lim_{(h,0)\to(0,0)} \frac{h}{h^3} = \lim_{(h,0)\to(0,0)} \frac{1}{h^2} = \infty,$$

$$\lim_{(0,h)\to(0,0)} f(x,y) = \lim_{(0,h)\to(0,0)} \frac{-h}{h^3} = \lim_{(0,h)\to(0,0)} \frac{-1}{h^2} = -\infty.$$

Skills

23. $\displaystyle\sum_{j=1}^{3}\sum_{k=1}^{2} j^k = \sum_{j=1}^{3}(j+j^2) = 1 + 1^2 + 2 + 2^2 + 3 + 3^2 = 20.$

25.

$$\sum_{j=1}^{3}\sum_{k=1}^{4}(3j-4k) = \sum_{j=1}^{3}\left(\sum_{k=1}^{4}3j - \sum_{k=1}^{4}4k\right) = \sum_{j=1}^{3}\left(12j - 4\sum_{k=1}^{4}k\right)$$

$$= \sum_{j=1}^{3}(12j-40) = 12\left(\sum_{j=1}^{3}j\right) - 120 = -48.$$

27. $\displaystyle\sum_{i=1}^{4}\sum_{j=1}^{3}\sum_{k=1}^{2} ij^2k^3 = \left(\sum_{i=1}^{4}i\right)\left(\sum_{j=1}^{3}j^2\right)\left(\sum_{k=1}^{2}k^3\right) = \frac{4(5)}{2}\cdot\frac{3(4)(7)}{6}\cdot(1+8) = 1260.$

29. We let our points $(x_j, y_k) = (0 + j\Delta x, 1 + k\Delta y)$, where there are m subintervals in the x direction and n in the y direction. Then $\Delta x = \frac{2-0}{m} = \frac{2}{m}$ and $\Delta y = \frac{4-1}{n} = \frac{3}{n}$, so that $\Delta A = \frac{6}{mn}$. Choosing

$(x_j^*, y_k^*) = (x_j, y_k)$, the double integral is

$$\iint_{\mathcal{R}} xy\, dA = \lim_{\Delta \to 0} \sum_{j=1}^{m} \sum_{k=1}^{n} f(x_j^*, y_k^*) \Delta A$$

$$= \lim_{m,n \to \infty} \frac{6}{mn} \sum_{j=1}^{m} \sum_{k=1}^{n} \left(\frac{2}{m}j\right)\left(1 + \frac{3}{n}k\right)$$

$$= \lim_{m,n \to \infty} \frac{6}{mn} \sum_{j=1}^{m} \left(\frac{2}{m}j\left(\sum_{k=1}^{n} 1 + \frac{3}{n}\sum_{k=1}^{n} k\right)\right)$$

$$= \lim_{m,n \to \infty} \frac{6}{mn} \sum_{j=1}^{m} \left(\frac{2}{m}j\left(n + \frac{3}{n}\cdot\frac{n(n+1)}{2}\right)\right)$$

$$= \lim_{m,n \to \infty} \left(\frac{12}{m^2 n}\left(n + \frac{3(n+1)}{2}\right)\sum_{j=1}^{m} j\right)$$

$$= \lim_{m,n \to \infty} \left(\frac{12}{m^2 n}\cdot\frac{5n+3}{2}\cdot\frac{m(m+1)}{2}\right)$$

$$= \lim_{m \to \infty} \left(\frac{3(m+1)}{m}\left(\lim_{n \to \infty} \frac{5n+3}{n}\right)\right)$$

$$= \lim_{m \to \infty} \left(\frac{3m+3}{m}\cdot 5\right) = 15.$$

31. We let our points $(x_j, y_k) = (-2 + j\Delta x, -1 + k\Delta y)$, where there are m subintervals in the x direction and n in the y direction. Then $\Delta x = \frac{2-(-2)}{m} = \frac{4}{m}$ and $\Delta y = \frac{1-(-1)}{n} = \frac{2}{n}$, so that $\Delta A = \Delta x \Delta y = \frac{8}{mn}$. Choosing $(x_j^*, y_k^*) = (x_j, y_k)$, the double integral is

$$\iint_{\mathcal{R}} xy^3\, dA = \lim_{\Delta \to 0} \sum_{j=1}^{m} \sum_{k=1}^{n} f(x_j^*, y_k^*) \Delta A$$

$$= \lim_{m,n \to \infty} \frac{8}{mn} \sum_{j=1}^{m} \sum_{k=1}^{n} \left(-2 + \frac{4}{m}j\right)\left(-1 + \frac{2}{n}k\right)^3$$

$$= \lim_{m,n \to \infty} \frac{8}{mn} \sum_{k=1}^{n} \sum_{j=1}^{m} \left(-1 + \frac{2}{n}k\right)^3\left(-2 + \frac{4}{m}j\right)$$

$$= \lim_{m,n \to \infty} \frac{8}{mn} \sum_{k=1}^{n} \left(\left(-1 + \frac{2}{n}k\right)^3 \sum_{j=1}^{m} \left(-2 + \frac{4}{m}j\right)\right)$$

$$= \lim_{m,n \to \infty} \frac{8}{mn} \sum_{k=1}^{n} \left(\left(-1 + \frac{2}{n}k\right)^3\left(-2m + \frac{4}{m}\sum_{j=1}^{m} j\right)\right)$$

$$= \lim_{m,n \to \infty} \frac{8}{mn} \sum_{k=1}^{n} \left(\left(-1 + \frac{2}{n}k\right)^3\left(-2m + \frac{4}{m}\cdot\frac{m(m+1)}{2}\right)\right)$$

$$= \lim_{m,n \to \infty} \frac{8}{mn}(-2m + 2m + 2) \sum_{k=1}^{n} \left(-1 + \frac{2}{n}k\right)^3$$

$$= \lim_{m,n \to \infty} \left(\frac{16}{m}\left(\frac{1}{n}\sum_{k=1}^{n} \left(-1 + \frac{2}{n}k\right)^3\right)\right)$$

$$= 0$$

since the limit as $m \to \infty$ of $\frac{16}{m}$ is zero.

33. Integrating first with respect to x, we get

$$\iint_{\mathcal{R}} xy\, dA = \int_1^4 \int_0^2 xy\, dx\, dy = \int_1^4 \left[\frac{1}{2}x^2 y\right]_{x=0}^{x=2} dy = \int_1^4 2y\, dy = \left[y^2\right]_1^4 = 15.$$

35. Integrating first with respect to x, we get

$$\iint_{\mathcal{R}} xy^3\, dA = \int_{-1}^1 \int_{-2}^2 xy^3\, dx\, dy = \int_{-1}^1 \left[\frac{1}{2}x^2 y^3\right]_{x=-2}^{x=2} dy = \int_{-1}^1 0\, dy = 0.$$

37. Integrating first with respect to x, we get

$$\iint_{\mathcal{R}} (3 - x + 4y)\, dA = \int_{-1}^3 \int_0^1 (3 - x + 4y)\, dx\, dy = \int_{-1}^3 \left[3x - \frac{1}{2}x^2 + 4xy\right]_{x=0}^{x=1} dy$$

$$= \int_{-1}^3 \left(\frac{5}{2} + 4y\right) dy = \left[\frac{5}{2}y + 2y^2\right]_{-1}^3 = 26.$$

39. Integrating first with respect to x gives

$$\iint_{\mathcal{R}} (2 - 3x^2 + y^2)\, dA = \int_3^5 \int_{-3}^2 (2 - 3x^2 + y^2)\, dx\, dy = \int_3^5 \left[2x - x^3 + xy^2\right]_{x=-3}^{x=2} dy$$

$$= \int_3^5 (-25 + 5y^2)\, dy = \left[-25y + \frac{5}{3}y^3\right]_3^5 = \frac{340}{3}.$$

41. Integrating first with respect to x gives

$$\iint_{\mathcal{R}} \sin(x + 2y)\, dA = \int_0^{\pi/2} \int_0^{\pi} \sin(x + 2y)\, dx\, dy = \int_0^{\pi/2} [-\cos(x + 2y)]_{x=0}^{x=\pi}\, dy$$

$$= \int_0^{\pi/2} (\cos 2y - \cos(\pi + 2y))\, dy = \left[\frac{1}{2}(\sin 2y - \sin(\pi + 2y))\right]_0^{\pi/2} = 0.$$

43. Integrating first with respect to y gives

$$\iint_{\mathcal{R}} xe^{xy}\, dA = \int_0^1 \int_0^{\ln 5} xe^{xy}\, dy\, dx.$$

Do the inner integration using the substitution $u = xy$, so that $du = x\, dy$ and thus

$$\int_0^{\ln 5} xe^{xy}\, dy = \int_{y=0}^{y=\ln 5} e^u\, du = [e^{xy}]_{y=0}^{y=\ln 5} = 5^x - 1.$$

Then the original problem becomes

$$\iint_{\mathcal{R}} xe^{xy}\, dA = \int_0^1 (5^x - 1)\, dx = \left[\frac{1}{\ln 5}5^x - x\right]_0^1 = \frac{5}{\ln 5} - 1 - \frac{1}{\ln 5} = \frac{4}{\ln 5} - 1.$$

45. Integrate first with respect to y to get

$$\iint_{\mathcal{R}} \frac{x}{x + y}\, dA = \int_1^4 \int_0^3 \frac{x}{x + y}\, dy\, dx = \int_1^4 [x \ln|x + y|]_{y=0}^{y=3}\, dx = \int_1^4 (x \ln|x + 3| - x \ln|x|)\, dx.$$

Integrate each of these using integration by parts with $u = \ln|x|$ (or $\ln|x+3|$) and $dv = x\,dx$, so that $du = \frac{1}{x}$ (or $\frac{1}{x+3}$) and $v = \frac{1}{2}x^2$. Then

$$\int_1^4 (x \ln|x+3| - x \ln|x|)\,dx = \left(\left[\frac{1}{2}x^2 \ln|x+3|\right]_1^4 - \frac{1}{2}\int_1^4 \frac{x^2}{x+3}\,dx\right) - \left(\left[\frac{1}{2}x^2 \ln|x|\right]_1^4 - \frac{1}{2}\int_1^4 x\,dx\right)$$

$$= 8\ln 7 - \frac{1}{2}\ln 4 - \frac{1}{2}\int_1^4 \frac{x^2}{x+3}\,dx - 8\ln 4 + \left[\frac{1}{4}x^2\right]_1^4$$

$$= 8\ln 7 - \frac{17}{2}\ln 4 + 4 - \frac{1}{4} - \frac{1}{2}\int_1^4\left(x - 3 + \frac{9}{x+3}\right)dx$$

$$= 8\ln 7 - \frac{17}{2}\ln 4 + \frac{15}{4} - \frac{1}{2}\left[\frac{1}{2}x^2 - 3x + 9\ln|x+3|\right]_1^4$$

$$= 8\ln 7 - \frac{17}{2}\ln 4 + \frac{15}{4} - 4 + 6 - \frac{9}{2}\ln 7 + \frac{1}{4} - \frac{3}{2} + \frac{9}{2}\ln 4$$

$$= \frac{9}{2} + \frac{7}{2}\ln 7 - 4\ln 4.$$

47. Integrate first with respect to x, giving

$$\iint_{\mathcal{R}} y \sin x\,dA = \int_0^1 \int_0^{\pi/2} y \sin x\,dx\,dy = \int_0^1 [-y\cos x]_0^{\pi/2}\,dy = \int_0^1 y\,dy = \left[\frac{1}{2}y^2\right]_0^1 = \frac{1}{2}.$$

49. Integrate first with respect to y, giving

$$\iint_{\mathcal{R}} y^2 \sin x\,dA = \int_0^\pi \int_0^3 y^2 \sin x\,dy\,dx = \int_0^\pi \left[\frac{1}{3}y^3 \sin x\right]_{y=0}^{y=3}\,dx$$

$$= \int_0^\pi 9\sin x\,dx = [-9\cos x]_0^\pi = 18.$$

51. Integrate first with respect to x, giving

$$\iint_{\mathcal{R}} y\cos(xy)\,dA = \int_0^1 \int_0^{\pi/2} y\cos(xy)\,dx\,dy = \int_0^1 \left[y \cdot \frac{1}{y}\sin(xy)\right]_{x=0}^{x=\pi/2}\,dy$$

$$= \int_0^1 \sin\left(\frac{\pi}{2}y\right)dy = \frac{2}{\pi}\left[-\cos\left(\frac{\pi}{2}y\right)\right]_0^1 = \frac{2}{\pi}.$$

53. Writing $x^2 e^{xy} = x \cdot (xe^{xy})$, the second factor is the y derivative of e^{xy}. So integrate first with respect to y, giving

$$\iint_{\mathcal{R}} x^2 e^{xy} = \int_0^1 \int_0^1 x \cdot (xe^{xy})\,dy\,dx = \int_0^1 x\,[e^{xy}]_{y=0}^{y=1}\,dx = \int_0^1 (xe^x - x)\,dx.$$

To integrate xe^x, use integration by parts with $u = x$ and $dv = e^x\,dx$, so that $du = dx$ and $v = e^x$. Then

$$\int_0^1 (xe^x - x)\,dx = [xe^x]_0^1 - \int_0^1 e^x\,dx - \left[\frac{1}{2}x^2\right]_0^1$$

$$= e - \frac{1}{2} - [e^x]_0^1 = \frac{1}{2}.$$

55. The signed volume is just the integral of the function over the rectangle, which is

$$
\iint_{\mathcal{R}} (3x - 2y^5 + 1) \, dA = \int_0^7 \int_{-4}^6 (3x - 2y^5 + 1) \, dx \, dy
$$

$$
= \int_0^7 \left[\frac{3}{2} x^2 - 2xy^5 + x \right]_{x=-4}^{x=6} dy = \int_0^7 \left(30 - 20y^5 + 10 \right) dy
$$

$$
= \int_0^7 (40 - 20y^5) \, dy = \left[40y - \frac{10}{3} y^6 \right]_0^7 = 280 - \frac{10}{3} \cdot 7^6 = -\frac{1175650}{3}.
$$

57. The signed volume is just the integral of the function over the rectangle. We can rewrite the function as $y^3 e^{xy^2} = y \cdot \left(y^2 e^{xy^2} \right)$, and then the second term is the x derivative of e^{xy^2}. So integrate with respect to x first:

$$
\iint_{\mathcal{R}} y^3 e^{xy^2} \, dA = \int_{-2}^3 \int_0^2 y \cdot \left(y^2 e^{xy^2} \right) dx \, dy = \int_{-2}^3 \left[y e^{xy^2} \right]_{x=0}^{x=2} dy = \int_{-2}^3 \left(y e^{2y^2} - y \right) dy.
$$

The factor of ye^{2y^2} is equal to $\frac{1}{4} \left(4 y e^{2y^2} \right)$, and then the second factor is the y derivative of e^{2y^2}, so that we get

$$
\int_{-2}^3 \left(y e^{2y^2} - y \right) dy = \left[\frac{1}{4} e^{2y^2} - \frac{1}{2} y^2 \right]_{-2}^3 = \frac{1}{4} (e^{18} - e^8) - \frac{5}{2}.
$$

59. $f(x, y) \le 0$ on $[-2, 0] \times [0, 5] \cup [0, 3] \times [-1, 0]$, and $f(x, y) \ge 0$ on $[0, 3] \times [0, 5] \cup [-2, 0] \times [-1, 0]$. So we need four integrals to compute the total volume:

$$
V = \int_0^3 \int_0^5 xy \, dy \, dx + \int_{-2}^0 \int_{-1}^0 xy \, dy \, dx - \int_{-2}^0 \int_0^5 xy \, dy \, dx - \int_0^3 \int_{-1}^0 xy \, dy \, dx
$$

$$
= \int_0^3 \left[\frac{1}{2} xy^2 \right]_{y=0}^{y=5} dx + \int_{-2}^0 \left[\frac{1}{2} xy^2 \right]_{y=-1}^{y=0} dx - \int_{-2}^0 \left[\frac{1}{2} xy^2 \right]_{y=0}^{y=5} dx - \int_0^3 \left[\frac{1}{2} xy^2 \right]_{y=-1}^{y=0} dx
$$

$$
= \int_0^3 \frac{25}{2} x \, dx - \int_{-2}^0 \frac{1}{2} x \, dx - \int_{-2}^0 \frac{25}{2} x \, dx + \int_0^3 \frac{1}{2} x \, dx
$$

$$
= \left[\frac{25}{4} x^2 \right]_0^3 - \left[\frac{1}{4} x^2 \right]_{-2}^0 - \left[\frac{25}{4} x^2 \right]_{-2}^0 + \left[\frac{1}{4} x^2 \right]_0^3
$$

$$
= \frac{225}{4} + 1 + 25 + \frac{9}{4} = 26 + \frac{117}{2} = \frac{169}{2}.
$$

61. Since $\sin x \ge 0$ on $[0, \pi]$, we see that $f(x, y) \ge 0$ on $[0, \pi] \times \left[0, \frac{\pi}{2} \right]$ and $f(x, y) \le 0$ on $[0, \pi] \times \left[\frac{\pi}{2}, \pi \right]$. So we need two integrals to compute the total volume:

$$
V = \int_0^\pi \int_0^{\pi/2} \sin x \cos y \, dy \, dx - \int_0^\pi \int_{\pi/2}^\pi \sin x \cos y \, dy \, dx
$$

$$
= \int_0^\pi [\sin x \sin y]_{y=0}^{y=\pi/2} \, dx - \int_0^\pi [\sin x \sin y]_{y=\pi/2}^{y=\pi} \, dx
$$

$$
= \int_0^\pi \sin x \, dx - \int_0^\pi (-\sin x) \, dx = 2 \int_0^\pi \sin x \, dx
$$

$$
= 2 \left[-\cos x \right]_0^\pi = 4.
$$

63. Since $f(x,y)$ is positive everywhere in $\mathcal{R}$, the volume is simply the integral over the region:

$$
\begin{aligned}
V &= \int_1^3 \int_1^5 \left(\frac{x}{y} + \frac{y}{x} \right) dy\, dx \\
&= \int_1^3 \left[x \ln |y| + \frac{1}{2x} y^2 \right]_{y=1}^{y=5} dx \\
&= \int_1^3 \left(x \ln 5 + \frac{25}{2x} - \frac{1}{2x} \right) dx = \int_1^3 \left(x \ln 5 + \frac{12}{x} \right) dx \\
&= \left[\frac{\ln 5}{2} x^2 + 12 \ln |x| \right]_1^3 \\
&= 4 \ln 5 + 12 \ln 3.
\end{aligned}
$$

65. Since the rectangles are $\frac{1}{2}$ unit squares, with area $\Delta A = \frac{1}{4}$, there are three subdivisions in the x direction and two in the y direction. The midpoints are at $x = \frac{1}{4}$, $x = \frac{3}{4}$, and $x = \frac{5}{4}$, and at $y = \frac{1}{4}$ and $y = \frac{3}{4}$. Thus the Riemann sum is

$$
\begin{aligned}
\sum_{j=1}^{3} \sum_{k=1}^{2} f(x_j^*, y_k^*) \Delta A &= \frac{1}{4} \sum_{j=1}^{3} \sum_{k=1}^{2} e^{x_j^* y_k^*} = \frac{1}{4} \left(e^{\frac{1}{4} \cdot \frac{1}{4}} + e^{\frac{1}{4} \cdot \frac{3}{4}} + e^{\frac{3}{4} \cdot \frac{1}{4}} + e^{\frac{3}{4} \cdot \frac{3}{4}} + e^{\frac{5}{4} \cdot \frac{1}{4}} + e^{\frac{5}{4} \cdot \frac{3}{4}} \right) \\
&= \frac{1}{4} \left(e^{1/16} + e^{3/16} + e^{3/16} + e^{9/16} + e^{5/16} + e^{15/16} \right) \approx 2.28811.
\end{aligned}
$$

67. In the x direction, there are three subintervals of width $\frac{\pi}{6}$, with midpoints $\frac{\pi}{12}$, $\frac{3\pi}{12} = \frac{\pi}{4}$, and $\frac{5\pi}{12}$. In the y direction there are two subintervals of width $\frac{1}{2}$, with midpoints $\frac{1}{4}$ and $\frac{3}{4}$. Thus the area of each rectangle is $\Delta A = \frac{\pi}{6} \cdot \frac{1}{2} = \frac{\pi}{12}$. The Riemann sum is

$$
\begin{aligned}
\sum_{j=1}^{3} \sum_{k=1}^{2} f(x_j^*, y_k^*) \Delta A &= \frac{\pi}{12} \sum_{j=1}^{3} \sum_{k=1}^{2} \cos \left(x_j^* y_k^* \right) \\
&= \frac{\pi}{12} \left(\cos \frac{\pi}{48} + \cos \frac{3\pi}{48} + \cos \frac{\pi}{16} + \cos \frac{3\pi}{16} + \cos \frac{5\pi}{48} + \cos \frac{15\pi}{48} \right) \approx 1.38581.
\end{aligned}
$$

Applications

69. The volume of a tank is clearly four times the volume of the fourth of the tank for $x \in [0, 40]$, $y \in [0, 40]$. Placing the surface of the liquid at $z = 0$, the z coordinate of the bottom of the tank at (x,y) is

$$
\begin{aligned}
d(x,y) &= -12 r(x) r(y) \\
&= -12 \begin{cases} \left(\frac{-(x-10)^2}{100} + 1 \right) \left(\frac{-(y-10)^2}{100} + 1 \right), & 0 \le x, y \le 10, \\ \frac{-(x-10)^2}{100} + 1, & 0 \le x \le 10, 10 \le y \le 40, \\ \frac{-(y-10)^2}{100} + 1, & 0 \le y \le 10, 10 \le x \le 40, \\ 1, & 10 \le x, y \le 40. \end{cases}
\end{aligned}
$$

Thus the volume of the tank is 4 times the integral of $-d(x, y)$ over $[0, 40] \times [0, 40]$, which is

$$V = 4 \int_0^{40} \int_0^{40} (12r(x)r(y)) \, dx \, dy$$

$$= 48 \left(\int_0^{10} \int_0^{10} \left(\frac{-(x-10)^2}{100} + 1 \right) \left(\frac{-(y-10)^2}{100} + 1 \right) \, dx \, dy + \right.$$

$$\int_{10}^{40} \int_0^{10} \left(\frac{-(x-10)^2}{100} + 1 \right) \, dx \, dy + \int_0^{10} \int_{10}^{40} \left(\frac{-(y-10)^2}{100} + 1 \right) \, dx \, dy +$$

$$\left. \int_{10}^{40} \int_{10}^{40} 1 \, dx \, dy \right)$$

Evaluate each integral separately. For the first, when performing the inner integral, treat the factor involving y as a constant; we get

$$\int_0^{10} \int_0^{10} \left(\frac{-(x-10)^2}{100} + 1 \right) \left(\frac{-(y-10)^2}{100} + 1 \right) \, dx \, dy$$

$$= \int_0^{10} \left(\left(\frac{-(y-10)^2}{100} + 1 \right) \left[\frac{-(x-10)^3}{300} + x \right]_{x=0}^{x=10} \right) \, dy$$

$$= \int_0^{10} \left(\left(10 - \frac{10}{3} \right) \left(\frac{-(y-10)^2}{100} + 1 \right) \right) \, dy$$

$$= \frac{20}{3} \int_0^{10} \left(\frac{-(y-10)^2}{100} + 1 \right) \, dy$$

$$= \frac{20}{3} \left[\frac{-(y-10)^3}{300} + y \right]_{y=0}^{y=10}$$

$$= \frac{400}{9}.$$

For the second, we get

$$\int_{10}^{40} \int_0^{10} \left(\frac{-(x-10)^2}{100} + 1 \right) \, dx \, dy = \int_{10}^{40} \left[\frac{-(x-10)^3}{300} + x \right]_{x=0}^{x=10} \, dy = \int_{10}^{40} \frac{20}{3} \, dy = 30 \cdot \frac{20}{3} = 200.$$

For the third,

$$\int_0^{10} \int_{10}^{40} \left(\frac{-(y-10)^2}{100} + 1 \right) \, dx \, dy = \int_0^{10} \left(30 \left(\frac{-(y-10)^2}{100} + 1 \right) \right) \, dy = 30 \cdot \frac{20}{3} = 200.$$

Finally, the value of the fourth integral is obviously $30 \cdot 30 = 900$. Thus, the total volume of the tank is

$$48 \left(\frac{400}{9} + 200 + 200 + 900 \right) = \frac{193600}{3} \approx 64533 \text{ ft}^3.$$

Proofs

71. We have

$$\sum_{j=1}^{m} \sum_{k=1}^{n} a_{j,k} = \sum_{j=1}^{m} (a_{j,1} + a_{j,2} + \cdots + a_{j,n})$$

$$= (a_{1,1} + a_{1,2} + \cdots + a_{1,n}) + (a_{2,1} + a_{2,2} + \cdots + a_{2,n}) + \cdots + (a_{m,1} + a_{m,2} + \cdots + a_{m,n}).$$

Since there are a finite number of terms and addition is commutative, we can rearrange them:

$$\sum_{j=1}^{m} \sum_{k=1}^{n} a_{j,k} = (a_{1,1} + a_{1,2} + \cdots + a_{1,n}) + (a_{2,1} + a_{2,2} + \cdots + a_{2,n}) + \cdots + (a_{m,1} + a_{m,2} + \cdots + a_{m,n})$$

$$= (a_{1,1} + a_{2,1} + \cdots + a_{m,1}) + (a_{1,2} + a_{2,2} + \cdots + a_{m,2}) + \cdots + (a_{1,n} + a_{2,n} + \cdots + a_{m,n})$$

$$= \sum_{k=1}^{n} (a_{1,k} + a_{2,k} + \cdots + a_{m,k})$$

$$= \sum_{k=1}^{n} \sum_{j=1}^{m} a_{j,k}.$$

73. By Fubini's Theorem,

$$\iint_{\mathcal{R}} g(x)h(y)\, dA = \int_{a}^{b} \int_{c}^{d} g(x)h(y)\, dy\, dx.$$

Looking at the inner integral, we see that $g(x)$ is a constant, so we can pull it out of the inner integral to get

$$\int_{a}^{b} \left(g(x) \int_{c}^{d} h(y)\, dy \right)\, dx.$$

Looking now at the outer integral, we see that $\int_{c}^{d} h(y)\, dy$ is a constant, so we can pull it out of the outer integral to get

$$\left(\int_{c}^{d} h(y)\, dy \right) \int_{a}^{b} g(x)\, dx = \left(\int_{a}^{b} g(x)\, dx \right) \left(\int_{c}^{d} h(y)\, dy \right).$$

75. Since 1 is a continuous function, Fubini's Theorem applies, and

$$\iint_{\mathcal{R}} 1\, dA = \int_{a}^{b} \int_{c}^{d} 1\, dy\, dx = \int_{a}^{b} [y]_{y=c}^{y=d}\, dx = \int_{a}^{b} (d-c)\, dx = [(d-c)x]_{a}^{b} = (b-a)(d-c).$$

The result of the integral is the area of the rectangle, $(b-a)(d-c)$. This is analogous to the fact that in a single variable,

$$\int_{a}^{b} 1\, dx = b - a,$$

the length of the interval $[a, b]$.

Thinking Forward

Riemann sums for functions of three variables

Let $a_1 < a_2$, $b_1 < b_2$, and $c_1 < c_2$ be real numbers, let $\mathcal{S}$ be the rectangular box defined by

$$\mathcal{S} = \{(x, y, z) \mid a_1 \leq x \leq a_2,\ b_1 \leq y \leq b_2,\ c_1 \leq z \leq c_2\},$$

and let $f(x, y, z)$ be a function defined on $\mathcal{S}$. Divide $[a_1, a_2]$ into l equal subintervals of width Δx, and choose a point x_i^* in the ith subinterval. Divide $[b_1, b_2]$ into m equal subintervals of width Δy, and choose a point y_j^* in the jth subinterval. Divide $[c_1, c_2]$ into n equal subintervals of width Δz, and choose a point z_k^* in the kth subinterval. Let $\Delta A = \Delta x \Delta y \Delta z$ be the volume of each piece of the solid defined by the subintervals. The sum

$$\sum_{i=1}^{l} \sum_{j=1}^{m} \sum_{k=1}^{n} f(x_i^*, y_j^*, z_k^*) \Delta A$$

is a Riemann sum for f on $\mathcal{S}$. By Exercise 72, the sums may be given in any order.

Triple integrals

Let $a_1 < a_2$, $b_1 < b_2$, and $c_1 < c_2$ be real numbers, let S be the rectangular box defined by

$$S = \{(x,y,z) \mid a_1 \le x \le a_2, \ b_1 \le y \le b_2, \ c_1 \le z \le c_2\},$$

and let $f(x,y,z)$ be a function defined on S. Provided that the limit exists, the triple integral of f over S is

$$\iiint_S f(x,y,z)\,dV = \lim_{\Delta \to 0} \sum_{i=1}^{l} \sum_{j=1}^{m} \sum_{k=1}^{n} f(x_i^*, y_j^*, z_k^*)\Delta A$$

where the triple sums are Riemann sums, and where $\Delta = \sqrt{(\Delta x)^2 + (\Delta y)^2 + (\Delta z)^2}$. When the limit exists, the function f is said to be integrable on S. By Exercise 72, the sums in the Riemann sum may be given in any order.

13.2   Double Integrals over General Regions

Thinking Back

Finding the area between two curves

Since $g_1(x) \le g_2(x)$, the curves do not cross, so that the area between them is

$$\int_a^b (g_2(x) - g_1(x))\,dx.$$

Finding the area between two curves

Since $h_1(y) \le h_2(y)$, the curves do not cross, so that the area between them is

$$\int_c^d (h_2(y) - h_1(y))\,dy.$$

Concepts

1. (a) True. If the rectangle is $[a,b] \times [c,d]$, then it is the Type I region bounded on the left by $x = a$, on the right by $x = b$, and below and above by $c \le y \le d$. It is also the Type II region bounded above by $y = d$, below by $y = c$, and to the left and right by $a \le x \le b$.

    (b) False. If the rectangle does not have its sides parallel to the axes, then the functions describing its edges are not simple functions of either x or y (unless you allow functions described piecewise).

    (c) False. For example, see item (b). Such a rectangle can be described as multiple Type I or Type II regions, however.

    (d) True. As a collection of Type I regions, choose the left and right boundaries to be the smallest and largest x coordinates $[a,b]$ of a point in the region. Subdivide $[a,b]$ at each point where there is a "corner" of the region. Then on each subdivision, the upper and lower boundaries will have descriptions as lines, so each one will form a Type I region. The decomposition is similar for Type II regions.

    (e) True. The integral is $\int_a^b \int_{g_1(x)}^{g_2(x)} f(x,y)\,dy\,dx$.

    (f) True. The first integral, for any particular value of y, integrates $f(x,y)$ along the horizontal line between the two edges of the circle at height y. The second integral sums those for all values of y.

(g) False. This is true only if Ω_1 and Ω_2 are disjoint; see Theorem 13.11.

(h) True. Since f is positive, $\iint_{\Omega-\Lambda} f(x,y)\,dA \geq 0$, so by Theorem 13.11,

$$\iint_{\Omega} f(x,y)\,dA = \iint_{\Lambda} f(x,y)\,dA + \iint_{\Omega-\Lambda} f(x,y)\,dA \geq \iint_{\Lambda} f(x,y)\,dA.$$

3. A Type I region is bounded to the left and right by lines $x = a$ and $x = b$ with $a < b$, while the upper and lower boundaries may be arbitrary functions of x on $[a,b]$. A Type II region is a type region rotated on its side: it is bounded above and below by lines $y = c$ and $y = d$ with $c < d$, while the left and right boundaries may be arbitrary functions of y on $[c,d]$.

5. This integral is correct. The inner integral is the integral from $g_1(x)$ to $g_2(x)$ in the expression of $\mathcal{R}$ as a Type I region, and the outer integral is the integral between the x bounds.

7. This integral is incorrect. The inner integral integrates x, rather than y, from 2 to 6, so it is integrating over the rectangle $[2,6] \times [1,4]$ rather than $\mathcal{R} = [1,4] \times [2,6]$. To correct the integral, either reverse dy and dx or reverse the two integral signs.

9. This integral is correct. It says: for each x between 0 and 2, integrate f from $y = 0$ to $y = -x + 2$. This is precisely the description of the region as a Type I region.

11. This integral is correct. Since $y = -x + 2$, we also have $x = -y + 2$, and the integral says: for each y between 0 and 2, integrate f from $x = 0$ to $x = -y + 2$. This is precisely the description of the region as a Type II region.

13. The two curves intersect for $\frac{1}{2}x = \sqrt{x}$, so that $x^2 = 4x$ and $x = 0$ or $x = 4$. As a Type I region, it is bounded by $x = 0$ and $x = 4$ (so that $a = 0$ and $b = 4$), and below and above by $g_1(x) = \frac{1}{2}x$ and $g_2(x) = \sqrt{x}$. Since the curves intersect at $x = 0$ and $x = 4$, these correspond to $y = 0$ and $y = 2$. Also, $y = \frac{1}{2}x$ is the same as $x = 2y$, and $y = \sqrt{x}$ is the same as $x = y^2$. Thus, as a Type II region, it is bounded by $y = 0$ and $y = 2$ (so that $c = 0$ and $d = 2$), and to the left and right by $h_1(y) = y^2$ and $h_2(y) = 2y$.

15. Treating Ω as a Type I region, we get for the area of Ω

$$\iint_{\Omega} 1\,dA = \int_a^b \int_{g_1(x)}^{g_2(x)} 1\,dy\,dx.$$

This is exactly the same as the computation of area in Chapter 4 using a definite integral, since

$$\int_a^b \int_{g_1(x)}^{g_2(x)} 1\,dy\,dx = \int_a^b [y]_{y=g_1(x)}^{y=g_2(x)}\,dx = \int_a^b (g_2(x) - g_1(x))\,dx.$$

17. As a Type I region, we have

$$A = \iint_{\Omega} 1\,dA = \int_0^4 \int_{x/2}^{\sqrt{x}} 1\,dy\,dx = \int_0^4 [y]_{y=x/2}^{y=\sqrt{x}}\,dx = \int_0^4 \left(\sqrt{x} - \frac{x}{2}\right)\,dx = \left[\frac{2}{3}x^{3/2} - \frac{1}{4}x^2\right]_0^4 = \frac{4}{3}.$$

As a Type II region, we have

$$A = \iint_{\Omega} 1\,dA = \int_0^2 \int_{y^2}^{2y} 1\,dx\,dy = \int_0^2 [x]_{x=y^2}^{x=2y}\,dy = \int_0^2 (2y - y^2)\,dy = \left[y^2 - \frac{1}{3}y^3\right]_0^2 = \frac{4}{3}.$$

19. Integrating first with respect to x means that we are regarding this as a Type II region. Solving $y = x^3$ for x gives $x = y^{1/3}$. As a Type II region, it splits up into two pieces, one from $y = -3^3 = -27$ to $y = 0$ and the other from $y = 0$ to $y = 3^3 = 27$. On each piece, the curve is $x = y^{1/3}$, and the area is bounded on one side by this curve and on the other by $x = \pm 3$. and the integral is

$$
\begin{aligned}
A &= \int_{-27}^{0} \int_{-3}^{y^{1/3}} 1 \, dx \, dy + \int_{0}^{27} \int_{y^{1/3}}^{3} 1 \, dx \, dy \\
&= \int_{-27}^{0} [x]_{x=-3}^{x=y^{1/3}} \, dy + \int_{0}^{27} [x]_{x=y^{1/3}}^{x=3} \, dy \\
&= \int_{-27}^{0} \left(y^{1/3} + 3 \right) dy + \int_{0}^{27} \left[3 - y^{1/3} \right] dy \\
&= \left[\frac{3}{4} y^{4/3} + 3y \right]_{-27}^{0} + \left[3y - \frac{3}{4} y^{4/3} \right]_{0}^{27} \\
&= -\frac{3}{4} \cdot (-27)^{4/3} - 3 \cdot (-27) + 3 \cdot 27 - \frac{3}{4} \cdot 27^{4/3} = 162 - \frac{3}{2} \cdot 27^{4/3} = \frac{81}{2}.
\end{aligned}
$$

Integrating first with respect to y, look at this as a Type I region. It splits up into two pieces, one from $x = -3$ to $x = 0$ and the other from $x = 0$ to $x = 3$. On each piece, we integrate from 0 to x^3 (note that since $x^3 < 0$ for $x < 0$ that this integral will be negative on the range $[-3, 0]$). The integral is

$$
\begin{aligned}
A &= \int_{-3}^{0} \int_{x^3}^{0} 1 \, dy \, dx + \int_{0}^{3} \int_{0}^{x^3} 1 \, dy \, dx \\
&= \int_{-3}^{0} [y]_{y=x^3}^{y=0} \, dx + \int_{0}^{3} [y]_{y=0}^{y=x^3} \, dx \\
&= -\int_{-3}^{0} x^3 \, dx + \int_{0}^{3} x^3 \, dx \\
&= -\left[\frac{1}{4} x^4 \right]_{-3}^{0} + \left[\frac{1}{4} x^4 \right]_{0}^{3} \\
&= \frac{1}{4}(-3)^4 + \frac{1}{4} 3^4 = \frac{81}{2}.
\end{aligned}
$$

Note that this is the unsigned area under the curve.

Skills

21. (a) Integrating first with respect to y (and treating this as a Type I region) gives bounds of $x = 0$ and $x = 1$ on the left and right, and of $y = 0$ and $y = e^x$ on the bottom and top, so we get the integral

$$
\iint_{\Omega} f(x, y) \, dA = \int_{0}^{1} \int_{0}^{e^x} f(x, y) \, dy \, dx.
$$

(b) Integrating first with respect to x (and treating this as a Type II region) gives bounds of $y = 0$ and $y = e$ on the bottom and top. Solving $y = e^x$ for x gives $x = \ln y$. Now, the region must be split up into two separate pieces as a Type II region, one from $y = 0$ to $y = 1$, where the left bound is $x = 0$, and one from $y = 1$ to $y = e$, where the left bound is $x = \ln y$. The right bound is $x = 1$ in both cases. Thus we get the integral

$$
\iint_{\Omega} f(x, y) \, dA = \int_{0}^{1} \int_{0}^{1} f(x, y) \, dx \, dy + \int_{1}^{e} \int_{\ln y}^{1} f(x, y) \, dx \, dy.
$$

23.　(a) Integrating first with respect to y (and treating this as a Type I region) gives bounds of $x = -2$ and $x = 2$ on the left and right. We will require two integrals, since from $x = -2$ to $x = 0$ the lower boundary is $y = -\sqrt{4 - x^2}$ and the upper is $y = \sqrt{4 - x^2}$, while from $x = 0$ to $x = 2$ the lower boundary is $y = 0$ and the upper is $y = \sqrt{4 - x^2}$. The integral is

$$\iint_\Omega f(x,y)\, dA = \int_{-2}^{0} \int_{-\sqrt{4-x^2}}^{\sqrt{4-x^2}} f(x,y)\, dy\, dx + \int_{0}^{2} \int_{0}^{\sqrt{4-x^2}} f(x,y)\, dy\, dx.$$

(b) Integrating first with respect to x (and treating this as a Type II region) gives bounds of $y = -2$ and $y = 2$ on the bottom and top. Again the region must be split up into two separate pieces as a Type II region, one from $y = -2$ to $y = 0$, where the left bound is $x = -\sqrt{4 - y^2}$ and the right bound is $x = 0$, and one from $y = 0$ to $y = 2$, where the left bound is $x = -\sqrt{4 - y^2}$ and the right bound is $x = \sqrt{4 - y^2}$. The integral is

$$\iint_\Omega f(x,y)\, dA = \int_{-2}^{0} \int_{-\sqrt{4-y^2}}^{0} f(x,y)\, dx\, dy + \int_{0}^{2} \int_{-\sqrt{4-y^2}}^{\sqrt{4-y^2}} f(x,y)\, dx\, dy.$$

25.　A sketch of the given region is

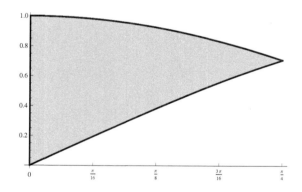

(a) Integrating first with respect to y (and treating this as a Type I region) gives bounds of $x = 0$ and $x = \frac{\pi}{4}$ on the left and right, and of $\sin x$ and $\cos x$ on the bottom and top, so the integral is

$$\iint_\Omega f(x,y)\, dA = \int_{0}^{\pi/4} \int_{\sin x}^{\cos x} f(x,y)\, dy\, dx.$$

(b) Integrating first with respect to x (and treating this as a Type II region) gives bounds of $y = 0$ and $y = 1$ on the bottom and top. The two curves cross at $x = \frac{\pi}{4}$, where $y = \sin x = \frac{\sqrt{2}}{2}$. Solving both equations for x gives $x = \sin^{-1} y$ and $x = \cos^{-1} y$. Thus we get two integrals:

$$\iint_\Omega f(x,y)\, dA = \int_{0}^{\sqrt{2}/2} \int_{0}^{\sin^{-1} y} f(x,y)\, dx\, dy + \int_{\sqrt{2}/2}^{1} \int_{0}^{\cos^{-1} y} f(x,y)\, dx\, dy.$$

27. A sketch of the given region is

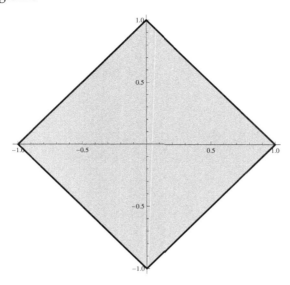

(a) Integrating first with respect to y (and treating this as a Type I region) gives bounds of $x = -1$ and $x = 1$ on the left and right. From $x = -1$ to $x = 0$, the bottom bound is $y = -1 - x$ and the top is $y = 1 + x$, while from $x = 0$ to $x = 1$ the bottom bound is $y = -1 + x$ and the top is $y = 1 - x$. So we get two integrals:

$$\iint_\Omega f(x,y)\,dA = \int_{-1}^0 \int_{-1-x}^{1+x} f(x,y)\,dy\,dx + \int_0^1 \int_{-1+x}^{1-x} f(x,y)\,dy\,dx.$$

(b) Integrating first with respect to x (and treating this as a Type II region) gives bounds of $y = -1$ and $y = 1$ on the bottom and top. From $y = -1$ to $y = 0$, the left bound is $x = -1 - y$ and the right is $x = 1 + y$, while from $y = 0$ to $y = 1$ the left bound is $x = -1 + y$ and the right is $x = 1 - y$. So we get two integrals:

$$\iint_\Omega f(x,y)\,dA = \int_{-1}^0 \int_{-1-y}^{1+y} f(x,y)\,dx\,dy + \int_0^1 \int_{-1+y}^{1-y} f(x,y)\,dx\,dy.$$

29. A sketch of the given region is

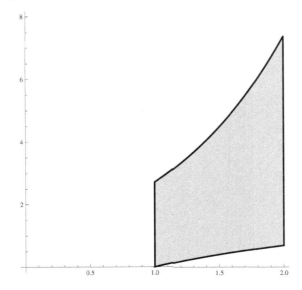

Integrating first with respect to x, and treating this as a Type II region, gives a lower bound of $y = 0$ and an upper bound of $y = e^2$. The lower curve, $y = \ln x$, intersects the right boundary at $(2, \ln 2)$, while the upper curve $y = e^x$ intersects the left boundary at $(1, e)$. The integral must thus be split into three separate integrals. In the first, the left boundary is $x = 1$ while the right is $x = e^y$. In the second, the left boundary is $x = 1$ and the right is $x = 2$. In the third, the left boundary is $x = \ln y$ and the right is $x = 2$. The integral is:

$$\int_0^{\ln 2} \int_1^{e^y} f(x,y)\, dx\, dy + \int_{\ln 2}^{e} \int_1^2 f(x,y)\, dx\, dy + \int_e^{e^2} \int_{\ln y}^2 f(x,y)\, dx\, dy.$$

31. A sketch of the given region is

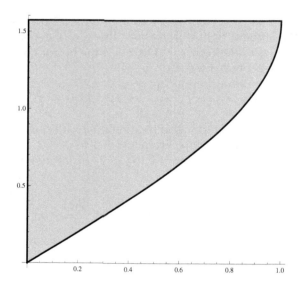

Integrating first with respect to y, and treating this as a Type I region, gives a left bound of $x = 0$. The right bound is the x coordinate corresponding to $y = \frac{\pi}{2}$, and it is at the point $\left(\sin \frac{\pi}{2}, \frac{\pi}{2}\right)$. Thus the right boundary is $x = \sin \frac{\pi}{2} = 1$. The curved boundary on the region is $x = \sin y$, or $y = \sin^{-1} x$, so the lower bound on the integral is $y = \sin^{-1} x$ and the upper bound is $y = \frac{\pi}{2}$, and the integral is

$$\int_0^1 \int_{\sin^{-1} x}^{\pi/2} f(x,y)\, dy\, dx.$$

33. A sketch of the given region is

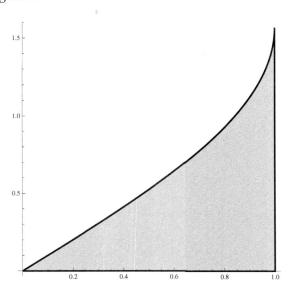

Integrating first with respect to y, and treating this as a Type I region, gives a left bound of $x = 0$. The right bound is the x coordinate corresponding to $y = \frac{\pi}{2}$, and it is at the point $\left(\sin\frac{\pi}{2}, \frac{\pi}{2}\right)$. Thus the right boundary is $x = \sin\frac{\pi}{2} = 1$. The curved boundary on the region is $x = \sin y$, or $y = \sin^{-1} x$, so the lower bound on the integral is $y = 0$ and the upper bound is $y = \sin^{-1} x$, and the integral is

$$\int_0^1 \int_0^{\sin^{-1} x} f(x, y) \, dy \, dx.$$

35. Using Exercise 21(a), we get

$$V = \int_0^1 \int_0^{e^x} (10 - 2x + y) \, dy \, dx = \int_0^1 \left[10y - 2xy + \frac{1}{2}y^2\right]_{y=0}^{y=e^x} dx = \int_0^1 \left(10e^x - 2xe^x + \frac{1}{2}e^{2x}\right) dx.$$

The terms in this integral may be easily handled except for the middle one, which requires integration by parts with $u = x$ and $dv = e^x \, dx$, so that $du = dx$ and $v = e^x$. Then we get

$$V = \int_0^1 \left(10e^x - 2xe^x + \frac{1}{2}e^{2x}\right) dx$$

$$= \int_0^1 10e^x \, dx - 2\int_0^1 xe^x \, dx + \frac{1}{2}\int_0^1 e^{2x} \, dx$$

$$= [10e^x]_0^1 - 2\left([xe^x]_0^1 - \int_0^1 e^x \, dx\right) + \frac{1}{4}\left[e^{2x}\right]_0^1$$

$$= 10e - 10 - 2\left(e - [e^x]_0^1\right) + \frac{1}{4}e^2 - \frac{1}{4}$$

$$= 8e + \frac{1}{4}e^2 - \frac{41}{4} + 2e - 2 = 10e + \frac{1}{4}e^2 - \frac{49}{4}.$$

37. Using Exercise 23(a) gives

$$V = \int_{-2}^0 \int_{-\sqrt{4-x^2}}^{\sqrt{4-x^2}} \sqrt{4 - x^2 - y^2} \, dy \, dx + \int_0^2 \int_0^{\sqrt{4-x^2}} \sqrt{4 - x^2 - y^2} \, dy \, dx.$$

We first find an antiderivative with respect to y of $\sqrt{4 - x^2 - y^2}$; we can then use that for each of the above integrals. Use the substitution $y = \sqrt{4 - x^2}\sin u$; then $dy = \sqrt{4 - x^2}\cos u\, du$ and we get

$$\int \sqrt{4 - x^2 - y^2}\, dy = \int \sqrt{(4 - x^2) - (4 - x^2)\sin^2 u} \cdot \sqrt{4 - x^2}\cos u\, du$$

$$= \int \sqrt{(4 - x^2)\cos^2 u} \cdot \sqrt{4 - x^2}\cos u\, du$$

$$= \int (4 - x^2)\cos^2 u\, du = (4 - x^2)\int \cos^2 u\, du$$

$$= (4 - x^2)\int \frac{1}{2}(1 + \cos 2u)\, du$$

$$= (4 - x^2) \cdot \frac{1}{2}\left(u + \frac{1}{2}\sin 2u\right) + C$$

$$= \frac{1}{2}(4 - x^2)\left(\sin^{-1}\frac{y}{\sqrt{4 - x^2}} + \sin u \cos u\right) + C$$

$$= \frac{1}{2}(4 - x^2)\left(\sin^{-1}\frac{y}{\sqrt{4 - x^2}} + \frac{y}{\sqrt{4 - x^2}} \cdot \frac{\sqrt{4 - x^2 - y^2}}{\sqrt{4 - x^2}}\right) + C$$

$$= \frac{1}{2}(4 - x^2)\sin^{-1}\frac{y}{\sqrt{4 - x^2}} + \frac{1}{2}y\sqrt{4 - x^2 - y^2} + C.$$

Evaluating this at $y = \sqrt{4 - x^2}$ gives

$$\frac{1}{2}(4 - x^2)\sin^{-1}\frac{\sqrt{4 - x^2}}{\sqrt{4 - x^2}} + \frac{1}{2}\sqrt{4 - x^2}\sqrt{4 - x^2 - (4 - x^2)} = (4 - x^2)\frac{\pi}{4},$$

and evaluating at $y = -\sqrt{4 - x^2}$ gives

$$\frac{1}{2}(4 - x^2)\sin^{-1}\frac{-\sqrt{4 - x^2}}{\sqrt{4 - x^2}} - \frac{1}{2}\sqrt{4 - x^2}\sqrt{4 - x^2 - (4 - x^2)} = -(4 - x^2)\frac{\pi}{4},$$

where we choose $C = 0$. Note also that evaluating at $y = 0$ gives zero.

Now, returning to the original integral, we get

$$V = \int_{-2}^{0}\int_{-\sqrt{4 - x^2}}^{\sqrt{4 - x^2}} \sqrt{4 - x^2 - y^2}\, dy\, dx + \int_{0}^{2}\int_{0}^{\sqrt{4 - x^2}} \sqrt{4 - x^2 - y^2}\, dy\, dx$$

$$= \int_{-2}^{0}\left((4 - x^2)\frac{\pi}{4} + (4 - x^2)\frac{\pi}{4}\right)dx + \int_{0}^{2}\left((4 - x^2)\frac{\pi}{4}\right)dx$$

$$= \frac{\pi}{2}\int_{-2}^{0}(4 - x^2)\, dx + \frac{\pi}{4}\int_{0}^{2}(4 - x^2)\, dx$$

$$= \frac{\pi}{2}\left[4x - \frac{1}{3}x^3\right]_{-2}^{0} + \frac{\pi}{4}\left[4x - \frac{1}{3}x^3\right]_{0}^{2}$$

$$= \frac{\pi}{2}\left(8 - \frac{8}{3}\right) + \frac{\pi}{4}\left(8 - \frac{8}{3}\right)$$

$$= 4\pi.$$

39. Using the integral from part (b) of Exercise 26 gives

$$V = \int_{0}^{2}\int_{-y}^{y} \sin x \cos y\, dx\, dy = \int_{0}^{2}\left[-\cos x \cos y\right]_{x=-y}^{x=y}\, dx = \int_{0}^{2}\left(-\cos^2 y + \cos^2 y\right)dx = 0.$$

41. The projection of this solid onto the xy plane can be found by setting $z = 0$; it is the triangle bounded by the x and y axes and the line $3x + 4y = 12$. Regarding this as a Type I region, we have left and right bounds $x = 0$ and $x = 4$, a lower bound of $y = 0$, and an upper bound of $y = \frac{1}{4}(12 - 3x) = 3 - \frac{3}{4}x$. Further, the upper plane of the surface is $z = \frac{1}{6}(12 - 3x - 4y) = 2 - \frac{1}{2}x - \frac{2}{3}y$, so the volume is

$$
\begin{aligned}
V &= \int_0^4 \int_0^{3-(3/4)x} \left(2 - \frac{1}{2}x - \frac{2}{3}y \right) dy\, dx \\
&= \int_0^4 \left[2y - \frac{1}{2}xy - \frac{1}{3}y^2 \right]_{y=0}^{y=3-(3/4)x} dx \\
&= \int_0^4 \left[2\left(3 - \frac{3}{4}x\right) - \frac{1}{2}x\left(3 - \frac{3}{4}x\right) - \frac{1}{3}\left(3 - \frac{3}{4}x\right)^2 \right] dx \\
&= \int_0^4 \left(\frac{3}{16}x^2 - \frac{3}{2}x + 3 \right) dx \\
&= \left[\frac{1}{16}x^3 - \frac{3}{4}x^2 + 3x \right]_0^4 \\
&= 4 - 12 + 12 = 4.
\end{aligned}
$$

43. Integrate first with respect to x to get

$$
\begin{aligned}
V &= \int_0^2 \int_1^2 (8 - x^2 - y^2)\, dx\, dy \\
&= \int_0^2 \left[8x - \frac{1}{3}x^3 - xy^2 \right]_{x=1}^{x=2} dy \\
&= \int_0^2 \left(16 - \frac{8}{3} - 2y^2 - 8 + \frac{1}{3} + y^2 \right) dy \\
&= \int_0^2 \left(\frac{17}{3} - y^2 \right) dy \\
&= \left[\frac{17}{3}y - \frac{1}{3}y^3 \right]_0^2 \\
&= \frac{26}{3}.
\end{aligned}
$$

45. The inner integral as given cannot be evaluated, since we do not know how to integrate $\sqrt{1 + x^3}$ with respect to x. However, the left boundary of this region is the curve $x = \sqrt{y}$, or $y = x^2$. A diagram of the region is

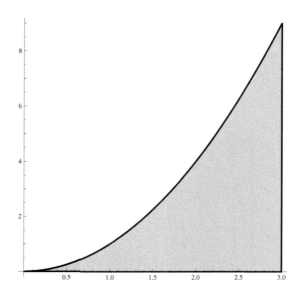

We can also consider this as a Type I region with left and right edges $x = 0$ and $x = 3$, lower edge $y = 0$, and upper edge $y = x^2$. This results in a simpler inner integral, and then an outer integral which we know how to evaluate:

$$\int_0^3 \int_0^{x^2} \sqrt{1 + x^3}\, dy\, dx = \int_0^3 \left[y\sqrt{1 + x^3} \right]_{y=0}^{y=x^2} dx = \int_0^3 x^2 \sqrt{1 + x^3}\, dx$$

$$= \left[\frac{2}{9}(1 + x^3)^{3/2} \right]_0^3 = \frac{2}{9}28^{3/2} - \frac{2}{9} = \frac{2}{9}\left(56\sqrt{7} - 1 \right).$$

47. The integral as written gives a rather ugly expression for the result of the inner integral, as an antiderivative of $\sec y$ is $\ln |\sec y + \tan y|$. However, the region is

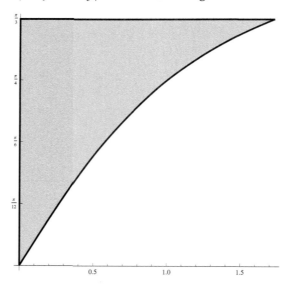

which has as its curved boundary $y = \tan^{-1} x$, or $x = \tan y$. Treating this as a Type II region results in a much simpler integration problem:

$$\int_0^{\pi/3} \int_0^{\tan y} \sec y\, dx\, dy = \int_0^{\pi/3} \left[x \sec y \right]_{x=0}^{x=\tan y} dy = \int_0^{\pi/3} \sec y \tan y\, dy = \left[\sec y \right]_0^{\pi/3} = 2 - 1 = 1.$$

49. The region is

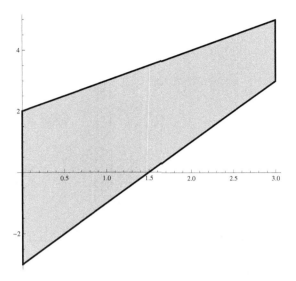

Keeping this as a Type I region, we have

$$\int_0^3 \int_{x+2}^{2x-3} (x^2 + 3xy)\, dy\, dx = \int_0^3 \left[x^2 y + \frac{3}{2} xy^2 \right]_{y=x+2}^{y=2x-3} dx$$

$$= \int_0^3 \left(x^2(2x-3) + \frac{3}{2} x(2x-3)^2 - x^2(x+2) - \frac{3}{2} x(x+2)^2 \right) dx$$

$$= \int_0^3 \left(\frac{11}{2} x^3 - 29x^2 + \frac{15}{2} x \right) dx$$

$$= \left[\frac{11}{8} x^4 - \frac{29}{3} x^3 + \frac{15}{4} x^2 \right]_0^3$$

$$= -\frac{927}{8}.$$

51. The region is

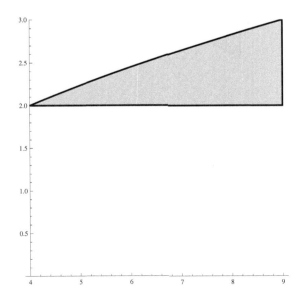

The curved edge of this region is $y = \sqrt{x}$, or $x = y^2$. While this integration can be performed without too much trouble as is, the integration bounds are somewhat simpler if we recast this as a Type II region with boundaries $y = 2$, $y = 3$, $x = y^2$, and $x = 9$:

$$\int_2^3 \int_{y^2}^9 (x^3 + y^2)\, dx\, dy = \int_2^3 \left[\frac{1}{4}x^4 + xy^2\right]_{x=y^2}^{x=9} dy$$

$$= \int_2^3 \left(\frac{6561}{4} + 9y^2 - \frac{1}{4}y^8 - y^4\right) dy$$

$$= \left[\frac{6561}{4}y + 3y^3 - \frac{1}{36}y^9 - \frac{1}{5}y^5\right]_2^3$$

$$= \frac{101027}{90}.$$

53. The region is

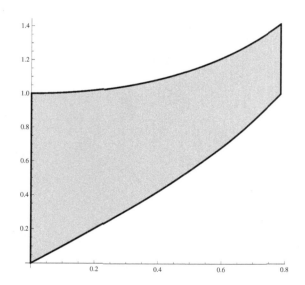

Leaving it as a Type I region, we get

$$\int_0^{\pi/4} \int_{\tan x}^{\sec x} y\, dy\, dx = \int_0^{\pi/4} \left[\frac{1}{2}y^2\right]_{y=\tan x}^{y=\sec x} dx = \frac{1}{2}\int_0^{\pi/4} (\sec^2 x - \tan^2 x)\, dx$$

$$= \frac{1}{2}\int_0^{\pi/4} 1\, dx = \frac{\pi}{8},$$

using the identity $\tan^2 x + 1 = \sec^2 x$.

55. The region is

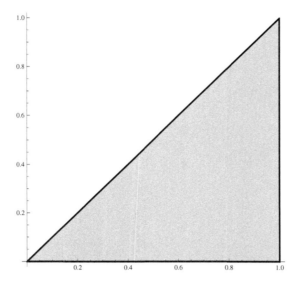

The region is a triangle; reverse the order of integration, giving a Type I region, in order to get an inner integral that we can evaluate:

$$\int_0^1 \int_0^x e^{x^2}\, dy\, dx = \int_0^1 \left[y e^{x^2} \right]_{y=0}^{y=x} dx = \int_0^1 x e^{x^2}\, dx = \left[\frac{1}{2} e^{x^2} \right]_0^1 = \frac{1}{2}(e-1).$$

57. The region is

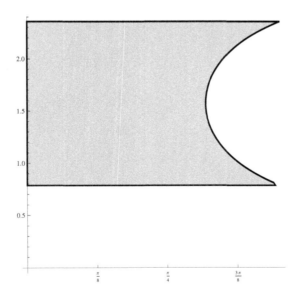

Leaving it as a Type II region, we get

$$\int_{\pi/4}^{3\pi/4} \int_0^{\csc y} \csc y\, dx\, dy = \int_{\pi/4}^{3\pi/4} \left[x \csc y \right]_{x=0}^{x=\csc y} dy = \int_{\pi/4}^{3\pi/4} \csc^2 y\, dy = \left[-\cot y \right]_{\pi/4}^{3\pi/4} = 2.$$

59. Since we do not know how to integrate xe^{x^3} with respect to x, we try a Type I region so that the first integration is with respect to y. As a Type I region, the left and right boundaries are $x = 0$ and $x = 2$, the bottom is $y = 0$, and the top is $y = x$. Then the integral is

$$\int_0^2 \int_0^x xe^{x^3}\, dy\, dx = \int_0^2 \left[yxe^{x^3} \right]_{y=0}^{y=x} dx = \int_0^2 x^2 e^{x^3}\, dx = \left[\frac{1}{3} e^{x^3} \right]_0^2 = \frac{1}{3}(e^8 - 1).$$

61. The curve $x = y^2$ in the first quadrant is the graph of $y = \sqrt{x}$; this and the graph of $y = x^2$ intersect at $x = 0$ and $x = 1$, and $x^2 \le \sqrt{x}$ everywhere on $[0, 1]$. So as a Type I region, this integral is

$$\int_0^1 \int_{x^2}^{\sqrt{x}} x^2 y^3\, dy\, dx = \int_0^1 \left[\frac{1}{4} x^2 y^4 \right]_{y=x^2}^{y=\sqrt{x}} dx$$

$$= \frac{1}{4} \int_0^1 (x^4 - x^{10})\, dx$$

$$= \frac{1}{4} \left[\frac{1}{5} x^5 - \frac{1}{11} x^{11} \right]_0^1$$

$$= \frac{1}{4} \cdot \frac{6}{55} = \frac{3}{110}.$$

Applications

63. This region intersects the plane $z = 0$ on the curve $\frac{1}{10}|x| + \frac{1}{10}|y| = 3$, or $|x| + |y| = 30$. Since this region is symmetric about both axes and the origin, we need only find its volume in the first quadrant and multiply by 4. There, the region is bounded by the two axes and the line $x + y = 30$, and the depth is $3 - \frac{1}{10}x - \frac{1}{10}y$ meters. Thus the total volume of the lagoon is

$$4 \int_0^{30} \int_0^{30-x} \left(3 - \frac{1}{10}x - \frac{1}{10}y \right) dy\, dx = 4 \int_0^{30} \left[3y - \frac{1}{10}xy - \frac{1}{20}y^2 \right]_{y=0}^{y=30-x} dx$$

$$= 4 \int_0^{30} \left(90 - 3x - \frac{1}{10}x(30 - x) - \frac{1}{20}(30 - x)^2 \right) dx$$

$$= 4 \int_0^{30} \left(45 - 3x + \frac{1}{20}x^2 \right) dx$$

$$= 4 \left[45x - \frac{3}{2}x^2 + \frac{1}{60}x^3 \right]_0^{30}$$

$$= 4\,(1350 - 1350 + 450) = 1800 \text{ cubic meters.}$$

Proofs

65. From Definition 13.4, the double integral is

$$\iint_{\mathcal{R}} \alpha f(x, y)\, dA = \lim_{\Delta \to 0} \sum_{j=1}^m \sum_{k=1}^n \alpha f(x_j^*, y_k^*) \Delta A$$

$$= \lim_{\Delta \to 0} \left(\alpha \sum_{j=1}^m \sum_{k=1}^n f(x_j^*, y_k^*) \Delta A \right)$$

$$= \alpha \lim_{\Delta \to 0} \sum_{j=1}^m \sum_{k=1}^n f(x_j^*, y_k^*) \Delta A$$

$$= \alpha \iint_{\mathcal{R}} f(x, y)\, dA.$$

67. Let $\mathcal{R} = \{(x,y) \mid a \leq x \leq b \text{ and } c \leq y \leq d\}$ be a rectangle containing Ω, and let

$$F(x,y) = \begin{cases} f(x,y), & (x,y) \in \Omega, \\ 0, & (x,y) \notin \Omega. \end{cases}$$

Then by Definition 13.9 and Exercise 65, since $\alpha F = \alpha f$ on Ω and $\alpha F = 0$ outside of Ω,

$$\iint_\Omega \alpha f(x,y)\, dA = \iint_\mathcal{R} \alpha F(x,y)\, dA = \alpha \iint_\mathcal{R} F(x,y)\, dA = \alpha \iint_\Omega f(x,y)\, dA.$$

69. The volume of this pyramid is a certain integral over the region bounded by $x = 0$ and $x = a$, and by $y = 0$ and $y = b - \frac{b}{a}x$. The function being integrated is the equation of the plane passing through the three points $(a,0,0)$, $(0,b,0)$ and $(0,0,c)$. That plane has normal vector

$$\langle -a, b, 0 \rangle \times \langle -a, 0, c \rangle = \langle bc, ac, ab \rangle,$$

so that its equation is $bc(x - a) + acy + abz = 0$, or $z = -\frac{bcx + acy}{ab} + c$. Thus the volume of the pyramid is

$$\int_0^a \int_0^{b-(b/a)x} \left(c - \frac{c}{a}x - \frac{c}{b}y \right) dy\, dx = \int_0^a \left[cy - \frac{c}{a}xy - \frac{c}{2b}y^2 \right]_{y=0}^{y=b-(b/a)x} dx$$

$$= \int_0^a \left(c\left(b - \frac{b}{a}x\right) - \frac{c}{a}x\left(b - \frac{b}{a}x\right) - \frac{c}{2b}\left(b - \frac{b}{a}x\right)^2 \right) dx$$

$$= \int_0^a \left(\frac{bc}{2a^2}x^2 - \frac{bc}{a}x + \frac{bc}{2} \right) dx$$

$$= \left[\frac{bc}{6a^2}x^3 - \frac{bc}{2a}x^2 + \frac{bc}{2}x \right]_0^a$$

$$= \frac{a^3 bc}{6a^2} - \frac{a^2 bc}{2a} + \frac{abc}{2}$$

$$= \frac{1}{6}abc.$$

Thinking Forward

Three iterated integrals

The innermost integral, $\int_{c_1}^{c_2} dz$, evaluates to $[z]_{c_1}^{c_2} = c_2 - c_1$. This is a constant which we can pull out of the integration, leaving

$$(c_2 - c_1) \int_{a_1}^{a_2} \int_{b_1}^{b_2} dy\, dx.$$

We know that the remaining integral is $(b_2 - b_1)(a_2 - a_1)$, so that the original integral is $(a_2 - a_1)(b_2 - b_1)(c_2 - c_1)$, which is the volume of the given rectangular solid.

Three more iterated integrals

The value of this integral is

$$
\begin{aligned}
V &= \int_0^2 \int_0^{-\frac{3}{2}x+3} \int_0^{4-2x-\frac{4}{3}y} dz\, dy\, dx \\
&= \int_0^2 \int_0^{-\frac{3}{2}x+3} [z]_0^{4-2x-\frac{4}{3}y} dy\, dx \\
&= \int_0^2 \int_0^{-\frac{3}{2}x+3} \left(4 - 2x - \frac{4}{3}y\right) dy\, dx \\
&= \int_0^2 \left[4y - 2xy - \frac{2}{3}y^2\right]_0^{-\frac{3}{2}x+3} dx \\
&= \int_0^2 \left(4\left(3 - \frac{3}{2}x\right) - 2x\left(3 - \frac{3}{2}x\right) - \frac{2}{3}\left(3 - \frac{3}{2}x\right)^2\right) dx \\
&= \int_0^2 \left(\frac{3}{2}x^2 - 6x + 6\right) dx \\
&= \left[\frac{1}{2}x^3 - 3x^2 + 6x\right]_0^2 \\
&= 4 - 12 + 12 = 4.
\end{aligned}
$$

This is the volume of the pyramid bounded by the plane $z = 4 - 2x - \frac{4}{3}y$, or $6x + 4y + 3z = 12$, and the coordinate planes. The range of integration for y is found by looking at the line of intersection of the plane with the plane $z = 0$, which is $4 - 2x - \frac{4}{3}y = 0$, or $y = -\frac{3}{2}x + 3$.

13.3   Double Integrals using Polar Coordinates

Thinking Back

▷ By Theorem 9.13, this is the area bounded by the function $\cos 3\theta$ for $\theta \in [0, \pi]$. This is a three-petaled rose, and we can find its area by tripling the area of one petal. A graph of this rose is

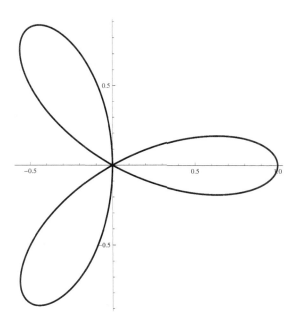

A single petal is traced by (for example) $\frac{\pi}{6} \leq \theta \leq \frac{\pi}{2}$, so the area of the rose is

$$3 \cdot \frac{1}{2} \int_{\pi/6}^{\pi/2} \cos^2 3\theta \, d\theta = \frac{3}{4} \int_{\pi/6}^{\pi/2} (1 + \cos 6\theta) \, d\theta$$

$$= \frac{3}{4} \left[\theta + \frac{1}{6} \sin 6\theta \right]_{\pi/6}^{\pi/2}$$

$$= \frac{3}{4} \left(\frac{\pi}{2} + \frac{1}{6} \sin 3\pi - \frac{\pi}{6} - \frac{1}{6} \sin \pi \right)$$

$$= \frac{\pi}{4}.$$

▷ By Theorem 9.13, this is eight times the area bounded by the curve $\cos 2\theta$ for $0 \leq \theta \leq \frac{\pi}{4}$. Now, this region is one half of a single leaf of the four-leaved rose $\cos 2\theta$, so that the integral represents the total area of the rose. A graph of this rose is

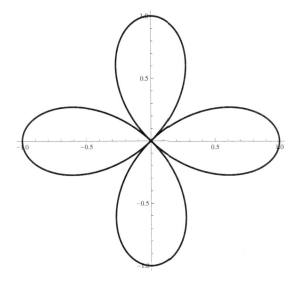

Its area is

$$4 \cdot \int_0^{\pi/4} \cos^2 2\theta \, d\theta = 2 \int_0^{\pi/4} (1 + \cos 4\theta) \, d\theta = 2 \left[\theta + \frac{1}{4} \sin 4\theta \right]_0^{\pi/4} = 2 \left(\frac{\pi}{4} + \frac{1}{4} \sin \pi \right) = \frac{\pi}{2}.$$

▷ The graph of $\frac{1}{2} + \cos \theta$ is a limaçon with an inner loop. From 0 to $\frac{2\pi}{3}$ we trace the upper half of the outer loop, and $\pi \le \theta \le \frac{4\pi}{3}$ traces the upper half of the inner loop. Thus the difference of these two integrals is the area between the inner and outer loop (note that there is no factor of $\frac{1}{2}$, so this is double the area of the difference in just the upper halves, so is the total area between the loops). A graph of the limaçon is

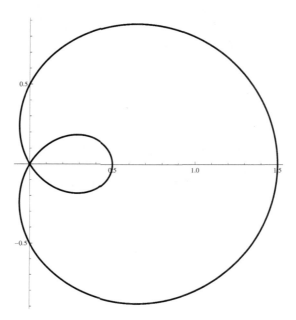

The area between the loops is

$$\int_0^{2\pi/3} \left(\frac{1}{2} + \cos \theta \right)^2 d\theta - \int_\pi^{4\pi/3} \left(\frac{1}{2} + \cos \theta \right)^2 d\theta$$

$$= \int_0^{2\pi/3} \left(\frac{1}{4} + \cos \theta + \cos^2 \theta \right) d\theta - \int_\pi^{4\pi/3} \left(\frac{1}{4} + \cos \theta + \cos^2 \theta \right) d\theta$$

$$= \int_0^{2\pi/3} \left(\frac{3}{4} + \cos \theta + \frac{1}{2} \cos 2\theta \right) d\theta - \int_\pi^{4\pi/3} \left(\frac{3}{4} + \cos \theta + \frac{1}{2} \cos 2\theta \right) d\theta$$

$$= \left[\frac{3}{4}\theta + \sin \theta + \frac{1}{4} \sin 2\theta \right]_0^{2\pi/3} - \left[\frac{3}{4}\theta + \sin \theta + \frac{1}{4} \sin 2\theta \right]_\pi^{4\pi/3}$$

$$= \frac{\pi}{2} + \frac{\sqrt{3}}{2} + \frac{1}{4} \cdot \left(-\frac{\sqrt{3}}{2} \right) - \pi + \frac{\sqrt{3}}{2} - \frac{1}{4} \frac{\sqrt{3}}{2} + \frac{3}{4}\pi$$

$$= \frac{\pi}{4} + \frac{3\sqrt{3}}{4}.$$

▷ The two curves $3 \sin \theta$ and $1 + \sin \theta$ intersect at $\theta = \frac{\pi}{6}$ and $\theta = \frac{5\pi}{6}$, so this is the area of the lune bounded by the two graphs (the graph of $1 + \sin \theta$ is in gray):

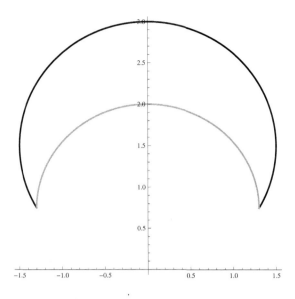

This area is

$$\frac{1}{2} \int_{\pi/6}^{5\pi/6} \left((3\sin\theta)^2 - (1+\sin\theta)^2 \right) d\theta = \frac{1}{2} \int_{\pi/6}^{5\pi/6} \left(8\sin^2\theta - 2\sin\theta - 1 \right) d\theta$$

$$= \frac{1}{2} \int_{\pi/6}^{5\pi/6} \left(4 - 4\cos 2\theta - 2\sin\theta - 1 \right) d\theta$$

$$= \frac{1}{2} \int_{\pi/6}^{5\pi/6} \left(3 - 4\cos 2\theta - 2\sin\theta \right) d\theta$$

$$= \frac{1}{2} \left[3\theta - 2\sin 2\theta + 2\cos\theta \right]_{\pi/6}^{5\pi/6}$$

$$= \frac{1}{2} \left(\frac{5\pi}{2} + \sqrt{3} - \sqrt{3} - \frac{\pi}{2} + \sqrt{3} - \sqrt{3} \right)$$

$$= \pi.$$

Concepts

1. (a) True. These circle sectors correspond to rectangles in cartesian coordinates. See the discussion at the beginning of the section.

   (b) True. Since the width of $\left[\frac{\pi}{4}, \frac{\pi}{2} \right]$ is $\frac{\pi}{2} - \frac{\pi}{4} = \frac{\pi}{4}$, dividing it into four equal subintervals gives each a width of $\frac{\pi}{16}$.

   (c) False. It is $\int_\alpha^\beta \int_0^{f(\theta)} r \, dr \, d\theta$, since the "volume element" in polar coordinates is $r \, dr \, d\theta$. See the discussion at the start of the section.

   (d) False. It is $\int_\alpha^\beta \int_0^{f(\theta)} r^2 \cdot r \, dr \, d\theta = \int_\alpha^\beta \int_0^{f(\theta)} r^3 \, dr \, d\theta$, since the "volume element" in polar coordinates is $r \, dr \, d\theta$. See the discussion in "Double Integrals in Polar Coordinates over General Regions" in this section.

   (e) False. The graph $2\cos\theta$ is traced twice from 0 to 2π. What is desired is $\int_0^\pi \int_0^{2\cos\theta} r \, dr \, d\theta$ for the area of the circle.

   (f) False. The period of this function is 2π, so it is traced once on $[0, 2\pi]$. It has eight petals each of which is traced as θ varies from one multiple of $\frac{\pi}{4}$ to the next.

   (g) True. This is just a change in notation. (But it would be wrong to interpret the second integral as the volume of the solid $\theta^2 + r^2$ lying over the given region, since the factor of r is missing).

(h) True. The region in the first integral is the first quadrant of the circle of radius 2 around the origin. In polar coordinates, this becomes θ from 0 to $\frac{\pi}{2}$ and r from 0 to 2. The integrand is $(x^2 + y^2)^3 = (r^2)^3 = r^6$, so to compute the volume of this solid over the given range we must include a factor of r in the polar integral. This gives r^7 as the integrand.

3. A sketch of a polar rectangle is given in the first figure in this section, along with subdivisions of that rectangle. This is the basic region for integration in the polar plane since when we integrate between constants bounds $0 \leq a \leq r \leq b$ and $\alpha \leq \theta \leq \beta$, we get a region of this form, where the angular width of the region is $\beta - \alpha$ and the difference between the inner and outer radii is $b - a$.

5.

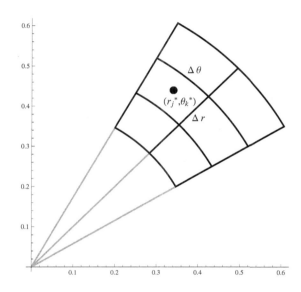

7. Evaluating the inner integral gives

$$\int_\alpha^\beta \int_0^{f(\theta)} dr\, d\theta = \int_\alpha^\beta [r]_0^{f(\theta)}\, d\theta = \int_\alpha^\beta f(\theta)\, d\theta.$$

In a rectangular coordinate system, this represents the area under the graph of $r = f(\theta)$ (here θ is thought of as the horizontal axis and r as the vertical axis). In polar coordinates, however, comparing to Theorem 9.13, we see that this is twice the area under the curve $\sqrt{f(\theta)}$ from $\theta = \alpha$ to $\theta = \beta$, since the area under that curve is

$$\frac{1}{2}\int_\alpha^\beta \left(\sqrt{f(\theta)}\right)^2 d\theta = \frac{1}{2}\int_\alpha^\beta f(\theta)\, d\theta.$$

9. Since the rose is traced once for $\theta \in [0, \pi]$, it is traced twice over the region of integration, so we get twice the area.

11. This is n times the area of the region bounded by $\theta = 0$ and $\theta = \frac{2\pi}{n}$, and by $r = 0$ and $r = R$, so it is n times the area of a wedge of the circle of radius R with angle $\frac{2\pi}{n}$. Such a wedge has an area equal to $\frac{1}{n}$ times the entire area of the circle, so n times it gives the area of the circle. The argument above never used the assumption that n is an integer, so as long as it is a positive real number, the statement is true.

13. Consider $z = x^2$ on $[0, 2]$:

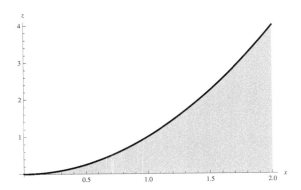

Using the shell method to rotate around the z axis, we get shells of height x^2 and radius x, for $x \in [0, 2]$, so the volume is

$$2\pi \int_0^2 x(x^2)\, dx = 2\pi \left[\frac{1}{4}x^4 \right]_0^2 = 8\pi.$$

15. Since the surface lies over the circle $x^2 + y^2 = 4$, the volume of the solid between the surface and the xy plane is

$$\int_{-2}^{2} \int_{-\sqrt{4-x^2}}^{\sqrt{4-x^2}} \left(\sqrt{x^2 + y^2} \right)^2 dy\, dx = \int_{-2}^{2} \int_{-\sqrt{4-x^2}}^{\sqrt{4-x^2}} (x^2 + y^2)\, dy\, dx.$$

By symmetry, we see that it suffices to integrate over the first quadrant and multiply by 4, so we must evaluate

$$4 \int_0^2 \int_0^{\sqrt{4-x^2}} (x^2 + y^2)\, dy\, dx = \int_0^2 \left[x^2 y + \frac{1}{3}y^3 \right]_{y=0}^{y=\sqrt{4-x^2}} dx$$

$$= 4 \int_0^2 \left(x^2 \sqrt{4 - x^2} + \frac{1}{3}(4 - x^2)^{3/2} \right) dx$$

$$= 4 \int_0^2 \left(x^2 \sqrt{4 - x^2} + \frac{1}{3}(4 - x^2)\sqrt{4 - x^2} \right) dx$$

$$= \frac{4}{3} \int_0^2 \left((2x^2 + 4)\sqrt{4 - x^2} \right) dx.$$

17. The integral from Exercise 16 is straightforward:

$$\int_0^{2\pi} \int_0^2 r^2 \cdot r\, dr\, d\theta = \int_0^{2\pi} \left[\frac{1}{4}r^4 \right]_0^2 d\theta = \int_0^{2\pi} 4\, d\theta = 8\pi.$$

The results from Exercises 13 and 16 are identical.

19. Using polar coordinates, and referring to Exercise 14, note that $g(x, y) = f\left(\sqrt{x^2 + y^2} \right) = f(r)$. Thus we want the volume enclosed by $f(r)$ for $a \le r \le b$, which is

$$\int_0^{2\pi} \int_a^b r f(r)\, dr\, d\theta.$$

Skills

21. This is the area enclosed by the cardioid $r = 1 + \sin\theta$, since that curve is traced once from $\theta = 0$ to $\theta = 2\pi$. A plot of the region is

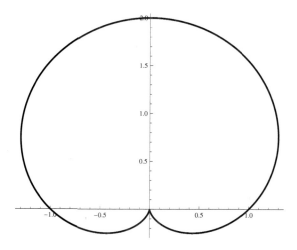

The integral is

$$\int_0^{2\pi} \int_0^{1+\sin\theta} r \, dr \, d\theta = \int_0^{2\pi} \left[\frac{1}{2}r^2\right]_{r=0}^{r=1+\sin\theta} d\theta$$

$$= \frac{1}{2}\int_0^{2\pi} (1 + \sin\theta)^2 \, d\theta$$

$$= \frac{1}{2}\int_0^{2\pi} \left(1 + 2\sin\theta + \sin^2\theta\right) d\theta$$

$$= \frac{1}{2}\int_0^{2\pi} \left(1 + 2\sin\theta + \frac{1}{2}\left(1 - \cos 2\theta\right)\right) d\theta$$

$$= \frac{1}{2}\left[\frac{3}{2}\theta - 2\cos\theta - \frac{1}{4}\sin 2\theta\right]_0^{2\pi}$$

$$= \frac{3}{2}\pi.$$

23. The curve $r = 2 - \sin\theta$ is a limaçon that is traced once from 0 to 2π, or alternatively from $-\frac{\pi}{2}$ to $\frac{3\pi}{2}$. It is symmetric around the y axis, since if (r, θ) is on the curve, then $2 - \sin(\pi - \theta) = 2 - \sin\theta$, so that $(r, \pi - \theta) = (-r, -\theta)$ is also on the curve. So the integral given is just the area under the curve, since it is twice the integral from $-\frac{\pi}{2}$ to $\frac{\pi}{2}$. A plot of the region is

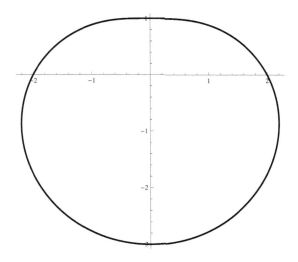

The integral is

$$2 \int_{-\pi/2}^{\pi/2} \int_{0}^{2-\sin\theta} r \, dr \, d\theta = 2 \int_{-\pi/2}^{\pi/2} \left[\frac{1}{2} r^2 \right]_{r=0}^{r=2-\sin\theta} d\theta$$

$$= \int_{-\pi/2}^{\pi/2} (2 - \sin\theta)^2 \, d\theta$$

$$= \int_{-\pi/2}^{\pi/2} \left(4 - 4\sin\theta + \sin^2\theta \right) d\theta$$

$$= \int_{-\pi/2}^{\pi/2} \left(4 - 4\sin\theta + \frac{1}{2}(1 - \cos 2\theta) \right) d\theta$$

$$= \left[\frac{9}{2}\theta + 4\cos\theta - \frac{1}{4}\sin 2\theta \right]_{-\pi/2}^{\pi/2}$$

$$= \frac{9}{2}\pi.$$

25. $r = \sin 3\theta$ is a three-petaled rose, and it is traced once from 0 to π. So this is twice the integral of the area under half of the rose, thus it is the area under the entire rose. A plot of the region is

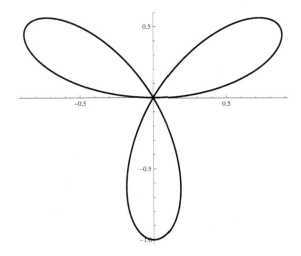

Note that the trace from 0 to $\frac{\pi}{2}$ consists of the loop in the first quadrant plus the half loop in the third quadrant. The integral is

$$2\int_0^{\pi/2}\int_0^{\sin 3\theta} r\,dr\,d\theta = 2\int_0^{\pi/2}\left[\frac{1}{2}r^2\right]_{r=0}^{r=\sin 3\theta} d\theta$$

$$= \int_0^{\pi/2} \sin^2 3\theta\,d\theta$$

$$= \int_0^{\pi/2}\left(\frac{1}{2} - \frac{1}{2}\cos 6\theta\right) d\theta$$

$$= \left[\frac{1}{2}\theta - \frac{1}{12}\sin 6\theta\right]_0^{\pi/2}$$

$$= \frac{\pi}{4}.$$

27. A plot of $r = \frac{\sqrt{2}}{2} + \sin\theta$ is

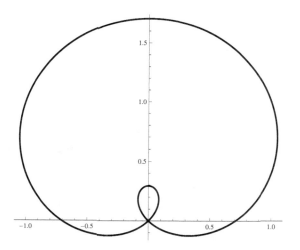

The entire curve is traced once for $\theta \in [0, 2\pi]$, or from $\left[-\frac{\pi}{2}, \frac{3\pi}{2}\right]$. It is symmetric around the y axis, since if (r, θ) is on the curve, then $\frac{\sqrt{2}}{2} + \sin(\pi - \theta) = \frac{\sqrt{2}}{2} + \sin\theta$ and thus $(r, \pi - \theta) = (-r, -\theta)$ is also on the curve. Now, $\theta = -\frac{\pi}{4}$ to $\frac{\pi}{2}$ traces the half of the large loop to the right of the y axis, while $\theta = -\frac{\pi}{2}$ to $-\frac{\pi}{4}$ traces the half of the small loop to the left of the y axis, which has the same area as the half of the small loop to the right of the y axis. So the formula given computes the difference between those two over the entire curve (due to the factor of 2), so it is the area inside the curve but outside the inner loop. This integral is

$$2 \int_{-\pi/4}^{\pi/2} \int_0^{\sqrt{2}/2+\sin\theta} r \, dr \, d\theta - 2 \int_{-\pi/2}^{-\pi/4} \int_0^{\sqrt{2}/2+\sin\theta} r \, dr \, d\theta$$

$$= 2 \int_{-\pi/4}^{\pi/2} \left[\frac{1}{2}r^2 \right]_{r=0}^{r=\sqrt{2}/2+\sin\theta} d\theta - 2 \int_{-\pi/2}^{-\pi/4} \left[\frac{1}{2}r^2 \right]_{r=0}^{r=\sqrt{2}/2+\sin\theta} d\theta$$

$$= \int_{-\pi/4}^{\pi/2} \left(\sqrt{2}/2 + \sin\theta \right)^2 d\theta - \int_{-\pi/2}^{-\pi/4} \left(\sqrt{2}/2 + \sin\theta \right)^2 d\theta$$

$$= \int_{-\pi/4}^{\pi/2} \left(\frac{1}{2} + \sqrt{2}\sin\theta + \sin^2\theta \right) d\theta - \int_{-\pi/2}^{-\pi/4} \left(\frac{1}{2} + \sqrt{2}\sin\theta + \sin^2\theta \right) d\theta$$

$$= \int_{-\pi/4}^{\pi/2} \left(1 + \sqrt{2}\sin\theta - \frac{1}{2}\cos 2\theta \right) d\theta$$

$$\qquad - \int_{-\pi/2}^{-\pi/4} \left(1 + \sqrt{2}\sin\theta - \frac{1}{2}\cos 2\theta \right) d\theta$$

$$= \left[\theta - \sqrt{2}\cos\theta - \frac{1}{4}\sin 2\theta \right]_{-\pi/4}^{\pi/2} - \left[\theta - \sqrt{2}\cos\theta - \frac{1}{4}\sin 2\theta \right]_{-\pi/2}^{-\pi/4}$$

$$= \left(\frac{3\pi}{4} + 1 - \frac{1}{4} \right) - \left(\frac{\pi}{4} - 1 + \frac{1}{4} \right) = \frac{\pi}{2} + \frac{3}{2}.$$

29. Since at $\theta = 0$ and $\theta = \pi$ the curve lies along the x axis, the area bounded by this portion of the curve and the x axis is just the area swept out by the curve between those two angles, so it is

$$\int_0^\pi \int_0^\theta r \, dr \, d\theta = \int_0^\pi \left[\frac{1}{2}r^2 \right]_{r=0}^{r=\theta} d\theta = \frac{1}{2} \int_0^\pi \theta^2 \, d\theta = \frac{1}{2} \left[\frac{1}{3}\theta^3 \right]_0^\pi = \frac{1}{6}\pi^3.$$

31. This limaçon is symmetric around the x axis, since if (r, θ) lies on the curve, then $1 + \sqrt{2}\cos(-\theta) = 1 + \sqrt{2}\cos\theta$, so that $(r, -\theta)$ also lies on the curve. The upper half of the outer loop is traced from $\theta = 0$ to $\theta = \frac{3\pi}{4}$, and the lower half of the inner loop from $\theta = \frac{3\pi}{4}$ to $\theta = \pi$. Thus the total area between the two loops is

$$2 \int_0^{3\pi/4} \int_0^{1+\sqrt{2}\cos\theta} r \, dr \, d\theta - 2 \int_{3\pi/4}^{\pi} \int_0^{1+\sqrt{2}\cos\theta} r \, dr \, d\theta$$

$$= 2 \int_0^{3\pi/4} \left[\frac{1}{2}r^2 \right]_{r=0}^{r=1+\sqrt{2}\cos\theta} d\theta - 2 \int_{3\pi/4}^{\pi} \left[\frac{1}{2}r^2 \right]_{r=0}^{r=1+\sqrt{2}\cos\theta} d\theta$$

$$= \int_0^{3\pi/4} (1 + \sqrt{2}\cos\theta)^2 d\theta - \int_{3\pi/4}^{\pi} (1 + \sqrt{2}\cos\theta)^2 d\theta$$

$$= \int_0^{3\pi/4} (1 + 2\sqrt{2}\cos\theta + 2\cos^2\theta) d\theta - \int_{3\pi/4}^{\pi} (1 + 2\sqrt{2}\cos\theta + 2\cos^2\theta) d\theta$$

$$= \int_0^{3\pi/4} \left(2 + 2\sqrt{2}\cos\theta + \cos 2\theta \right) d\theta - \int_{3\pi/4}^{\pi} \left(2 + 2\sqrt{2}\cos\theta + \cos 2\theta \right) d\theta$$

$$= \left[2\theta + 2\sqrt{2}\sin\theta + \frac{1}{2}\sin 2\theta \right]_0^{3\pi/4} - \left[2\theta + 2\sqrt{2}\sin\theta + \frac{1}{2}\sin 2\theta \right]_{3\pi/4}^{\pi}$$

$$= \left(\frac{3\pi}{2} + 2 - \frac{1}{2} \right) - \left(2\pi - \frac{3\pi}{2} - 2 + \frac{1}{2} \right) = 3 + \pi.$$

33. A plot of the two curves is below, with $r = 3 - 3\sin\theta$ in black:

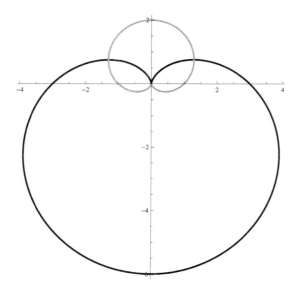

Both curves are symmetric around the y axis, since they both involve sine, for which $\sin(\pi - \theta) = \sin\theta$. They intersect where $3 - 3\sin\theta = 1 + \sin\theta$, or $2 = 4\sin\theta$. This happens when $\theta = \frac{\pi}{6}, \frac{5\pi}{6}$. Now, both curves trace out their right-hand loops for $\theta \in \left[-\frac{\pi}{2}, \frac{\pi}{2}\right]$. So to compute the area inside $3 - 3\sin\theta$ and outside $1 + \sin\theta$, we compute the area inside $3 - 3\sin\theta$ from $-\frac{\pi}{2}$ to $\frac{\pi}{6}$ and subtract the area inside $1 + \sin\theta$ over the same range:

$$2\int_{-\pi/2}^{\pi/6}\int_0^{3-3\sin\theta} r\,dr\,d\theta - 2\int_{-\pi/2}^{\pi/6}\int_0^{1+\sin\theta} r\,dr\,d\theta$$

$$= 2\int_{-\pi/2}^{\pi/6}\left[\frac{1}{2}r^2\right]_{r=0}^{r=3-3\sin\theta} d\theta - 2\int_{-\pi/2}^{\pi/6}\left[\frac{1}{2}r^2\right]_{r=0}^{r=1+\sin\theta} d\theta$$

$$= \int_{-\pi/2}^{\pi/6}(3-3\sin\theta)^2\,d\theta - \int_{-\pi/2}^{\pi/6}(1+\sin\theta)^2\,d\theta$$

$$= \int_{-\pi/2}^{\pi/6}(9 - 18\sin\theta + 9\sin^2\theta)\,d\theta - \int_{-\pi/2}^{\pi/6}(1 + 2\sin\theta + \sin^2\theta)\,d\theta$$

$$= \int_{-\pi/2}^{\pi/6}\left(\frac{27}{2} - 18\sin\theta - \frac{9}{2}\cos 2\theta\right)d\theta - \int_{-\pi/2}^{\pi/6}\left(\frac{3}{2} + 2\sin\theta - \frac{1}{2}\cos 2\theta\right)d\theta$$

$$= \left[\frac{27}{2}\theta + 18\cos\theta - \frac{9}{4}\sin 2\theta\right]_{-\pi/2}^{\pi/6} - \left[\frac{3}{2}\theta - 2\cos\theta - \frac{1}{4}\sin 2\theta\right]_{-\pi/2}^{\pi/6}$$

$$= \left(9\pi + 9\sqrt{3} - \frac{9}{8}\sqrt{3}\right) - \left(\pi - \sqrt{3} - \frac{\sqrt{3}}{8}\right)$$

$$= 8\pi + 9\sqrt{3}.$$

35. Thinking in polar coordinates, the circle $x^2 + y^2 = 1$ is the circle $r = 1$. This circle intersects the line $x = \frac{1}{2}$, which is $r\cos\theta = \frac{1}{2}$, or $r = \frac{1}{2}\sec\theta$, at $(x, y) = \left(\frac{1}{2}, \pm\frac{\sqrt{3}}{2}\right)$, which is for $\theta = \pm\frac{\pi}{3}$. So the

area inside the circle but to the right of the vertical line is:

$$\int_{-\pi/3}^{\pi/3} \int_0^1 r\,dr\,d\theta - \int_{-\pi/3}^{\pi/3} \int_0^{\frac{1}{2}\sec\theta} r\,dr\,d\theta$$

$$= \int_{-\pi/3}^{\pi/3} \left[\frac{1}{2}r^2\right]_{r=0}^{r=1} d\theta - \int_{-\pi/3}^{\pi/3} \left[\frac{1}{2}r^2\right]_{r=0}^{r=\frac{1}{2}\sec\theta} d\theta$$

$$= \int_{-\pi/3}^{\pi/3} \frac{1}{2}\,d\theta - \frac{1}{8}\int_{-\pi/3}^{\pi/3} \sec^2\theta\,d\theta$$

$$= \frac{\pi}{3} - \frac{1}{8}[\tan\theta]_{-\pi/3}^{\pi/3}$$

$$= \frac{\pi}{3} - \frac{\sqrt{3}}{4}.$$

37. The limaçon is traced once from $\theta = 0$ to 2π, so its area is

$$\int_0^{2\pi} \int_0^{1+k\sin\theta} r\,dr\,d\theta = \int_0^{2\pi} \left[\frac{1}{2}r^2\right]_{r=0}^{r=1+k\sin\theta} d\theta$$

$$= \frac{1}{2}\int_0^{2\pi} (1+k\sin\theta)^2\,d\theta$$

$$= \frac{1}{2}\int_0^{2\pi} (1 + 2k\sin\theta + k^2\sin^2\theta)\,d\theta$$

$$= \frac{1}{2}\int_0^{2\pi} \left(1 + \frac{k^2}{2} + 2k\sin\theta - \frac{k^2}{2}\cos 2\theta\right) d\theta$$

$$= \frac{1}{2}\left[\theta + \frac{k^2}{2}\theta - 2k\cos\theta - \frac{k^2}{4}\sin 2\theta\right]_0^{2\pi}$$

$$= \frac{1}{2}\left(2\pi + k^2\pi\right) = \left(1 + \frac{k^2}{2}\right)\pi.$$

Clearly this approaches π as $k \to 0$, which makes sense, since when $k = 0$ the equation is $r = 1$, which is a circle of radius 1 having area $\pi \cdot 1^2 = \pi$.

39. The integrand is r times $\sqrt{16 - r^2}$, so $\sqrt{16 - r^2}$ is the equation of the surface over which we are integrating; this is the hemisphere lying over the disc $r \le 4$, which is the region of integration. Since there is a factor of 2 in front, this is the volume of the entire sphere (so it should be $\frac{4}{3}\pi \cdot 4^3 = \frac{256}{3}\pi$):

$$2\int_0^{2\pi} \int_0^4 r\sqrt{16 - r^2}\,dr\,d\theta = 2\int_0^{2\pi} \left[-\frac{1}{3}(16 - r^2)^{3/2}\right]_{r=0}^{r=4} d\theta = \frac{2}{3}\int_0^{2\pi} 64\,d\theta = \frac{256}{3}\pi.$$

41. The integrand is $4r - r^3 = (4 - r^2)r$, so $4 - r^2$ is the equation of the paraboloid surface over which we are integrating (it is a paraboloid since in rectangular coordinates it is $4 - x^2 - y^2$). This surface intersects the xy plane along the circle of radius 2; since we are integrating from $r = 0$ to $r = 2$, the integration determines the volume of the solid bounded by this surface and the xy plane. Its volume is

$$\int_0^{2\pi} \int_0^2 (4r - r^3)\,dr\,d\theta = \int_0^{2\pi} \left[2r^2 - \frac{1}{4}r^4\right]_{r=0}^{r=2} d\theta = \int_0^{2\pi} 4\,d\theta = 8\pi.$$

43. The integrand is $6r - 2r^2 = (6 - 2r)r$, so $6 - 2r$ is the equation of the conical surface over which we are integrating (it is a cone since in rectangular coordinates it is $2(3 - \sqrt{x^2 + y^2})$). This surface

intersects the xy plane along the circle of radius 3, which is the range of integration. Since we are integrating from 0 to 2π, the integral determines the volume of the solid bounded by $6 - 2r$ and the xy plane. Its volume is

$$\int_0^{2\pi} \int_0^3 (6r - 2r^2)\, dr\, d\theta = \int_0^{2\pi} \left[3r^2 - \frac{2}{3}r^3 \right]_0^3 d\theta = \int_0^{2\pi} 9\, d\theta = 18\pi.$$

45. The curve $\frac{\sqrt{2}}{2} + \sin\theta$ is a cardioid that is traced once from $\theta = 0$ to $\theta = 2\pi$, or from $\theta = -\frac{\pi}{4}$ to $\frac{7\pi}{4}$. The inner loop is traced out from $\theta = \frac{5\pi}{4}$ to $\theta = \frac{7\pi}{4}$, so that the given integral is the area inside the outer loop, which is the same as the volume under the surface $z = 1$ lying over the outer loop of the cardioid. This volume is

$$\int_{-\pi/4}^{5\pi/4} \int_0^{\sqrt{2}/2 + \sin\theta} r\, dr\, d\theta = \int_{-\pi/4}^{5\pi/4} \left[\frac{1}{2}r^2 \right]_{r=0}^{r=\sqrt{2}/2 + \sin\theta} d\theta$$

$$= \frac{1}{2} \int_{-\pi/4}^{5\pi/4} \left(\sqrt{2}/2 + \sin\theta \right)^2 d\theta$$

$$= \frac{1}{2} \int_{-\pi/4}^{5\pi/4} \left(\frac{1}{2} + \sqrt{2}\sin\theta + \sin^2\theta \right) d\theta$$

$$= \frac{1}{2} \int_{-\pi/4}^{5\pi/4} \left(1 + \sqrt{2}\sin\theta - \frac{1}{2}\cos 2\theta \right) d\theta$$

$$= \frac{1}{2} \left[\theta - \sqrt{2}\cos\theta - \frac{1}{4}\sin 2\theta \right]_{-\pi/4}^{5\pi/4}$$

$$= \frac{1}{2} \left(\left(\frac{5\pi}{4} + 1 - \frac{1}{4} \right) - \left(-\frac{\pi}{4} - 1 + \frac{1}{4} \right) \right) = \frac{3\pi}{4} + \frac{3}{4}.$$

47. These two surfaces intersect where $x^2 + y^2 = 16 - x^2 - y^2$, or on the circle $x^2 + y^2 = 8$ of radius $2\sqrt{2}$. In polar coordinates, then, the region of integration is $\theta \in [0, 2\pi]$ and $r \in [0, 2\sqrt{2}]$, and the volume is the volume contained between the surfaces $16 - r^2$ and r^2. Thus the volume is

$$\int_0^{2\pi} \int_0^{2\sqrt{2}} (16 - r^2 - r^2)r\, dr\, d\theta = \int_0^{2\pi} \int_0^{2\sqrt{2}} (16r - 2r^3)\, dr\, d\theta$$

$$= \int_0^{2\pi} \left[8r^2 - \frac{1}{2}r^4 \right]_{r=0}^{r=2\sqrt{2}} d\theta$$

$$= \int_0^{2\pi} 32\, d\theta = 64\pi.$$

49. The given cone and the sphere of radius 1 (which has equation $x^2 + y^2 + z^2 = 1$) intersect above the xy plane (i.e., for positive z), so they intersect where $\sqrt{x^2 + y^2} = \sqrt{1 - x^2 - y^2}$. This reduces to $x^2 + y^2 = \frac{1}{2}$, so they intersect along a circle of radius $\frac{1}{\sqrt{2}}$ at height $z = \frac{1}{2}$. Thus (in polar coordinates) the region of integration is $\theta \in [0, 2\pi]$ and $r \in \left[0, \frac{1}{\sqrt{2}} \right]$, and the volume is the volume contained between the two surfaces r and $\sqrt{1 - r^2}$. Thus the volume is

$$\int_0^{2\pi} \int_0^{1/\sqrt{2}} \left(\sqrt{1 - r^2} - r \right) r\, dr\, d\theta = \int_0^{2\pi} \left[-\frac{1}{3}(1 - r^2)^{3/2} - \frac{1}{3}r^3 \right]_{r=0}^{r=1/\sqrt{2}} d\theta$$

$$= \int_0^{2\pi} \left(-\frac{1}{6\sqrt{2}} - \frac{1}{6\sqrt{2}} + \frac{1}{3} \right) d\theta$$

$$= 2\pi \left(\frac{1}{3} - \frac{\sqrt{2}}{6} \right) = \frac{2 - \sqrt{2}}{3}\pi.$$

51. These two surfaces intersect where $x^2 + y^2 + \left(\frac{3}{5}\right)^2 = 1$, so where $x^2 + y^2 = \frac{16}{25}$. This is a circle of radius $\frac{4}{5}$ centered on the z axis. Thus (in polar coordinates) the region of integration is $\theta \in [0, 2\pi]$ and $r \in \left[0, \frac{4}{5}\right]$, and the volume is the volume contained between the surfaces $\sqrt{1 - r^2}$ and $\frac{3}{5}$. Thus the volume is

$$\int_0^{2\pi} \int_0^{4/5} \left(\sqrt{1 - r^2} - \frac{3}{5}\right) r\, dr\, d\theta = \int_0^{2\pi} \left[-\frac{1}{3}(1 - r^2)^{3/2} - \frac{3}{10}r^2\right]_0^{4/5} d\theta$$
$$= \int_0^{2\pi} \left(-\frac{27}{375} - \frac{48}{250} + \frac{1}{3}\right) d\theta$$
$$= \frac{52}{375}\pi.$$

53. Place one sphere at the origin, with equation $x^2 + y^2 + z^2 = 1$, and the other at $(0, 0, 1)$, with equation $x^2 + y^2 + (z - 1)^2 = 1$. The two spheres intersect where the two equations are equal, which is for $z^2 = (z - 1)^2$, or $2z - 1 = 0$. So they intersect at $z = \frac{1}{2}$. At that height, the first equation gives $x^2 + y^2 = \frac{3}{4}$, which is a circle of radius $\frac{\sqrt{3}}{2}$. Thus (in polar coordinates) the region of integration is $\theta \in [0, 2\pi]$ and $r \in \left[0, \frac{\sqrt{3}}{2}\right]$. A diagram of the situation is

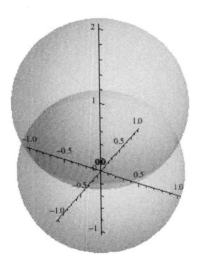

Note that the region of intersection is composed of two identical regions: one, above the plane $z = \frac{1}{2}$, consists of the volume between $x^2 + y^2 + z^2 = 1$ and $z = \frac{1}{2}$; the other, below that plane, consists of the volume between $x^2 + y^2 + (z - 1)^2 = 1$ and $z = \frac{1}{2}$. Writing $x^2 + y^2 + z^2 = 1$ in polar coordinates gives $z = \sqrt{1 - r^2}$, so the total volume is

$$2 \int_0^{2\pi} \int_0^{\sqrt{3}/2} \left(\sqrt{1 - r^2} - \frac{1}{2}\right) r\, dr\, d\theta = 2 \int_0^{2\pi} \left[-\frac{1}{3}(1 - r^2)^{3/2} - \frac{1}{4}r^2\right]_0^{\sqrt{3}/2} d\theta$$
$$= 2 \int_0^{2\pi} \left[-\frac{1}{24} - \frac{3}{16} + \frac{1}{3}\right] = \frac{5}{12}\pi.$$

55. The two surfaces intersect where $\sqrt{x^2 + y^2} = h$, or $x^2 + y^2 = h^2$; this is a circle of radius h. Thus (in cylindrical coordinates) the region of integration is $\theta \in [0, 2\pi]$ and $r \in [0, h]$. The volume is the volume between h and $\sqrt{r^2} = r$, so it is

$$\int_0^{2\pi} \int_0^h (h - r) r \, dr \, d\theta = \int_0^{2\pi} \left[\frac{1}{2} h r^2 - \frac{1}{3} r^3 \right]_0^h d\theta = \int_0^{2\pi} \left(\frac{1}{2} h^3 - \frac{1}{3} h^3 \right) d\theta = \frac{1}{3} \pi h^3.$$

57. The region of integration is

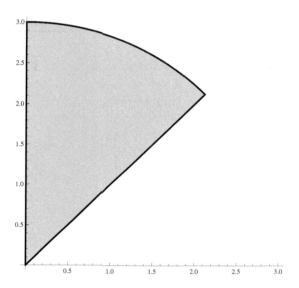

This is a sector of a circle of radius 3 between $\theta = \frac{\pi}{4}$ and $\theta = \frac{\pi}{2}$. To see this, note that the outer edge of the region satisfies $y = \sqrt{9 - x^2}$, or $x^2 + y^2 = 9$; the left edge is where $x = 0$, so that y ranges from 0 to 3, and the right edge is along the line from $(0, 0)$ to $\left(\frac{3\sqrt{2}}{2}, \frac{3\sqrt{2}}{2} \right)$, which is an angle whose tangent is 1, or $\frac{\pi}{4}$. The value of the integral is

$$\int_{\pi/4}^{\pi/2} \int_0^3 r \, dr \, d\theta = \int_{\pi/4}^{\pi/2} \left[\frac{1}{2} r^2 \right]_0^3 d\theta = \int_{\pi/4}^{\pi/2} \frac{9}{2} \, d\theta = \frac{9}{2} \cdot \frac{\pi}{4} = \frac{9}{8} \pi.$$

Note that this is the correct answer, since the integral is just the area of the region, and this region is one eighth of a circle of radius 3.

59. The region of integration is

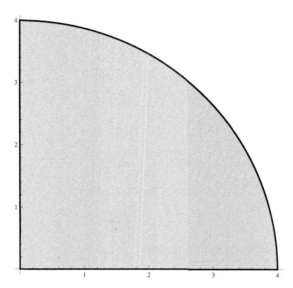

This is the quarter circle of radius 4 in the first quadrant. To see this, note that the outer edge of the region is $y = \sqrt{16 - x^2}$, or $x^2 + y^2 = 16$, and that x ranges from 0 to 4, restricting to the first quadrant. Converting the integrand to polar coordinates gives $e^{x^2} e^{y^2} = e^{x^2 + y^2} = e^{r^2}$, so the integral is

$$\int_0^{\pi/2} \int_0^4 r \cdot e^{r^2} \, dr \, d\theta = \int_0^{\pi/2} \left[\frac{1}{2} e^{r^2} \right]_{r=0}^{r=4} d\theta = \int_0^{\pi/2} \frac{1}{2} \left(e^{16} - 1 \right) d\theta = \frac{\pi}{4} \left(e^{16} - 1 \right).$$

Applications

61. The number of wolves within 12 miles of both herds is the number of wolves contained in the intersection of the circles of radius 12 miles centered at $(0,0)$ and $(0, 12)$. These circles have equations $x^2 + y^2 = 144$ and $x^2 + (y - 12)^2 = 144$. Substituting $x^2 = 144 - y^2$ into the second equation and solving gives $-24y + 144 = 0$, so that $y = 6$ and then $x = \pm\sqrt{144 - 36} = \pm\sqrt{108} = \pm 6\sqrt{3}$. The two points of intersection are $(\pm 6\sqrt{3}, 6) = \left(\pm 12 \cos \frac{\pi}{6}, 12 \sin \frac{\pi}{6} \right)$. A diagram of the region is:

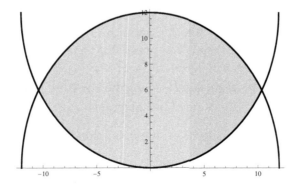

Since the density of wolves is constant in this region, the number of wolves is 0.08 times the area of the region. We can compute the area of the region as twice the area of the region to the right of the x axis, which in turn is the sum of two polar integrals, one extending from $\theta = 0$ to $\theta = \frac{\pi}{6}$, where the two circles intersect, and the second extending from $\theta = \frac{\pi}{6}$ to $\theta = \frac{\pi}{2}$. The polar equation of the circle centered at $(0, 12)$ is $r = 24 \sin \theta$ and the polar equation of the circle centered at the origin is $r = 12$. Thus the number of wolves is (recalling that $\int \sin^2 \theta \, d\theta = \frac{\theta}{2} - \frac{1}{4} \sin 2\theta$)

$$2 \cdot 0.08 \left(\int_0^{\pi/6} \int_0^{24 \sin \theta} r \, dr \, d\theta + \int_{\pi/6}^{\pi/2} \int_0^{12} r \, dr \, d\theta \right)$$

$$= 0.16 \left(\int_0^{\pi/6} \left[\frac{1}{2} r^2 \right]_0^{24 \sin \theta} d\theta + \int_{\pi/6}^{\pi/2} \left[\frac{1}{2} r^2 \right]_0^{12} d\theta \right)$$

$$= 0.16 \left(\int_0^{\pi/6} 288 \sin^2 \theta \, d\theta + \int_{\pi/6}^{\pi/2} 72 \, d\theta \right)$$

$$= 0.16 \left([144\theta - 72 \sin 2\theta]_0^{\pi/6} + 72 \left(\frac{\pi}{2} - \frac{\pi}{6} \right) \right)$$

$$= 0.16 \left(24\pi - 36\sqrt{3} + 24\pi \right)$$

$$= 7.68\pi - 5.76\sqrt{3} \approx 14.15.$$

Thus there are 14 or 15 wolves in that region.

Proofs

63. This circle is traced once from $\theta = 0$ to $\theta = \pi$, so the area enclosed by the circle is

$$\int_0^\pi \int_0^{2R \cos \theta} r \, dr \, d\theta = \int_0^\pi \left[\frac{1}{2} r^2 \right]_{r=0}^{r=2R \cos \theta} d\theta$$

$$= \frac{1}{2} \int_0^\pi 4R^2 \cos^2 \theta \, d\theta$$

$$= \frac{1}{2} \int_0^\pi 4R^2 \cdot \frac{1}{2} (1 + \cos 2\theta) \, d\theta$$

$$= R^2 \int_0^\pi (1 + \cos 2\theta) \, d\theta$$

$$= R^2 \left[\theta + \frac{1}{2} \sin 2\theta \right]_0^\pi = \pi R^2.$$

65. The volume of the entire sphere is twice the volume of the upper hemisphere, which is the solid lying over the disc $x^2 + y^2 \leq R^2$ and under the sphere $x^2 + y^2 + z^2 = R^2$. In polar coordinates, then, we must integrate from $\theta = 0$ to 2π, and from $r = 0$ to R; the region is $\sqrt{R^2 - r^2}$. Thus the volume of the sphere is

$$2 \int_0^{2\pi} \int_0^R r \sqrt{R^2 - r^2} \, dr \, d\theta = 2 \int_0^{2\pi} \left[-\frac{1}{3} (R^2 - r^2)^{3/2} \right]_0^R d\theta = 2 \int_0^{2\pi} \frac{R^3}{3} \, d\theta = \frac{4}{3} \pi R^3.$$

67. Position the angle so that one edge is on the polar axis and the other is on the polar line $\theta = \phi$. Then the area of the sector is

$$\int_0^\phi \int_0^R r \, dr \, d\theta = \int_0^\phi \left[\frac{1}{2} r^2 \right]_0^R d\theta = \frac{1}{2} \int_0^\phi R^2 \, d\theta = \frac{1}{2} \phi R^2.$$

69. The rose $r = \cos 2n\theta$ has $4n$ leaves, and is traced once from $\theta = 0$ to $\theta = 2\pi$. The first two positive values of θ for which $r = \cos 2n\theta = 0$ are when $2n\theta = \frac{\pi}{2}$ and $2n\theta = \frac{3\pi}{2}$, so for $\theta = \frac{\pi}{4n}$ and $\theta = \frac{3\pi}{4n}$.

So these two values bound a single leaf, and the total area of the rose is $2n$ times the area of that leaf:

$$2n \int_{\frac{\pi}{4n}}^{\frac{3\pi}{4n}} \int_0^{\cos 2n\theta} r \, dr \, d\theta = 2n \int_{\frac{\pi}{4n}}^{\frac{3\pi}{4n}} \left[\frac{1}{2} r^2 \right]_{r=0}^{r=\cos 2n\theta} d\theta$$

$$= n \int_{\frac{\pi}{4n}}^{\frac{3\pi}{4n}} \cos^2 2n\theta \, d\theta$$

$$= \frac{n}{2} \int_{\frac{\pi}{4n}}^{\frac{3\pi}{4n}} (1 + \cos 4n\theta) \, d\theta$$

$$= \frac{n}{2} \left[\theta + \frac{1}{4n} \sin 4n\theta \right]_{\frac{\pi}{4n}}^{\frac{3\pi}{4n}}$$

$$= \frac{n}{2} \left(\frac{3\pi}{4n} - \frac{\pi}{4n} \right) = \frac{\pi}{4}.$$

This area is independent of n.

Thinking Forward

A triple integral using cylindrical coordinates

The value of the integral is

$$\int_0^{2\pi} \int_0^R \int_0^h r \, dz \, dr \, d\theta = \int_0^{2\pi} \int_0^R [rz]_{z=0}^{z=h} \, dr \, d\theta$$

$$= \int_0^{2\pi} \int_0^R hr \, dr \, d\theta$$

$$= h \int_0^{2\pi} \left[\frac{1}{2} r^2 \right]_{r=0}^{r=R} d\theta$$

$$= \frac{h}{2} \int_0^{2\pi} R^2 \, d\theta$$

$$= \pi R^2 h.$$

Since all three integrals have constant bounds, and since the functions are continuous, we can reorder the integrals to get

$$\int_0^h \int_0^{2\pi} \int_0^R r \, dr \, d\theta \, dz.$$

Now the inner two integrals represent the area of a circle of radius R, computed using polar coordinates. The outer integral multiplies that area over the height of the cylinder, giving the total volume of the cylinder.

A triple integral using spherical coordinates

The value of the integral is

$$\int_0^\pi \int_0^{2\pi} \int_0^R \rho^2 \sin\phi \, d\rho \, d\theta \, d\phi = \int_0^\pi \int_0^{2\pi} \left[\frac{1}{3}\rho^3 \sin\phi \right]_{\rho=0}^{\rho=R} d\theta \, d\phi$$

$$= \int_0^\pi \int_0^{2\pi} \frac{1}{3} R^3 \sin\phi \, d\theta \, d\phi$$

$$= \int_0^\pi \frac{2}{3}\pi R^3 \sin\phi \, d\phi$$

$$= \frac{2}{3}\pi R^3 \left[-\cos\phi \right]_0^\pi = \frac{4}{3}\pi R^3.$$

13.4   Applications of Double Integrals

Thinking Back

Mass of a rod

Since the density varies only with the x coordinate, the mass of the rod is the integral over the length of the rod of the cross-sectional area times the density, which is

$$\int_0^{20} \pi \cdot (10 + 0.01x^2) \, dx = \pi \left[10x + \frac{0.01}{3}x^3 \right]_0^{20} = \left(200 + \frac{80}{3} \right)\pi = \frac{680}{3}\pi.$$

More questions about the rod

Since the cross-sectional area of the rod is π, in order to apply the material in the text, we think of the rod as a flat rectangle Ω in the xy plane with height π and width 20, with density function $\rho(x) = 10 + 0.01x^2$. This has the same mass as the rod itself, and also the same x coordinate for its center of mass. This coordinate is given, from Definition 13.12 and the surrounding discussion, by $\frac{M_y}{m}$ where m is the mass of the lamina (see the previous problem) and M_y is the first moment of the lamina about the y axis. This is

$$M_y = \iint_\Omega x\rho(x,y) \, dA = \int_0^\pi \int_0^{20} x \left(10 + 0.01x^2 \right) dx \, dy$$

$$= \int_0^\pi \left[5x^2 + \frac{0.01}{4}x^4 \right]_0^{20} dy = \int_0^\pi 2400 \, dy = 2400\pi.$$

Thus the x coordinate of the center of mass is

$$\frac{M_y}{m} = \frac{2400\pi}{(680/3)\pi} = \frac{2400 * 3}{680} = \frac{180}{17} \approx 10.588 \text{ cm.}$$

By Definition 13.13, the moment of inertia is

$$\iint_\Omega x^2\rho(x,y) \, dA = \int_0^\pi \int_0^{20} x^2 \left(10 + 0.01x^2 \right) dx \, dy$$

$$= \int_0^\pi \left[\frac{10}{3}x^3 + \frac{0.01}{5}x^5 \right]_0^{20} dy = \int_0^\pi \frac{99200}{3} \, dy = \frac{99200}{3}\pi,$$

so that the radius of gyration is

$$R_y = \sqrt{\frac{I_y}{m}} = \sqrt{\frac{(99200/3)\pi}{(680/3)\pi}} = \sqrt{\frac{99200}{680}} = \sqrt{\frac{2480}{17}} \approx 12.078 \text{ cm.}$$

Concepts

1. (a) False. For example, if Ω is an annulus, say $\{(x,y) \mid 1 \leq \sqrt{x^2 + y^2} \leq 2\}$ with constant density, its center of mass is the origin, which is not in Ω.

   (b) False. The same example as in part (a) shows this.

   (c) False. For example, consider the unit circle centered at the origin, with the density function $\rho(x,y) = x + 2$. Then clearly the density is higher the further right you move, so that the x coordinate of the center of mass will be to the right of the origin.

   (d) True. Consider the unit circle centered at the origin. Since the circle is assumed to have constant density, say $\rho = 1$, we have

   $$M_y = \iint_\Omega x\rho(x,y)\,dA = \int_{-1}^1 \int_{-\sqrt{1-x^2}}^{\sqrt{1-x^2}} x\,dy\,dx = \int_{-1}^1 2x\sqrt{1-x^2}\,dx$$

   $$M_x = \iint_\Omega y\rho(x,y)\,dA = \int_{-1}^1 \int_{-\sqrt{1-y^2}}^{\sqrt{1-y^2}} y\,dx\,dy = \int_{-1}^1 2y\sqrt{1-y^2}\,dy.$$

   But both of these integrals are zero, since the integrand is an odd function and the region of integration is symmetric around 0. Thus the centroid is at the origin.

   (e) False. By Definition 13.12, it is given by $M_x = \iint_\Omega y\rho(x,y)\,dA$.

   (f) False. By Definitions 13.12 and 13.13,

   $$M_x = \iint_\Omega y\rho(x,y)\,dA, \qquad I_x = \iint_\Omega y^2\rho(x,y)\,dA.$$

   There is no reason to expect that $xM_x = I_x$, or even that $yM_x = I_x$.

   (g) True. By definition 13.13,

   $$I_o = \iint_\Omega (x^2 + y^2)\rho(x,y)\,dA = \iint_\Omega x^2\rho(x,y)\,dA + \iint_\Omega y^2\rho(x,y)\,dA = I_y + I_x.$$

   (h) True. The definition of the radius of gyration is that $R_y = \sqrt{\frac{I_y}{m}}$, so that $R_y^2 = \frac{I_y}{m}$.

3. This is the mass of the region Ω assuming it has constant density 1; its units are square centimeters times grams per square centimeter, or grams.

5. These are respectively the x and y coordinates, $\bar{x}$ and $\bar{y}$, of the centroid of Ω where the density is constant at 1, since the numerators are the first moments (see Exercise 4) and the denominators are each equal to the mass of Ω. Since the numerators have units gram-centimeters and the denominators have units grams, the quotient is measured in centimeters.

7. These are respectively the first moments around the y and x axes, M_y and M_x, of the region Ω assuming that the lamina Ω has density $\rho(x,y)$. See Definition 13.12. The units of x are centimeters, while the units of ρ are grams per square centimeter, and thus the units of the integrand are grams/centimeter. Integrating multiplies by area, or square centimeters, so that the units of $\iint_\Omega x\rho(x,y)\,dA$ are gram-centimeters. The units of the other integral are also gram-centimeters by the same reasoning.

9. These are respectively the moment of inertia of Ω about the y axis and the moment of inertia of Ω about the x axis assuming that Ω has density $\rho(x,y)$; see Definition 13.13. The units of x^2 are square centimeters, while the units of ρ are grams per square centimeter, and thus the units of the integrand are grams. Integrating multiplies by area, or square centimeters, so that the units of $\iint_\Omega x^2\rho(x,y)\,dA$ are grams times square centimeters. The units of the other integral are also grams times square centimeters by the same reasoning.

11. These are respectively the radius of gyration of Ω about the y axis and the x axis, R_y and R_x. From Exercise 9, both numerators have units grams times square centimeters, while from Exercise 6, both denominators have units grams, so the quotient has units square centimeters and thus its square root, which is R_x or R_y, has units centimeters.

13. The triangle is bounded below by the line joining $(1,1)$ and $(2,0)$, which has equation $y - 0 = \frac{1-0}{1-2}(x-2)$, or $y = -x + 2$. It is bounded above by the line joining $(1,1)$ with $(2,3)$, which has equation $y - 1 = \frac{3-1}{2-1}(x-1)$, or $y = 2x - 1$. The region is thus described by $1 \leq x \leq 2$ and $-x + 2 \leq y \leq 2x - 1$, so that by Definition 13.12

$$M_y = \iint_\Omega x\rho(x,y)\,dA = \int_1^2 \int_{-x+2}^{2x-1} x\,dy\,dx$$
$$= \int_1^2 [xy]_{y=-x+2}^{y=2x-1}\,dx$$
$$= \int_1^2 (x(2x-1) - x(-x+2))\,dx$$
$$= \int_1^2 (3x^2 - 3x)\,dx$$
$$= \left[x^3 - \frac{3}{2}x^2\right]_1^2 = \frac{5}{2}.$$

15. Let the density of the region be $\rho(x,y) = kx$. The mass is the integral of the density over the region, and the given integral correctly describes the region (see Exercise 13), so the mass is

$$M_y = \iint_\Omega \rho(x,y)\,dA = \int_1^2 \int_{-x+2}^{2x-1} kx\,dy\,dx$$
$$= \int_1^2 [kxy]_{y=-x+2}^{y=2x-1}\,dx$$
$$= k\int_1^2 (x(2x-1) - x(-x+2))\,dx$$
$$= k\int_1^2 (3x^2 - 3x)\,dx$$
$$= k\left[x^3 - \frac{3}{2}x^2\right]_1^2 = \frac{5}{2}k.$$

17. Let the density of the region be $\rho(x,y) = kx$. The first moment about the x axis is the integral over the region of y times the density. The given integral correctly describes the region (see Exercise 13), so the first moment about the x axis is

$$M_x = \iint_\omega y\rho(x,y)\,dA = \int_1^2 \int_{-x+2}^{2x-1} kxy\,dy\,dx$$
$$= \int_1^2 \left[\frac{1}{2}kxy^2\right]_{y=-x+2}^{y=2x-1}\,dx$$
$$= \frac{1}{2}k\int_1^2 \left(x(2x-1)^2 - x(-x+2)^2\right)\,dx$$
$$= \frac{1}{2}k\int_1^2 \left(3x^3 - 3x\right)\,dx$$
$$= \frac{1}{2}k\left[\frac{3}{4}x^4 - \frac{3}{2}x^2\right]_1^2 = \frac{27}{8}k.$$

19. Let the density of the region be $\rho(x,y) = kx$. The moment of inertia about the x axis is the integral over the region of y^2 times the density. The given integral correctly describes the region (see Exercise 13), so the moment of inertia about the x axis is

$$I_x = \iint_\omega y^2 \rho(x,y)\, dA = \int_1^2 \int_{-x+2}^{2x-1} kxy^2\, dy\, dx$$

$$= \int_1^2 \left[\frac{1}{3} kxy^3 \right]_{y=-x+2}^{y=2x-1} dx$$

$$= \frac{1}{3} k \int_1^2 \left(x(2x-1)^3 - x(-x+2)^3 \right) dx$$

$$= \frac{1}{3} k \int_1^2 \left(9x^4 - 18x^3 + 18x^2 - 9x \right) dx$$

$$= 3k \left[\frac{1}{5} x^5 - \frac{1}{2} x^4 + \frac{2}{3} x^3 - \frac{1}{2} x^2 \right]_1^2 = \frac{28}{5} k.$$

21. Definition 13.13 gives, after converting to polar coordinates as in Example 2 (using kr for the density and $r\, dr\, d\theta$ for the area element)

$$I_y = \iint_\Omega x^2 \rho(x,y)\, dA = \int_{-\pi/4}^{\pi/4} \int_{\sec\theta}^{2\cos\theta} (r\cos\theta)^2 \cdot kr \cdot r\, dr\, d\theta$$

$$= k \int_{-\pi/4}^{\pi/4} \int_{\sec\theta}^{2\cos\theta} r^4 \cos^2\theta\, dr\, d\theta$$

$$= k \int_{-\pi/4}^{\pi/4} \left[\frac{1}{5} r^5 \cos^2\theta \right]_{\sec\theta}^{2\cos\theta} d\theta$$

$$= \frac{1}{5} k \int_{-\pi/4}^{\pi/4} \left(32\cos^7\theta - \sec^5\theta \cos^2\theta \right) d\theta$$

$$= \frac{1}{5} k \int_{-\pi/4}^{\pi/4} \left(32\cos\theta(1 - \sin^2\theta)^3 - \sec^3\theta \right) d\theta$$

$$= \frac{1}{5} k \int_{-\pi/4}^{\pi/4} \left(32\cos\theta - 96\cos\theta\sin^2\theta + 96\cos\theta\sin^4\theta - 32\cos\theta\sin^6\theta - \sec^3\theta \right) d\theta$$

$$= \frac{1}{5} k \left[32\sin\theta - 32\sin^3\theta + \frac{96}{5}\sin^5\theta - \frac{32}{7}\sin^7\theta \right.$$

$$\left. - \frac{1}{2}(\sec\theta\tan\theta + \ln|\sec\theta + \tan\theta|) \right]_{-\pi/4}^{\pi/4}$$

$$= \frac{1}{5} k \left(32\sqrt{2} - 16\sqrt{2} + \frac{24}{5}\sqrt{2} - \frac{4}{7}\sqrt{2} \right.$$

$$\left. - \frac{1}{2}\left(\sqrt{2} + \ln\left|\sqrt{2} + 1\right|\right) + \frac{1}{2}\left(-\sqrt{2} + \ln\left|\sqrt{2} - 1\right|\right) \right)$$

$$= \frac{1}{5} k \left(\frac{673}{35}\sqrt{2} + \frac{1}{2}\ln\left|\frac{\sqrt{2}-1}{\sqrt{2}+1}\right| \right) = \left(\frac{673}{175}\sqrt{2} + \frac{1}{10}\ln\left|\frac{\sqrt{2}-1}{\sqrt{2}+1}\right| \right) k.$$

For I_x we get

$$I_x = \iint_\Omega y^2 \rho(x,y)\, dA = \int_{-\pi/4}^{\pi/4} \int_{\sec\theta}^{2\cos\theta} (r\sin\theta)^2 \cdot kr \cdot r\, dr\, d\theta$$

$$= k \int_{-\pi/4}^{\pi/4} \int_{\sec\theta}^{2\cos\theta} r^4 \sin^2\theta\, dr\, d\theta$$

$$= k \int_{-\pi/4}^{\pi/4} \left[\frac{1}{5} r^5 \sin^2\theta \right]_{\sec\theta}^{2\cos\theta} d\theta$$

$$= \frac{1}{5} k \int_{-\pi/4}^{\pi/4} \left(32\cos^5\theta \sin^2\theta - \sec^5\theta \sin^2\theta \right) d\theta$$

$$= \frac{1}{5} k \int_{-\pi/4}^{\pi/4} \left(32\cos\theta(1 - \sin^2\theta)^2 \sin^2\theta - \sec^3\theta \tan^2\theta \right) d\theta$$

$$= \frac{1}{5} k \int_{-\pi/4}^{\pi/4} \left(32\cos\theta \sin^2\theta - 64\cos\theta \sin^4\theta + 32\cos\theta \sin^6\theta - \sec^3\theta \tan^2\theta \right) d\theta$$

$$= \frac{1}{5} k \left(\left[\frac{32}{3} \sin^3\theta - \frac{64}{5} \sin^5\theta + \frac{32}{7} \sin^7\theta \right]_{-\pi/4}^{\pi/4} - \int_{-\pi/4}^{\pi/4} \sec^3\theta \tan^2\theta\, d\theta \right)$$

$$= \frac{1}{5} k \left(\frac{16}{3} \sqrt{2} - \frac{16}{5} \sqrt{2} + \frac{4}{7} \sqrt{2} - \int_{-\pi/4}^{\pi/4} \sec^3\theta \tan^2\theta\, d\theta \right)$$

$$= \frac{1}{5} k \left(\frac{284}{105} \sqrt{2} - \int_{-\pi/4}^{\pi/4} \sec^3\theta \tan^2\theta\, d\theta \right).$$

To integrate $\sec^3\theta \tan^2\theta$, use the Pythagorean identity $\tan^2\theta = \sec^2\theta - 1$ to convert it to $\sec^5\theta - \sec^3\theta$. By the reduction formula from Exercise 86 in Section 5.4, we have

$$\int_{-\pi/4}^{\pi/4} \sec^3\theta \tan^2\theta\, d\theta = \int_{-\pi/4}^{\pi/4} (\sec^5\theta - \sec^3\theta)\, d\theta$$

$$= \frac{1}{4} \left[\sec^3\theta \tan\theta \right]_{-\pi/4}^{\pi/4} + \frac{3}{4} \int_{-\pi/4}^{\pi/4} \sec^3\theta\, d\theta - \int_{-\pi/4}^{\pi/4} \sec^3\theta\, d\theta$$

$$= \sqrt{2} - \frac{1}{4} \int_{-\pi/4}^{\pi/4} \sec^3\theta\, d\theta$$

$$= \sqrt{2} - \frac{1}{8} \left[\sec\theta \tan\theta + \ln|\sec\theta + \tan\theta| \right]_{-\pi/4}^{\pi/4}$$

$$= \sqrt{2} - \frac{1}{8} \left(2\sqrt{2} + \ln\left|\sqrt{2} + 1\right| - \ln\left|\sqrt{2} - 1\right| \right)$$

$$= \frac{3}{4} \sqrt{2} + \frac{1}{8} \ln\left| \frac{\sqrt{2} - 1}{\sqrt{2} + 1} \right|.$$

Putting this all together gives

$$I_x = \frac{1}{5} k \left(\frac{284}{105} \sqrt{2} - \left(\frac{3}{4} \sqrt{2} + \frac{1}{8} \ln\left| \frac{\sqrt{2} - 1}{\sqrt{2} + 1} \right| \right) \right)$$

$$= \frac{1}{5} k \left(\frac{821}{420} \sqrt{2} - \frac{1}{8} \ln\left| \frac{\sqrt{2} - 1}{\sqrt{2} + 1} \right| \right)$$

$$= \left(\frac{821}{2100} \sqrt{2} - \frac{1}{40} \ln\left| \frac{\sqrt{2} - 1}{\sqrt{2} + 1} \right| \right) k.$$

23. We get

$$\int_{-\pi/4}^{\pi/4} \int_{\sec\theta}^{2\cos\theta} kr^3 \cos\theta \, dr \, d\theta = \int_{-\pi/4}^{\pi/4} \left[\frac{1}{4}kr^4 \cos\theta\right]_{r=\sec\theta}^{r=2\cos\theta} d\theta$$

$$= \frac{1}{4}k \int_{-\pi/4}^{\pi/4} \left(16\cos^5\theta - \sec^4\theta\cos\theta\right) d\theta$$

$$= \frac{1}{4}k \int_{-\pi/4}^{\pi/4} \left(16\cos\theta(1-\sin^2\theta)^2 - \sec^3\theta\right) d\theta$$

$$= \frac{1}{4}k \int_{-\pi/4}^{\pi/4} \left(16\cos\theta - 32\cos\theta\sin^2\theta + 16\cos\theta\sin^4\theta - \sec^3\theta\right) d\theta$$

$$= \frac{1}{4}k \left[16\sin\theta - \frac{32}{3}\sin^3\theta + \frac{16}{5}\sin^5\theta - \frac{1}{2}(\sec\theta\tan\theta + \ln|\sec\theta + \tan\theta|)\right]_{-\pi/4}^{\pi/4}$$

$$= \frac{1}{4}k \left(16\sqrt{2} - \frac{16}{3}\sqrt{2} + \frac{4}{5}\sqrt{2} - \frac{1}{2}(\sqrt{2} + \ln\left|\sqrt{2}+1\right|) + \frac{1}{2}(-\sqrt{2} + \ln\left|\sqrt{2}-1\right|)\right)$$

$$= \frac{1}{4}k \left(\frac{157}{15}\sqrt{2} + \frac{1}{2}\ln\frac{\sqrt{2}-1}{\sqrt{2}+1}\right)$$

$$= \frac{k}{60} \left(157\sqrt{2} + \frac{15}{2}\ln\left(\frac{\sqrt{2}-1}{\sqrt{2}+1} \cdot \frac{\sqrt{2}-1}{\sqrt{2}-1}\right)\right)$$

$$= \frac{k}{60} \left(157\sqrt{2} + \frac{15}{2}\ln\left(\frac{(\sqrt{2}-1)^2}{1}\right)\right)$$

$$= \frac{k}{60} \left(157\sqrt{2} + 15\ln(\sqrt{2}-1)\right).$$

Skills

For Exercises 24–30, note that the triangle T is described by $0 \le x \le 1$ and $-x \le y \le x$.

25. Assume the density is $\rho(x,y) = kx$. Then the mass of T is

$$\iint_T \rho(x,y) \, dA = \int_0^1 \int_{-x}^x kx \, dy \, dx = \int_0^1 [kxy]_{y=-x}^{y=x} \, dx = 2k \int_0^1 x^2 \, dx = 2k \left[\frac{1}{3}x^3\right]_0^1 = \frac{2}{3}k.$$

27. Assume the density is $\rho(x,y) = kx$. Then the moments of inertia are

$$I_y = \iint_T x^2 \rho(x,y) \, dA = \int_0^1 \int_{-x}^x kx^3 \, dy \, dx = k \int_0^1 \left[x^3 y\right]_{y=-x}^{y=x} dx = k \int_0^1 2x^4 \, dx = k \left[\frac{2}{5}x^5\right]_0^1 = \frac{2}{5}k$$

$$I_x = \iint_T y^2 \rho(x,y) \, dA = \int_0^1 \int_{-x}^x kxy^2 \, dy \, dx = k \int_0^1 \left[\frac{1}{3}xy^3\right]_{y=-x}^{y=x} dx = \frac{2}{3}k \int_0^1 x^4 \, dx = \frac{2}{3}k \left[\frac{1}{5}x^5\right]_0^1$$

$$= \frac{2}{15}k.$$

Since the mass of T is $\frac{2}{3}k$, the radii of gyration are

$$R_y = \sqrt{\frac{I_y}{m}} = \sqrt{\frac{\frac{2}{5}k}{\frac{2}{3}k}} = \sqrt{\frac{3}{5}} = \frac{\sqrt{15}}{5}, \qquad R_x = \sqrt{\frac{I_x}{m}} = \sqrt{\frac{\frac{2}{15}k}{\frac{2}{3}k}} = \sqrt{\frac{1}{5}} = \frac{\sqrt{5}}{5}.$$

29. Assume the density is $\rho(x, y) = k\,|y|$. Since both the region and the density function are symmetric around the x axis, the y coordinate of the center of mass will be zero. For the x coordinate, we get

$$\bar{x} = \frac{M_y}{m} = \frac{\iint_T x\rho(x, y)\,dA}{\frac{1}{3}k}$$

$$= \frac{3}{k} \int_0^1 \int_{-x}^{x} kx\,|y|\,dy\,dx$$

$$= \frac{3}{k} \left(\int_0^1 \int_{-x}^0 (-kxy)\,dy\,dx + \int_0^1 \int_0^x kxy\,dy\,dx \right)$$

$$= \frac{3}{k} \left(\int_0^1 \left[-\frac{1}{2}kxy^2 \right]_{y=-x}^{y=0} dx + \int_0^1 \left[\frac{1}{2}kxy^2 \right]_{y=0}^{y=x} dx \right)$$

$$= \frac{3}{k} \left(\frac{1}{2}k \int_0^1 x^3\,dx + \frac{1}{2}k \int_0^1 x^3\,dx \right)$$

$$= 3 \int_0^1 x^3\,dx$$

$$= 3 \left[\frac{1}{4}x^4 \right]_0^1 = \frac{3}{4}.$$

Thus the center of mass is $\left(\frac{3}{4},\, 0 \right)$.

For Exercises 31–37, note that the triangle $\mathcal{T}$ is described by $1 \le x \le 2$ and $-x + 1 \le y \le x - 1$.

31. By symmetry, the y coordinate of the centroid is 0. For the x coordinate, we have

$$\bar{x} = \frac{M_y}{m} = \frac{\iint_{T_2} x\rho(x, y)\,dA}{\iint_{T_2} \rho(x, y)\,dA} = \frac{\int_1^2 \int_{-x+1}^{x-1} x\,dy\,dx}{\int_1^2 \int_{-x+1}^{x-1} 1\,dy\,dx} = \frac{\int_1^2 [xy]_{y=-x+1}^{y=x-1}\,dx}{\int_1^2 [y]_{y=-x+1}^{y=x-1}\,dx}$$

$$= \frac{\int_1^2 x(x - 1 - (-x + 1))\,dx}{\int_1^2 (x - 1 - (-x + 1))\,dx} = \frac{\int_1^2 (2x^2 - 2x)\,dx}{\int_1^2 (2x - 2)\,dx} = \frac{\left[\frac{2}{3}x^3 - x^2 \right]_1^2}{[x^2 - 2x]_1^2} = \frac{5}{3}.$$

So the centroid is at $\left(\frac{5}{3},\, 0 \right)$ (as we would expect since this triangle is simply translated to the right by one unit from the triangle in Exercise 24).

33. Assume the density is $\rho(x, y) = kx$. Then by symmetry, since the density is not dependent on the y coordinate, the y coordinate of the center of mass is zero. For the x coordinate we have (using the previous exercise)

$$\bar{x} = \frac{M_y}{m} = \frac{\iint_{T_2} x\rho(x, y)\,dA}{\iint_{T_2} \rho(x, y)\,dA} = \frac{3}{5k} \int_1^2 \int_{-x+1}^{x-1} kx^2\,dy\,dx = \frac{3}{5} \int_1^2 \left[x^2 y \right]_{y=-x+1}^{y=x-1} dx$$

$$= \frac{3}{5} \int_1^2 x^2(x - 1 - (-x + 1))\,dx = \frac{3}{5} \int_1^2 (2x^3 - 2x^2)\,dx$$

$$= \frac{3}{5} \left[\frac{1}{2}x^4 - \frac{2}{3}x^3 \right]_1^2 = \frac{3}{5} \cdot \frac{17}{6} = \frac{17}{10}.$$

35. Assume the density is $\rho(x,y) = kx^2$. Then the mass is

$$\iint_{T_2} \rho(x,y)\,dA = \int_1^2 \int_{-x+1}^{x-1} kx^2\,dy\,dx = k\int_1^2 \left[x^2 y\right]_{y=-x+1}^{y=x-1}\,dx$$

$$= k\int_1^2 \int_1^2 x^2(x-1-(-x+1))\,dx = k\int_1^2 (2x^3 - 2x^2)\,dx$$

$$= k\left[\frac{1}{2}x^4 - \frac{2}{3}x^3\right]_1^2 = \frac{17}{6}k.$$

37. Assume the density is $\rho(x,y) = kx^2$. Then the moments of inertia are

$$I_y = \iint_{T_2} x^2\rho(x,y)\,dA = \int_1^2 \int_{-x+1}^{x-1} kx^4\,dy\,dx = k\int_1^2 \left[x^4 y\right]_{y=-x+1}^{y=x-1}\,dx$$

$$= k\int_1^2 x^4(x-1-(-x+1))\,dx = k\int_1^2 (2x^5 - 2x^4)\,dx$$

$$= k\left[\frac{1}{3}x^6 - \frac{2}{5}x^5\right]_1^2 = \frac{43}{5}k$$

$$I_x = \iint_{T_2} y^2\rho(x,y)\,dA = \int_1^2 \int_{-x+1}^{x-1} kx^2y^2\,dy\,dx = k\int_1^2 \left[\frac{1}{3}x^2 y^3\right]_{y=-x+1}^{y=x-1}\,dx$$

$$= \frac{1}{3}k\int_1^2 x^2((x-1)^3 - (-x+1)^3)\,dx = \frac{2}{3}k\int_1^2 x^2(x-1)^3\,dx$$

$$= \frac{2}{3}k\int_1^2 (x^5 - 3x^4 + 3x^3 - x^2)\,dx$$

$$= \frac{2}{3}k\left[\frac{1}{6}x^6 - \frac{3}{5}x^5 + \frac{3}{4}x^4 - \frac{1}{3}x^3\right]_1^2 = \frac{2}{3}k\cdot\frac{49}{60} = \frac{49}{90}k.$$

Since the mass of T_2 is $\frac{17}{6}k$, the radii of gyration are

$$R_y = \sqrt{\frac{I_y}{m}} = \sqrt{\frac{\frac{43}{5}k}{\frac{17}{6}k}} = \sqrt{\frac{258}{85}}, \qquad R_x = \sqrt{\frac{I_x}{m}} = \sqrt{\frac{\frac{49}{90}k}{\frac{17}{6}k}} = \sqrt{\frac{49}{255}} = \frac{7}{\sqrt{255}}.$$

39. Let the density function be $\rho(x,y) = kx$, which is proportional to the distance from the y axis. Then the mass of $\mathcal{R}$ is

$$m = \iint_{\mathcal{R}} \rho(x,y)\,dA = \int_0^b \int_0^h kx\,dy\,dx = \int_0^b [kxy]_{y=0}^{y=h}\,dx = \int_0^b khx\,dx = \left[\frac{1}{2}khx^2\right]_0^b = \frac{1}{2}khb^2.$$

41. Let the density function be $\rho(x,y) = kx$, which is proportional to the distance from the y axis. Then the moments of inertia are

$$I_y = \iint_{\mathcal{R}} x^2\rho(x,y)\,dA = \int_0^b \int_0^h kx^3\,dy\,dx = k\int_0^b \left[x^3 y\right]_{y=0}^{y=h}\,dx = k\int_0^b x^3 h\,dx$$

$$= kh\left[\frac{1}{4}x^4\right]_0^b = \frac{1}{4}khb^4$$

$$I_x = \iint_{\mathcal{R}} y^2\rho(x,y)\,dA = \int_0^b \int_0^h kxy^2\,dy\,dx = k\int_0^b \left[\frac{1}{3}xy^3\right]_{y=0}^{y=h}\,dx = \frac{1}{3}k\int_0^b xh^3\,dx$$

$$= \frac{1}{3}kh^3\left[\frac{1}{2}x^2\right]_0^b = \frac{1}{6}kb^2h^3.$$

Thus the radii of gyration are, since the mass is $\frac{1}{2}khb^2$,

$$R_y = \sqrt{\frac{I_y}{m}} = \sqrt{\frac{\frac{1}{4}khb^4}{\frac{1}{2}khb^2}} = \frac{\sqrt{2}}{2}b, \qquad R_x = \sqrt{\frac{I_x}{m}} = \sqrt{\frac{\frac{1}{6}kb^2h^3}{\frac{1}{2}khb^2}} = \frac{\sqrt{3}}{3}h.$$

43. Let the density function be $\rho(x,y) = kx^2$. Then the coordinates of the center of mass are

$$\bar{x} = \frac{M_y}{m} = \frac{3}{kb^3h} \iint_{\mathcal{R}} x\rho(x,y)\,dA = \frac{3}{kb^3h} \int_0^b \int_0^h kx^3\,dy\,dx = \frac{3}{b^3h} \int_0^b \left[x^3 y\right]_{y=0}^{y=h}\,dx$$

$$= \frac{3}{b^3h} \int_0^b hx^3\,dx = \frac{3}{b^3}\left[\frac{1}{4}x^4\right]_0^b = \frac{3}{4}b$$

$$\bar{y} = \frac{M_x}{m} = \frac{3}{kb^3h} \iint_{\mathcal{R}} y\rho(x,y)\,dA = \frac{3}{kb^3h} \int_0^b \int_0^h kx^2 y\,dy\,dx = \frac{3}{b^3h} \int_0^b \left[\frac{1}{2}x^2 y^2\right]_{y=0}^{y=h}\,dx$$

$$= \frac{3}{b^3h} \int_0^b \frac{1}{2}h^2 x^2\,dx = \frac{3h}{2b^3}\left[\frac{1}{3}x^3\right]_0^b = \frac{1}{2}h.$$

Therefore the center of mass is $\left(\frac{3}{4}b,\ \frac{1}{2}h\right)$.

45. Since this region is the unit disc, it is symmetric around the origin, so its centroid is $(0,0)$.

47. The density function is $\rho(x,y) = k\,|x|$, which in polar coordinates is $\rho(r) = kr\,|\cos\theta|$. Then the first moments are

$$M_y = \iint_C x\rho(r)\,dA = \int_{-\pi/2}^{3\pi/2} \int_0^1 (r\cos\theta)\cdot kr\,|\cos\theta|\cdot r\,dr\,d\theta$$

$$= k\int_{-\pi/2}^{3\pi/2} \int_0^1 r^3 \cos\theta\,|\cos\theta|\,dr\,d\theta$$

$$= k\int_{-\pi/2}^{3\pi/2} \left[\frac{1}{4}r^4 \cos\theta\,|\cos\theta|\right]_{r=0}^{r=1}\,d\theta$$

$$= \frac{1}{4}k\int_{-\pi/2}^{3\pi/2} \cos\theta\,|\cos\theta|\,d\theta$$

$$= \frac{1}{4}k\left(\int_{-\pi/2}^{\pi/2} \cos^2\theta\,d\theta - \int_{\pi/2}^{3\pi/2} \cos^2\theta\,d\theta\right)$$

$$= \frac{1}{4}k\left(\int_{-\pi/2}^{\pi/2} \frac{1}{2}(1+\cos 2\theta)\,d\theta - \int_{\pi/2}^{3\pi/2} \frac{1}{2}(1+\cos 2\theta)\,d\theta\right)$$

$$= \frac{1}{4}k\left(\left[\frac{1}{2}\theta + \frac{1}{4}\sin 2\theta\right]_{-\pi/2}^{\pi/2} - \left[\frac{1}{2}\theta + \frac{1}{4}\sin 2\theta\right]_{\pi/2}^{3\pi/2}\right)$$

$$= 0$$

$$M_x = \iint_C y\rho(r)\,dA = \int_{-\pi/2}^{3\pi/2} \int_0^1 (r\sin\theta)\cdot kr\,|\cos\theta|\cdot r\,dr\,d\theta$$

$$= k\int_{-\pi/2}^{3\pi/2} \int_0^1 r^3 \sin\theta\,|\cos\theta|\,dr\,d\theta$$

$$= k\int_{-\pi/2}^{3\pi/2} \left[\frac{1}{4}r^4 \sin\theta\,|\cos\theta|\right]_{r=0}^{r=1}\,d\theta$$

$$= \frac{1}{4}k\int_{-\pi/2}^{3\pi/2} \sin\theta\,|\cos\theta|\,d\theta$$

$$= \frac{1}{4}k\left(\int_{-\pi/2}^{\pi/2} \sin\theta\cos\theta\,d\theta - \int_{\pi/2}^{3\pi/2} \sin\theta\cos\theta\,d\theta\right)$$

$$= \frac{1}{4}k\left(\left[-\frac{1}{2}\cos^2\theta\right]_{-\pi/2}^{\pi/2} - \left[-\frac{1}{2}\cos^2\theta\right]_{\pi/2}^{3\pi/2}\right)$$

$$= 0.$$

Thus the centroid is at $(0,0)$. This is the expected result: since the density function does not depend on the y coordinate, and the region is symmetric around the x axis, the y coordinate of the center of mass should be zero. Since the density function and the region are both symmetric around the y axis, the x coordinate should be zero as well.

49. The density function is $\rho(x,y) = k\sqrt{x^2 + y^2}$, which in polar coordinates is $\rho(r) = kr$. Then the mass is

$$m = \iint_C \rho(r)\,dA = \int_0^{2\pi}\int_0^1 kr\cdot r\,dr\,d\theta = k\int_0^{2\pi}\left[\frac{1}{3}r^3\right]_{r=0}^{r=1}\,d\theta = \frac{1}{3}k\int_0^{2\pi} 1\,d\theta = \frac{2}{3}\pi k.$$

51. The density function is $\rho(x,y) = k\sqrt{x^2 + y^2}$, which in polar coordinates is $\rho(r) = kr$. Then the moments of inertia are

$$I_y = \iint_C x^2\rho(r)\,dA = \int_0^{2\pi}\int_0^1 (r\cos\theta)^2\cdot kr\cdot r\,dr\,d\theta$$

$$= k\int_0^{2\pi}\int_0^1 r^4\cos^2\theta\,dr\,d\theta$$

$$= k\int_0^{2\pi}\left[\frac{1}{5}r^5\cos^2\theta\right]_{r=0}^{r=1}\,d\theta$$

$$= \frac{1}{5}k\int_0^{2\pi}\cos^2\theta\,d\theta$$

$$= \frac{1}{10}k\int_0^{2\pi}(1+\cos 2\theta)\,d\theta$$

$$= \frac{1}{10}k\left[\theta + \frac{1}{2}\sin 2\theta\right]_0^{2\pi} = \frac{1}{5}\pi k$$

$$I_x = \iint_C y^2 \rho(r)\, dA = \int_0^{2\pi} \int_0^1 (r\sin\theta)^2 \cdot kr \cdot r\, dr\, d\theta$$

$$= k \int_0^{2\pi} \int_0^1 r^4 \sin^2\theta\, dr\, d\theta$$

$$= k \int_0^{2\pi} \left[\frac{1}{5} r^5 \sin^2\theta\right]_{r=0}^{r=1} d\theta$$

$$= \frac{1}{5} k \int_0^{2\pi} \sin^2\theta\, d\theta$$

$$= \frac{1}{10} k \int_0^{2\pi} (1 - \cos 2\theta)\, d\theta$$

$$= \frac{1}{10} k \left[\theta - \frac{1}{2}\sin 2\theta\right]_0^{2\pi} = \frac{1}{5}\pi k.$$

Thus the radii of gyration are, since the mass is $\frac{2}{3}\pi k$,

$$R_y = \sqrt{\frac{I_y}{m}} = \sqrt{\frac{\frac{1}{5}\pi k}{\frac{2}{3}\pi k}} = \sqrt{\frac{3}{10}} = \frac{\sqrt{30}}{10}, \qquad R_x = \sqrt{\frac{I_x}{m}} = \sqrt{\frac{\frac{1}{5}\pi k}{\frac{2}{3}\pi k}} = \sqrt{\frac{3}{10}} = \frac{\sqrt{30}}{10}.$$

53. The density function is $\rho(x,y) = k|x|$, which in polar coordinates is $\rho(r) = kr|\cos\theta|$. Since $x \geq 0$ everywhere in the region, $\cos\theta \geq 0$, so that the density is simply $\rho(r) = kr\cos\theta$. The mass is then

$$m = \iint_S \rho(r)\, dA = \int_{-\pi/2}^{\pi/2} \int_0^1 kr\cos\theta \cdot r\, dr\, d\theta$$

$$= k \int_{-\pi/2}^{\pi/2} \int_0^1 r^2 \cos\theta\, dr\, d\theta$$

$$= k \int_{-\pi/2}^{\pi/2} \left[\frac{1}{3} r^3 \cos\theta\right]_{r=0}^{r=1} d\theta$$

$$= \frac{1}{3} k \int_{-\pi/2}^{\pi/2} \cos\theta\, d\theta$$

$$= \frac{1}{3} k [\sin\theta]_{-\pi/2}^{\pi/2} = \frac{2}{3} k.$$

55. The density function is $\rho(x,y) = k|x|$, which in polar coordinates is $\rho(r) = kr|\cos\theta|$. Since $x \geq 0$ everywhere in the region, $\cos\theta \geq 0$, so that the density is simply $\rho(r) = kr\cos\theta$. The moments of

inertia are

$$
\begin{aligned}
I_y = \iint_C x^2 \rho(r)\, dA &= \int_{-\pi/2}^{\pi/2} \int_0^1 (r\cos\theta)^2 \cdot kr\cos\theta \cdot r\, dr\, d\theta \\
&= k \int_{-\pi/2}^{\pi/2} \int_0^1 r^4 \cos^3\theta\, dr\, d\theta \\
&= k \int_{-\pi/2}^{\pi/2} \left[\frac{1}{5} r^5 \cos^3\theta \right]_{r=0}^{r=1} d\theta \\
&= \frac{1}{5} k \int_{-\pi/2}^{\pi/2} \cos^3\theta\, d\theta \\
&= \frac{1}{5} k \int_{-\pi/2}^{\pi/2} \cos\theta(1-\sin^2\theta)\, d\theta \\
&= \frac{1}{5} k \left[\sin\theta - \frac{1}{3}\sin^3\theta \right]_{-\pi/2}^{\pi/2} \\
&= \frac{4}{15} k
\end{aligned}
$$

$$
\begin{aligned}
I_x = \iint_C y^2 \rho(r)\, dA &= \int_{-\pi/2}^{\pi/2} \int_0^1 (r\sin\theta)^2 \cdot kr\cos\theta \cdot r\, dr\, d\theta \\
&= k \int_{-\pi/2}^{\pi/2} \int_0^1 r^4 \sin^2\theta \cos\theta\, dr\, d\theta \\
&= k \int_{-\pi/2}^{\pi/2} \left[\frac{1}{5} r^5 \sin^2\theta \cos\theta \right]_{r=0}^{r=1} d\theta \\
&= \frac{1}{5} k \int_{-\pi/2}^{\pi/2} \sin^2\theta \cos\theta\, d\theta \\
&= \frac{1}{5} k \left[\frac{1}{3}\sin^3\theta \right]_{-\pi/2}^{\pi/2} \\
&= \frac{2}{15} k.
\end{aligned}
$$

Thus the radii of gyration are, since the mass is $\frac{2}{3} k$,

$$
R_y = \sqrt{\frac{I_y}{m}} = \sqrt{\frac{\frac{4}{15} k}{\frac{2}{3} k}} = \sqrt{\frac{2}{5}} = \frac{\sqrt{10}}{5}, \qquad
R_x = \sqrt{\frac{I_x}{m}} = \sqrt{\frac{\frac{2}{15} k}{\frac{2}{3} k}} = \sqrt{\frac{1}{5}} = \frac{\sqrt{5}}{5}.
$$

57. The density function is $\rho(x,y) = k\sqrt{x^2 + y^2}$, which in polar coordinates is simply $\rho(r) = kr$. The first moments are

$$
\begin{aligned}
M_y = \iint_S x\rho(r)\,dA &= \int_{-\pi/2}^{\pi/2} \int_0^1 (r\cos\theta) \cdot kr \cdot r\,dr\,d\theta \\
&= k \int_{-\pi/2}^{\pi/2} \int_0^1 r^3 \cos\theta\,dr\,d\theta \\
&= k \int_{-\pi/2}^{\pi/2} \left[\frac{1}{4} r^4 \cos\theta \right]_{r=0}^{r=1} d\theta \\
&= \frac{1}{4} k \int_{-\pi/2}^{\pi/2} \cos\theta\,d\theta \\
&= \frac{1}{4} k \left[\sin\theta \right]_{-\pi/2}^{\pi/2} = \frac{1}{2} k
\end{aligned}
$$

$$
\begin{aligned}
M_x = \iint_S y\rho(r)\,dA &= \int_{-\pi/2}^{\pi/2} \int_0^1 (r\sin\theta) \cdot kr \cdot r\,dr\,d\theta \\
&= k \int_{-\pi/2}^{\pi/2} \int_0^1 r^3 \sin\theta\,dr\,d\theta \\
&= k \int_{-\pi/2}^{\pi/2} \left[\frac{1}{4} r^4 \sin\theta \right]_{r=0}^{r=1} d\theta \\
&= \frac{1}{4} k \int_{-\pi/2}^{\pi/2} \sin\theta\,d\theta \\
&= \frac{1}{4} k \left[-\cos\theta \right]_{-\pi/2}^{\pi/2} = 0.
\end{aligned}
$$

Since the mass is $\frac{1}{3}\pi k$, the center of mass is

$$
(\bar{x}, \bar{y}) = \left(\frac{M_y}{m}, \frac{M_x}{m} \right) = \left(\frac{3}{2\pi}, 0 \right).
$$

We needn't have actually computed $\bar{y}$; we could have known it was zero from symmetry considerations.

59. By Definition 13.12 and the material that follows, the centers of mass of Ω_1 and Ω_2 are

$$
(\bar{x}_1, \bar{y}_1) = \left(\frac{M_y}{m_1}, \frac{M_x}{m_1} \right) = \left(\frac{\iint_{\Omega_1} x\rho(x,y)\,dA}{m_1}, \frac{\iint_{\Omega_1} y\rho(x,y)\,dA}{m_1} \right)
$$

$$
(\bar{x}_2, \bar{y}_2) = \left(\frac{M_y}{m_2}, \frac{M_x}{m_2} \right) = \left(\frac{\iint_{\Omega_2} x\rho(x,y)\,dA}{m_2}, \frac{\iint_{\Omega_2} y\rho(x,y)\,dA}{m_2} \right)
$$

where $\rho(x,y)$ is the density of the lamina $\Omega = \Omega_1 \cup \Omega_2$. But then

$$
(m_1\bar{x}_1, m_1\bar{y}_1) = \left(\iint_{\Omega_1} x\rho(x,y)\,dA, \iint_{\Omega_1} y\rho(x,y)\,dA \right)
$$

$$
(m_2\bar{x}_2, m_2\bar{y}_2) = \left(\iint_{\Omega_2} x\rho(x,y)\,dA, \iint_{\Omega_2} y\rho(x,y)\,dA \right),
$$

so that

$$(m_1\bar{x}_1 + m_2\bar{x}_2, m_1\bar{y}_1 + m_2\bar{y}_2) = \left(\iint_{\Omega_1} x\rho(x,y)\,dA + \iint_{\Omega_2} x\rho(x,y)\,dA, \right.$$

$$\iint_{\Omega_1} y\rho(x,y)\,dA + \iint_{\Omega_2} y\rho(x,y)\,dA \bigg)$$

$$= \left(\iint_{\Omega} x\rho(x,y)\,dA, \iint_{\Omega} y\rho(x,y)\,dA \right).$$

Divide through by $m_1 + m_2$ to get

$$\left(\frac{m_1\bar{x}_1 + m_2\bar{x}_2}{m_1 + m_2}, \frac{m_1\bar{y}_1 + m_2\bar{y}_2}{m_1 + m_2} \right) = \left(\frac{\iint_{\Omega} x\rho(x,y)\,dA}{m_1 + m_2}, \frac{\iint_{\Omega} y\rho(x,y)\,dA}{m_1 + m_2} \right) = (\bar{x}, \bar{y}).$$

The last equality follows since the mass of Ω is $m_1 + m_2$ and $\rho(x,y)$ is the density of Ω.

61. The region is composed of three squares of side length 1; since the density is constant, say equal to ρ, the centroid of the lower left square is at

$$(\bar{x}, \bar{y}) = \left(\frac{\iint_{[0,1]\times[0,1]} x\rho\,dA}{\iint_{[0,1]\times[0,1]} \rho\,dA}, \frac{\iint_{[0,1]\times[0,1]} y\rho\,dA}{\iint_{[0,1]\times[0,1]} \rho\,dA} \right)$$

$$= \left(\int_0^1 \int_0^1 x\,dx\,dy, \int_0^1 \int_0^1 y\,dx\,dy \right)$$

$$= \left(\int_0^1 \left[\frac{1}{2}x^2 \right]_0^1 dy, \int_0^1 [xy]_0^1\,dy \right)$$

$$= \left(\int_0^1 \frac{1}{2}\,dy, \int_0^1 y\,dy \right)$$

$$= \left(\frac{1}{2}, \frac{1}{2} \right),$$

a fact that is also apparent from symmetry. Similarly, the centers of mass of the other two squares are at $\left(\frac{1}{2}, \frac{3}{2} \right)$ and $\left(\frac{3}{2}, \frac{1}{2} \right)$. Since the density is constant, the masses of the three squares are equal, say to m, so that by Exercise 60, the center of mass of the lamina is at

$$\left(\frac{\frac{1}{2}m + \frac{1}{2}m + \frac{3}{2}m}{3m}, \frac{\frac{1}{2}m + \frac{3}{2}m + \frac{1}{2}m}{3m} \right) = \left(\frac{5}{6}, \frac{5}{6} \right).$$

63. Since the density is constant, and constant densities factor out of the calculations, assume the density is 1. Then the mass of each piece is the same as its area. The lower left rectangle (R_1) has area $b_1 h_1$; the upper left (R_2) has area $b_1(h_2 - h_1)$, and the lower right (R_3) has area $h_1(b_2 - b_1)$.

Then we have

$$(\bar{x}_1, \bar{y}_1) = \left(\frac{\int_0^{h_1} \int_0^{b_1} x \, dx \, dy}{b_1 h_1}, \frac{\int_0^{h_1} \int_0^{b_1} y \, dx \, dy}{b_1 h_1} \right)$$

$$= \frac{1}{b_1 h_1} \left(\int_0^{h_1} \left[\frac{1}{2} x^2 \right]_{x=0}^{x=b_1} dy, \int_0^{h_1} b_1 y \, dy \right)$$

$$= \frac{1}{b_1 h_1} \left(\int_0^{h_1} \frac{b_1^2}{2} \, dy, \int_0^{h_1} b_1 y \, dy \right)$$

$$= \frac{1}{b_1 h_1} \left(\left[\frac{b_1^2}{2} y \right]_0^{h_1}, \left[\frac{b_1}{2} y^2 \right]_0^{h_1} \right)$$

$$= \left(\frac{b_1}{2}, \frac{h_1}{2} \right),$$

$$(\bar{x}_2, \bar{y}_2) = \left(\frac{\int_{h_1}^{h_2} \int_0^{b_1} x \, dx \, dy}{b_1 (h_2 - h_1)}, \frac{\int_{h_1}^{h_2} \int_0^{b_1} y \, dx \, dy}{b_1 (h_2 - h_1)} \right)$$

$$= \frac{1}{b_1 (h_2 - h_1)} \left(\int_{h_1}^{h_2} \left[\frac{1}{2} x^2 \right]_{x=0}^{x=b_1} dy, \int_{h_1}^{h_2} b_1 y \, dy \right)$$

$$= \frac{1}{b_1 (h_2 - h_1)} \left(\int_{h_1}^{h_2} \frac{b_1^2}{2} \, dy, \int_{h_1}^{h_2} b_1 y \, dy \right)$$

$$= \frac{1}{b_1 (h_2 - h_1)} \left(\left[\frac{b_1^2}{2} y \right]_{h_1}^{h_2}, \left[\frac{1}{2} b_1 y^2 \right]_{h_1}^{h_2} \right)$$

$$= \left(\frac{b_1}{2}, \frac{h_2 + h_1}{2} \right),$$

$$(\bar{x}_3, \bar{y}_3) = \left(\frac{\int_0^{h_1} \int_{b_1}^{b_2} x \, dx \, dy}{(b_2 - b_1) h_1}, \frac{\int_0^{h_1} \int_{b_1}^{b_2} y \, dx \, dy}{(b_2 - b_1) h_1} \right)$$

$$= \frac{1}{(b_2 - b_1) h_1} \left(\int_0^{h_1} \left[\frac{1}{2} x^2 \right]_{x=b_1}^{x=b_2} dy, \int_0^{h_1} (b_2 - b_1) y \, dy \right)$$

$$= \frac{1}{(b_2 - b_1) h_1} \left(\int_0^{h_1} \frac{(b_2 - b_1)^2}{2} \, dy, \int_0^{h_1} (b_2 - b_1) y \, dy \right)$$

$$= \frac{1}{(b_2 - b_1) h_1} \left(\left[\frac{(b_2 - b_1)^2}{2} y \right]_0^{h_1}, \left[\frac{b_2 - b_1}{2} y^2 \right]_0^{h_1} \right)$$

$$= \left(\frac{b_2 + b_1}{2}, \frac{h_1}{2} \right).$$

Thus by Exercise 60, the center of mass of the entire lamina is

$$\frac{b_1 h_1 \left(\frac{b_1}{2}, \frac{h_1}{2}\right) + b_1(h_2 - h_1)\left(\frac{b_1}{2}, \frac{h_2 + h_1}{2}\right) + (b_2 - b_1)h_1 \left(\frac{b_2 + b_1}{2}, \frac{h_1}{2}\right)}{b_1 h_1 + b_1(h_2 - h_1) + (b_2 - b_1)h_1},$$

which simplifies to

$$\left(\frac{b_2^2 h_1 + b_1^2(h_2 - h_1)}{2(b_2 h_1 + b_1(h_2 - h_1))}, \frac{b_2 h_1^2 + b_1(h_2^2 - h_1^2)}{2(b_2 h_1 + b_1(h_2 - h_1))}\right).$$

65. Using the same rectangles as in exercise 64, we have for the masses (assuming the constant factor of proportionality is 1, since it will cancel in any case)

$$m_1 = \int_{-3}^{3} \int_{4}^{6} y \, dy \, dx = \int_{-3}^{3} \left[\frac{1}{2}y^2\right]_{y=4}^{y=6} dx = \int_{-3}^{3} 10 \, dx = 60$$

$$m_2 = \int_{-1}^{1} \int_{0}^{4} y \, dy \, dx = \int_{-1}^{1} \left[\frac{1}{2}y^2\right]_{y=0}^{y=4} dx = \int_{-1}^{1} 8 \, dx = 16.$$

Then the centers of mass are

$$(\bar{x}_1, \bar{y}_1) = \frac{1}{60}\left(\int_{-3}^{3} \int_{4}^{6} xy \, dy \, dx, \int_{-3}^{3} \int_{4}^{6} y^2 \, dy \, dx\right)$$

$$= \frac{1}{60}\left(\int_{-3}^{3} \left[\frac{1}{2}xy^2\right]_{y=4}^{y=6} dx, \int_{-3}^{3} \left[\frac{1}{3}y^3\right]_{y=4}^{y=6} dx\right)$$

$$= \frac{1}{60}\left(\int_{-3}^{3} 10x \, dx, \int_{-3}^{3} \frac{152}{3} dx\right)$$

$$= \frac{1}{60}\left(\left[5x^2\right]_{-3}^{3}, \left[\frac{152}{3}x\right]_{-3}^{3}\right)$$

$$= \left(0, \frac{76}{15}\right),$$

$$(\bar{x}_2, \bar{y}_2) = \frac{1}{16}\left(\int_{-1}^{1} \int_{0}^{4} xy \, dy \, dx, \int_{-1}^{1} \int_{0}^{4} y^2 \, dy \, dx\right)$$

$$= \frac{1}{16}\left(\int_{-1}^{1} \left[\frac{1}{2}xy^2\right]_{y=0}^{y=4} dx, \int_{-1}^{1} \left[\frac{1}{3}y^3\right]_{y=0}^{y=4} dx\right)$$

$$= \frac{1}{16}\left(\int_{-1}^{1} 8x \, dx, \int_{-1}^{1} \frac{64}{3} dx\right)$$

$$= \frac{1}{16}\left(\left[4x^2\right]_{-1}^{1}, \left[\frac{64}{3}x\right]_{-1}^{1}\right)$$

$$= \left(0, \frac{8}{3}\right).$$

Hence the center of mass of the lamina is, by Exercise 59 or 60,

$$(\bar{x}, \bar{y}) = \left(\frac{60 \cdot 0 + 16 \cdot 0}{60 + 16}, \frac{60 \cdot \frac{76}{15} + 16 \cdot \frac{8}{3}}{60 + 16}\right) = \left(0, \frac{260}{57}\right)$$

67. Using the same division as in Exercise 66, we get for the masses of the two regions

$$m_R = \int_0^a \int_{-a}^a y\,dx\,dy = \int_0^a [xy]_{x=-a}^{x=a}\,dy = \int_0^a 2ay\,dy = \left[ay^2\right]_0^a = a^3$$

$$m_T = \int_a^{2a} \int_{y-2a}^{2a-y} y\,dx\,dy = \int_a^{2a} [xy]_{x=y-2a}^{x=2a-y}\,dy = \int_a^{2a} (4ay - 2y^2)\,dy$$

$$= \left[2ay^2 - \frac{2}{3}y^3\right]_a^{2a} = 8a^3 - \frac{16}{3}a^3 - 2a^3 + \frac{2}{3}a^3 = \frac{4}{3}a^3.$$

Then the centers of mass are

$$(\bar{x}_R, \bar{y}_R) = \frac{1}{a^3} \left(\int_0^a \int_{-a}^a xy\,dx\,dy, \int_0^a \int_{-a}^a y^2\,dx\,dy \right)$$

$$= \frac{1}{a^3} \left(\int_0^a \left[\frac{1}{2}x^2 y\right]_{x=-a}^{x=a}\,dy, \int_0^a \left[xy^2\right]_{x=-a}^{x=a}\,dy \right)$$

$$= \frac{1}{a^3} \left(\int_0^a 0\,dy, \int_0^a 2ay^2\,dy \right)$$

$$= \left(0, \frac{1}{a^3}\left[\frac{2}{3}ay^3\right]_0^a\right) = \left(0, \frac{2}{3}a\right)$$

$$(\bar{x}_T, \bar{y}_T) = \frac{3}{4a^3} \left(\int_a^{2a} \int_{y-2a}^{2a-y} xy\,dx\,dy, \int_a^{2a} \int_{y-2a}^{2a-y} y^2\,dx\,dy \right)$$

$$= \frac{3}{4a^3} \left(\int_a^{2a} \left[\frac{1}{2}x^2 y\right]_{x=y-2a}^{x=2a-y}\,dy, \int_a^{2a} \left[xy^2\right]_{x=y-2a}^{x=2a-y}\,dy \right)$$

$$= \frac{3}{4a^3} \left(\int_a^{2a} 0\,dy, \int_a^{2a} (4ay^2 - 2y^3)\,dy \right)$$

$$= \left(0, \frac{3}{4a^3}\left[\frac{4}{3}ay^3 - \frac{1}{2}y^4\right]_a^{2a}\right)$$

$$= \left(0, \frac{3}{4a^3}\left(\frac{32}{3}a^4 - 8a^4 - \frac{4}{3}a^4 + \frac{1}{2}a^4\right)\right)$$

$$= \left(0, \frac{11}{8}a\right).$$

Then by Exercise 59 or 60, the center of mass of the entire lamina is

$$(\bar{x}, \bar{y}) = \left(\frac{a^3 \cdot 0 + \frac{4}{3}a^3 \cdot 0}{\frac{7}{3}a^3}, \frac{a^3 \cdot \frac{2}{3}a + \frac{4}{3}a^3 \cdot \frac{11}{8}a}{\frac{7}{3}a^3} \right) = \left(0, \frac{15}{14}a\right).$$

Applications

69. The expected value is the first moment about the y axis (in the x direction) with respect to the given probability density function, so it is

$$\int_0^{4.3/0.8} \int_0^{4.3-0.8x} 0.0873x \, dy \, dx = \int_0^{5.375} [0.0873xy]_{y=0}^{y=4.3-0.8x}$$

$$= \int_0^{5.375} 0.0873x(4.3 - 0.8x) \, dx$$

$$= 0.0873 \int_0^{5.375} (4.3x - 0.8x^2) \, dx$$

$$= 0.0873 \left[2.15x^2 - \frac{0.8}{3}x^3 \right]_0^{5.375}$$

$$\approx 1.79.$$

Proofs

71. Since the triangle has base a and height d, its area is $\frac{1}{2}ad$, so assuming the density is $\rho(x,y) = 1$, its mass is also $m = \frac{1}{2}ad$. Now, regarding $\mathcal{T}$ as a Type II region, it is bounded by $y = 0$ and $y = d$, and by the lines AC and BC, which have equations

$$y = \frac{d}{c}x, \qquad y = \frac{d}{c-a}(x-a) = \frac{d}{c-a}x - \frac{ad}{c-a}.$$

As functions of y, these become

$$x = \frac{c}{d}y, \qquad x = \frac{c-a}{d}y + a.$$

Thus the first moments are

$$M_y = \iint_{\mathcal{T}} x\rho(x,y) \, dA = \int_0^d \int_{\frac{c}{d}y}^{\frac{c-a}{d}y+a} x \, dx \, dy$$

$$= \int_0^d \left[\frac{1}{2}x^2 \right]_{x=\frac{c}{d}y}^{x=\frac{c-a}{d}y+a} dy$$

$$= \frac{1}{2} \int_0^d \left(\left(\frac{c-a}{d}y + a \right)^2 - \left(\frac{c}{d}y \right)^2 \right) dy$$

$$= \frac{1}{2} \int_0^d \left(\frac{(c-a)^2}{d^2}y^2 + \frac{2a(c-a)}{d}y + a^2 - \frac{c^2}{d^2}y^2 \right) dy$$

$$= \frac{1}{2} \int_0^d \left(\frac{a^2 - 2ac}{d^2}y^2 + \frac{2a(c-a)}{d}y + a^2 \right) dy$$

$$= \frac{1}{2} \left[\frac{a^2 - 2ac}{3d^2}y^3 + \frac{a(c-a)}{d}y^2 + a^2 y \right]_0^d$$

$$= \frac{1}{6}ad(a+c)$$

$$M_x = \iint_T y\rho(x,y)\,dA = \int_0^d \int_{\frac{c}{d}y}^{\frac{c-a}{d}y+a} y\,dx\,dy$$

$$= \int_0^d [xy]_{x=\frac{c}{d}y}^{x=\frac{c-a}{d}y+a}\,dy$$

$$= \int_0^d y\left(\frac{c-a}{d}y + a - \frac{c}{d}y\right)\,dy$$

$$= \int_0^d \left(-\frac{a}{d}y^2 + ay\right)\,dy$$

$$= \left[-\frac{a}{3d}y^3 + \frac{a}{2}y^2\right]_0^d$$

$$= \frac{1}{6}ad^2.$$

Thus the centroid is

$$(\bar{x}, \bar{y}) = \left(\frac{M_y}{m}, \frac{M_x}{m}\right) = \left(\frac{\frac{1}{6}ad(a+c)}{\frac{1}{2}ad}, \frac{\frac{1}{6}ad^2}{\frac{1}{2}ad}\right) = \left(\frac{a+c}{3}, \frac{d}{3}\right),$$

which matches the result from the previous exercise.

73. Assume the circle has radius R. We may as well assume it is centered at the origin, since translating it around the plane does not change whether or not its centroid is at its center. Then letting $\rho(x,y) = 1$, and we have for the first moments, using polar coordinates,

$$M_y = \iint_C x\rho(x,y)\,dA = \int_0^{2\pi} \int_0^R r\cos\theta \cdot r\,dr\,d\theta$$

$$= \int_0^{2\pi} \int_0^R r^2 \cos\theta\,dr\,d\theta$$

$$= \int_0^{2\pi} \left[\frac{1}{3}r^3 \cos\theta\right]_{r=0}^{r=R}\,d\theta$$

$$= \frac{1}{3}R^3 \int_0^{2\pi} \cos\theta\,d\theta$$

$$= \frac{1}{3}R^3 [\sin\theta]_0^{2\pi} = 0$$

$$M_x = \iint_C y\rho(x,y)\,dA = \int_0^{2\pi} \int_0^R r\sin\theta \cdot r\,dr\,d\theta$$

$$= \int_0^{2\pi} \int_0^R r^2 \sin\theta\,dr\,d\theta$$

$$= \int_0^{2\pi} \left[\frac{1}{3}r^3 \sin\theta\right]_{r=0}^{r=R}\,d\theta$$

$$= \frac{1}{3}R^3 \int_0^{2\pi} \sin\theta\,d\theta$$

$$= \frac{1}{3}R^3 [-\cos\theta]_0^{2\pi} = 0.$$

Since both first moments are zero, the centroid is at the origin, which is the center of the circle.

Thinking Forward

A triple iterated integral

We have

$$\int_\alpha^\beta \int_\gamma^\delta \int_\epsilon^\zeta dz\,dy\,dx = \int_\alpha^\beta \int_\gamma^\delta [z]_{z=\epsilon}^{z=\zeta}\,dy\,dx$$

$$= \int_\alpha^\beta \int_\gamma^\delta (\zeta - \epsilon)\,dy\,dx$$

$$= \int_\alpha^\beta [(\zeta - \epsilon)y]_{y=\gamma}^{y=\delta}\,dx$$

$$= \int_\alpha^\beta (\zeta - \epsilon)(\delta - \gamma)\,dx$$

$$= [(\zeta - \epsilon)(\delta - \gamma)]_{x=\alpha}^{x=\beta}$$

$$= (\zeta - \epsilon)(\delta - \gamma)(\beta - \alpha).$$

This integral represents the volume of the rectangular parallelepiped with opposite vertices $(\alpha, \gamma, \epsilon)$ and (β, δ, ζ).

A triple iterated integral of a density function

This triple integral represents the mass of the rectangular parallelepiped from the previous part that has density given by the function $\rho(x, y, z)$.

13.5   Triple Integrals

Thinking Back

Using iterated double integrals to compute area

As a Type I region, Ω is bounded on the left by $x = 0$ and on the right by $x = 1$, below by $y = 0$ and above by $y = x$, so the double integral representing its area is

$$\int_0^1 \int_0^x 1\,dy\,dx.$$

As a Type II region, Ω is bounded below by $y = 0$ and above by $y = 1$, on the left by $y = x$ (or $x = y$) and on the right by $x = 1$. Thus the double integral representing its area is

$$\int_0^1 \int_x^1 1\,dy\,dx.$$

Using iterated double integrals to compute mass

The two lines $y = x$ and $y = 2 - x$ meet at the point $(1, 1)$. So as a Type I region, Ω is bounded on the left by $x = 0$ and on the right by $x = 1$. It is bounded below by $y = x$ and above by $y = 2 - x$. So the double integral representing its mass is

$$\int_0^1 \int_x^{2-x} \rho(x, y)\,dy\,dx.$$

As a Type II region, Ω is bounded below by $y = 0$ and above by $y = 2$. It is bounded on the left by $x = 0$. On the right, it is bounded by two different functions: $y = x$ (or $x = y$) for $0 \leq y \leq 1$, and $y = 2 - x$ (or $x = 2 - y$) for $1 \leq y \leq 2$. Thus the corresponding double integral is actually the sum of two integrals:

$$\int_0^1 \int_0^y \rho(x, y)\, dx\, dy + \int_1^2 \int_0^{2-y} \rho(x, y)\, dx\, dy.$$

Concepts

1. (a) False. The integration order is incorrect. As written, for example, x corresponds to the interval from c_1 to c_2. The correct integral is $\int_{a_1}^{a_2} \int_{b_1}^{b_2} \int_{c_1}^{c_2} f(x, y, z)\, dz\, dy\, dx$.

    (b) True. According to the order of integration, y corresponds to the interval $[b_1, b_2]$, z to $[c_1, c_2]$, and x to $[a_1, a_2]$, which is correct.

    (c) False. This is only the case if Ω_1 and Ω_2 are disjoint.

    (d) True. This is the definition of a bounded subset of $\mathbb{R}^3$.

    (e) True. Since f is positive, we have

$$\iiint_\Omega f(x, y, z)\, dV = \iiint_\Lambda f(x, y, z)\, dV + \iiint_{\Omega-\Lambda} f(x, y, z)\, dV \geq \iiint_\Lambda f(x, y, z)\, dV.$$

    (f) False. By Definition 13.19, it is $M_{xy} = \iiint_\Omega z\rho(x, y, z)\, dV$.

    (g) True. See Definition 13.20(a).

    (h) False. We must integrate with respect to z first. The correct integral is $\int_0^2 \int_0^2 \int_{g_1(x,y)}^{g_2(x,y)} dV$.

3. The difference between the two expressions is simply that the summations have been reordered; the thing being summed is the same on both sides. Since expanding gives a finite number of terms, those terms may be written in any order. Thus the two sides are the same.

5. Using the procedure in Exercise 4, instead of selecting an arbitrary point $(x_j^*, y_k^*, z_l^*) \in [x_{j-1}, x_j] \times [y_{k-1}, y_k] \times [z_{l-1}, z_l]$, select the centroid of that region, which is

$$(x_j^*, y_k^*, z_l^*) = \left(\frac{x_{j-1} + x_j}{2}, \frac{y_{k-1} + y_k}{2}, \frac{z_{l-1} + z_l}{2} \right).$$

    Then the Riemann sum will be a midpoint sum.

7. A triple integral is the limit of a three-dimensional Riemann sum as the number of subintervals in all three directions goes to infinity at the same time. An iterated integral breaks that process up into three one-dimensional processes: for a fixed number of subintervals in two directions, let the number of subintervals go to infinity in the other direction and compute the limit. Obviously this limit may depend on the number of chosen subintervals in the other directions. Then let the number of subintervals in one of the remaining directions go to infinity. Finally, let the number of subintervals in the third direction go to infinity. Thus an iterated integral consist of an iterated evaluation of limits.

9. The region over which we are integrating in the yz plane is a Type I region bounded by $a \leq y \leq b$ and $h_1(y) \leq z \leq h_2(y)$, and the range of values over which we integrate for each point (y, z) is $g_1(y, z) \leq x \leq g_2(y, z)$. Thus the integral is

$$\int_a^b \int_{h_1(y)}^{h_2(y)} \int_{g_1(y,z)}^{g_2(y,z)} f(x, y, z)\, dx\, dz\, dy.$$

In the second case, the region is described instead as a Type II region bounded by $a \leq z \leq b$ and $h_1(z) \leq y \leq h_2(z)$, and the range of values over which we integrate is the same as above. Thus the integral is

$$\int_a^b \int_{h_1(z)}^{h_2(z)} \int_{g_1(y,z)}^{g_2(y,z)} f(x,y,z)\, dx\, dy\, dz.$$

11. Since the solid $\mathcal{R}$ is rectangular, changing the order of integration involves just moving the integral signs and exchanging the order of the differentials. The six orders are

$$\int_{-1}^{3} \int_0^2 \int_2^7 \rho(x,y,z)\, dz\, dy\, dx, \qquad \int_{-1}^{3} \int_2^7 \int_0^2 \rho(x,y,z)\, dy\, dz\, dx, \qquad \int_0^2 \int_{-1}^{3} \int_2^7 \rho(x,y,z)\, dz\, dx\, dy,$$

$$\int_2^7 \int_0^2 \int_{-1}^{3} \rho(x,y,z)\, dx\, dy\, dz, \qquad \int_0^2 \int_2^7 \int_{-1}^{3} \rho(x,y,z)\, dx\, dz\, dy, \qquad \int_2^7 \int_{-1}^{3} \int_0^2 \rho(x,y,z)\, dy\, dx\, dz.$$

13. The equation of the slanted plane of the tetrahedron is $\frac{x}{2} + \frac{y}{4} + \frac{z}{3} = 1$, or $6x + 3y + 4z = 12$. The projection of this tetrahedron on the xy plane is bounded by (setting $z = 0$ in the equation of the plane) the line $6x + 3y = 12$, or $2x + y = 4$. We can treat this as a Type I or a Type II region, which gives the two integrals

$$\int_0^2 \int_0^{4-2x} \int_0^{\frac{12-6x-3y}{4}} \rho(x,y,z)\, dz\, dy\, dx, \qquad \int_0^4 \int_0^{\frac{4-y}{2}} \int_0^{\frac{12-6x-3y}{4}} \rho(x,y,z)\, dz\, dx\, dy.$$

The projection of the tetrahedron on the yz plane is bounded by (setting $x = 0$ in the equation of the plane) $3y + 4z = 12$. We can treat this as a Type I region or a Type II region, which gives the two integrals

$$\int_0^4 \int_0^{\frac{12-3y}{4}} \int_0^{\frac{12-3y-4z}{6}} \rho(x,y,z)\, dx\, dz\, dy, \qquad \int_0^3 \int_0^{\frac{12-4z}{3}} \int_0^{\frac{12-3y-4z}{6}} \rho(x,y,z)\, dx\, dy\, dz.$$

The projection on the xz plane is bounded by (setting $y = 0$ in the equation of the plane) $6x + 4z = 12$, or $3x + 2z = 6$. We can treat this as a Type I or a Type II region, which gives the two integrals

$$\int_0^2 \int_0^{\frac{6-3x}{2}} \int_0^{\frac{12-6x-4z}{3}} \rho(x,y,z)\, dy\, dz\, dx, \qquad \int_0^3 \int_0^{\frac{6-2z}{3}} \int_0^{\frac{12-6x-4z}{3}} \rho(x,y,z)\, dy\, dx\, dz.$$

15. Integrating gives

$$I_x = \int_0^1 \int_0^{-x+1} \int_0^{-x-y+1} k(y^2 + z^2)\, dz\, dy\, dx = k \int_0^1 \int_0^{-x+1} \left[y^2 z + \frac{1}{3} z^3 \right]_{z=0}^{z=-x-y+1} dy\, dx$$

$$= k \int_0^1 \int_0^{-x+1} \left(y^2(-x-y+1) + \frac{1}{3}(-x-y+1)^3 \right) dy\, dx$$

$$= k \int_0^1 \int_0^{-x+1} \left((-x+1)y^2 - y^3 + \frac{1}{3}(-x-y+1)^3 \right) dy\, dx$$

$$= k \int_0^1 \left[\frac{1}{3}(-x+1)y^3 - \frac{1}{4}y^4 - \frac{1}{12}(-x-y+1)^4 \right]_{y=0}^{y=-x+1} dx$$

$$= k \int_0^1 \left(\frac{1}{3}(-x+1)^4 - \frac{1}{4}(-x+1)^4 - \frac{1}{12}(-x-(-x+1)+1)^4 + \frac{1}{12}(-x+1)^4 \right) dx$$

$$= k \int_0^1 \left(\frac{1}{6}(-x+1)^4 \right) dx$$

$$= k \left[-\frac{1}{30}(-x+1)^5 \right]_0^1 = \frac{1}{30}k.$$

Thus the radius of gyration around the x axis is

$$R_x = \sqrt{\frac{I_x}{m}} = \sqrt{\frac{\frac{1}{30}k}{\frac{1}{6}k}} = \frac{\sqrt{5}}{5}.$$

17. Integrating gives

$$I_z = \int_0^1 \int_0^{-x+1} \int_0^{-x-y+1} k(x^2+y^2)\,dz\,dy\,dx = k\int_0^1 \int_0^{-x+1} \left[x^2 z + y^2 z\right]_{z=0}^{z=-x-y+1} dy\,dx$$

$$= k\int_0^1 \int_0^{-x+1} \left(x^2(-x-y+1) + y^2(-x-y+1)\right) dy\,dx$$

$$= k\int_0^1 \int_0^{-x+1} \left(x^2(-x+1) + y^2(-x+1) - x^2 y - y^3\right) dy\,dx$$

$$= k\int_0^1 \left[x^2(-x+1)y + \frac{1}{3}y^3(-x+1) - \frac{1}{2}x^2 y^2 - \frac{1}{4}y^4\right]_{y=0}^{y=-x+1} dx$$

$$= k\int_0^1 \left(x^2(-x+1)^2 + \frac{1}{3}(-x+1)^4 - \frac{1}{2}x^2(-x+1)^2 - \frac{1}{4}(-x+1)^4\right) dx$$

$$= k\int_0^1 \left(\frac{1}{12}(-x+1)^4 + \frac{1}{2}x^4 - x^3 + \frac{1}{2}x^2\right) dx$$

$$= k\left[-\frac{1}{60}(-x+1)^5 + \frac{1}{10}x^5 - \frac{1}{4}x^4 + \frac{1}{6}x^3\right]_0^1$$

$$= k\left(\frac{2}{5} - \frac{3}{4} + \frac{1}{6} + \frac{1}{60}\right) = \frac{1}{30}k.$$

Thus the radius of gyration around the z axis is

$$R_z = \sqrt{\frac{I_x}{m}} = \sqrt{\frac{\frac{1}{30}k}{\frac{1}{6}k}} = \frac{\sqrt{5}}{5}.$$

19. This triple integral represents the volume of Ω (or the mass of Ω if $\rho(x,y,z) = 1$). The units of the expression are cubic centimeters (or, if interpreting it as a mass, grams per cubic centimeter times cubic centimeters, or grams). See the discussion following Definition 13.15.

21. This represents the first moment of Ω, with mass density function $\rho(x,y,z)$, with respect to the yz plane. See Definition 13.19(a). Since $\rho(x,y,z)$ has units grams per cubic centimeter and x has units centimeters, the units of the integrand are grams per square centimeter. Since each integral multiplies by centimeters, the result is a unit of gram-centimeters.

23. This represents the moment of inertia about the y axis of the mass Ω with mass density function $\rho(x,y,z)$. See Definition 13.20(b). Now, $\rho(x,y,z)$ has units grams per cubic centimeter, while x^2 and z^2 have units square centimeters, so $(x^2+z^2)\rho(x,y,z)$ has units grams per centimeter. Each integration multiplies by centimeters, so the integral has units gram-square centimeters.

Skills

25. We have

$$\int_0^1 \int_2^4 \int_{-1}^5 (x + yz^2)\, dx\, dy\, dz = \int_0^1 \int_2^4 \left[\frac{1}{2}x^2 + xyz^2\right]_{x=-1}^{x=5}\, dx\, dy\, dz$$

$$= \int_0^1 \int_2^4 \left(12 + 6yz^2\right)\, dy\, dz$$

$$= \int_0^1 \left[12y + 3y^2z^2\right]_{y=2}^{y=4}\, dz$$

$$= \int_0^1 \left(24 + 36z^2\right)\, dz = \left[24z + 12z^3\right]_0^1 = 36.$$

27. We have

$$\int_0^1 \int_0^{6-y} \int_0^{3-3y-(1/2)z} (y - z)\, dx\, dz\, dy = \int_0^1 \int_0^{6-y} [x(y-z)]_{x=0}^{x=3-3y-(1/2)z}\, dz\, dy$$

$$= \int_0^1 \int_0^{6-y} \left(3 - 3y - \frac{1}{2}z\right)(y - z)\, dz\, dy$$

$$= \int_0^1 \int_0^{6-y} \left(3y - 3z - 3y^2 + 3yz - \frac{1}{2}yz + \frac{1}{2}z^2\right)\, dz\, dy$$

$$= \int_0^1 \int_0^{6-y} \left(3y - 3z - 3y^2 + \frac{5}{2}yz + \frac{1}{2}z^2\right)\, dz\, dy$$

$$= \int_0^1 \left[3yz - \frac{3}{2}z^2 - 3y^2z + \frac{5}{4}yz^2 + \frac{1}{6}z^3\right]_{z=0}^{z=6-y}\, dy$$

$$= \int_0^1 \left(3y(6 - y) - \frac{3}{2}(6 - y)^2 - 3y^2(6 - y) + \frac{5}{4}y(6 - y)^2 + \frac{1}{6}(6 - y)^3\right)\, dy$$

$$= \int_0^1 \left(\frac{49}{12}y^3 - \frac{69}{2}y^2 + 63y - 18\right)\, dy$$

$$= \left[\frac{49}{48}y^4 - \frac{23}{2}y^3 + \frac{63}{2}y^2 - 18y\right]_0^1 = \frac{145}{48}.$$

29. We have

$$\int_0^{\pi/2} \int_0^{\cos y} \int_0^{\sin y} (2x + y)\, dz\, dx\, dy = \int_0^{\pi/2} \int_0^{\cos y} [(2x + y)z]_{z=0}^{z=\sin y}\, dx\, dy$$

$$= \int_0^{\pi/2} \int_0^{\cos y} (2x \sin y + y \sin y)\, dx\, dy$$

$$= \int_0^{\pi/2} \left[x^2 \sin y + xy \sin y\right]_{x=0}^{x=\cos y}\, dy$$

$$= \int_0^{\pi/2} \left(\sin y \cos^2 y + y \sin y \cos y\right)\, dy.$$

To integrate $y \sin y \cos y$, use integration by parts with $u = y$ and $dv = \sin y \cos y\, dy$; then $du = dy$

and $v = \frac{1}{2}\sin^2 y$. Then the integral becomes

$$\int_0^{\pi/2}\left(\sin y\cos^2 y + y\sin y\cos y\right)dy = \left[-\frac{1}{3}\cos^3 y\right]_0^{\pi/2} + \left[\frac{1}{2}y\sin^2 y\right]_0^{\pi/2} - \frac{1}{2}\int_0^{\pi/2}\sin^2 y\,dy$$

$$= \frac{1}{3} + \frac{\pi}{4} - \frac{1}{4}\int_0^{\pi/2}(1-\cos 2y)\,dy$$

$$= \frac{1}{3} + \frac{\pi}{4} - \frac{1}{4}\left[y - \frac{1}{2}\sin 2y\right]_0^{\pi/2}$$

$$= \frac{1}{3} + \frac{\pi}{8}.$$

31. We get

$$\iiint_{\mathcal{R}} x^2 yz\,dV = \int_{-2}^{1}\int_{0}^{3}\int_{1}^{5} x^2 yz\,dz\,dy\,dx$$

$$= \int_{-2}^{1}\int_{0}^{3}\left[\frac{1}{2}x^2 yz^2\right]_{z=1}^{z=5} dy\,dx$$

$$= \int_{-2}^{1}\int_{0}^{3} 12x^2 y\,dy\,dx$$

$$= \int_{-2}^{1}\left[6x^2 y^2\right]_{y=0}^{y=3} dx$$

$$= \int_{-2}^{1} 54x^2\,dx = \left[18x^3\right]_{-2}^{1} = 162.$$

33. We get

$$\iiint_{\mathcal{R}} z\sin x\cos y\,dV = \int_{0}^{\pi}\int_{3\pi/2}^{2\pi}\int_{1}^{3} z\sin x\cos y\,dz\,dy\,dx$$

$$= \int_{0}^{\pi}\int_{3\pi/2}^{2\pi}\left[\frac{1}{2}z^2\sin x\cos y\right]_{z=1}^{z=3} dy\,dx$$

$$= \int_{0}^{\pi}\int_{3\pi/2}^{2\pi} 4\sin x\cos y\,dy\,dx$$

$$= \int_{0}^{\pi}\left[4\sin x\sin y\right]_{y=3\pi/2}^{y=2\pi} dx$$

$$= 4\int_{0}^{\pi}\sin x\,dx = -4\left[\cos x\right]_0^{\pi} = 8.$$

35. This is a rectangular solid described by $0 \le y \le 5$, $2 \le x \le 6$, and $-2 \le z \le 4$.

37. The x and y bounds determine the rectangle $[0,3] \times [0,3]$ in the xy plane; that forms the lower bound of the integration with respect to z. The upper bound is the plane $z = 3 - y$. A diagram of the region is

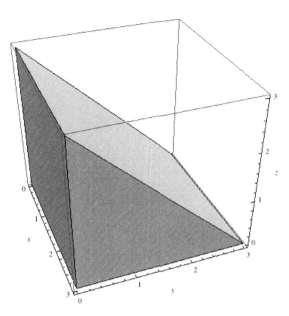

39. The range of integration for y is from 0 to 3. For each value of y, the z range is from 0 to $1 - \frac{y}{3}$, so the region in the yz plane is bounded by the coordinate axes and the line $z = 1 - \frac{y}{3}$, or $y + 3z = 3$. Over that triangle, the values of x range from the yz plane to the plane $x = 2 - \frac{2}{3}y - 2z$. Thus this is a tetrahedron, with vertices $(0,0,0)$, $(2,0,0)$, $(0,3,0)$, and $(0,0,1)$. A diagram of the region is

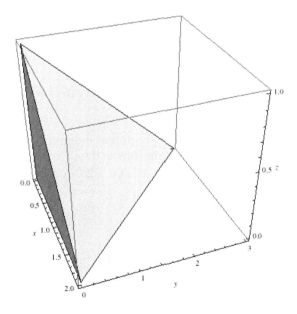

41. In the plane $z = 0$, the range of integration of x is from -3 to 3, and y is from $-\sqrt{9 - x^2}$ to $\sqrt{9 - x^2}$. Thus this is a circle of radius 3. Over that circle, z ranges from $-\sqrt{9 - x^2 - y^2}$ to $\sqrt{9 - x^2 - y^2}$, so that the range of integration is $x^2 + y^2 + z^2 \le 9$. Thus this is the ball of radius 3 around the origin.

43. In the yz plane, the range of integration is a circle of radius 3 centered at the origin. Above that circle, integration ranges from $x = 0$ to $x = 9 - y^2 - z^2$, which is a paraboloid. Thus the region is the interior of the paraboloid $x = 9 - y^2 - z^2$ for $x \ge 0$. A diagram of the region is

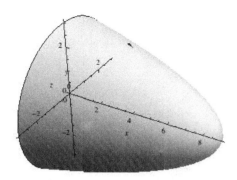

45. Integrating first with respect to x, then with respect to y and finally with respect to z gives

$$\int_{-2}^{4} \int_{0}^{5} \int_{2}^{6} f(x,y,z)\, dx\, dy\, dz.$$

47. Looking at the diagram in Exercise 37, the region of integration in the yz plane is bounded by the axes and the line $y + z = 3$, so that the integration bounds on z (the outermost integral) are from 0 to 3, and then the bounds on y are from 0 to $3 - z$. Finally, x ranges from 0 to 3. So the integral is

$$\int_{0}^{3} \int_{0}^{3-z} \int_{0}^{3} f(x,y,z)\, dx\, dy\, dz.$$

49. Looking at the diagram in Exercise 39, the region of integration in the xy plane is bounded by the axes and the line $3x + 2y = 6$. Thus x ranges from 0 to 2 and y from 0 to $\frac{6-3x}{2}$. Finally, the upper bound of z is determined by the plane $x = 2 - \frac{2}{3}y - 2z$, or $z = 1 - \frac{1}{3}y - \frac{1}{2}x$. So the integral is

$$\int_{0}^{2} \int_{0}^{(6-3x)/2} \int_{0}^{1-(1/3)y-(1/2)x} f(x,y,z)\, dz\, dy\, dx.$$

51. Since this is a sphere, and is thus symmetric around the origin, in the new integration order $z \in [-3,3]$, then x goes from $-\sqrt{9 - z^2}$ to $\sqrt{9 - z^2}$, and y goes from $-\sqrt{9 - x^2 - z^2}$ to $\sqrt{9 - x^2 - z^2}$ (this is just the integral from Exercise 41 with the axes relabeled, by symmetry):

$$\int_{-3}^{3} \int_{-\sqrt{9-z^2}}^{\sqrt{9-z^2}} \int_{-\sqrt{9-x^2-z^2}}^{\sqrt{9-x^2-z^2}} f(x,y,z)\, dy\, dx\, dz.$$

53. The projection of the solid in the xy plane is bounded by the axes and by the line $3x + 4y = 12$, so we will integrate x from 0 to 4 and y from 0 to $\frac{12-3x}{4} = 3 - \frac{3}{4}x$. Finally, z ranges from 0 to $\frac{12-3x-4y}{6} = 2 - \frac{1}{2}x - \frac{2}{3}y$. The density, which is proportional to the distance from the xz plane, is

$\rho = ky$. Hence the mass of the solid is

$$\iiint_\mathcal{S} \rho(x,y,z)\, dV = \int_0^4 \int_0^{3-(3/4)x} \int_0^{2-(1/2)x-(2/3)y} ky\, dz\, dy\, dx$$

$$= k \int_0^4 \int_0^{3-(3/4)x} [yz]_{z=0}^{z=2-(1/2)x-(2/3)y}\, dy\, dx$$

$$= k \int_0^4 \int_0^{3-(3/4)x} y\left(2 - \frac{1}{2}x - \frac{2}{3}y\right)\, dy\, dx$$

$$= k \int_0^4 \left[y^2 - \frac{1}{4}xy^2 - \frac{2}{9}y^3\right]_{y=0}^{y=3-(3/4)x}\, dx$$

$$= k \int_0^4 \left(\left(3 - \frac{3}{4}x\right)^2 - \frac{1}{4}x\left(3 - \frac{3}{4}x\right)^2 - \frac{2}{9}\left(3 - \frac{3}{4}x\right)^3\right)\, dx$$

$$= k \int_0^4 \left(-\frac{3}{64}x^3 + \frac{9}{16}x^2 - \frac{9}{4}x + 3\right)\, dx$$

$$= k \left[-\frac{3}{256}x^4 + \frac{3}{16}x^3 - \frac{9}{8}x^2 + 3x\right]_0^4 = 3k.$$

55. The innermost integral will be with respect to z, from 0 to $8 - x^2 - y^2$; the other two integrals will cover the given rectangle. The density function given is $\rho(x,y,z) = k(x^2 + y^2 + z^2)$. Thus the mass is

$$\iiint_\mathcal{S} \rho(x,y,z)\, dV = \int_1^2 \int_0^2 \int_0^{8-x^2-y^2} k(x^2 + y^2 + z^2)\, dz\, dy\, dx$$

$$= k \int_1^2 \int_0^2 \left[x^2 z + y^2 z + \frac{1}{3}z^3\right]_{z=0}^{z=8-x^2-y^2}\, dy\, dx$$

$$= k \int_1^2 \int_0^2 \left(x^2(8 - x^2 - y^2) + y^2(8 - x^2 - y^2) + \frac{1}{3}(8 - x^2 - y^2)^3\right)\, dy\, dx$$

$$= k \int_1^2 \int_0^2 \left(8x^2 - x^4 + 8y^2 - y^4 - 2x^2y^2\right.$$

$$+ \frac{1}{3}\left(512 - 192x^2 + 24x^4 - x^6 - 192y^2 + 48x^2y^2\right.$$

$$\left.\left. -3x^4y^2 + 24y^4 - 3x^2y^4 - y^6\right)\right)\, dy\, dx$$

$$= k \int_1^2 \int_0^2 \left(-\frac{1}{3}y^6 - x^2y^4 + 7y^4 - x^4y^2 + 14x^2y^2\right.$$

$$\left. -56y^2 - \frac{1}{3}x^6 + 7x^4 - 56x^2 + \frac{512}{3}\right)\, dy\, dx$$

$$= k \int_1^2 \left[-\frac{1}{21}y^7 - \frac{1}{5}x^2y^5 + \frac{7}{5}y^5 - \frac{1}{3}x^4y^3 + \frac{14}{3}x^2y^3\right.$$

$$\left. -\frac{56}{3}y^3 - \frac{1}{3}x^6y + 7x^4y - 56x^2y + \frac{512}{3}y\right]_{y=0}^{y=2}\, dx$$

$$= k \int_1^2 \left(-\frac{128}{21} - \frac{32}{5}x^2 + \frac{224}{5} - \frac{8}{3}x^4 + \frac{112}{3}x^2 \right.$$

$$\left. - \frac{448}{3} - \frac{2}{3}x^6 + 14x^4 - 112x^2 + \frac{1024}{3} \right) dx$$

$$= k \int_1^2 \left(-\frac{2}{3}x^6 + \frac{34}{3}x^4 - \frac{1216}{15}x^2 + \frac{24224}{105} \right) dx$$

$$= k \left[-\frac{2}{21}x^7 + \frac{34}{15}x^5 - \frac{1216}{45}x^3 + \frac{24224}{105}x \right]_{x=1}^{x=2}$$

$$= \frac{31412}{315}k.$$

Applications

57. (a) Since the density is uniform and the solid is symmetric about the point $\left(2, \frac{3}{2}, 1\right)$, this must be the center of mass since as much of the solid lies on one side of that point as on the other side, in any direction.

(b) Assuming the density is 1, since constants will cancel out of the computation, the mass (volume) of the solid is $4 \cdot 3 \cdot 2 = 24$. Then the center of mass is

$$\bar{x} = \frac{1}{24} \int_0^4 \int_0^3 \int_0^2 x \, dz \, dy \, dx = \frac{1}{24} \int_0^4 6x \, dx = \frac{1}{24} \left[3x^2 \right]_0^4 = 2$$

$$\bar{y} = \frac{1}{24} \int_0^4 \int_0^3 \int_0^2 y \, dz \, dy \, dx = \frac{1}{24} \int_0^4 \int_0^3 2y \, dy \, dx = \frac{1}{24} \int_0^4 \left[y^2 \right]_0^3 dx = \frac{1}{24} \int_0^4 9 \, dx = \frac{36}{24} = \frac{3}{2}$$

$$\bar{z} = \frac{1}{24} \int_0^4 \int_0^3 \int_0^2 z \, dz \, dy \, dx = \frac{1}{24} \int_0^4 \int_0^3 \left[\frac{1}{2}z^2 \right]_{z=0}^{z=2} dy \, dx = \frac{1}{24} \int_0^4 \int_0^3 2 \, dy \, dx$$

$$= \frac{1}{24}(4 \cdot 3 \cdot 2) = 1.$$

59. (a) The distance from the xy plane is simply the z coordinate. Mentally decompose the solid into a bunch of tubes, each running parallel to the x axis. Each of these tubes has a roughly constant z coordinate, so the x coordinate of the center of mass is at $x = 2$. Thus the x coordinate of the center of mass of the whole solid is at $x = 2$. An identical argument applies to the y coordinate, since all points in any small tube parallel to the y axis has constant z coordinate as well. However, this will not work for the z coordinate, since if you try decomposing the solid, each tube has a varying density depending on its distance to the xy plane.

(b) Assume the constant of proportionality is 1, since it will appear in the numerator and denominator in any case, so will simply cancel out. The z coordinate of the center of mass is

$$\bar{z} = \frac{\int_0^4 \int_0^3 \int_0^2 z^2 \, dz \, dy \, dx}{\int_0^4 \int_0^3 \int_0^2 z \, dz \, dy \, dx} = \frac{\int_0^4 \int_0^3 \left[\frac{1}{3}z^3 \right]_{z=0}^{z=2} dy \, dx}{\int_0^4 \int_0^3 \left[\frac{1}{2}z^2 \right]_{z=0}^{z=2} dy \, dx} = \frac{\int_0^4 \int_0^3 \frac{8}{3} \, dy \, dx}{\int_0^4 \int_0^3 2 \, dy \, dx} = \frac{4}{3}.$$

61. (a) Since the density is uniform and the solid is symmetric about the point $\left(\frac{a_1+a_2}{2}, \frac{b_1+b_2}{2}, \frac{c_1+c_2}{2} \right)$, this must be the center of mass since as much of the solid lies on one side of that point as on the other side, in any direction.

(b) Assuming the density is 1, since constants will cancel out of the computation, the mass (volume) of the solid is $(a_2 - a_1)(b_2 - b_1)(c_2 - c_1)$. Then the center of mass is

$$
\begin{aligned}
\bar{x} &= \frac{1}{(a_2 - a_1)(b_2 - b_1)(c_2 - c_1)} \int_{a_1}^{a_2} \int_{b_1}^{b_2} \int_{c_1}^{c_2} x \, dz \, dy \, dx \\
&= \frac{1}{(a_2 - a_1)(b_2 - b_1)(c_2 - c_1)} \int_{a_1}^{a_2} (b_2 - b_1)(c_2 - c_1) x \, dx \\
&= \frac{1}{a_2 - a_1} \left[\frac{1}{2} x^2 \right]_{a_1}^{a_2} = \frac{a_2^2 - a_1^2}{2(a_2 - a_1)} = \frac{a_1 + a_2}{2}
\end{aligned}
$$

$$
\begin{aligned}
\bar{y} &= \frac{1}{(a_2 - a_1)(b_2 - b_1)(c_2 - c_1)} \int_{a_1}^{a_2} \int_{b_1}^{b_2} \int_{c_1}^{c_2} y \, dz \, dy \, dx \\
&= \frac{1}{(a_2 - a_1)(b_2 - b_1)(c_2 - c_1)} \int_{a_1}^{a_2} \int_{b_1}^{b_2} (c_2 - c_1) y \, dy \, dx \\
&= \frac{1}{(a_2 - a_1)(b_2 - b_1)} \int_{a_1}^{a_2} \left[\frac{1}{2} y^2 \right]_{b_1}^{b_2} dx \\
&= \frac{1}{(a_2 - a_1)(b_2 - b_1)} \int_{a_1}^{a_2} \frac{1}{2} (b_2^2 - b_1^2) \, dx \\
&= \frac{1}{(a_2 - a_1)(b_2 - b_1)} \cdot \frac{1}{2} (b_2^2 - b_1^2)(a_2 - a_1) = \frac{b_1 + b_2}{2}
\end{aligned}
$$

$$
\begin{aligned}
\bar{z} &= \frac{1}{(a_2 - a_1)(b_2 - b_1)(c_2 - c_1)} \int_{a_1}^{a_2} \int_{b_1}^{b_2} \int_{c_1}^{c_2} z \, dz \, dy \, dx \\
&= \frac{1}{(a_2 - a_1)(b_2 - b_1)(c_2 - c_1)} \int_{a_1}^{a_2} \int_{b_1}^{b_2} \left[\frac{1}{2} z^2 \right]_{z=c_1}^{z=c_2} dy \, dx \\
&= \frac{1}{(a_2 - a_1)(b_2 - b_1)(c_2 - c_1)} \int_{a_1}^{a_2} \int_{b_1}^{b_2} \frac{1}{2} (c_2^2 - c_1^2) \, dy \, dx \\
&= \frac{1}{(a_2 - a_1)(b_2 - b_1)(c_2 - c_1)} \cdot \frac{1}{2} (c_2^2 - c_1^2)(b_2 - b_1)(a_2 - a_1) = \frac{c_1 + c_2}{2}.
\end{aligned}
$$

63. (a) The distance from the xy plane is simply the z coordinate. Mentally decompose the solid into a bunch of tubes, each running parallel to the x axis. Each of these tubes has a roughly constant z coordinate, so the x coordinate is at $x = 2$. Thus the x coordinate of the center of mass of the whole solid is at $x = \frac{a_1 + a_2}{2}$. An identical argument applies to the y coordinate, since all points in any small tube parallel to the y axis has constant z coordinate as well. However, this will not work for the z coordinate, since if you try decomposing the solid, each tube has a varying density depending on its distance to the xy plane.

(b) Assuming the density is 1, since constants will cancel out of the computation, the z coordinate of the center of mass is

$$
\begin{aligned}
\bar{z} &= \frac{\int_{a_1}^{a_2} \int_{b_1}^{b_2} \int_{c_1}^{c_2} z^2 \, dz \, dy \, dx}{\int_{a_1}^{a_2} \int_{b_1}^{b_2} \int_{c_1}^{c_2} z \, dz \, dy \, dx} = \frac{\int_{a_1}^{a_2} \int_{b_1}^{b_2} \left[\frac{1}{3} z^3 \right]_{z=c_1}^{z=c_2} dy \, dx}{\int_{a_1}^{a_2} \int_{b_1}^{b_2} \left[\frac{1}{2} z^2 \right]_{z=c_1}^{z=c_2} dy \, dx} = \frac{\int_{a_1}^{a_2} \int_{b_1}^{b_2} \frac{1}{3} (c_2^3 - c_1^3) \, dy \, dx}{\int_{a_1}^{a_2} \int_{b_1}^{b_2} \frac{1}{2} (c_2^2 - c_1^2) \, dy \, dx} \\
&= \frac{2}{3} \cdot \frac{c_2^3 - c_1^3}{c_2^2 - c_1^2}.
\end{aligned}
$$

65. (a) First find the volume of the tetrahedron. The oblique planar face forming the upper boundary of the tetrahedron has equation $\frac{x}{a} + \frac{y}{b} + \frac{z}{c} = 1$. The diagonal line forming the boundary of

the tetrahedron in the xy plane is $\frac{x}{a} + \frac{y}{b} = 1$. Thus as x varies from 0 to a, y can vary from 0 to $b - \frac{b}{a}x$, and then z can vary from 0 to

$$c\left(1 - \frac{x}{a} - \frac{y}{b}\right)$$

Thus the volume of the tetrahedron is

$$
\begin{aligned}
V &= \int_0^a \int_0^{b-\frac{b}{a}x} \int_0^{c\left(1-\frac{x}{a}-\frac{y}{b}\right)} 1 \, dz \, dy \, dx \\
&= c \int_0^a \int_0^{b-\frac{b}{a}x} \left(1 - \frac{x}{a} - \frac{y}{b}\right) dy \, dx \\
&= c \int_0^a \left[y - \frac{x}{a}y - \frac{1}{2b}y^2\right]_{y=0}^{y=b-\frac{b}{a}x} dx \\
&= c \int_0^a \left(b - \frac{b}{a}x - \frac{x}{a}\left(b - \frac{b}{a}x\right) - \frac{1}{2b}\left(b - \frac{b}{a}x\right)^2\right) dx \\
&= bc \int_0^a \left(\frac{1}{2} - \frac{x}{a} + \frac{x^2}{2a^2}\right) dx \\
&= bc \left[\frac{1}{2}x - \frac{1}{2a}x^2 + \frac{1}{6a^2}x^3\right]_0^a \\
&= bc \left(\frac{1}{2}a - \frac{1}{2}a + \frac{1}{6}a\right) \\
&= \frac{1}{6}abc.
\end{aligned}
$$

The x coordinate of the center of mass is thus

$$
\begin{aligned}
\bar{x} &= \frac{6}{abc} \int_0^a \int_0^{b-\frac{b}{a}x} \int_0^{c\left(1-\frac{x}{a}-\frac{y}{b}\right)} x \, dz \, dy \, dx \\
&= \frac{6}{ab} \int_0^a \int_0^{b-\frac{b}{a}x} x\left(1 - \frac{x}{a} - \frac{y}{b}\right) dy \, dx \\
&= \frac{6}{ab} \int_0^a x\left[y - \frac{x}{a}y - \frac{1}{2b}y^2\right]_{y=0}^{y=b-\frac{b}{a}x} dx \\
&= \frac{6}{ab} \int_0^a x\left(b - \frac{b}{a}x - \frac{x}{a}\left(b - \frac{b}{a}x\right) - \frac{1}{2b}\left(b - \frac{b}{a}x\right)^2\right) dx \\
&= \frac{6}{a} \int_0^a \left(\frac{x}{2} - \frac{x^2}{a} + \frac{x^3}{2a^2}\right) dx \\
&= \frac{6}{a} \left[\frac{1}{4}x^2 - \frac{1}{3a}x^3 + \frac{1}{8a^2}x^4\right]_0^a \\
&= \frac{6}{a} \left(\frac{1}{4}a^2 - \frac{1}{3}a^2 + \frac{1}{8}a^2\right) \\
&= \frac{a}{4}.
\end{aligned}
$$

(b) Since the density is constant, we can relabel the vertices and the axes to find that the y coordinate of the center of mass will be $\frac{b}{4}$, and the z coordinate will be $\frac{c}{4}$.

67. The density function is $\rho(x, y, z) = ky$, but we may assume $k = 1$ since constants will cancel out in the calculations. Thus using Definition 13.19 and the integration bounds from Exercise 65, we get

$$\bar{x} = \frac{M_{yz}}{m} = \frac{\int_0^a \int_0^{b-\frac{b}{a}x} \int_0^{c\left(1-\frac{x}{a}-\frac{y}{b}\right)} xy \, dz \, dy \, dx}{\int_0^a \int_0^{b-\frac{b}{a}x} \int_0^{c\left(1-\frac{x}{a}-\frac{y}{b}\right)} y \, dz \, dy \, dx}$$

$$\bar{y} = \frac{M_{xz}}{m} = \frac{\int_0^a \int_0^{b-\frac{b}{a}x} \int_0^{c\left(1-\frac{x}{a}-\frac{y}{b}\right)} y^2 \, dz \, dy \, dx}{\int_0^a \int_0^{b-\frac{b}{a}x} \int_0^{c\left(1-\frac{x}{a}-\frac{y}{b}\right)} y \, dz \, dy \, dx}$$

$$\bar{z} = \frac{M_{xy}}{m} = \frac{\int_0^a \int_0^{b-\frac{b}{a}x} \int_0^{c\left(1-\frac{x}{a}-\frac{y}{b}\right)} yz \, dz \, dy \, dx}{\int_0^a \int_0^{b-\frac{b}{a}x} \int_0^{c\left(1-\frac{x}{a}-\frac{y}{b}\right)} y \, dz \, dy \, dx}$$

Proofs

69. By Fubini's Theorem,

$$\iiint_{\mathcal{R}} \alpha(x)\beta(y)\gamma(z) \, dA = \int_{a_1}^{a_2} \int_{b_1}^{b_2} \int_{c_1}^{c_2} \alpha(x)\beta(y)\gamma(z) \, dz \, dy \, dx.$$

Looking at the inner integral, we see that $\alpha(x)\beta(y)$ is a constant, so we can pull it out of the inner integral to get

$$\int_{a_1}^{a_2} \int_{b_1}^{b_2} \left(\alpha(x)\beta(y) \int_{c_1}^{c_2} \gamma(z) \, dz \, dy \, dx \right).$$

Similarly, $\alpha(x)$ is a constant in the integral with respect to y, as is $\int_{c_1}^{c_2} \gamma(z) \, dz$, so we can pull them out of the y integral to get

$$\int_{a_1}^{a_2} \left(\alpha(x) \int_{c_1}^{c_2} \gamma(z) \, dz \int_{b_1}^{b_2} \beta(y) \, dy \right) dx.$$

Finally, both of the inner integrals are constants in the x integral, so we can pull them out as well to get

$$\left(\int_{c_1}^{c_2} \gamma(z) \, dz \right) \left(\int_{b_1}^{b_2} \beta(y) \, dy \right) \left(\int_{a_1}^{a_2} \alpha(x) \, dx \right).$$

Reorder this product to get the desired answer.

71. By Exercise 69,

$$\iiint_{\mathcal{R}} dA = \left(\int_{a_1}^{a_2} dx \right) \left(\int_{b_1}^{b_2} dy \right) \left(\int_{c_1}^{c_2} dz \right) = [x]_{a_1}^{a_2} [y]_{b_1}^{b_2} [z]_{c_1}^{c_2} = (a_2 - a_1)(b_2 - b_1)(c_2 - c_1).$$

This integral is clearly the volume of the rectangular parallelepiped $\mathcal{R}$.

73. From Definition 13.15, the double integral is

$$\iint_{\mathcal{R}} (f(x, y, z) + g(x, y, z)) \, dA = \lim_{\Delta \to 0} \sum_{i=1}^{l} \sum_{j=1}^{m} \sum_{k=1}^{n} (f(x_i^*, y_j^*, z_k^*) + g(x_i^*, y_j^*, z_k^*)) \Delta V$$

$$= \lim_{\Delta \to 0} \left(\sum_{i=1}^{l} \sum_{j=1}^{m} \sum_{k=1}^{n} f(x_i^*, y_j^*, z_k^*) \Delta V + \sum_{i=1}^{l} \sum_{j=1}^{m} \sum_{k=1}^{n} g(x_i^*, y_j^*, z_k^*) \Delta V \right)$$

$$= \lim_{\Delta \to 0} \sum_{i=1}^{l} \sum_{j=1}^{m} \sum_{k=1}^{n} f(x_i^*, y_j^*, z_k^*) \Delta V + \lim_{\Delta \to 0} \sum_{i=1}^{l} \sum_{j=1}^{m} \sum_{k=1}^{n} g(x_i^*, y_j^*, z_k^*) \Delta V$$

$$= \iint_{\mathcal{R}} f(x, y, z) \, dV + \iint_{\mathcal{R}} g(x, y, z) \, dV.$$

Thinking Forward

Cylindrical coordinates

The first two integrals are equal by Fubini's Theorem. The integration is over the region Ω_{xy}, and goes from $g_1(x,y)$ to $g_2(x,y)$. If we regard the region Ω_{xy} as a region in polar coordinates, however, it is the region described by $\alpha \leq \theta \leq \beta$ and $h_1(\theta) \leq r \leq h_2(\theta)$, so converting the integral to polar coordinates, we get the two outer integrals in the third equation and the factor of r in the integrand for the volume element. Finally, substituting the polar equivalents for x and y in the inner integral and the integrand gives the third equation.

Cylindrical coordinates

The outer two integrals describe a polar circle of radius 3 centered at the origin ($\theta \in [0, 2\pi]$ and $r \in [0,3]$). So in rectangular coordinates, the integral is

$$\int_{-3}^{3} \int_{-\sqrt{9-x^2}}^{\sqrt{9-x^2}} \int_{0}^{9-x^2-y^2} k\sqrt{x^2 + y^2 + z^2}\, dz\, dy\, dx.$$

Since the paraboloid $z = 9 - x^2 - y^2$ intersects the xy plane in the circle of radius 3 around the origin, this integral exactly describes the volume under that paraboloid above the xy plane. The factor of $k\sqrt{x^2 + y^2 + z^2}$ in the integrand results in computing the mass of that solid given a density function proportional to the distance from the origin. Evaluating the integral, we get

$$\int_{0}^{2\pi} \int_{0}^{3} \int_{0}^{9-r^2} kr^2\, dz\, dr\, d\theta = \int_{0}^{2\pi} \int_{0}^{3} \left[kr^2 z \right]_{z=0}^{z=9-r^2} dr\, d\theta$$

$$= k \int_{0}^{2\pi} \int_{0}^{3} (9r^2 - r^4)\, dr\, d\theta$$

$$= k \int_{0}^{2\pi} \left[3r^3 - \frac{1}{5}r^5 \right]_{r=0}^{r=3} d\theta$$

$$= k \int_{0}^{2\pi} \frac{162}{5}\, d\theta = \frac{324}{5} k\pi.$$

13.6   Integration using Cylindrical and Spherical Coordinates

Thinking Back

Graphing with polar coordinates

▷ Since r is constant over all θ, this is a circle of radius $r = 3$ centered at the origin:

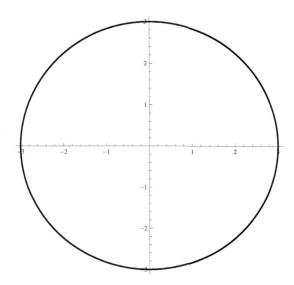

▷ Since θ is constant, this is a line through the origin at an angle of $\frac{\pi}{3}$, i.e., a line with slope $\sqrt{3}$:

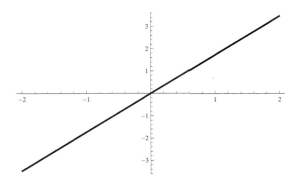

▷ $r = 2\cos\theta$ is a circle of radius 1 centered at $(1,0)$, since $r = 2\cos\theta$ gives $r^2 = 2r\cos\theta$; converting to rectangular coordinates gives $x^2 + y^2 = 2x$, or $(x^2 - 2x + 1) + y^2 = 1$, so that $(x - 1)^2 + y^2 = 1$. A graph is:

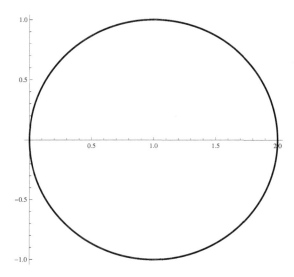

$\triangleright$ $r = 1 + 2\sin\theta$ is a limaçon with an inner loop; see Section 9.3:

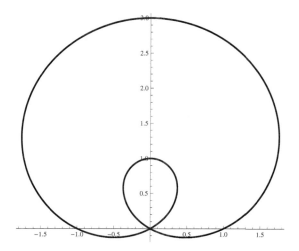

Rectangular versus polar coordinates

Many answers are possible. Polar coordinates tend to be easier when the graph of a function exhibits periodicity, i.e., when its graph repeats after a period of time, or sometimes (but not always) when the graph exhibits symmetries. When presented with a function described implicitly in rectangular coordinates, you should think about using a polar coordinate description if, for example, the function contains elements of the form $x^2 + y^2$ or $\sqrt{x^2 + y^2}$, as these convert naturally to polar coordinates.

Integrating with polar coordinates

The area element in polar coordinates is $r\,dr\,d\theta$, so if Ω is described by $\alpha \leq \theta \leq \beta$ and $g_1(\theta) \leq r \leq g_2(\theta)$, then the area of Ω is

$$\iint_\Omega dA = \int_\alpha^\beta \int_{g_1(\theta)}^{g_2(\theta)} r\,dr\,d\theta.$$

Concepts

1. (a) True. Converting to rectangular coordinates gives $z = \sqrt{2x - x^2 - y^2}$, so that $x^2 - 2x + y^2 + z^2 = 0$. Completing the square gives $(x - 1)^2 + y^2 + z^2 = 1$, which is the equation of a sphere. However, in the original equation, z must be nonnegative, so in fact we get only the upper hemisphere.

   (b) True. ϕ is the angle to the positive z axis; when that angle is $\frac{\pi}{2}$, the point lies on the xy plane. Here we are taking all points in that plane.

   (c) False. $\rho = 1$ is a sphere of radius 1, while $\phi = a$ is a cone; the angle that the slant edge of the cone makes with the z axis is ϕ. Thus when ϕ is close to zero, the intersection of the two is a small circle around the north pole of the sphere, while when ϕ is $\frac{\pi}{2}$, the intersection is a circle of radius 1. Thus the intersection is always a circle (except for $\phi = 0, \pi$, when it is a point), but it is not generally a circle of radius 1.

   (d) True. Converting $z = r$ to rectangular coordinates gives $z = \sqrt{x^2 + y^2}$, so that $z^2 = x^2 + y^2$ and $z \geq 0$. Now convert to spherical coordinates, giving

   $$\rho^2 \cos^2 \phi = \rho^2 \sin^2 \phi \cos^2 \theta + \rho^2 \sin^2 \phi \sin^2 \theta = \rho^2 \sin^2 \phi (\cos^2 \theta + \sin^2 \theta) = \rho^2 \sin^2 \phi.$$

   For $\rho > 0$, this simplifies to $\tan \phi = \pm 1$. But we are restricted to $0 \leq \phi \leq \frac{\pi}{2}$ since $z \geq 0$, so we must have $\phi = \frac{\pi}{4}$.

   (e) True. Since x, y, and z are dummy variables of integration, we may rename them to r, θ, and z if we wish. Also, since the integral is over the entire region, not expressed as an iterated integral with bounds, we may change the order of the differentials (it is $dz\, dy\, dx$ on the left, which would after substitution become $dz\, d\theta\, dr$ on the right).

   (f) True. This converts the integral to cylindrical coordinates, in which z remains z. The extra factor of r is to account for the volume element in polar/cylindrical coordinates.

   (g) False. This is indeed the integral for the volume of a sphere in spherical coordinates, since the integrand is the volume element $\rho^2 \sin \phi\, d\rho\, d\theta\, d\phi$. However, the integration limits are incorrect: the limits for ϕ should be from 0 to π.

   (h) False. There are many other possible coordinate systems in $\mathbb{R}^3$. For example, one could specify the location of a point by its x and y rectangular coordinates and the angle formed by the line from the origin to the point. This gives a representation with two lengths and one angle. Some other coordinate systems that arise in particular applications are the bipolar, parabolic, and toroidal systems.

3. These are planes. $x = x_0$ is a plane parallel to the yz plane at a distance of x_0. Also, $y = y_0$ is a plane parallel to the xz plane at a distance of y_0, and $z = z_0$ is a plane parallel to the xy plane at a distance of z_0.

5. $\rho = \rho_0$ is the set of points at a distance ρ_0 from the origin, so it is a sphere of radius ρ_0. Next, $\theta = \theta_0$ is, as it is in cylindrical coordinates, the plane through the origin, containing the z axis, that forms an angle of θ with the xz plane, since θ means the same thing in both coordinate systems. Finally, for most values of ϕ_0, the graph of $\phi = \phi_0$ is a cone through the origin; the slant face of the cone makes an angle ϕ with the positive z axis. The exceptions are when $\phi_0 = 0, \pi$, or $\frac{\pi}{2}$; see the next exercise for details.

7. $r = \sqrt{x^2 + y^2}$, $\tan \theta = \frac{y}{x}$, $z = z$. When computing θ from $\tan \theta$, it must be interpreted depending on the signs of x and y. For example, if x is negative but y is positive, we must choose the value for $\tan^{-1} \frac{y}{x}$ lying in the second quadrant. See Theorem 13.21(a).

9. $\rho = \sqrt{x^2 + y^2 + z^2}$, $\tan \theta = \frac{y}{x}$, $\cos \phi = \frac{z}{\sqrt{x^2 + y^2 + z^2}}$. When computing θ from its tangent, it must be interpreted depending on the signs of x and y, as in Exercise 7. See Theorem 13.21(c).

11. $\rho = \sqrt{r^2 + z^2}, \quad \theta = \theta, \quad \tan\phi = \frac{r}{z}$. See Theorem 13.21(b). Note that here ϕ must be between 0 and π, so that it is uniquely determined from its tangent.

13. The six possibilities are

$$dx\, dy\, dz, \quad dx\, dz\, dy, \quad dy\, dx\, dz, \quad dy\, dz\, dx, \quad dz\, dx\, dy, \quad dz\, dy\, dx.$$

The choice of form is determined by the shape of the region involved. Typically the outermost (rightmost) variable is the one in which the region has the simplest description.

15. The usual choice is $dV = \rho^2 \sin\phi\, d\rho\, d\theta\, d\phi$. Since the solid is usually symmetric under rotation around the z axis when we use spherical coordinates, ϕ usually has the simplest integration range, followed by θ and then ρ. So this integration order makes sense.

17. The triple integral describes some region of integration Ω; the same geometric considerations apply as in the previous exercise: if the projection onto the xy plane (or onto some other of the coordinate planes) has a natural expression in polar coordinates, then cylindrical coordinates may make the most sense, since Ω is likely easily expressible in that set of coordinates. If the boundaries of solids are spheres or parts of spheres, or if the solid has rotational symmetry around the z axis, spherical coordinates may be useful. Otherwise, rectangular coordinates are probably the best bet.

19. Evaluating the integral gives

$$
\begin{aligned}
\iiint_{\mathcal{E}} k\, dV &= \int_0^{\pi/2} \int_0^{\pi/2} \int_0^1 k\rho^2 \sin\phi\, d\rho\, d\theta\, d\phi \\
&= \int_0^{\pi/2} \int_0^{\pi/2} \left[\frac{1}{3} k\rho^3 \sin\phi\right]_{\rho=0}^{\rho=1} d\theta\, d\phi \\
&= \frac{1}{3} k \int_0^{\pi/2} \int_0^{\pi/2} \sin\phi\, d\theta\, d\phi \\
&= \frac{1}{6} \pi k \int_0^{\pi/2} \sin\phi\, d\phi \\
&= \frac{1}{6} \pi k \left[-\cos\phi\right]_{\phi=0}^{\phi=\pi/2} \\
&= \frac{1}{6} \pi k.
\end{aligned}
$$

21. Using the fact that $y = \rho \sin \phi \sin \theta$ and the fact that the volume element in spherical coordinates is $\rho^2 \sin \phi \, d\rho \, d\theta \, d\phi$ gives

$$
\begin{aligned}
M_{xz} = \iiint_{\mathcal{E}} ky \, dV &= \int_0^{\pi/2} \int_0^{\pi/2} \int_0^1 k\rho^3 \sin^2 \phi \sin \theta \, d\rho \, d\theta \, d\phi \\
&= \int_0^{\pi/2} \int_0^{\pi/2} \left[\frac{1}{4} k\rho^4 \sin^2 \phi \sin \theta \right]_{\rho=0}^{\rho=1} d\theta \, d\phi \\
&= \frac{1}{4} k \int_0^{\pi/2} \int_0^{\pi/2} \sin^2 \phi \sin \theta \, d\theta \, d\phi \\
&= \frac{1}{4} k \int_0^{\pi/2} \left[-\sin^2 \phi \cos \theta \right]_{\theta=0}^{\theta=\pi/2} d\phi \\
&= \frac{1}{4} k \int_0^{\pi/2} \sin^2 \phi \, d\phi \\
&= \frac{1}{8} k \int_0^{\pi/2} (1 - \cos 2\phi) \, d\phi \\
&= \frac{1}{8} k \left[\phi - \frac{1}{2} \sin 2\phi \right]_0^{\pi/2} = \frac{1}{16} \pi k.
\end{aligned}
$$

Skills

23. In cylindrical coordinates,

$$
r = \sqrt{x^2 + y^2} = \sqrt{1^2 + 0^2} = 1, \quad \tan \theta = \frac{0}{1} = 0, \quad z = 0.
$$

Thus $\theta = 0$ and the cylindrical coordinates are $(1, 0, 0)$.

In spherical coordinates,

$$
\rho = \sqrt{x^2 + y^2 + z^2} = \sqrt{1^2 + 0^2 + 0^2} = 1, \quad \tan \theta = \frac{0}{1} = 0, \quad \cos \phi = \frac{0}{\sqrt{1^2 + 0^2 + 0^2}} = 0.
$$

Thus $\theta = 0$ and $\phi = \frac{\pi}{2}$, so the spherical coordinates are $\left(1, 0, \frac{\pi}{2}\right)$.

25. In rectangular coordinates,

$$
x = r \cos \theta = \sqrt{48} \cos \frac{\pi}{3} = 2\sqrt{3}, \quad y = r \sin \theta = \sqrt{48} \sin \frac{\pi}{3} = 6, \quad z = 4,
$$

so the point is $(2\sqrt{3}, 6, 4)$.

In spherical coordinates,

$$
\rho = \sqrt{r^2 + z^2} = \sqrt{48 + 16} = 8, \quad \theta = \frac{\pi}{3}, \quad \tan \phi = \frac{r}{z} = \frac{4\sqrt{3}}{4} = \sqrt{3}.
$$

Thus $\phi = \frac{\pi}{3}$, and the point is $\left(8, \frac{\pi}{3}, \frac{\pi}{3}\right)$.

27. In rectangular coordinates,

$$
x = \rho \sin \phi \cos \theta = 0, \quad y = \rho \sin \phi \sin \theta = 0, \quad z = \rho \cos \phi = -8,
$$

so the point is $(0, 0, -8)$.

In cylindrical coordinates,

$$
r = \rho \sin \phi = 0, \quad \theta = \frac{\pi}{2}, \quad z = \rho \cos \phi = -8,
$$

so the point is $\left(0, \frac{\pi}{2}, -8\right)$ (which is the same as $(0, 0, -8)$ since $r = 0$).

29. $x = 4$ is a plane parallel to the yz plane through the point $(4,0,0)$. In cylindrical coordinates, $x = r \cos \theta$, so that $x = 4$ becomes $r \cos \theta = 4$, or $r = 4 \sec \theta$. In spherical coordinates, $x = \rho \sin \phi \cos \theta$, so we get $\rho \sin \phi \cos \theta = 4$, or $\rho = 4 \csc \phi \sec \theta$.

31. $r = 2$ in cylindrical coordinates is a cylinder of radius 2 whose axis of symmetry is the z axis. In rectangular coordinates, we get $\sqrt{x^2 + y^2} = 2$, or $x^2 + y^2 = 4$ (note that z is arbitrary). In spherical coordinates, substituting for r gives $\rho \sin \phi = 2$.

33. $\theta = \frac{\pi}{2}$ is the yz plane, since it is the plane forming an angle of $\frac{\pi}{2}$ with the x axis. In rectangular coordinates, this is the plane through $(0,0,0)$ with normal vector $(1,0,0)$, so it is the plane $x = 0$.

35. $\rho = 2$ is a sphere of radius 2. Converting to cylindrical coordinates, we get $\sqrt{r^2 + z^2} = 2$, or $r^2 + z^2 = 4$. Converting to rectangular coordinates gives $\sqrt{x^2 + y^2 + z^2} = 2$, or $x^2 + y^2 + z^2 = 4$.

37. $\phi = \frac{\pi}{2}$ is the xy plane, since it consists of points such that the line from $(0,0,0)$ to the point makes an angle of $\frac{\pi}{2}$ with the z axis. In rectangular coordinates, this is the plane $z = 0$. In cylindrical coordinates, it is also the plane $z = 0$.

39. $\theta \in [0, 2\pi]$ and $r \in [0, 3]$, so the projection of the solid on the xy ($r\theta$) plane is a circle of radius 3 centered at the origin. Over that circle, z ranges from 0 to r, which is everything *except* a cone over that circle. Thus the solid is the cylinder of height 3 and radius 3 with symmetry axis the z axis, minus the cone $z = r$. That is, it is bounded by the xy plane, the cylinder $r = 3$, and the cone $z = r$.

41. $\theta \in [0, \pi]$, so the solid lies over the upper half-plane. In fact, $r = 2 \sin \theta$ traces out a circle of radius 1 centered at $(0, 1)$, so this is the base of the solid. Over that base, z ranges from 0 to $\sqrt{16 - r^2}$, so the solid is the interior of the portion of the sphere of radius 4 centered at the origin that lies over the circle of radius 1 centered at $(0, 1)$.

43. Since ϕ ranges from $\frac{\pi}{2}$ to π, the solid lies entirely on or below the xy plane (i.e., nonpositive z). θ ranges from 0 to 2π, so it is rotationally symmetric around the z axis. Finally, for each such point, r varies from 0 to 2. Thus this is the lower half hemisphere of radius 2 centered at the origin. That is, it is the volume below the xy plane bounded by the xy plane and the sphere $\rho = 2$.

45. ρ ranges from 0 to $3 \sec \phi$. Now, $\rho = 3 \sec \phi$ means $\rho \cos \phi = 3$, so $z = 3$. Thus this region is bounded above by $z = 3$. Since θ ranges from 0 to 2π, it is rotationally symmetric around the z axis. Finally, ϕ ranges from 0 to $\frac{\pi}{4}$. So this is the solid above the cone $\phi = \frac{\pi}{4}$ and below the plane $z = 3$; in cartesian coordinates, it is the region above $z = \sqrt{x^2 + y^2}$ and below the plane $z = 3$.

47. The cylinder has cylindrical equation $r = 2 \sin \theta$, and the cylindrical equation of the sphere of radius 2 centered at the origin is $z^2 + r^2 = 4$ since $r^2 = x^2 + y^2$. The circular base of the solid is traced once for $\theta \in [0, \pi]$. By symmetry of both the cylinder and the sphere, it suffices to compute the integral for $\theta \in \left[0, \frac{\pi}{2}\right]$ and $r \in [0, 2 \sin \theta]$ and quadruple it. (This has the advantage that both

$\sin\theta$ and $\cos\theta$ will be nonnegative on the region of integration). Thus the volume of this solid is

$$
4\int_0^{\pi/2}\int_0^{2\sin\theta}\int_0^{\sqrt{4-r^2}} r\,dz\,dr\,d\theta = 4\int_0^{\pi/2}\int_0^{2\sin\theta} [rz]_{z=0}^{z=\sqrt{4-r^2}}\,dr\,d\theta
$$

$$
= 4\int_0^{\pi/2}\int_0^{2\sin\theta} r\sqrt{4-r^2}\,dr\,d\theta
$$

$$
= 4\int_0^{\pi/2}\left[-\frac{1}{3}(4-r^2)^{3/2}\right]_{r=0}^{r=2\sin\theta}\,d\theta
$$

$$
= 4\int_0^{\pi/2}\left(\frac{8}{3}-\frac{1}{3}(4-4\sin^2\theta)^{3/2}\right)\,d\theta
$$

$$
= \frac{16\pi}{3}-\frac{32}{3}\int_0^{\pi/2}\cos^3\theta\,d\theta
$$

$$
= \frac{16\pi}{3}-\frac{32}{3}\int_0^{\pi/2}\cos\theta(1-\sin^2\theta)\,d\theta
$$

$$
= \frac{16\pi}{3}-\frac{32}{3}\left[\sin\theta-\frac{1}{3}\sin^3\theta\right]_0^{\pi/2}
$$

$$
= \frac{16\pi}{3}-\frac{32}{3}\left(1-\frac{1}{3}\right) = \frac{16}{3}\pi-\frac{64}{9}.
$$

49. The cylinder has cylindrical equation $r = 1$, and the sphere has cylindrical equation $r^2 + z^2 = 4$. Since we want the volume outside the cylinder, we integrate from $r = 1$ to $r = 2$, and for $\theta \in [0, 2\pi]$. By symmetry, we can integrate from 0 to $\frac{\pi}{2}$ (assuring nonnegativity of both $\cos\theta$ and $\sin\theta$) and quadruple the result. Since we are looking for the volume above the xy plane, the integral is

$$
4\int_0^{\pi/2}\int_1^2\int_0^{\sqrt{4-r^2}} r\,dz\,dr\,d\theta = 4\int_0^{\pi/2}\int_1^2 [rz]_0^{\sqrt{4-r^2}}\,dr\,d\theta
$$

$$
= 4\int_0^{\pi/2}\int_1^2 r\sqrt{4-r^2}\,dr\,d\theta
$$

$$
= 4\int_0^{\pi/2}\left[-\frac{1}{3}(4-r^2)^{3/2}\right]_{r=1}^{r=2}\,d\theta
$$

$$
= 4\int_0^{\pi/2}\frac{1}{3}3^{3/2}\,d\theta = 2\sqrt{3}\,\pi.
$$

51. Use cylindrical coordinates. The equation of the paraboloid is $z = r^2$, and the equation of the plane is $z = r\cos\theta$. The two intersect for $r^2 = r\cos\theta$, so on the circle $r = \cos\theta$ of radius $\frac{1}{2}$ centered at

$\left(\frac{1}{2}, 0\right)$. The circle is traced once for $\theta \in [0, \pi]$. Thus the volume is

$$
\begin{aligned}
\int_0^\pi \int_0^{\cos\theta} \int_{r^2}^{r\cos\theta} r \, dz \, dr \, d\theta &= \int_0^\pi \int_0^{\cos\theta} [rz]_{z=r^2}^{z=r\cos\theta} \, dr \, d\theta \\
&= \int_0^\pi \int_0^{\cos\theta} \left(r^2 \cos\theta - r^3 \right) dr \, d\theta \\
&= \int_0^\pi \left[\frac{1}{3} r^3 \cos\theta - \frac{1}{4} r^4 \right]_{r=0}^{r=\cos\theta} d\theta \\
&= \int_0^\pi \frac{1}{12} \cos^4\theta \, d\theta \\
&= \int_0^\pi \frac{1}{12} (\cos^2\theta)^2 \, d\theta \\
&= \int_0^\pi \frac{1}{48} (1 + \cos 2\theta)^2 \, d\theta \\
&= \frac{1}{48} \int_0^\pi (1 + 2\cos 2\theta + \cos^2 2\theta) \, d\theta \\
&= \frac{1}{48} \int_0^\pi \left(1 + 2\cos 2\theta + \frac{1}{2}(1 + \cos 4\theta) \right) d\theta \\
&= \frac{1}{48} \left[\frac{3}{2}\theta + \sin 2\theta + \frac{1}{8}\sin 4\theta \right]_0^\pi = \frac{1}{32}\pi.
\end{aligned}
$$

53. Use cylindrical coordinates. The cylinder has equation $r = \sqrt{5}$, and the hyperboloid has equation $r^2 - z^2 = 1$, so that $z^2 = r^2 - 1$. The hyperboloid intersects the xy plane when $z = 0$, so on the circle $r = 1$ of radius 1. Thus the volume of the solid is

$$
\begin{aligned}
\int_0^{2\pi} \int_1^{\sqrt{5}} \int_0^{\sqrt{r^2-1}} r \, dz \, dr \, d\theta &= \int_0^{2\pi} \int_1^{\sqrt{5}} [rz]_{z=0}^{z=\sqrt{r^2-1}} \, dr \, d\theta \\
&= \int_0^{2\pi} \int_1^{\sqrt{5}} r\sqrt{r^2-1} \, dr \, d\theta \\
&= \int_0^{2\pi} \left[\frac{1}{3}(r^2-1)^{3/2} \right]_{r=1}^{r=\sqrt{5}} d\theta \\
&= \int_0^{2\pi} \frac{8}{3} \, d\theta = \frac{16\pi}{3}.
\end{aligned}
$$

55. Use spherical coordinates. For $\phi \in [0, \alpha]$, we integrate over all $\theta \in [0, 2\pi]$ and all ρ from 0 to R. Thus the integral is

$$
\begin{aligned}
\int_0^{2\pi} \int_0^\alpha \int_0^R \rho^2 \sin\phi \, d\rho \, d\phi \, d\theta &= \int_0^{2\pi} \int_0^\alpha \left[\frac{1}{3}\rho^3 \sin\phi \right]_{\rho=0}^{\rho=R} d\phi \, d\theta \\
&= \frac{R^3}{3} \int_0^{2\pi} \int_0^\alpha \sin\phi \, d\phi \, d\theta \\
&= \frac{R^3}{3} \int_0^{2\pi} [-\cos\phi]_{\phi=0}^{\phi=\alpha} \, d\theta \\
&= \frac{R^3}{3} \int_0^{2\pi} (1 - \cos\alpha) \, d\theta = \frac{2}{3}\pi R^3(1 - \cos\alpha).
\end{aligned}
$$

When $\alpha = 0$, the "cone" is the positive z axis, so there is no area inside the cone, so the result should be zero (and in fact the integral is zero). When $\alpha = \pi$, the "cone" is the negative z axis, so everything inside the sphere is also inside the cone, so the result should be the volume of the sphere (and in fact the integral is $\frac{4}{3}\pi R^3$, the volume of the sphere).

57. Integrating using cylindrical coordinates (see Exercise 47) gives (with $\rho(x, y, z) = k(x^2 + y^2) = kr^2$)

$$m = 4 \int_0^{\pi/2} \int_0^{2\sin\theta} \int_0^{\sqrt{4-r^2}} r \cdot kr^2 \, dz \, dr \, d\theta = 4k \int_0^{\pi/2} \int_0^{2\sin\theta} \left[r^3 z \right]_{z=0}^{z=\sqrt{4-r^2}} dr \, d\theta$$

$$= 4k \int_0^{\pi/2} \int_0^{2\sin\theta} r^3 \sqrt{4 - r^2} \, dr \, d\theta.$$

Integrate using integration by parts with $u = r^2$ and $dv = r\sqrt{4-r^2}$, so that $du = 2r\,dr$ and $v = -\frac{1}{3}(4 - r^2)^{3/2}$. Then

$$4k \int_0^{\pi/2} \int_0^{2\sin\theta} r^3 \sqrt{4 - r^2} \, dr \, d\theta$$

$$= 4k \int_0^{\pi/2} \left(\left[-\frac{1}{3} r^2 (4 - r^2)^{3/2} \right]_{r=0}^{r=2\sin\theta} + \frac{2}{3} \int_0^{2\sin\theta} r(4 - r^2)^{3/2} \, dr \right) d\theta$$

$$= 4k \int_0^{\pi/2} \left(-\frac{4}{3} \sin^2\theta (4 - 4\sin^2\theta)^{3/2} + \frac{2}{3} \left[-\frac{1}{5}(4 - r^2)^{5/2} \right]_0^{2\sin\theta} \right) d\theta$$

$$= 4k \int_0^{\pi/2} \left(-\frac{32}{3} \sin^2\theta \cos^3\theta - \frac{2}{15}(4 - 4\sin^2\theta)^{5/2} + \frac{2}{15} 4^{5/2} \right) d\theta$$

$$= 4k \int_0^{\pi/2} \left(-\frac{32}{3} \sin^2\theta \cos^3\theta - \frac{64}{15} \cos^5\theta + \frac{64}{15} \right) d\theta$$

$$= 4k \int_0^{\pi/2} \left(-\frac{32}{3}(\sin^2\theta - \sin^4\theta) \cos\theta - \frac{64}{15}(1 - \sin^2\theta)^2 \cos\theta + \frac{64}{15} \right) d\theta$$

$$= 4k \int_0^{\pi/2} \left(-\frac{32}{3}(\sin^2\theta - \sin^4\theta) \cos\theta - \frac{64}{15}(1 - 2\sin^2\theta + \sin^4\theta) \cos\theta + \frac{64}{15} \right) d\theta$$

$$= 4k \int_0^{\pi/2} \left(\frac{64}{15} - \frac{64}{15} \cos\theta - \frac{32}{15} \sin^2\theta \cos\theta + \frac{32}{5} \sin^4\theta \cos\theta \right) d\theta$$

$$= 4k \left[\frac{64}{15}\theta - \frac{64}{15} \sin\theta - \frac{32}{45} \sin^3\theta + \frac{32}{25} \sin^5\theta \right]_0^{\pi/2}$$

$$= 4k \left(\frac{32}{15}\pi - \frac{64}{15} - \frac{152}{45} + \frac{52}{25} \right) = \left(\frac{128}{15}\pi - \frac{3328}{225} \right) k.$$

This integral can also be evaluated using the substitution $u = 4 - r^2$; it is arguable which method is easier.

59. Use cylindrical coordinates (see Exercise 49 for details). The distance from the z axis is r, so the density is $\rho(x, y, z) = kr$. We assume $k = 1$ since the constant of proportionality will cancel out in the final calculation of the center of mass. Then the mass of the region is

$$4 \int_0^{\pi/2} \int_1^2 \int_0^{\sqrt{4-r^2}} r \cdot r \, dz \, dr \, d\theta = 4 \int_0^{\pi/2} \int_1^2 \left[r^2 z \right]_{z=0}^{z=\sqrt{4-r^2}} dr \, d\theta$$

$$= 4 \int_0^{\pi/2} \int_1^2 r^2 \sqrt{4 - r^2} \, dr \, d\theta.$$

Now substitute $r = 2 \sin u$, so that $dr = 2 \cos u \, du$. Then we get

$$4 \int_0^{\pi/2} \int_1^2 r^2 \sqrt{4-r^2} \, dr \, d\theta = 4 \int_0^{\pi/2} \int_{r=1}^{r=2} 4 \sin^2 u \sqrt{4 - 4\sin^2 u} \cdot 2 \cos u \, du \, d\theta$$

$$= 64 \int_0^{\pi/2} \int_{r=1}^{r=2} \sin^2 u \cos^2 u \, du \, d\theta$$

$$= 16 \int_0^{\pi/2} \int_{r=1}^{r=2} \sin^2 2u \, du \, d\theta$$

$$= 16 \int_0^{\pi/2} \int_{r=1}^{r=2} \frac{1}{2}(1 - \cos 4u) \, du \, d\theta$$

$$= 8 \int_0^{\pi/2} \left[u - \frac{1}{4} \sin 4u \right]_{r=1}^{r=2} d\theta$$

$$= 8 \int_0^{\pi/2} \left[u - \frac{1}{2} \sin 2u \cos 2u \right]_{r=1}^{r=2} d\theta$$

$$= 8 \int_0^{\pi/2} \left[u - \sin u \cos u (\cos^2 u - \sin^2 u) \right]_{r=1}^{r=2} d\theta$$

$$= 8 \int_0^{\pi/2} \left[\sin^{-1} \frac{r}{2} - \frac{r}{2} \cdot \frac{\sqrt{4-r^2}}{2} \left(\frac{4-r^2}{4} - \frac{r^2}{4} \right) \right]_{r=1}^{r=2} d\theta$$

$$= 8 \int_0^{\pi/2} \left(\frac{\pi}{2} - \frac{\pi}{6} + \frac{1}{2} \cdot \frac{\sqrt{3}}{2} \left(\frac{3}{4} - \frac{1}{4} \right) \right) d\theta$$

$$= 8 \int_0^{\pi/2} \left(\frac{\pi}{3} + \frac{\sqrt{3}}{8} \right) d\theta = \frac{4}{3} \pi^2 + \frac{\sqrt{3}}{2} \pi.$$

By symmetry, since both the region and the density function are symmetric around the z axis, both $\bar{x}$ and $\bar{y}$ are zero. The first moment around the xy plane is

$$M_{xy} = 4 \int_0^{\pi/2} \int_1^2 \int_0^{\sqrt{4-r^2}} z \cdot r \cdot r \, dz \, dr \, d\theta = 4 \int_0^{\pi/2} \int_1^2 \left[\frac{1}{2} r^2 z^2 \right]_{z=0}^{z=\sqrt{4-r^2}} dr \, d\theta$$

$$= 2 \int_0^{\pi/2} \int_1^2 (4r^2 - r^4) \, dr \, d\theta$$

$$= 2 \int_0^{\pi/2} \left[\frac{4}{3} r^3 - \frac{1}{5} r^5 \right]_{r=1}^{r=2} d\theta$$

$$= 2 \int_0^{\pi/2} \frac{47}{15} \, d\theta = \frac{47}{15} \pi.$$

Thus

$$\bar{z} = \frac{\frac{47}{15}\pi}{\frac{4}{3}\pi^2 + \frac{\sqrt{3}}{2}\pi} = \frac{94}{40\pi + 15\sqrt{3}},$$

so the center of mass is

$$\left(0, 0, \frac{94}{40\pi + 15\sqrt{3}} \right).$$

61. Use cylindrical coordinates (see Exercise 51 for details). The density is $\rho(x, y, z) = kz^2$, so the mass

is

$$\int_0^\pi \int_0^{\cos\theta} \int_{r^2}^{r\cos\theta} kz^2 \cdot r \, dz \, dr \, d\theta = k \int_0^\pi \int_0^{\cos\theta} \left[\frac{1}{3}rz^3\right]_{z=r^2}^{z=r\cos\theta} dr \, d\theta$$

$$= \frac{k}{3} \int_0^\pi \int_0^{\cos\theta} (r^4 \cos^3\theta - r^7) \, dr \, d\theta$$

$$= \frac{k}{3} \int_0^\pi \left[\frac{1}{5}r^5 \cos^3\theta - \frac{1}{8}r^8\right]_{r=0}^{r=\cos\theta} d\theta$$

$$= \frac{k}{3} \int_0^\pi \left(\frac{1}{5}\cos^8\theta - \frac{1}{8}\cos^8\theta\right) d\theta$$

$$= \frac{k}{40} \int_0^\pi \cos^8\theta \, d\theta = \frac{k}{40} \int_0^\pi \left(\cos^2\theta\right)^4 d\theta$$

$$= \frac{k}{40} \int_0^\pi \left(\frac{1}{2}(1+\cos 2\theta)\right)^4 d\theta$$

$$= \frac{k}{640} \int_0^\pi (1 + 4\cos 2\theta + 6\cos^2 2\theta + 4\cos^3 2\theta + \cos^4 2\theta) \, d\theta$$

$$= \frac{k}{640} \int_0^\pi \Big(1 + 4\cos 2\theta + 3(1 + \cos 4\theta) + 4\cos 2\theta(1 - \sin^2 2\theta)$$
$$+ \left(\frac{1}{2}(1+\cos 4\theta)\right)^2\Big) \, d\theta$$

$$= \frac{k}{640} \int_0^\pi \Big(1 + 4\cos 2\theta + 3(1 + \cos 4\theta) + 4\cos 2\theta(1 - \sin^2 2\theta)$$
$$+ \frac{1}{4} + \frac{1}{2}\cos 4\theta + \frac{1}{4}\cos^2 4\theta\Big) \, d\theta$$

$$= \frac{k}{640} \int_0^\pi \Big(\frac{17}{4} + 8\cos 2\theta - 4\cos 2\theta \sin^2 2\theta + \frac{1}{2}\cos 4\theta$$
$$+ \frac{1}{4}\left(\frac{1}{2} + \frac{1}{2}\cos 8\theta\right)\Big) \, d\theta$$

$$= \frac{k}{640} \int_0^\pi \left(\frac{35}{8} + 8\cos 2\theta - 4\cos 2\theta \sin^2 2\theta + \frac{1}{2}\cos 4\theta + \frac{1}{8}\cos 8\theta\right) d\theta$$

$$= \frac{k}{640} \left[\frac{35}{8}\theta + 4\sin 2\theta - \frac{2}{3}\sin^3 2\theta + \frac{1}{8}\sin 4\theta + \frac{1}{64}\sin 8\theta\right]_0^\pi$$

$$= \frac{k}{640} \left(\frac{35}{8}\pi\right) = \frac{7}{1024}k\pi.$$

63. Use cylindrical coordinates (see Exercise 53 for details). The density function is $\frac{k}{\sqrt{x^2+y^2}} = \frac{k}{r}$. Thus the moment of inertia about the z axis is

$$I_z = \iiint_\Omega (x^2 + y^2)\rho(x, y, z)\, dV$$

$$= k \iiint_\Omega r^2 \cdot \frac{1}{r}\, dV$$

$$= k \int_0^{2\pi} \int_1^{\sqrt{5}} \int_0^{\sqrt{r^2-1}} r \cdot r\, dz\, dr\, d\theta$$

$$= k \int_0^{2\pi} \int_1^{\sqrt{5}} \left[r^2 z \right]_{z=0}^{z=\sqrt{r^2-1}}\, dr\, d\theta$$

$$= k \int_0^{2\pi} \int_1^{\sqrt{5}} r^2 \sqrt{r^2 - 1}\, dr\, d\theta.$$

Use the trigonometric substitution $r = \sec u$, so that $\sqrt{r^2 - 1} = \tan u$ and $dr = \sec u \tan u\, du$; then the integral becomes

$$k \int_0^{2\pi} \int_{r=1}^{r=\sqrt{5}} \sec^2 u \tan u \sec u \tan u\, du\, d\theta = k \int_0^{2\pi} \int_{r=1}^{r=\sqrt{5}} \sec^3 u (\sec^2 u - 1)\, du\, d\theta$$

$$= k \int_0^{2\pi} \int_{r=1}^{r=\sqrt{5}} (\sec^5 u - \sec^3 u)\, du\, d\theta.$$

Since θ is not involved in the inner integral, we may rewrite this is $2k\pi \int_{r=1}^{r=\sqrt{5}} (\sec^5 u - \sec^3 u)\, du$. Now, using the reduction formula from Exercise 86 in Section 5.4, we have

$$\int_{r=1}^{r=\sqrt{5}} \sec^5 u\, du = \frac{1}{4} \left[\sec^3 u \tan u \right]_{r=1}^{r=\sqrt{5}} + \frac{3}{4} \int_{r=1}^{r=\sqrt{5}} \sec^3 u\, du$$

$$= \frac{1}{4} \left[r^3 \sqrt{r^2 - 1} \right]_1^{\sqrt{5}} + \frac{3}{4} \int_{r=1}^{r=\sqrt{5}} \sec^3 u\, du$$

$$= \frac{5}{2}\sqrt{5} + \frac{3}{4} \int_{r=1}^{r=\sqrt{5}} \sec^3 u\, du.$$

Thus the original integral becomes (using the formula for $\int \sec^3 u\, du$ from Section 5.4)

$$I_z = 2k\pi \int_{r=1}^{r=\sqrt{5}} (\sec^5 u - \sec^3 u)\, du$$

$$= 2k\pi \left(\frac{5}{2}\sqrt{5} - \frac{1}{4} \int_{r=1}^{r=\sqrt{5}} du \right)$$

$$= 2k\pi \left(\frac{5}{2}\sqrt{5} - \frac{1}{8} \left[\sec u \tan u + \ln|\sec u + \tan u| \right]_{r=1}^{r=\sqrt{5}} \right)$$

$$= 2k\pi \left(\frac{5}{2}\sqrt{5} - \frac{1}{8} \left[r\sqrt{r^2 - 1} + \ln\left| r + \sqrt{r^2 - 1} \right| \right]_1^{\sqrt{5}} \right)$$

$$= 2k\pi \left(\frac{5}{2}\sqrt{5} - \frac{1}{8} \left(2\sqrt{5} + \ln(2 + \sqrt{5}) \right) \right)$$

$$= \left(\frac{9}{2}\sqrt{5} - \frac{1}{4} \ln(2 + \sqrt{5}) \right) k\pi.$$

65. Use spherical coordinates (see Exercise 55 for details). Since the region is rotationally symmetric around the z axis and the density is constant, we have $\bar{x} = \bar{y} = 0$. With $\rho(x, y, z) = 1$, the mass

was computed in Exercise 55 to be $\frac{2}{3}\pi R^3(1 - \cos\alpha)$, and we get for the first moment around the xy plane

$$M_{xy} = \int_0^{2\pi} \int_0^\alpha \int_0^R \rho\cos\phi \cdot \rho^2 \sin\phi \, d\rho \, d\phi \, d\theta = \int_0^{2\pi} \int_0^\alpha \left[\frac{1}{4}\rho^4 \sin\phi\cos\phi\right]_{\rho=0}^{\rho=R} d\phi \, d\theta$$

$$= \frac{1}{4}R^4 \int_0^{2\pi} \int_0^\alpha \sin\phi\cos\phi \, d\phi \, d\theta$$

$$= \frac{1}{4}R^4 \int_0^{2\pi} \left[\frac{1}{2}\sin^2\phi\right]_{\phi=0}^{\phi=\alpha} d\theta$$

$$= \frac{1}{8}R^4 \int_0^{2\pi} \sin^2\alpha \, d\theta = \frac{1}{4}\pi R^4 \sin^2\alpha.$$

Then

$$\bar{z} = \frac{M_{xy}}{m} = \frac{\frac{1}{4}\pi R^4 \sin^2\alpha}{\frac{2}{3}\pi R^3(1 - \cos\alpha)} = \frac{3}{8}\pi R\frac{\sin^2\alpha}{1 - \cos\alpha} = \frac{3}{8}\pi R\frac{1 - \cos^2\alpha}{1 - \cos\alpha} = \frac{3}{8}\pi R(1 + \cos\alpha),$$

so that the center of mass is

$$\left(0, 0, \frac{3}{8}\pi R(1 + \cos\alpha)\right).$$

Applications

67. The tank is composed of two pieces. The upper piece is a cylindrical shell with height 8, outer radius 75, and inner radius 8, so the volume of this portion of the tank is $8\pi(75^2 - 8^2)$. The remaining piece of the tank is a portion of the sphere of radius 150 that lies outside of a cylinder of radius 8 but inside a cylinder of radius 75. Place the origin at the center of that sphere, and point the positive z axis *downwards*. Then using cylindrical coordinates, $\theta \in [0, 2\pi]$ and $r \in [8, 75]$. Then $(75, 0, 75\sqrt{3})$ is a point on the circle where the sphere intersects the vertical tank wall, so that we want to integrate for $z \in [75\sqrt{3}, \sqrt{150^2 - r^2}]$. Thus the volume of the spherical portion is

$$\int_0^{2\pi} \int_8^{75} \int_{75\sqrt{3}}^{\sqrt{150^2 - r^2}} r \, dz \, dr \, d\theta = \int_0^{2\pi} \int_8^{75} [rz]_{z=75\sqrt{3}}^{z=\sqrt{150^2 - r^2}} dr \, d\theta$$

$$= \int_0^{2\pi} \int_8^{75} \left(r\sqrt{150^2 - r^2} - 75r\sqrt{3}\right) dr \, d\theta$$

$$= \int_0^{2\pi} \left[-\frac{1}{3}(150^2 - r^2)^{3/2} - \frac{75}{2}r^2\sqrt{3}\right]_{r=8}^{r=75} d\theta$$

$$= \int_0^{2\pi} \left(-\frac{1}{3}(150^2 - 75^2)^{3/2} - \frac{75^3}{2}\sqrt{3} + \frac{1}{3}(150^2 - 8^2)^{3/2} + \frac{75}{2}64\sqrt{3}\right) d\theta$$

$$= 2\pi \left(-\frac{1}{3}(75\sqrt{3})^3 - \frac{1}{2}75^3\sqrt{3} + \frac{1}{3}(2\sqrt{5609})^3 + 2400\sqrt{3}\right)$$

$$= 2\pi \left(\left(-\frac{5}{6}\cdot 75^3 + 2400\right)\sqrt{3} + \frac{44872}{3}\sqrt{5609}\right).$$

Thus the total volume of the tank is

$$8\pi(75^2 - 8^2) + 2\pi \left(\left(-\frac{5}{6}\cdot 75^3 + 2400\right)\sqrt{3} + \frac{44872}{3}\sqrt{5609}\right) \approx 317558 \text{ cubic feet.}$$

Proofs

69. In cylindrical coordinates, $x = r\cos\theta$, so $x = a$ is $r\cos\theta = a$, or $r = a\sec\theta$. Note that this is not defined for $\theta = \frac{\pi}{2}$, which makes sense since the line $x = a$ (for $a \neq 0$) never intersects the y axis.

71. In spherical coordinates, $x = \rho\sin\phi\cos\theta$, so $x = a$ is $\rho\sin\phi\cos\theta = a$, or $\rho = a\csc\phi\sec\theta$. Note that this plane never intersects the yz plane, which corresponds to $\phi = 0$ and to $\theta = \frac{\pi}{2} + n\pi$, which is exactly where this equation is undefined.

73. Place the origin at the tip of the cone, and let the cone open upwards. Then the height of the cone when the radius is r is $r\frac{h}{R}$, so using cylindrical coordinates, the volume of the cone is

$$\int_0^{2\pi}\int_0^R\int_{rh/R}^h r\,dz\,dr\,d\theta = \int_0^{2\pi}\int_0^R [rz]_{z=rh/R}^{z=h}\,dr\,d\theta = \int_0^{2\pi}\int_0^R \left(rh - \frac{h}{R}r^2\right)dr\,d\theta$$

$$= \int_0^{2\pi}\left[\frac{h}{2}r^2 - \frac{h}{3R}r^3\right]_{r=0}^{r=R}d\theta = \int_0^{2\pi}\frac{h}{6}R^2\,d\theta = \frac{1}{3}\pi R^2 h.$$

75. Place the origin at the tip of the cone, and let the cone open upwards. Then the angle ϕ of the slant side of the cone to the z axis is such that $\tan\phi = \frac{R}{h}$, so the equation of the cone is $\phi = \tan^{-1}\frac{R}{h}$. Thus we want to integrate from $\phi = 0$ to $\phi = \tan^{-1}\frac{R}{h}$ and from $\theta = 0$ to 2π. Since $z = \rho\cos\phi$, the range of integration for ρ is from 0 to $h\sec\phi$. Thus the volume is

$$\int_0^{2\pi}\int_0^{\tan^{-1}(R/h)}\int_0^{h\sec\phi} \rho^2\sin\phi\,d\rho\,d\phi\,d\theta = \int_0^{2\pi}\int_0^{\tan^{-1}(R/h)}\left[\frac{1}{3}\rho^3\sin\phi\right]_{\rho=0}^{\rho=h\sec\phi}d\phi\,d\theta$$

$$= \frac{1}{3}h^3\int_0^{2\pi}\int_0^{\tan^{-1}(R/h)}\sin\phi\sec^3\phi\,d\phi\,d\theta$$

$$= \frac{1}{3}h^3\int_0^{2\pi}\int_0^{\tan^{-1}(R/h)}\tan\phi\cdot\sec^2\phi\,d\phi\,d\theta$$

$$= \frac{1}{3}h^3\int_0^{2\pi}\left[\frac{1}{2}\tan^2\phi\right]_{\phi=0}^{\phi=\tan^{-1}(R/h)}d\theta$$

$$= \frac{1}{6}h^3\int_0^{2\pi}\frac{R^2}{h^2}\,d\theta = \frac{1}{3}\pi R^2 h.$$

Thinking Forward

A non-standard coordinate system in $\mathbb{R}^2$

▷ Every point has unique coordinates since the two vectors representing a unit length in each of the u and v directions are not parallel. Thus to find the coordinates of any point in these coordinates, draw a line parallel to u and one parallel to v from the point until it meets the two axes. The intersection points are the coordinates of the point.

▷ Let $\mathbf{u}$ be a unit vector in the u direction, and $\mathbf{v}$ a unit vector in the v direction. Thus if the angle between the u and v axes is θ, then by Corollary 10.33, the area of the parallelogram is $\|\mathbf{u} \times \mathbf{v}\| = \|\mathbf{u}\|\,\|\mathbf{v}\|\sin\theta = \sin\theta$.

▷ Since $\mathbf{a}$ is parallel to the u axis, it is $a_1\mathbf{u}$ for some real a_1; similarly $\mathbf{b} = b_1\mathbf{v}$ for some real b_1. Then the area of the parallelogram determined by $\mathbf{a}$ and $\mathbf{b}$ is

$$\|\mathbf{a} \times \mathbf{b}\| = \|\mathbf{a}\| \times \|\mathbf{b}\|\sin\theta = a_1 b_1\sin\theta.$$

A non-standard coordinate system in $\mathbb{R}^3$

▷ Let $a = \frac{u}{\|u\|}$, $b = \frac{v}{\|v\|}$, $c = \frac{w}{\|w\|}$. Let θ be the angle between b and c, and let ϕ be the angle between a and the plane determined by b and c. Then $\frac{\pi}{2} - \phi$ is the angle between a and $b \times c$, so that

$$|a \cdot (b \times c)| = \|a\| \, \|b \times c\| \cos\left(\frac{\pi}{2} - \phi\right) = \|a\| \, \|b\| \, \|c\| \sin\theta \sin\phi = \sin\theta \sin\phi,$$

since a, b, and c are all unit vectors.

▷ By Theorem 10.36, the volume of the parallelepiped determined by $\mathbf{a}$, $\mathbf{b}$, and $\mathbf{c}$ is

$$|\mathbf{a} \cdot (\mathbf{b} \times \mathbf{c})|.$$

13.7 Jacobians and Change of Variables

Thinking Back

Integration by substitution

The integration-by-substitution formula is

$$\int_a^b f(g(x))g'(x)\,dx = \int_{g(a)}^{g(b)} f(u)\,du,$$

where $u = g(x)$. This is simply an application of the chain rule to $f(u) = f(g(x))$. Suppose $F(u)$ is an antiderivative of $f(u)$. Then $F'(u) = F'(g(x)) = f(g(x))g'(x)$ and thus $F(u)$ is also an antiderivative of $f(g(x))g'(x)$. So by the Fundamental Theorem of Calculus, we have

$$\int_a^b f(g(x))g'(x)\,dx = F(u)\Big|_{x=a}^{x=b} = F(g(x))\Big|_{x=a}^{x=b} = F(g(b)) - F(g(a)) = \int_{g(a)}^{g(b)} F(u)\,du.$$

Integrating with polar coordinates

The integral is

$$\iint_\Omega 1\,dA = \iint_\Omega r\,dr\,d\theta.$$

Integrating with cylindrical coordinates

The integral is

$$\iint_\Omega 1\,dV = \iint_\Omega r\,dz\,dr\,d\theta.$$

Integrating with spherical coordinates

The integral is

$$\iint_\Omega 1\,dA = \iint_\Omega \rho^2 \sin\phi\,d\rho\,d\phi\,d\theta.$$

Concepts

1. (a) True. Such a transformation might "flatten" each of the four quadrants of the circle to get a rectangle (or a square). Such a transformation would not be differentiable at the corners of the rectangle, but it would be a continuous transformation.

    (b) True. The equations of two of the parallel sides are $ax + by = u_1$ and $ax + by = u_2$; let $u = ax + by$ so that $u = u_1$ and $u = u_2$ are two sides of the transformed parallelogram. Similarly, if the other two sides of the parallelogram are $cx + dy = v_1$ and $cx + dy = v_2$, let $v = cx + dy$ so that $u = v_1$ and $u = v_2$ are the other two sides. If the resulting figure is a rectangle rather than a square, apply an appropriate scaling factor to the definition of v so that all four sides are equal.

    (c) True. Use the transformation described in part (a), flattening each of the four quadrants of the cylinder, to get a rectangular solid. Again this will not be differentiable along the vertical edges of the cylinder.

    (d) True. Similarly to part (a), flatten each of six portions of the sphere to get a rectangular solid.

    (e) True. Multiply $v = x - 2y$ by 2 to get $2v = 2x - 4y$. Subtract from $u = 2x + y$ to get $u - 2v = 5y$ so that $y = \frac{1}{5}u - \frac{2}{5}v$. Substitute into $v = x - 2y$ and simplify to get $x = \frac{2}{5}u + \frac{1}{5}v$.

    (f) False. From the result of part (e), we have

$$\frac{\partial(x, y)}{\partial(u, v)} = \det \begin{bmatrix} \frac{\partial x}{\partial u} & \frac{\partial y}{\partial u} \\ \frac{\partial x}{\partial v} & \frac{\partial y}{\partial v} \end{bmatrix} = \det \begin{bmatrix} \frac{2}{5} & \frac{1}{5} \\ \frac{1}{5} & \frac{-2}{5} \end{bmatrix} = -\frac{1}{5}.$$

    (g) True. Simply apply a transformation that scales one pair of sides so that they are equal in length to the other pair. For example, first apply a transformation that rotates $\mathcal{R}$ so that its sides are parallel to the x and y axes. If the horizontal side length is a and the vertical side length is b, apply the transformation $u = x$ and $v = \frac{a}{b}y$.

    (h) False. For example, in the example in part (g), you could use the transformation given there, or you could use the transformation $u = \frac{b}{a}x$ and $v = y$. There are many other possibilities as well.

3. The Jacobian of the transformation represents a scaling factor on the increment of area. For example, the identity transformation, corresponding to rectangular coordinates, has a Jacobian of 1, and the increment of area is $dx\,dy$. Polar coordinates have a Jacobian of r, and the increment of area is $r\,dr\,d\theta$. A transformation that simply stretches by a factor of c in the x direction and leaves the y direction alone would have a Jacobian of c and an area increment of $c\,dx\,dy$.

5. Using the diagram from Exercise 4, as a Type II region $\mathcal{R}$ is bounded below by $y = 0$ and above by $y = 3$. From $y = 0$ to $y = 1$ it is bounded on the left by $x = 1 - y$ and on the right by $x = y + 1$. From $y = 1$ to $y = 2$ it is bounded on the left by $x = y - 1$ and on the right by $x = y + 1$. Finally, from $y = 2$ to $y = 3$ it is bounded on the left by $x = y - 1$ and on the right by $x = 5 - y$. Thus the

integral is

$$\iint_{\mathcal{R}} x^2 y \, dA = \int_0^1 \int_{1-y}^{y+1} x^2 y \, dx \, dy + \int_1^2 \int_{y-1}^{y+1} x^2 y \, dx \, dy + \int_2^3 \int_{y-1}^{5-y} x^2 y \, dx \, dy$$

$$= \int_0^1 \left[\frac{1}{3} x^3 y \right]_{x=1-y}^{x=y+1} dy + \int_1^2 \left[\frac{1}{3} x^3 y \right]_{x=y-1}^{x=y+1} dy + \int_2^3 \left[\frac{1}{3} x^3 y \right]_{x=y-1}^{x=5-y} dy$$

$$= \frac{1}{3} \left(\int_0^1 ((y+1)^3 - (1-y)^3) y \, dy + \int_1^2 ((y+1)^3 - (y-1)^3) y \, dy \right.$$

$$\left. + \int_2^3 ((5-y)^3 - (y-1)^3) y \, dy \right)$$

$$= \frac{1}{3} \left(\int_0^1 (2y^4 + 6y^2) \, dy + \int_1^2 (6y^3 + 2y) \, dy + \int_2^3 (-2y^4 + 18y^3 - 78y^2 + 126y) \, dy \right)$$

$$= \frac{1}{3} \left(\left[\frac{2}{5} y^5 + 2y^3 \right]_0^1 + \left[\frac{3}{2} y^4 + y^2 \right]_1^2 + \left[-\frac{2}{5} y^5 + \frac{9}{2} y^4 - 26y^3 + 63y^2 \right]_2^3 \right)$$

$$= \frac{1}{3} \left(\frac{12}{5} + 28 - \frac{5}{2} + \frac{1323}{10} - \frac{516}{5} \right) = 19.$$

7. $u = x + y$ and $v = x - y$ gives, adding, $2x = u + v$, or $x = \frac{u+v}{2}$. Substituting into the first equation and solving for y gives $y = \frac{u-v}{2}$. Then the Jacobian is

$$\frac{\partial(x,y)}{\partial(u,v)} = \det \begin{bmatrix} \frac{\partial x}{\partial u} & \frac{\partial y}{\partial u} \\ \frac{\partial x}{\partial v} & \frac{\partial y}{\partial v} \end{bmatrix} = \det \begin{bmatrix} \frac{1}{2} & \frac{1}{2} \\ \frac{1}{2} & -\frac{1}{2} \end{bmatrix} = -\frac{1}{2}.$$

9. In the uv plane, the region becomes $\mathcal{R}' = [1,5] \times [-1,1]$, and the integrand is

$$x^2 y = \left(\frac{u+v}{2} \right)^2 \left(\frac{u-v}{2} \right) = \frac{1}{8} (u+v)^2 (u-v) = \frac{1}{8} (u+v)(u^2 - v^2) = \frac{1}{8} (u^3 + u^2 v - uv^2 - v^3).$$

So the integral is

$$\int_{-1}^1 \int_1^5 \frac{1}{8} (u^3 + u^2 v - uv^2 - v^3) \cdot \left| \frac{\partial(x,y)}{\partial(u,v)} \right| du \, dv = \frac{1}{16} \int_{-1}^1 \int_1^5 (u^3 + u^2 v - uv^2 - v^3) \, du \, dv$$

$$= \frac{1}{16} \int_{-1}^1 \left[\frac{1}{4} u^4 + \frac{1}{3} u^3 v - \frac{1}{2} u^2 v^2 - uv^3 \right]_{u=1}^{u=5} dv$$

$$= \frac{1}{16} \int_{-1}^1 \left(156 + \frac{124}{3} v - 12v^2 - 4v^3 \right) dv$$

$$= \frac{1}{16} \left[156v + \frac{62}{3} v^2 - 4v^3 - v^4 \right]_{-1}^1 = \frac{304}{16} = 19.$$

11. The image of any region Ω in the xy plane under this transformation is a region in the uv plane that is congruent to Ω but that is shifted by a in the u direction and b in the v direction from its position in the xy plane. In the specific case given, if $a = 3$ and $b = 4$, and if Ω is the rectangle $0 \leq x \leq 1$ and $0 \leq y \leq 2$, we see from the previous exercise that $x = 0$ becomes $u = 3$ and $x = 1$ becomes $u = 4$, so that $0 \leq x \leq 1$ becomes $3 \leq u \leq 4$. Similarly, $y = 0$ becomes $v = 4$ and $y = 2$ becomes $v = 6$, so that $0 \leq y \leq 2$ becomes $4 \leq v \leq 6$. Thus the new rectangle in the uv plane is $[3,4] \times [4,6]$, which is the original rectangle translated by $(3,4)$.

13. Since all this transformation does is shift a region around in the plane, it is not going to be useful in evaluating an integral. The purpose of these transformations is to simplify either the integrand or the bounds, and this transformation does neither.

15. The image of any region Ω in the xy plane under this transformation is a region in the uv plane that is similar (in the sense of geometry) to Ω but that has been stretched (or compressed) by a factor of a in the u direction and b in the v direction from its size in the xy plane. In the specific case given, if $a = 3$ and $b = \frac{1}{2}$, and if Ω is defined by $0 \leq x \leq 1$ and $0 \leq y \leq 1$, we see from the previous exercise that $x = 0$ becomes $u = 0$, $x = 1$ becomes $u = 3$, $y = 0$ becomes $v = 0$, and $y = 1$ becomes $v = \frac{1}{2}$. Thus Ω' is defined by $0 \leq u \leq 3, 0 \leq v \leq \frac{1}{2}$, so it is the rectangle $[0,3] \times \left[0, \frac{1}{2}\right]$ in the uv plane.

17. Exercise 14 will remain the same if a and/or b is negative (although then the lines will be on the opposite side of the corresponding axis in the uv plane from where they were in the xy plane). In Exercise 16, the Jacobian will still be $\frac{1}{ab}$, although this may end up being a negative number. However, since the integral in the uv plane uses the absolute value of the Jacobian, this will not affect the fact that the volume element is still positive.

19. Multiply the first equation by $\sin \theta$ and the second by $\cos \theta$ and add to get

$$u \sin \theta + v \cos \theta = (\sin^2 \theta + \cos^2 \theta)x = x,$$

so that $x = u \sin \theta + v \cos \theta$. Now multiply the first equation by $-\cos \theta$ and the second by $\sin \theta$ and add to get

$$-u \cos \theta + v \sin \theta = (\sin^2 \theta + \cos^2 \theta)y = y,$$

so that $y = -u \cos \theta + v \sin \theta$. Then the Jacobian is

$$\frac{\partial(x,y)}{\partial(u,v)} = \det \begin{bmatrix} \frac{\partial x}{\partial u} & \frac{\partial y}{\partial u} \\ \frac{\partial x}{\partial v} & \frac{\partial y}{\partial v} \end{bmatrix} = \det \begin{bmatrix} \sin \theta & -\cos \theta \\ \cos \theta & \sin \theta \end{bmatrix} = \sin^2 \theta + \cos^2 \theta = 1.$$

Since the transformation is a rotation, it makes sense that the Jacobian is 1, since the volume element does not change, so Exercise 3 implies that the Jacobian is 1.

21. Treating Ω as a union of two Type I regions, the first goes from $x = 0$ to $x = 1$ and the second from $x = 1$ to $x = 4$. For the first region, the lower bound is $y = 1 - x$ and the upper bound is $y = 4 - x$, while for the second, the lower bound is $y = 0$ and the upper is $y = 4 - x$. Then the integral becomes

$$\iint_\Omega \frac{1}{(x+y)^2} \, dA = \int_0^1 \int_{1-x}^{4-x} \frac{1}{(x+y)^2} \, dy \, dx + \int_1^4 \int_0^{4-x} \frac{1}{(x+y)^2} \, dy \, dx$$

$$= \int_0^1 \left[-\frac{1}{x+y}\right]_{y=1-x}^{y=4-x} dx + \int_1^4 \left[-\frac{1}{x+y}\right]_{y=0}^{y=4-x} dx$$

$$= \int_0^1 \left(-\frac{1}{4} + 1\right) dx + \int_1^4 \left(-\frac{1}{4} + \frac{1}{x}\right) dx$$

$$= \frac{3}{4} + \left[-\frac{1}{4}x + \ln|x|\right]_1^4 = \ln 4.$$

23. The original trapezoid has one base between $(1,0)$ and $(0,1)$, with length $b_1 = \sqrt{2}$. The other base is from $(4,0)$ to $(0,4)$, so it has length $b_2 = 4\sqrt{2}$. The height is the perpendicular distance

between the two bases. Draw a line from $(1,0)$ perpendicular to the other base, which has equation $y = 4 - x$; it meets that line at the point $\left(\frac{5}{2}, \frac{3}{2}\right)$, so the length of the perpendicular is

$$h = \sqrt{\left(1 - \frac{5}{2}\right)^2 + \left(0 - \frac{3}{2}\right)^2} = \frac{1}{2}\sqrt{18} = \frac{3\sqrt{2}}{2}.$$

Then the area of the trapezoid is

$$\frac{b_1 + b_2}{2}h = \frac{5\sqrt{2}}{2} \cdot \frac{3\sqrt{2}}{2} = \frac{15}{2}.$$

The transformed trapezoid has one base from $(1, -1)$ to $(1, 1)$, with length 2, and one from $(4, -4)$ to $(4, 4)$, with length $b_2 = 8$. The height is $h = 4 - 1 = 3$. Then the area of the trapezoid is

$$\frac{b_1 + b_2}{2}h = \frac{2 + 8}{2} \cdot 3 = 15.$$

Since the Jacobian of the transformation is $\frac{1}{2}$, this all makes sense, since a Jacobian of $\frac{1}{2}$ means that the area of the new region must be multiplied by $\frac{1}{2}$ to get the area of the old region.

25. Since $v = \frac{y}{x}$, we have $y = vx$. Substitute for y in the equation for u to get $u = x^2 v$, so that $x = \sqrt{\frac{u}{v}} = u^{1/2}v^{-1/2}$. Substitution $x = \frac{y}{v}$ and simplify to get $y = \sqrt{uv} = u^{1/2}v^{1/2}$. Then the Jacobian is

$$\frac{\partial(x, y)}{\partial(u, v)} = \det \begin{bmatrix} \frac{\partial x}{\partial u} & \frac{\partial y}{\partial u} \\ \frac{\partial x}{\partial v} & \frac{\partial y}{\partial v} \end{bmatrix} = \det \begin{bmatrix} \frac{1}{2}u^{-1/2}v^{-1/2} & \frac{1}{2}u^{-1/2}v^{1/2} \\ -\frac{1}{2}u^{1/2}v^{-3/2} & \frac{1}{2}u^{1/2}v^{-1/2} \end{bmatrix} = \frac{1}{4}v^{-1} + \frac{1}{4}v^{-1} = \frac{1}{2v}.$$

Skills

27. Adding the two equations gives $2u = x + y$, so that $u = \frac{x+y}{2}$; subtracting them gives $v = \frac{y-x}{2}$. Thus level curves of u and of v correspond to lines in the xy plane, so that the region in the xy plane is a quadrilateral. The slope of $x + y = c$ is -1, while the slope of $y - x = d$ is 1. Since the slopes are negative reciprocals, the lines are perpendicular. Thus the region in the xy plane is a rectangle, with vertices at the points (x, y) corresponding to the vertices in the uv plane. These vertices are:

$$(0 - 0, 0 + 0) = (0, 0), \quad (2 - 0, 2 + 0) = (2, 2),$$
$$(2 - 1, 2 + 1) = (1, 3), \quad (0 - 1, 0 + 1) = (-1, 1).$$

The Jacobian of the transformation is

$$\frac{\partial(x, y)}{\partial(u, v)} = \det \begin{bmatrix} \frac{\partial x}{\partial u} & \frac{\partial y}{\partial u} \\ \frac{\partial x}{\partial v} & \frac{\partial y}{\partial v} \end{bmatrix} = \det \begin{bmatrix} 1 & 1 \\ -1 & 1 \end{bmatrix} = 2.$$

29. Substitute $u = xv$ into the second equation to get $y = xv^2$ so that $v = \sqrt{\frac{y}{x}} = y^{1/2}x^{-1/2}$. Multiply this equation by x to get $u = vx = y^{1/2}x^{-1/2}x = x^{1/2}y^{1/2}$. So level curves of v correspond to curves of the form $y^{1/2}x^{-1/2} = c$, or $y = c^2x$, which is a line. Level curves of u correspond to $x^{1/2}y^{1/2} = d$, or $xy = d^2$, which is a hyperbola. So the line in the uv plane from $(1, 1)$ to $(2, 1)$ corresponds to the level curve $c = 1$ for v, so the line $y = x$ in the xy plane. Similarly, the line from $(1, 3)$ to $(2, 3)$ corresponds to the level curve $c = 3$ for v, so the line $y = 9x$ in the xy plane. The line from $(1, 1)$ to $(1, 3)$ in the uv plane corresponds to the level curve for $u = 1$, so the hyperbola

$xy = 1$, while the line from $(2,1)$ to $(2,3)$ corresponds to the level curve for $u = 2$, so to the hyperbola $xy = 4$. The four points $(1,1)$, $(2,1)$, $(1,3)$, and $(2,3)$ correspond to the points $(1,1)$, $(2,2)$, $\left(\frac{1}{3},3\right)$ and $\left(\frac{2}{3},6\right)$. A diagram of the region in the xy plane is

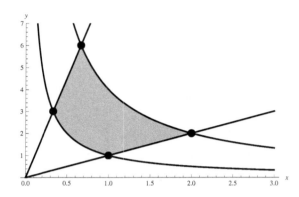

The Jacobian is

$$\frac{\partial(x,y)}{\partial(u,v)} = \det \begin{bmatrix} \frac{\partial x}{\partial u} & \frac{\partial y}{\partial u} \\ \frac{\partial x}{\partial v} & \frac{\partial y}{\partial v} \end{bmatrix} = \det \begin{bmatrix} \frac{1}{v} & v \\ -\frac{u}{v^2} & u \end{bmatrix} = \frac{u}{v} + \frac{u}{v} = \frac{2u}{v}.$$

31. Square both equations to get $x^2 = u^2 \sin^2 v$ and $y^2 = u^2 \cos^2 v$, so that $x^2 + y^2 = u^2(\sin^2 v + \cos^2 v) = u^2$. Thus $u = \sqrt{x^2 + y^2}$ ($0 \le u \le 2$, so we can take the positive square root). It follows that $\sin v = \frac{x}{\sqrt{x^2+y^2}}$ and $v = \sin^{-1} \frac{x}{\sqrt{x^2+y^2}} = \tan^{-1} \frac{x}{y}$. So level curves of u are $\sqrt{x^2 + y^2} = c$, or $x^2 + y^2 = c^2$, which are circles centered at the origin. Level curves of v are $d = \tan^{-1} \frac{x}{y}$, or $x = y \tan d$, so these are lines through the origin. The level curve corresponding to $u = 0$ is the origin itself, and the level curve corresponding to $u = 2$ is the circle of radius 2. The level curve corresponding to $v = 0$ is the line $x = y \tan 0 = 0$, and the level curve corresponding to $v = \pi$ is $x = y \tan \pi$, which is also $x = 0$. However, in the first case, y is positive, while in the second y is negative. So the first level curve is the positive y axis and the second level curve is the negative y axis. Thus the region in the xy plane is the semicircle of radius 2 centered at the origin and lying to the right of the y axis. The Jacobian is

$$\frac{\partial(x,y)}{\partial(u,v)} = \det \begin{bmatrix} \frac{\partial x}{\partial u} & \frac{\partial y}{\partial u} \\ \frac{\partial x}{\partial v} & \frac{\partial y}{\partial v} \end{bmatrix} = \det \begin{bmatrix} \sin v & \cos v \\ u \cos v & -u \sin v \end{bmatrix} = -u(\sin^2 v + \cos^2 v) = -u.$$

33. (a) The given trapezoid Ω is

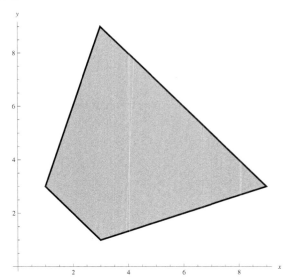

(b) Under the given transformation, the lower left boundary in the original trapezoid, which is $x + y = 4$, becomes $u = 4$, while the upper right boundary, which is $x + y = 12$, becomes $u = 12$. For the other two boundaries, first solve the given equations for x and y to get $x = \frac{u+v}{2}$ and $y = \frac{u-v}{2}$. Then if a is a real number, $y = ax$ becomes $\frac{u-v}{2} = a\frac{u+v}{2}$, so that $u - v = au + av$, or $(1 + a)v = (1 - a)u$. This is the equation $v = \frac{1-a}{1+a}u$. The lower right boundary is $y = \frac{1}{3}x$, which corresponds to

$$v = \frac{1 - \frac{1}{3}}{1 + \frac{1}{3}}u, \text{ or } v = \frac{1}{2}u.$$

The upper left boundary is $y = 3x$, which corresponds to

$$v = \frac{1 - 3}{1 + 3}u, \text{ or } v = -\frac{1}{2}u.$$

Thus the transformed region is

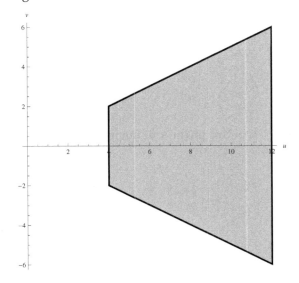

(c) The Jacobian of this transformation is

$$\frac{\partial(x,y)}{\partial(u,v)} = \det \begin{bmatrix} \frac{\partial x}{\partial u} & \frac{\partial y}{\partial u} \\ \frac{\partial x}{\partial v} & \frac{\partial y}{\partial v} \end{bmatrix} = \det \begin{bmatrix} \frac{1}{2} & \frac{1}{2} \\ \frac{1}{2} & -\frac{1}{2} \end{bmatrix} = -\frac{1}{2}.$$

Since $x + y = u$, the integrand $(x+y)^2$ becomes u^2 in the uv plane, so the integral in the uv plane is

$$\iint_{\Omega'} u^2 \, dA = \int_4^{12} \int_{-(1/2)u}^{(1/2)u} u^2 \cdot \frac{1}{2} \, dv \, du = \frac{1}{2} \int_4^{12} \left[u^2 v \right]_{v=-(1/2)u}^{v=(1/2)u} \, du$$

$$= \frac{1}{2} \int_4^{12} u^3 \, du = \frac{1}{2} \left[\frac{1}{4} u^4 \right]_4^{12} = 2560.$$

35. (a) The given trapezoid Ω is shown in Exercise 33(a).

(b) Under the given transformation, the lower left boundary in the original trapezoid, which is $x + y = 4$, becomes $u = 4$, while the upper right boundary, which is $x + y = 12$, becomes $u = 12$. For the other two boundaries, solve the given equations for x and y. The second equation gives $x = vy$; substitute for x in the first equation to get $u = vy + y$ so that $y = \frac{u}{v+1}$; then substituting that value into $x = vy$ gives $x = \frac{uv}{v+1}$. Then if a is a real number, $y = ax$ becomes $\frac{u}{v+1} = a \frac{uv}{v+1}$, so that $u = auv$ and $v = \frac{1}{a}$. Then the lower right boundary is $y = \frac{1}{3}x$, which corresponds to $v = 3$; the upper left boundary is $y = 3x$, which corresponds to $v = \frac{1}{3}$. Thus the transformed region is

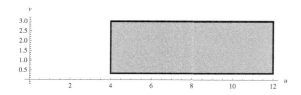

(c) Note that

$$\frac{\partial x}{\partial v} = \frac{(v+1)u - uv(1)}{(v+1)^2} = \frac{u}{(v+1)^2},$$

so that the Jacobian of this transformation is

$$\frac{\partial(x,y)}{\partial(u,v)} = \det \begin{bmatrix} \frac{\partial x}{\partial u} & \frac{\partial y}{\partial u} \\ \frac{\partial x}{\partial v} & \frac{\partial y}{\partial v} \end{bmatrix} = \det \begin{bmatrix} \frac{v}{v+1} & \frac{1}{v+1} \\ \frac{u}{(v+1)^2} & -\frac{u}{(v+1)^2} \end{bmatrix} = -\frac{uv}{(v+1)^3} - \frac{u}{(v+1)^3} = -\frac{u}{(v+1)^2}.$$

Note that u and v are positive, so that the absolute value of the Jacobian is $\frac{u}{(v+1)^2}$. Since $x + y = u$, the integrand $(x+y)^2$ becomes u^2 in the uv plane, so the integral in the uv plane is

$$\iint_{\Omega'} u^2 \, dA = \int_4^{12} \int_{1/3}^3 u^2 \cdot \left(\frac{u}{(v+1)^2} \right) dv \, du$$

$$= \int_4^{12} \int_{1/3}^3 \frac{u^3}{(v+1)^2} \, dv \, du$$

$$= \int_4^{12} \left[-\frac{u^3}{v+1} \right]_{v=1/3}^{v=3} du$$

$$= \int_4^{12} \left(-\frac{u^3}{4} + \frac{3}{4}u^3 \right) du$$

$$= \int_4^{12} \frac{1}{2} u^3 \, du$$

$$= \left[\frac{1}{8} u^4 \right]_4^{12} = 2560.$$

37. (a) The given trapezoid Ω is shown in Exercise 36(a).

(b) Under the given transformation, the lower left boundary in the original trapezoid, which is $x + y = 5$, becomes $u = 5$, while the upper right boundary, which is $x + y = 8$, becomes $u = 8$. For the other two boundaries, first solve the given equations for x and y to get $x = \frac{2u-v}{3}$ and $y = \frac{u+v}{3}$. Then if a is a real number, $y = ax + b$ becomes $\frac{u+v}{3} = a\frac{2u-v}{3} + b$, so that $u + v = 2au - av + 3b$. Rearranging terms gives $(a+1)v = (2a-1)u + 3b$, or $v = \frac{2a-1}{a+1}u + \frac{3b}{a+1}$. The lower right boundary is $y - 2 = \frac{3-2}{5-3}(x - 3)$, or $y = \frac{1}{2}x + \frac{1}{2}$. Then $a = b = \frac{1}{2}$, so this corresponds to

$$v = \frac{2 \cdot \frac{1}{2} - 1}{\frac{1}{2} + 1} u + \frac{3 \cdot \frac{1}{2}}{\frac{1}{2} + 1}, \text{ or } v = 1.$$

The upper left boundary is $y - 3 = \frac{5-3}{3-2}(x - 2)$, or $y = 2x - 1$. then $a = 2$ and $b = -1$, so this corresponds to

$$v = \frac{4-1}{2+1}u + \frac{-3}{2+1}, \text{ or } v = u - 1.$$

Thus the transformed region is

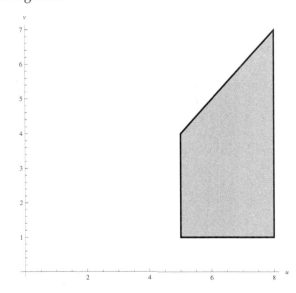

(c) The Jacobian of this transformation is

$$\frac{\partial(x,y)}{\partial(u,v)} = \det\begin{bmatrix} \frac{\partial x}{\partial u} & \frac{\partial y}{\partial u} \\ \frac{\partial x}{\partial v} & \frac{\partial y}{\partial v} \end{bmatrix} = \det\begin{bmatrix} \frac{2}{3} & \frac{1}{3} \\ -\frac{1}{3} & \frac{1}{3} \end{bmatrix} = \frac{1}{3}.$$

The integrand xy becomes $\frac{2u-v}{3} \cdot \frac{u+v}{3} = \frac{1}{9}(2u^2 + uv - v^2)$ in the uv plane, so the integral is

$$\iint_{\Omega'} \frac{1}{9}(2u^2 + uv - v^2)\, dA = \frac{1}{9}\int_5^8 \int_1^{u-1} (2u^2 + uv - v^2) \cdot \frac{1}{3}\, dv\, du$$

$$= \frac{1}{27}\int_5^8 \left[2u^2 v + \frac{1}{2}uv^2 - \frac{1}{3}v^3 \right]_{v=1}^{v=u-1} du$$

$$= \frac{1}{27}\int_5^8 \left(2u^2(u-2) + \frac{1}{2}u((u-1)^2 - 1) - \frac{1}{3}((u-1)^3 - 1) \right) du$$

$$= \frac{1}{27}\int_5^8 \left(2u^3 - 4u^2 + \frac{1}{2}u^3 - u^2 - \frac{1}{3}u^3 + u^2 - u + \frac{2}{3} \right) du$$

$$= \frac{1}{27}\int_5^8 \left(\frac{13}{6}u^3 - 4u^2 - u + \frac{2}{3} \right) du$$

$$= \frac{1}{27}\left[\frac{13}{24}u^4 - \frac{4}{3}u^3 - \frac{1}{2}u^2 + \frac{2}{3}u \right]_5^8$$

$$= \frac{1}{27} \cdot \frac{10773}{8} = \frac{399}{8}.$$

39. Use the transformation $u = x - y$, $v = y$. Then the bottom and top edges, which are $y = 0$ and $y = 2$, become $v = 0$ and $v = 2$. The left and right edges, which are $y = x$ and $y = x - 3$, become $u = 0$ and $u = 3$. Solving these equations for x and y gives $y = v$ and thus $x = u + v$. The Jacobian of this transformation is

$$\frac{\partial(x,y)}{\partial(u,v)} = \det\begin{bmatrix} \frac{\partial x}{\partial u} & \frac{\partial y}{\partial u} \\ \frac{\partial x}{\partial v} & \frac{\partial y}{\partial v} \end{bmatrix} = \det\begin{bmatrix} 1 & 0 \\ 1 & 1 \end{bmatrix} = 1.$$

The integrand $(x - y)^3$ becomes u^3, so the integral is

$$\int_0^3 \int_0^2 u^3 \cdot 1\, dv\, du = \int_0^3 \left[u^3 v \right]_{v=0}^{v=2} du = \int_0^3 2u^3\, du = \left[\frac{1}{2}u^4 \right]_0^3 = \frac{81}{2}.$$

41. The bottom edge is the line from $(0,0)$ to $(2,1)$, which is $y = \frac{1}{2}x$. The top edge is the line from $(-1,3)$ to $(1,4)$, which is $y = \frac{1}{2}x + \frac{7}{2}$. So let $v = y - \frac{1}{2}x$; then these two edges becomes $v = 0$ and $v = \frac{7}{2}$. The left edge is the line from $(0,0)$ to $(-1,3)$, which is $y = -3x$, and the right edge is the line from $(2,1)$ to $(1,4)$, which is $y = -3x + 7$. So let $u = 3x + y$; then these two edges become $u = 0$ and $u = 7$. Solve these equations for x and y in terms of u and v: subtracting gives $u - v = \frac{7}{2}x$, so that $x = \frac{2}{7}(u - v)$. Substituting into the equation for v gives $v = y - \frac{1}{7}(u - v)$, so that $y = \frac{1}{7}u + \frac{6}{7}v$. Then the Jacobian is

$$\frac{\partial(x,y)}{\partial(u,v)} = \det\begin{bmatrix} \frac{\partial x}{\partial u} & \frac{\partial y}{\partial u} \\ \frac{\partial x}{\partial v} & \frac{\partial y}{\partial v} \end{bmatrix} = \det\begin{bmatrix} \frac{2}{7} & \frac{1}{7} \\ -\frac{2}{7} & \frac{6}{7} \end{bmatrix} = \frac{14}{49} = \frac{2}{7}.$$

The integrand becomes

$$\frac{2y - x}{3x + y + 1} = \frac{\frac{2}{7}u + \frac{12}{7}v - (\frac{2}{7}u - \frac{2}{7}v)}{\frac{6}{7}u - \frac{6}{7}v + \frac{1}{7}u + \frac{6}{7}v + 1} = \frac{2v}{u + 1}.$$

Thus the integral is

$$\int_0^7 \int_0^{7/2} \frac{2v}{u+1} \cdot \frac{2}{7} \, dv \, du = \frac{2}{7} \int_0^7 \left[\frac{v^2}{u+1} \right]_{v=0}^{v=7/2} du$$

$$= \frac{2}{7} \cdot \frac{49}{4} \int_0^7 \frac{1}{u+1} \, du = \frac{7}{2} \left[\ln |u+1| \right]_0^7 = \frac{7}{2} \ln 8.$$

43. Since two of the edges are $xy = 3$ and $xy = 27$, let $u = xy$; then those edges become $u = 3$ and $u = 27$. The other two edges are $y = 3x$ and $y = \frac{1}{3}x$; let $v = \frac{y}{x}$ so that these edges become $v = 3$ and $v = \frac{1}{3}$. Solve these equations for x and y: the equation for v gives $y = vx$; substitute into the other equation to get $u = vx^2$, so that $x = u^{1/2}v^{-1/2}$ (note that u and v are positive on the range of interest). Then using the equation for u, we get $y = u^{1/2}v^{1/2}$. The Jacobian is

$$\frac{\partial(x,y)}{\partial(u,v)} = \det \begin{bmatrix} \frac{\partial x}{\partial u} & \frac{\partial y}{\partial u} \\ \frac{\partial x}{\partial v} & \frac{\partial y}{\partial v} \end{bmatrix} = \det \begin{bmatrix} \frac{1}{2}u^{-1/2}v^{-1/2} & \frac{1}{2}u^{-1/2}v^{1/2} \\ -\frac{1}{2}u^{1/2}v^{-3/2} & \frac{1}{2}u^{1/2}v^{-1/2} \end{bmatrix} = \frac{1}{4v} + \frac{1}{4v} = \frac{1}{2v}.$$

The integrand becomes

$$\frac{x^2}{y^2} + x^2y^2 = \frac{1}{v^2} + u^2.$$

Thus the integral is

$$\int_3^{27} \int_{1/3}^3 \left(\frac{1}{v^2} + u^2 \right) \cdot \frac{1}{2v} \, dv \, du = \int_3^{27} \int_{1/3}^3 \left(\frac{1}{2v^3} + \frac{u^2}{2v} \right) dv \, du$$

$$= \int_3^{27} \left[-\frac{1}{4v^2} + \frac{u^2}{2} \ln |v| \right]_{v=1/3}^{v=3} du$$

$$= \int_3^{27} \left(-\frac{1}{36} + \frac{9}{4} + \frac{u^2}{2} \ln 3 - \frac{u^2}{2} \ln \frac{1}{3} \right) du$$

$$= \int_3^{27} \left(\frac{20}{9} + u^2 \ln 3 \right) du$$

$$= \left[\frac{20}{9}u + \frac{1}{3}u^3 \ln 3 \right]_3^{27}$$

$$= \frac{160}{3} + 6552 \ln 3.$$

45. Use the transformation $u = \frac{y}{x^2}, v = \frac{y}{x}$. Then the left and right edges becomes $u = 2$ and $u = 1$, while the top and bottom become $v = 2$ and $v = 1$. Solve these equations for x and y: the equation for v gives $y = vx$; substitute into the equation for u to get $u = \frac{vx}{x^2}$, so that $x = \frac{v}{u}$. Then $y = \frac{v^2}{u}$. The Jacobian is

$$\frac{\partial(x,y)}{\partial(u,v)} = \det \begin{bmatrix} \frac{\partial x}{\partial u} & \frac{\partial y}{\partial u} \\ \frac{\partial x}{\partial v} & \frac{\partial y}{\partial v} \end{bmatrix} = \det \begin{bmatrix} -\frac{v}{u^2} & -\frac{v^2}{u^2} \\ \frac{1}{u} & \frac{2v}{u} \end{bmatrix} = -\frac{2v^2}{u^3} + \frac{v^2}{u^3} = -\frac{v^2}{u^3}.$$

Since u and v are positive in the range of interest, the absolute value of the Jacobian is $\frac{v^2}{u^3}$. The integrand becomes

$$\frac{y^2}{x^3} = \frac{y}{x} \cdot \frac{y}{x^2} = uv.$$

Thus the integral is

$$\int_1^2 \int_1^2 uv \cdot \frac{v^2}{u^3} \, dv \, du = \int_1^2 \int_1^2 \frac{v^3}{u^2} \, dv \, du$$

$$= \int_1^2 \left[\frac{1}{4u^2} v^4 \right]_1^2 du$$

$$= \frac{15}{4} \int_1^2 \frac{1}{u^2} \, du$$

$$= \frac{15}{4} \left[-\frac{1}{u} \right]_1^2 = \frac{15}{8}.$$

47. Use the transformation $u = \frac{x}{y}$ and $v = y^2 - x^2$; then the left boundary is $u = 0$ and the right boundary is $u = \frac{1}{2}$, while the bottom boundary becomes $v = 1$ and the top becomes $v = 4$. Solve the equations for x and y: the equation for u gives $x = uy$; substitute into the equation for v to get $v = y^2 - u^2 y^2$, so that $y = \sqrt{\frac{v}{1-u^2}}$ (note that this is defined for the relevant range of both u and v). From $x = uy$ we then get $x = u\sqrt{\frac{v}{1-u^2}}$. Note that

$$\frac{\partial}{\partial u} \sqrt{\frac{v}{1 - u^2}} = \sqrt{v} \frac{d}{du} \frac{1}{\sqrt{1 - u^2}} = \frac{u\sqrt{v}}{(1 - u^2)^{3/2}},$$

$$\frac{\partial}{\partial u} \left(u \sqrt{\frac{v}{1 - u^2}} \right) = \sqrt{v} \frac{d}{du} \frac{u}{\sqrt{1 - u^2}} = \frac{\sqrt{v}}{(1 - u^2)^{3/2}},$$

so the Jacobian is

$$\frac{\partial(x, y)}{\partial(u, v)} = \det \begin{bmatrix} \frac{\partial x}{\partial u} & \frac{\partial y}{\partial u} \\ \frac{\partial x}{\partial v} & \frac{\partial y}{\partial v} \end{bmatrix} = \det \begin{bmatrix} \frac{\sqrt{v}}{(1-u^2)^{3/2}} & \frac{u\sqrt{v}}{(1-u^2)^{3/2}} \\ \frac{u}{2\sqrt{v}\sqrt{1-u^2}} & \frac{1}{2\sqrt{v}\sqrt{1-u^2}} \end{bmatrix}$$

$$= \frac{1}{2(1 - u^2)^2} - \frac{u^2}{2(1 - u^2)^2} = \frac{1}{2(1 - u^2)}.$$

Since $0 \leq u < 1$ in the range of interest, this is positive, so it is the absolute value of the Jacobian. The integrand becomes

$$\frac{x^3}{y^3} = u^3.$$

Thus the integral is

$$\int_0^{1/2} \int_1^4 u^3 \cdot \frac{1}{2 - 2u^2} \, dv \, du = \frac{1}{2} \int_0^{1/2} \left[\frac{u^3}{1 - u^2} v \right]_{v=1}^{v=4} du$$

$$= \frac{3}{2} \int_0^{1/2} \frac{u^3}{1 - u^2} \, du$$

$$= \frac{3}{2} \int_0^{1/2} \left(-u + \frac{u}{1 - u^2} \right) du$$

$$= \frac{3}{2} \left[-\frac{1}{2} u^2 - \frac{1}{2} \ln \left| 1 - u^2 \right| \right]_0^{1/2}$$

$$= \frac{3}{2} \left(-\frac{1}{8} - \frac{1}{2} \ln \frac{3}{4} \right) = \frac{3}{4} \ln \frac{4}{3} - \frac{3}{16}.$$

Applications

49. Use the transformation

$$u = x - \frac{y}{\sqrt{3}}, \qquad v = y,$$

which transforms the parallelogram into the rectangle $[0, 25] \times \left[0, \frac{15\sqrt{3}}{2}\right]$. The inverse of this transformation is

$$x = u + \frac{v}{\sqrt{3}}, \qquad y = v,$$

which has the Jacobian

$$\det \begin{pmatrix} \frac{\partial x}{\partial u} & \frac{\partial y}{\partial u} \\ \frac{\partial x}{\partial v} & \frac{\partial y}{\partial v} \end{pmatrix} = \det \begin{pmatrix} 1 & 0 \\ \frac{1}{\sqrt{3}} & 1 \end{pmatrix} = 1.$$

Let Ω represent the parallelogram and R the transformed rectangle. Let $d(x, y)$ be the function giving the depth of the water at the point (x, y). The transformed version of this function, $d(u, v)$, is a simple linear function that is 3 when $u = 0$ and 8 when $u = 25$, independent of v, so its equation is $d(u, v) = \frac{1}{5}u + 3$. Thus the volume of water in the pool is

$$\iint_\Omega d(x, y)\, dA = \iint_R d(u, v) \left| \frac{\partial(x, y)}{\partial(u, v)} \right| du\, dv$$

$$= \int_0^{\frac{15\sqrt{3}}{2}} \int_0^{25} \left(\frac{1}{5}u + 3 \right) du\, dv$$

$$= \int_0^{\frac{15\sqrt{3}}{2}} \left[\frac{1}{10}u^2 + 3u \right]_0^{25} dv$$

$$= \int_0^{\frac{15\sqrt{3}}{2}} \frac{275}{2}\, dv$$

$$= \frac{275 \cdot 15\sqrt{3}}{4} \approx 1786 \text{ cubic feet.}$$

Proofs

51. The Jacobian is

$$\frac{\partial(x, y, z)}{\partial(r, \theta, z)} = \det \begin{pmatrix} \frac{\partial x}{\partial r} & \frac{\partial y}{\partial r} & \frac{\partial z}{\partial r} \\ \frac{\partial x}{\partial \theta} & \frac{\partial y}{\partial \theta} & \frac{\partial z}{\partial \theta} \\ \frac{\partial x}{\partial z} & \frac{\partial y}{\partial z} & \frac{\partial z}{\partial z} \end{pmatrix} = \det \begin{pmatrix} \cos\theta & \sin\theta & 0 \\ -r\sin\theta & r\cos\theta & 0 \\ 0 & 0 & 1 \end{pmatrix} = r\cos^2\theta + r\sin^2\theta = r.$$

53. The Jacobian of the transformation is

$$\frac{\partial(x, y)}{\partial(u, v)} = \det \begin{pmatrix} \frac{\partial x}{\partial u} & \frac{\partial y}{\partial u} \\ \frac{\partial x}{\partial v} & \frac{\partial y}{\partial v} \end{pmatrix} = \det \begin{pmatrix} \alpha & \gamma \\ \beta & \delta \end{pmatrix} = \alpha\delta - \beta\gamma.$$

Substituting for x and y in the equation $ax + by = c$ gives

$$a(\alpha u + \beta v) + b(\gamma u + \delta v) = (a\alpha + b\gamma)u + (a\beta + b\delta)v = c.$$

This is a line as long as either $a\alpha + b\gamma \neq 0$ or $a\beta + b\delta \neq 0$. Suppose

$$a\alpha + b\gamma = r$$
$$a\beta + b\delta = s.$$

Multiply the first equation by β and the second by α to get

$$a\alpha\beta + b\beta\gamma = r\beta$$
$$a\alpha\beta + b\alpha\delta = s\alpha.$$

Then subtract to get $b(\alpha\delta - \beta\gamma) = s\alpha - r\beta$. Similarly, we get $a(\alpha\delta - \beta\gamma) = r\delta - s\gamma$. Either a or b is nonzero, since $ax + by = c$ is a line, and so if the Jacobian, $\alpha\delta - \beta\gamma$, is also nonzero, then one of $s\alpha - r\beta$ or $r\delta - s\gamma$ must also be nonzero, so that either r or s is nonzero. Thus the coefficient of either u or v in the transformed equation is nonzero, so the equation is a line.

55. Assume that the linear transformation has nonzero Jacobian $\alpha\delta - \beta\gamma$, and consider the equations

$$\alpha u + \beta v = x$$
$$\gamma u + \delta v = y.$$

Multiply the first equation by γ and the second by α and subtract to get $(\alpha\delta - \beta\gamma)v = \alpha y - \gamma x$. Similarly, multiply the first equation by δ and the second by β and subtract to get $(\alpha\delta - \beta\gamma)u = \delta x - \beta y$. Divide both of these through by $\alpha\delta - \beta\gamma \neq 0$ to get

$$u = \frac{\delta}{\alpha\delta - \beta\gamma}x - \frac{\beta}{\alpha\delta - \beta\gamma}y, \qquad v = -\frac{\gamma}{\alpha\delta - \beta\gamma}x + \frac{\alpha}{\alpha\delta - \beta\gamma}y.$$

57. Let $u = \frac{x}{a}$ and $v = \frac{y}{b}$; then $x = ua$ and $y = vb$. The Jacobian of this transformation is

$$\frac{\partial(x,y)}{\partial(u,v)} = \det\begin{pmatrix} \frac{\partial x}{\partial u} & \frac{\partial y}{\partial u} \\ \frac{\partial x}{\partial v} & \frac{\partial y}{\partial v} \end{pmatrix} = \det\begin{pmatrix} a & 0 \\ 0 & b \end{pmatrix} = ab.$$

The transformed equation is $u^2 + v^2 = 1$. Use polar coordinates to integrate. Then the equation $u^2 + v^2 = 1$ becomes $r = 1$; using the factor ab from the Jacobian as well as the volume element multiplier of r from the use of polar coordinates, we get for the area

$$\int_0^{2\pi}\int_0^1 1\cdot abr\,dr\,d\theta = ab\int_0^{2\pi}\left[\frac{1}{2}r^2\right]_{r=0}^{r=1}d\theta = ab\int_0^{2\pi}\frac{1}{2}d\theta = \pi ab.$$

59. Recall three facts about matrix multiplication: If A and B are 2×2 matrices, then

$$\det(AB) = \det(A)\cdot\det(B) = \det(B)\cdot\det(A), \qquad \det(A^T) = \det A, \qquad (AB)^T = B^T A^T,$$

where A^T is the transpose of the matrix A. Then we compute

$$\frac{\partial(x,y)}{\partial(u,v)}\frac{\partial(u,v)}{\partial(s,t)} = \det\begin{pmatrix} \frac{\partial x}{\partial u} & \frac{\partial y}{\partial u} \\ \frac{\partial x}{\partial v} & \frac{\partial y}{\partial v} \end{pmatrix}\det\begin{pmatrix} \frac{\partial u}{\partial s} & \frac{\partial v}{\partial s} \\ \frac{\partial u}{\partial t} & \frac{\partial v}{\partial t} \end{pmatrix}$$

$$= \det\begin{pmatrix} \frac{\partial u}{\partial s} & \frac{\partial v}{\partial s} \\ \frac{\partial u}{\partial t} & \frac{\partial v}{\partial t} \end{pmatrix}\det\begin{pmatrix} \frac{\partial x}{\partial u} & \frac{\partial y}{\partial u} \\ \frac{\partial x}{\partial v} & \frac{\partial y}{\partial v} \end{pmatrix}$$

$$= \det\left(\begin{pmatrix} \frac{\partial u}{\partial s} & \frac{\partial v}{\partial s} \\ \frac{\partial u}{\partial t} & \frac{\partial v}{\partial t} \end{pmatrix}\begin{pmatrix} \frac{\partial x}{\partial u} & \frac{\partial y}{\partial u} \\ \frac{\partial x}{\partial v} & \frac{\partial y}{\partial v} \end{pmatrix}\right)$$

$$= \det\left(\begin{pmatrix} \frac{\partial u}{\partial s} & \frac{\partial v}{\partial s} \\ \frac{\partial u}{\partial t} & \frac{\partial v}{\partial t} \end{pmatrix}\begin{pmatrix} \frac{\partial x}{\partial u} & \frac{\partial y}{\partial u} \\ \frac{\partial x}{\partial v} & \frac{\partial y}{\partial v} \end{pmatrix}\right)^T$$

$$= \det\left(\begin{pmatrix} \frac{\partial x}{\partial u} & \frac{\partial x}{\partial v} \\ \frac{\partial y}{\partial u} & \frac{\partial y}{\partial v} \end{pmatrix}\begin{pmatrix} \frac{\partial u}{\partial s} & \frac{\partial u}{\partial t} \\ \frac{\partial v}{\partial s} & \frac{\partial v}{\partial t} \end{pmatrix}\right)$$

$$= \det\begin{pmatrix} \frac{\partial x}{\partial u}\frac{\partial u}{\partial s} + \frac{\partial x}{\partial v}\frac{\partial v}{\partial s} & \frac{\partial x}{\partial u}\frac{\partial u}{\partial t} + \frac{\partial x}{\partial v}\frac{\partial v}{\partial t} \\ \frac{\partial y}{\partial u}\frac{\partial u}{\partial s} + \frac{\partial y}{\partial v}\frac{\partial v}{\partial s} & \frac{\partial y}{\partial u}\frac{\partial u}{\partial t} + \frac{\partial y}{\partial v}\frac{\partial v}{\partial t} \end{pmatrix}$$

$$= \det\begin{pmatrix} \frac{\partial x}{\partial s} & \frac{\partial x}{\partial t} \\ \frac{\partial y}{\partial s} & \frac{\partial y}{\partial t} \end{pmatrix} = \det\begin{pmatrix} \frac{\partial x}{\partial s} & \frac{\partial y}{\partial s} \\ \frac{\partial x}{\partial t} & \frac{\partial y}{\partial t} \end{pmatrix} = \frac{\partial(x,y)}{\partial(s,t)}.$$

Here we used the multivariable chain rule, Theorem 12.34.

Thinking Forward

Determinants of 3×3 matrices

▷ Evaluating the formula in the exercise gives

$$\det A = a_{11}\det\begin{bmatrix} a_{22} & a_{23} \\ a_{32} & a_{33} \end{bmatrix} - a_{12}\det\begin{bmatrix} a_{21} & a_{23} \\ a_{31} & a_{33} \end{bmatrix} + a_{13}\det\begin{bmatrix} a_{21} & a_{22} \\ a_{31} & a_{32} \end{bmatrix}$$

$$= a_{11}\left(a_{22}a_{33} - a_{23}a_{32}\right) - a_{12}\left(a_{21}a_{33} - a_{23}a_{31}\right) + a_{13}\left(a_{21}a_{32} - a_{22}a_{31}\right)$$

$$= a_{11}a_{22}a_{33} - a_{11}a_{23}a_{32} + a_{12}a_{23}a_{31} - a_{12}a_{21}a_{33} + a_{13}a_{21}a_{32} - a_{13}a_{22}a_{31}.$$

Except for the notation used in the definition of the matrix in Definition 10.24, the two answers are the same.

▷ Computing, we get

$$\det\begin{bmatrix} 1 & 3 & 4 \\ -2 & 5 & -1 \\ -4 & 2 & -3 \end{bmatrix} = 1\det\begin{bmatrix} 5 & -1 \\ 2 & -3 \end{bmatrix} - 3\det\begin{bmatrix} -2 & -1 \\ -4 & -3 \end{bmatrix} + 4\det\begin{bmatrix} -2 & 5 \\ -4 & 2 \end{bmatrix}$$

$$= 1\left(5\cdot(-3) - (-1)\cdot 2\right) - 3\left(-2\cdot(-3) - (-1)\cdot(-4)\right)$$
$$+ 4\left(-2\cdot 2 - 5\cdot(-4)\right)$$

$$= 45.$$

▷ A similar formula is

$$\det A = a_{11} \det \begin{bmatrix} a_{22} & a_{23} & a_{24} \\ a_{32} & a_{33} & a_{34} \\ a_{42} & a_{43} & a_{44} \end{bmatrix} - a_{12} \det \begin{bmatrix} a_{21} & a_{23} & a_{24} \\ a_{31} & a_{33} & a_{34} \\ a_{41} & a_{43} & a_{44} \end{bmatrix}$$

$$+ a_{13} \det \begin{bmatrix} a_{21} & a_{22} & a_{24} \\ a_{31} & a_{32} & a_{34} \\ a_{41} & a_{42} & a_{44} \end{bmatrix} - a_{14} \det \begin{bmatrix} a_{21} & a_{22} & a_{23} \\ a_{31} & a_{32} & a_{33} \\ a_{41} & a_{42} & a_{43} \end{bmatrix}.$$

Chapter Review and Self-Test

1. If there are m subintervals in the x direction and n in the y direction, then $\Delta x = \frac{2-0}{m} = \frac{2}{m}$ and $\Delta y = \frac{4-1}{n} = \frac{3}{n}$. Thus $\Delta A = \Delta x \Delta y = \frac{6}{mn}$. We get $x_j = \frac{2}{m}j$ for $j = 0, 1, \ldots, m$, and $y_k = 1 + \frac{3}{n}k$ for $k = 0, 1, \ldots, n$. In $[x_{j-1}, x_j] \times [y_{k-1}, y_k]$, choose $(x_j^*, y_k^*) = (x_j, y_k)$. Then the integral is

$$\iint_{\mathcal{R}} (x + 2y)\, dA = \lim_{\Delta \to 0} \sum_{j=1}^{m} \sum_{k=1}^{n} (x_j^* + 2y_k^*) \Delta A = \lim_{m,n \to \infty} \frac{6}{mn} \sum_{j=1}^{m} \sum_{k=1}^{n} (x_j + 2y_k)$$

$$= \lim_{m,n \to \infty} \frac{6}{mn} \sum_{j=1}^{m} \left(n\frac{2}{m}j + 2\sum_{k=1}^{n} \left(1 + \frac{3}{n}k\right)\right)$$

$$= \lim_{m,n \to \infty} \frac{6}{mn} \sum_{j=1}^{m} \left(n\frac{2}{m}j + 2n + \frac{6}{n}\sum_{k=1}^{n} k\right)$$

$$= \lim_{m,n \to \infty} \frac{6}{mn} \sum_{j=1}^{m} \left(n\frac{2}{m}j + 2n + \frac{6}{n} \cdot \frac{n(n+1)}{2}\right)$$

$$= \lim_{m,n \to \infty} \frac{6}{mn} \left(\frac{2n}{m}\left(\sum_{j=1}^{m} j\right) + 2mn + 3m(n+1)\right)$$

$$= \lim_{m,n \to \infty} \frac{6}{mn} \left(\frac{2n}{m} \cdot \frac{m(m+1)}{2} + 2mn + 3m(n+1)\right)$$

$$= \lim_{m,n \to \infty} \left(\frac{6(m+1)}{m} + 12 + \frac{18(n+1)}{n}\right)$$

$$= \lim_{m \to \infty} \left(\frac{6(m+1)}{m} + 12 + \lim_{n \to \infty} \frac{18(n+1)}{n}\right)$$

$$= \lim_{m \to \infty} \left(\frac{6(m+1)}{m} + 30\right) = 36.$$

3. By Fubini's Theorem,

$$\iint_{\mathcal{R}} (x + 2y)\, dA = \int_0^2 \int_1^4 (x + 2y)\, dy\, dx = \int_0^2 \left[xy + y^2 \right]_{y=1}^{y=4} dx = \int_0^2 (3x + 15)\, dx = \left[\frac{3}{2}x^2 + 15x \right]_0^2$$

$$= 36$$

$$\iint_{\mathcal{R}} (x + 2y)\, dA = \int_1^4 \int_0^2 (x + 2y)\, dx\, dy = \int_1^4 \left[\frac{1}{2}x^2 + 2xy \right]_{x=0}^{x=2} dy = \int_1^4 (2 + 4y)\, dy = \left[2y + 2y^2 \right]_1^4$$

$$= 36.$$

The two answers are the same.

5. For each $x \in [0, 1]$, the region is bounded below by $y = x^2$ and above by $y = \sqrt{x}$:

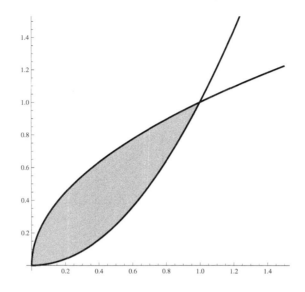

The integral is

$$\int_0^1 \int_{x^2}^{\sqrt{x}} x^2 y^3 \, dy \, dx = \int_0^1 \left[\frac{1}{4} x^2 y^4 \right]_{y=x^2}^{y=\sqrt{x}} dx$$

$$= \frac{1}{4} \int_0^1 \left(x^4 - x^{10} \right) dx$$

$$= \frac{1}{4} \int_0^1 \left[\frac{1}{5} x^5 - \frac{1}{11} x^{11} \right]_0^1 = \frac{3}{110}.$$

7. For each $y \in [0, 1]$, the region is bounded on the left by $-\sqrt{1 - y^2}$ and on the right by $\sqrt{1 - y^2}$. This is the upper half of the unit circle:

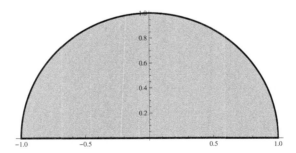

The integral is

$$\int_0^1 \int_{-\sqrt{1-y^2}}^{\sqrt{1-y^2}} \frac{x}{y+1} \, dx \, dy = \int_0^1 \left[\frac{x^2}{2(y+1)} \right]_{x=-\sqrt{1-y^2}}^{x=\sqrt{1-y^2}} dy = \int_0^1 0 \, dy = 0.$$

9. For each $y \in [1, 4]$, the region is bounded on the left by $x = 0$ and on the right by $x = \frac{1}{y}$:

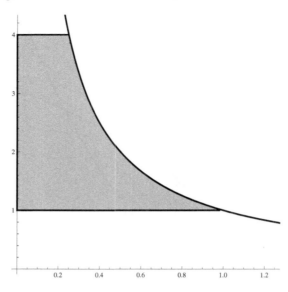

The integral is

$$\int_1^4 \int_0^{1/y} \frac{x}{y} \, dx \, dy = \int_1^4 \left[\frac{x^2}{2y} \right]_{x=0}^{x=1/y} dx = \int_1^4 \frac{1}{2y^3} \, dy = \left[-\frac{1}{4y^2} \right]_1^4 = \frac{15}{64}.$$

11. For each $x \in [0, 4]$, the range of y is $[\sqrt{x}, 2]$:

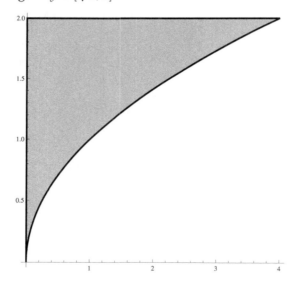

Reversing the order of integration, we have $y \in [0, 2]$, and x ranges from 0 to y^2, so the integral is

$$\int_0^2 \int_0^{y^2} y \cos x \, dx \, dy = \int_0^2 [y \sin x]_{x=0}^{x=y^2} \, dy$$

$$= \int_0^2 y \sin y^2 \, dy$$

$$= \left[-\frac{1}{2} \cos y^2 \right]_0^2 = \frac{1}{2} - \frac{1}{2} \cos 4 = \sin^2 2.$$

13. For each $y \in [0, 9]$, the range of x is $[\sqrt{y}, 3]$:

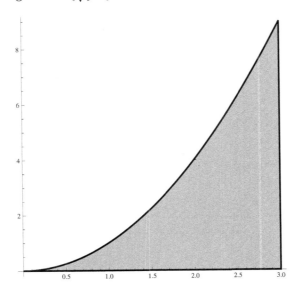

Reversing the order of integration, we have $x \in [0, 3]$, and y ranges from 0 to x^2, so the integral is

$$\int_0^3 \int_0^{x^2} \frac{1}{1+x^3} \, dy \, dx = \int_0^3 \left[\frac{y}{1+x^3} \right]_{y=0}^{y=x^2} dx = \int_0^3 \frac{x^2}{1+x^3} \, dx = \left[\frac{1}{3} \ln \left| 1+x^3 \right| \right]_0^3 = \frac{1}{3} \ln 28.$$

15. For each $y \in [0, 4]$, the range of x is $[y, 4]$:

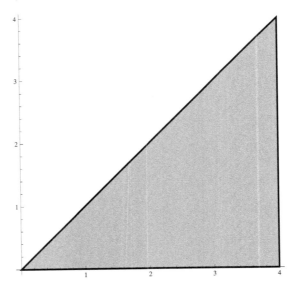

Reversing the order of integration, we have $x \in [0, 4]$, and y ranges from 0 to x, so the integral is

$$\int_0^4 \int_0^x e^{y/x} \, dy \, dx = \int_0^4 \left[xe^{y/x} \right]_{y=0}^{y=x} dx = \int_0^4 (ex - x) \, dx = \left[\frac{1}{2}(e-1)x^2 \right]_0^4 = 8(e-1).$$

17. Since $\theta \in \left[0, \frac{\pi}{2} \right]$ and r has constant bounds starting at zero and ending at 3, this is a quarter-circle of radius 3:

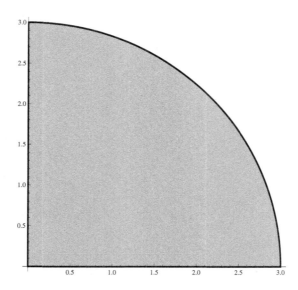

The integral is

$$\int_0^{\pi/2} \int_0^3 r^2 \, dr \, d\theta = \int_0^{\pi/2} \left[\frac{1}{3} r^3\right]_{r=0}^{r=3} d\theta = \int_0^{\pi/2} 9 \, d\theta = \frac{9}{2}\pi.$$

19. $r = 2\sin\theta$ is a circle of radius 1 centered at $(0,1)$, traced once for $\theta \in [0, \pi]$. So this is the right half of that circle:

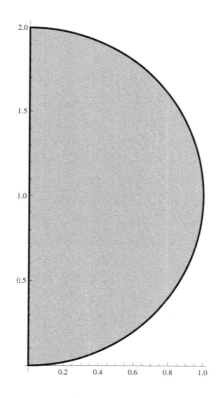

The integral is

$$\int_0^{\pi/2}\int_0^{2\sin\theta} r^3\, dr\, d\theta = \int_0^{\pi/2}\left[\frac{1}{4}r^4\right]_{r=0}^{r=2\sin\theta} d\theta$$

$$= \int_0^{\pi/2} 4\sin^4\theta\, d\theta$$

$$= \int_0^{\pi/2} (1-\cos 2\theta)^2\, d\theta$$

$$= \int_0^{\pi/2} (1-2\cos 2\theta + \cos^2 2\theta)\, d\theta$$

$$= \int_0^{\pi/2}\left(1-2\cos 2\theta + \frac{1}{2}(1+\cos 4\theta)\right) d\theta$$

$$= \left[\frac{3}{2}\theta - \sin 2\theta + \frac{1}{8}\sin 4\theta\right]_0^{\pi/2}$$

$$= \frac{3}{4}\pi.$$

21. This is a quarter-circle of radius 2:

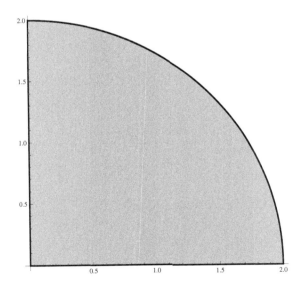

In polar coordinates, this is $\theta \in \left[0, \frac{\pi}{2}\right]$ and $r \in [0, 2]$. The integrand becomes e^{r^2}, so (remembering the factor of r), the integral is

$$\int_0^2\int_0^{\sqrt{4-y^2}} e^{x^2+y^2}\, dx\, dy = \int_0^{\pi/2}\int_0^2 re^{r^2}\, dr\, d\theta = \int_0^{\pi/2}\left[\frac{1}{2}e^{r^2}\right]_{r=0}^{r=2} d\theta$$

$$= \int_0^{\pi/2}\frac{1}{2}(e^4-1)\, d\theta = \frac{1}{4}(e^4-1)\pi.$$

23. This region is the full circle of radius 3 centered at the origin:

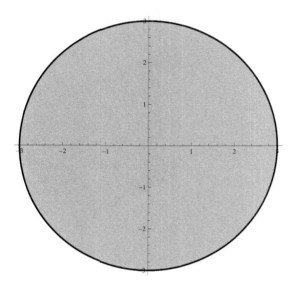

In polar coordinates, the region of integration is $\theta \in [0, 2\pi]$, $r \in [0, 3]$. The integrand becomes

$$\frac{x + 2y}{x^2 + y^2} = \frac{r\cos\theta + 2r\sin\theta}{r^2} = \frac{\cos\theta + 2\sin\theta}{r},$$

so (remembering the factor of r), the integral is

$$\int_{-3}^{3} \int_{-\sqrt{9-x^2}}^{\sqrt{9-x^2}} \frac{x + 2y}{x^2 + y^2}\, dy\, dx = \int_{0}^{2\pi} \int_{0}^{3} \frac{\cos\theta + 2\sin\theta}{r} \cdot r\, dr\, d\theta$$

$$= \int_{0}^{2\pi} \int_{0}^{3} (\cos\theta + 2\sin\theta)\, dr\, d\theta$$

$$= \int_{0}^{2\pi} [r(\cos\theta + 2\sin\theta)]_{r=0}^{r=3}\, d\theta$$

$$= 3 \int_{0}^{2\pi} (\cos\theta + 2\sin\theta)\, d\theta$$

$$= 3\,[\sin\theta - 2\cos\theta]_{0}^{2\pi} = 0.$$

25. This solid is a box:

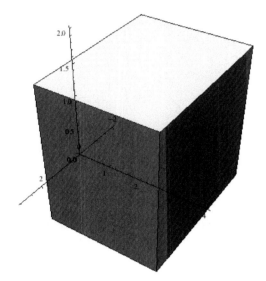

The volume of the box is $(3 - (-2))(4 - 1)(2 - 0) = 30$. Evaluating the integral, we also get

$$\int_0^2 \int_1^4 \int_{-2}^3 dx\, dy\, dz = \int_0^2 \int_1^4 [x]_{x=-2}^{x=3}\, dy\, dz = 5 \int_0^2 \int_1^4 dy\, dz$$

$$= 5 \int_0^2 [y]_{y=1}^{y=4}\, dz = 15 \int_0^2 dz = 15\, [z]_0^2 = 30.$$

27. The solid is a tetrahedron with one vertex at the origin. Another vertex is at $(4,0,0)$. When $x = y = 0$, we see that another vertex is at $z = 2$, and finally when $x = z = 0$, we find a vertex at $y = 3$:

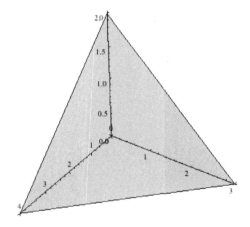

The volume of a tetrahedron is one third the area of the base times the height. In this case, the base is a right triangle with legs 3 and 4, so it has area 6, and the height is 2, so the volume should be

$\frac{1}{3} \cdot 6 \cdot 2 = 4$:

$$\int_0^4 \int_0^{3-(3/4)x} \int_0^{2-(1/2)x-(2/3)y} dz\, dy\, dx = \int_0^4 \int_0^{3-(3/4)x} [z]_{z=0}^{z=2-(1/2)x-(2/3)y} dy\, dx$$

$$= \int_0^4 \int_0^{3-(3/4)x} \left(2 - \frac{1}{2}x - \frac{2}{3}y\right) dy\, dx$$

$$= \int_0^4 \left[2y - \frac{1}{2}xy - \frac{1}{3}y^2\right]_{y=0}^{y=3-(3/4)x} dx$$

$$= \int_0^4 \left(2\left(3 - \frac{3}{4}x\right) - \frac{1}{2}x\left(3 - \frac{3}{4}x\right) - \frac{1}{3}\left(3 - \frac{3}{4}x\right)^2\right) dx$$

$$= \int_0^4 \left(3 - \frac{3}{2}x + \frac{3}{16}x^2\right) dx = \left[3x - \frac{3}{4}x^2 + \frac{1}{16}x^3\right]_0^4 = 4.$$

29. The projection in the xy plane of this solid is the semicircle of radius 2 above the x axis. The solid is a half-cylinder above the xy plane truncated by the plane $z = y$:

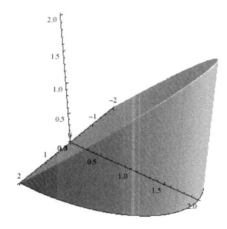

The volume of the solid is

$$\int_{-2}^2 \int_0^{\sqrt{4-x^2}} \int_0^y dz\, dy\, dx = \int_{-2}^2 \int_0^{\sqrt{4-x^2}} [z]_{z=0}^{z=y} dy\, dx$$

$$= \int_{-2}^2 \int_0^{\sqrt{4-x^2}} y\, dy\, dx$$

$$= \int_{-2}^2 \left[\frac{1}{2}y^2\right]_{y=0}^{y=\sqrt{4-x^2}} dx$$

$$= \int_{-2}^2 \left(2 - \frac{1}{2}x^2\right) dx$$

$$= \left[2x - \frac{1}{6}x^3\right]_{-2}^2 = \frac{16}{3}.$$

31. The base of this solid is the interior of the (polar) circle $r = 3$ for $\theta \in [0, 2\pi]$. It is bounded above by the hemisphere $z = \sqrt{9 - r^2}$:

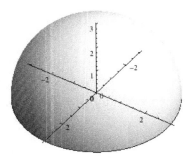

The volume of a hemisphere of radius 3 is $\frac{2}{3}\pi \cdot 3^3 = 18\pi$:

$$\int_0^{2\pi} \int_0^3 \int_0^{\sqrt{9-r^2}} r \, dz \, dr \, d\theta = \int_0^{2\pi} \int_0^3 [rz]_{z=0}^{z=\sqrt{9-r^2}} \, dr \, d\theta$$

$$= \int_0^{2\pi} \int_0^3 r\sqrt{9 - r^2} \, dr \, d\theta$$

$$= \int_0^{2\pi} \left[-\frac{1}{3}(9 - r^2)^{3/2} \right]_{r=0}^{r=3} \, d\theta$$

$$= \int_0^{2\pi} 9 \, d\theta = 18\pi.$$

33. The projection of this solid in the xy plane is the interior of the (polar) circle $r = 4$ for $\theta \in [0, 2\pi]$. The solid extends from $z = 0$ to $z = 4 - r$, so it is the interior of a downwards-opening cone:

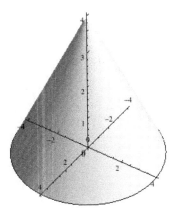

The volume is

$$\int_0^{2\pi} \int_0^4 \int_0^{4-r} r\, dz\, dr\, d\theta = \int_0^{2\pi} \int_0^4 [rz]_{z=0}^{z=4-r}\, dr\, d\theta$$

$$= \int_0^{2\pi} \int_0^4 (4r - r^2)\, dr\, d\theta$$

$$= \int_0^{2\pi} \left[2r^2 - \frac{1}{3}r^3\right]_{r=0}^{r=4}\, d\theta$$

$$= \int_0^{2\pi} \frac{32}{3}\, d\theta = \frac{64}{3}\pi.$$

35. Given the ranges of ρ, θ, and ϕ, we see that this is a sphere of radius 5 centered at the origin:

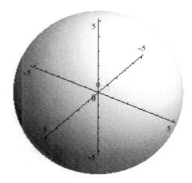

From the formula for the volume of a sphere, its volume should be $\frac{4}{3}\pi \cdot 5^3 = \frac{500}{3}\pi$:

$$\int_0^{\pi} \int_0^{2\pi} \int_0^5 \rho^2 \sin\phi\, d\rho\, d\theta\, d\phi = \int_0^{\pi} \int_0^{2\pi} \left[\frac{1}{3}\rho^3 \sin\phi\right]_{\rho=0}^{\rho=5}\, d\theta\, d\phi$$

$$= \int_0^{\pi} \int_0^{2\pi} \frac{125}{3}\sin\phi\, d\theta\, d\phi$$

$$= \frac{250}{3}\pi \int_0^{\pi} \sin\phi\, d\phi$$

$$= \frac{250}{3}\pi \left[-\cos\phi\right]_0^{\pi} = \frac{500}{3}\pi.$$

Chapter 14

Vector Analysis

14.1   Introduction and Vector Fields

Thinking Back

Work as an integral of force and distance

Since work is the application of force through a distance, the work is (using integration by parts with $u = x$ and $dv = \sin x\, dx$)

$$\int_0^{\pi/2} x \sin x\, dx = [-x \cos x]_0^{\pi/2} + \int_0^{\pi/2} \cos x\, dx = 0 + [\sin x]_0^{\pi/2} = 1.$$

Vector geometry

▷ Since $\mathbf{F}(x, y, z) = z(x - y)^2 \mathbf{i} + \sqrt{y}\, \mathbf{j} + \mathbf{k}$, we have

$$\mathbf{F}(3, 3, \sqrt{13}) = \sqrt{13}(3 - 3)^2 \mathbf{i} + \sqrt{3}\mathbf{j} + \mathbf{k} = \left\langle 0, \sqrt{3}, 1 \right\rangle.$$

The norm of this vector is $\sqrt{0 + 3 + 1} = 2$, so a unit vector in this direction is $\left\langle 0, \frac{\sqrt{3}}{2}, \frac{1}{2} \right\rangle$.

▷ A vector equation for the line is $\langle 0, 0, 0 \rangle + t \left\langle 0, \frac{\sqrt{3}}{2}, \frac{1}{2} \right\rangle$.

Calculus of vector-valued functions

▷ To differentiate, we differentiate each component:

$$\frac{d}{dt}(\mathbf{r}(t)) = \frac{d}{dt}(3\cos^2 t)\mathbf{i} + \frac{d}{dt}(5t)\mathbf{j} + \frac{d}{dt}\left(\frac{t}{t^2 + 1}\right)\mathbf{k} = -6\sin t \cos t\, \mathbf{i} + 5\mathbf{j} + \frac{1 - t^2}{(1 + t^2)^2}\mathbf{k}.$$

▷ To integrate, we integrate each component:

$$\int \mathbf{r}(t)\, dt = \left(\int e^t\, dt\right)\mathbf{i} + \left(\int t^3\, dt\right)\mathbf{j} + \left(\int -4\, dt\right)\mathbf{k} = e^t \mathbf{i} + \frac{1}{4}t^4 \mathbf{j} - 4t\mathbf{k} + (C_1\mathbf{i} + C_2\mathbf{j} + C_3\mathbf{k}).$$

Concepts

1. (a) False. A vector field is a function whose outputs are vectors. See Definition 14.1.

   (b) True. See Definition 14.1.

   (c) False. A vector field in $\mathbb{R}^2$ has as inputs points of $\mathbb{R}^2$, while a vector field in $\mathbb{R}^3$ has as inputs points of $\mathbb{R}^3$. See Definition 14.1.

   (d) True. See part (c), and Definition 14.1.

   (e) True. If f is a potential function for $\mathbf{F}$, then so is $f + \alpha$ for any real number α, since $\nabla(f + \alpha) = \nabla f + \nabla \alpha = \nabla f = \mathbf{F}$.

   (f) False. A vector field $\mathbf{F}$ is a gradient if and only if it is conservative (Definition 14.2), which is true if and only if (in the case of a vector field on $\mathbb{R}^2$) $\frac{\partial F_1}{\partial y} = \frac{\partial F_2}{\partial x}$. The analogous condition for vector fields on $\mathbb{R}^3$ is more complicated.

   (g) False. f and $f + \alpha$ for $\alpha \in \mathbb{R}$ always have the same gradient, since $\nabla(f + \alpha) = \nabla f + \nabla \alpha = \nabla f$.

   (h) True. This is the definition of work; see Definition 6.12 and Example 1 in Section 6.4.

3. By Definition 14.1, a vector field in the Cartesian plane, which is $\mathbb{R}^2$, is a function from points in $\mathbb{R}^2$ to vectors in $\mathbb{R}^2$. So its inputs are points in the Cartesian plane (or $\mathbb{R}^2$).

5. By Definition 14.1, a vector field in the Cartesian plane, which is $\mathbb{R}^2$, is a function from points in $\mathbb{R}^2$ to vectors in $\mathbb{R}^2$. So its outputs are vectors in $\mathbb{R}^2$.

7. A vector field is conservative if it is the gradient of some function. So, for example, a vector field $\langle F_1(x,y), F_2(x,y) \rangle$ in $\mathbb{R}^2$ is conservative if there is some function $f(x,y)$ such that $\frac{\partial f}{\partial x} = F_1$ and $\frac{\partial f}{\partial y} = F_2$.

9. The vectors $\mathbf{F}(x,y)$ and $\mathbf{G}(x,y)$ at any point (x,y) point in the same direction, since $\langle 2x, 2y \rangle = 2\langle x, y \rangle$. But the vectors in $\mathbf{F}(x,y)$ are twice as long at each point as the vectors of $\mathbf{G}(x,y)$.

11. Since $\mathbf{F}(x,y) = -\mathbf{G}(x,y)$ for any point (x,y), the vectors of $\mathbf{F}$ are the same length at each point as those in $\mathbf{G}$, but point in the opposite direction.

13. We have
$$2\mathbf{F}(x,y,z) = 2\langle yz, xz, xy \rangle = \langle 2yz, 2xz, 2xy \rangle,$$
so let $\mathbf{G}(x,y,z) = \langle 2yz, 2xz, 2xy \rangle$.

15. If $\mathbf{F} = \langle F_1(x,y), F_2(x,y) \rangle$, then if $\frac{\partial F_1}{\partial y} \neq \frac{\partial F_2}{\partial x}$, the vector field is not conservative. This follows from the fact that if it were conservative, then there would be some function $f(x,y)$ with $\frac{\partial f}{\partial x} = F_1$ and $\frac{\partial f}{\partial y} = F_2$. Since mixed partials are equal, we would have $\frac{\partial^2 f}{\partial x \partial y} = \frac{\partial F_1}{\partial y} = \frac{\partial^2 f}{\partial y \partial x} = \frac{\partial F_2}{\partial x}$.

Skills

17. Using the procedure in the text, integrate the first component with respect to x:
$$\int 3x^2 \cos y \, dx = x^3 \cos y + \alpha.$$

Since the only term the second component also depends on x, no further work is required, so $x^3 \cos y + \alpha$ is a potential function for any $\alpha \in \mathbb{R}$. As a check, $\frac{\partial}{\partial y}(x^3 \cos y + \alpha) = -x^3 \sin y$, which is the second component of the vector field.

19. Using the procedure in the text, integrate the first component with respect to x:

$$\int (5x^4 + y)\, dx = x^5 + xy + \alpha.$$

The term $-12y^3$ in the second component does not depend on x, so we must integrate it as well, with respect to y:

$$\int (-12y^3)\, dy = -3y^4 + \beta.$$

Add these results and set $\alpha + \beta = \gamma$, to give $x^5 + xy - 3y^4 + \gamma$ as a potential function for any $\gamma \in \mathbb{R}$. As a check,

$$\frac{\partial}{\partial x}(x^5 + xy - 3y^4 + \gamma) = 5x^4 + y, \qquad \frac{\partial}{\partial y}(x^5 + xy - 3y^4 + \gamma) = x - 12y^3.$$

21. Using the procedure in the text, integrate the first component with respect to x:

$$\int yz\, dx = xyz + \alpha.$$

Since the only term in each of the other components depends on x, no further work is required, and $xyz + \alpha$ is a potential function for any $\alpha \in \mathbb{R}$. As a check, $\frac{\partial}{\partial y}(xyz + \alpha) = xz$ and $\frac{\partial}{\partial z}(xyz + \alpha) = xy$.

23. Using the procedure in the text, but starting with the second component, integrate with respect to y:

$$\int (\sin z - x \sin y)\, dy = y \sin z + x \cos y + \alpha.$$

Since the only term in each of the other two components depends on y, no further work is required, and $y \sin z + x \cos y + \alpha$ is a potential function for any $\alpha \in \mathbb{R}$ (this is why we started with the second component — starting with either of the other components would have required at least two integrations). As a check,

$$\frac{\partial}{\partial x}(y \sin z + x \cos y + \alpha) = \cos y, \qquad \frac{\partial}{\partial z}(y \sin z + x \cos y + \alpha) = y \cos z.$$

25. These are horizontal vectors of constant length 2 pointing to the right:

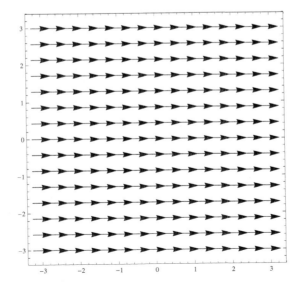

27. These are constant length vectors of magnitude $\sqrt{1+1} = \sqrt{2}$ pointing up and to the right:

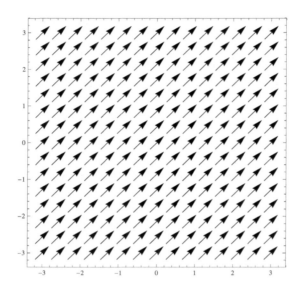

29. These are constant length vectors of magnitude $\sqrt{1+1} = \sqrt{2}$ pointing down and to the right:

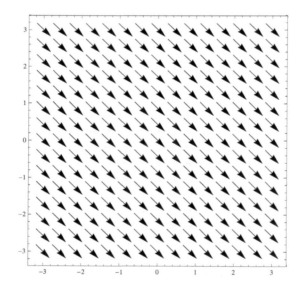

31. These are vectors of varying length pointing generally away from the origin (this vector field is similar to $\langle x, y \rangle$ except for the length of the y component):

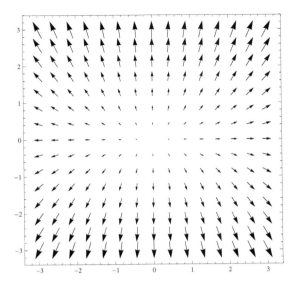

33. Since

$$\frac{\partial F_1}{\partial y} = \frac{\partial}{\partial y}(xy) = x, \qquad \frac{\partial F_2}{\partial x} = \frac{\partial}{\partial x}(-y) = 0,$$

the mixed partial condition is not satisfied, so this is not a conservative vector field.

35. Since

$$\frac{\partial G_1}{\partial y} = \frac{\partial}{\partial y}\frac{1}{x^2 + y} = -\frac{1}{(x^2 + y)^2}, \qquad \frac{\partial G_2}{\partial x} = \frac{\partial}{\partial x}\left(\frac{y}{x}\right) = -\frac{y}{x^2},$$

the mixed partial condition is not satisfied, so this is not a conservative vector field.

37. Since

$$\frac{\partial F_2}{\partial z} = \frac{\partial}{\partial z}(-z) = -1, \qquad \frac{\partial F_3}{\partial y} = \frac{\partial}{\partial y}\left(e^{yz}\right) = ze^{yz},$$

this pair of mixed partials is unequal, so this is not a conservative vector field.

39. Since

$$\frac{\partial G_2}{\partial z} = \frac{\partial}{\partial z}(yz) = y, \qquad \frac{\partial G_3}{\partial y} = \frac{\partial}{\partial y}(z + 12) = 0,,$$

this pair of mixed partials is unequal, so this is not a conservative vector field.

41. Since

$$\frac{\partial F_1}{\partial y} = \frac{\partial}{\partial y}\left(e^y\right) = e^y, \qquad \frac{\partial F_2}{\partial x} = \frac{\partial}{\partial x}(\sin y) = 0,$$

the mixed partial condition is not satisfied, so this is not a conservative vector field.

43. Since

$$\frac{\partial G_1}{\partial y} = \frac{\partial}{\partial y}\left(2x + y\cos(xy)\right) = \cos(xy) - xy\sin(xy),$$

$$\frac{\partial G_2}{\partial x} = \frac{\partial}{\partial x}(x\cos(xy) - 1) = \cos(xy) - xy\sin(xy),$$

the mixed partial condition is satisfied, so this is a conservative vector field. To find a potential function, integrate the first component with respect to x to get $x^2 + \sin(xy) + \alpha$. Since the second term in the second component does not depend on x, we must integrate it as well, with respect to y, to get $-y + \beta$. The sum of these, $x^2 - y + \sin(xy) + \gamma$, for $\gamma \in \mathbb{R}$, is a potential function.

45. Since

$$\frac{\partial F_1}{\partial y} = e^{2z}, \qquad \frac{\partial F_2}{\partial x} = e^{2z},$$

$$\frac{\partial F_1}{\partial z} = 2ye^{2z}, \qquad \frac{\partial F_3}{\partial x} = 2ye^{2z},$$

$$\frac{\partial F_2}{\partial z} = 2xe^{2z}, \qquad \frac{\partial F_3}{\partial y} = 2xe^{2z},$$

all three sets of mixed partials are equal, so this is a conservative vector field. To find a potential function, integrate the first component with respect to x to get $xye^{2z} + x + \alpha$. Since the only term in each of the other components depends on x, we are done, and this is a potential function for any $\alpha \in \mathbb{R}$.

47. Since

$$\frac{\partial G_1}{\partial y} = -1, \qquad \frac{\partial G_2}{\partial x} = -y,$$

$$\frac{\partial G_1}{\partial z} = 1, \qquad \frac{\partial G_3}{\partial x} = z,$$

$$\frac{\partial G_2}{\partial z} = 0, \qquad \frac{\partial G_3}{\partial y} = 1,$$

the mixed partials are not equal, so this is not a conservative field.

Applications

49. (a) Since the water flows from right to left, the arrows will be pointing leftwards; since it is at a constant rate, they will all be the same size. A possible vector field is $\langle -1, 0 \rangle$:

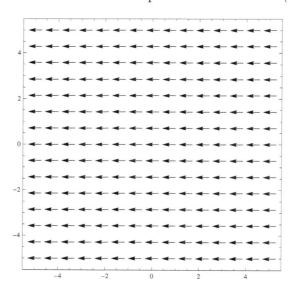

(b) Since the water flows from top left to bottom right, the arrows will be pointing southeast; since it is at a constant rate, they will all be the same size. A possible vector field is $\langle 1, -1 \rangle$.

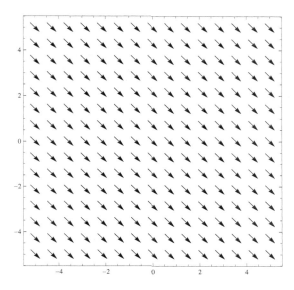

(c) Since the flow is at a constant rate, the arrows will all be the same size. A possible vector field is $\left\langle -\frac{y}{\sqrt{x^2+y^2}}, \ \frac{x}{\sqrt{x^2+y^2}} \right\rangle$, which is a unit vector field:

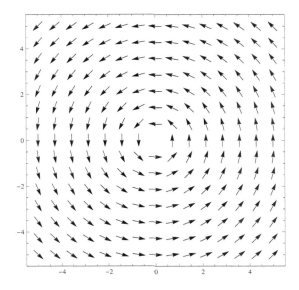

51. (a) The vector field is:

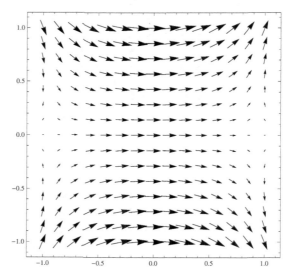

(b) Yes, it is conservative, since

$$\frac{\partial F_1}{\partial y} = \frac{\partial}{\partial y}\left(0.9 - x^2 + 0.5y^2\right) = y, \qquad \frac{\partial F_2}{\partial x} = \frac{\partial}{\partial x}(xy) = y,$$

and the mixed partials are equal. A potential function is found by integrating $0.9 - x^2 + 0.5y^2$ with respect to x to get $0.9x - \frac{1}{3}x^3 + 0.5xy^2 + \alpha$ for any $\alpha \in \mathbb{R}$. Since the only term in the second component depends on x, we are done, and this is a potential function. The potential function is a function from $\mathbb{R}^2$ to $\mathbb{R}$ whose gradient at any point is the current at that point.

(c) Looking at the diagram, it appears that there is land both above $y = 1$ and below $y = -1$; the current is flowing around those pieces of land and through the channel.

Proofs

53. Suppose that $\nabla f = \nabla g$. Then $\frac{\partial f}{\partial x} = \frac{\partial g}{\partial x}$, so that integrating both sides with respect to x gives $f(x, y) = g(x, y) + h(y)$. Thus

$$\frac{\partial f}{\partial y} = \frac{\partial}{\partial y}(g(x, y) + h(y)) = \frac{\partial g}{\partial y} + \frac{\partial h}{\partial y}.$$

But since $\frac{\partial f}{\partial y} = \frac{\partial g}{\partial y}$ because the gradients are equal, it must be the case that $\frac{\partial h}{\partial y} = 0$, so that $h(y)$ is a constant $\alpha \in \mathbb{R}$. Thus $f(x, y) = g(x, y) + \alpha$ and the two differ by a constant (which could be zero).

Thinking Forward

Integration of vector fields

▷ In computing the arc length, we divided up the curve into multiple subintervals corresponding to subintervals in the range of t, and added up the lengths of secants between the end points of those subintervals. Letting the number of subintervals go to infinity, the secants approximate the curve itself arbitrarily closely, so the limit of that sum was the length of the segment. To use that concept to compute the area of the fence, compute the limit of a Riemann sum in which each term is the length of that secant multiplied by the height of the fence over some point in that subinterval. Then once again the Riemann sum will be computing areas, and the limit of their sum is the area of the fence over the entire length of $\mathbf{r}(t)$.

▷ Given a curve $\mathbf{r}(t)$ and a vector field $\mathbf{F}$, perhaps the most obvious idea is to integrate the function $\|\mathbf{F}(\mathbf{r}(t))\|$ along the curve $\mathbf{r}(t)$ using the ideas in part (a). However, if this kind of integral is to give a usable definition of work, this idea is incorrect, for its value does not depend on the direction of the vectors in $\mathbf{F}$: the work would be the same regardless of whether $\mathbf{F}$ pointed in the same direction as $\mathbf{r}(t)$ moved for increasing values of t, or in the opposite direction. So a more reasonable definition is that we want to integrate only the magnitude of the portion of $\mathbf{F}$ that is parallel to $\mathbf{r}(t)$ at each point; that magnitude is given by $\mathbf{F}(\mathbf{r}(t)) \cdot T(t)$, where $T(t)$ is the unit tangent vector $\frac{\mathbf{r}'(t)}{\|\mathbf{r}'(t)\|}$ to $\mathbf{r}(t)$ at t.

14.2 Line Integrals

Thinking Back

Average values

▷ The average value is

$$\frac{1}{5-1}\int_1^5 (x^2+x+2)\,dx = \frac{1}{4}\left[\frac{1}{3}x^3 + \frac{1}{2}x^2 + 2x\right]_1^5 = \frac{46}{3}.$$

▷ The average value is

$$\frac{1}{2\pi - \pi}\int_\pi^{2\pi} \cos x\,dx = \frac{1}{\pi}[\sin x]_\pi^{2\pi} = 0.$$

Arc length

▷ By Theorem 11.18, the arc length of $\mathbf{r}(t) = \langle t, t^2 \rangle$ from $t=4$ to $t=6$ is

$$\int_4^6 \|\mathbf{r}'(t)\|\,dt = \int_4^6 \sqrt{1^2+(2t)^2}\,dt = \int_4^6 \sqrt{4t^2+1}\,dt.$$

To integrate, use the trigonometric substitution $t = \frac{1}{2}\tan u$, so that $dt = \frac{1}{2}\sec^2 u\,du$. Then

$$\int_4^6 \sqrt{4t^2+1}\,dt = \int_{t=4}^{t=6} \sqrt{\tan^2 u + 1} \cdot \frac{1}{2}\sec^2 u\,du$$

$$= \frac{1}{2}\int_{t=4}^{t=6} \sec^3 u\,du$$

$$= \frac{1}{4}[\sec u \tan u + \ln|\sec u + \tan u|]_{t=4}^{t=6}$$

$$= \frac{1}{4}\left[2t\sqrt{4t^2+1} + \ln\left|\sqrt{4t^2+1}+2t\right|\right]_4^6$$

$$= \frac{1}{4}\left(12\sqrt{145} + \ln\left|12+\sqrt{145}\right| - 8\sqrt{65} - \ln\left|8+\sqrt{65}\right|\right)$$

$$= 3\sqrt{145} - 2\sqrt{65} + \frac{1}{4}\ln\frac{12+\sqrt{145}}{8+\sqrt{65}}.$$

▷ By Theorem 11.18, the arc length of $\mathbf{r}(t) = \cos z\mathbf{i} + \sin z\mathbf{j} + z\mathbf{k}$ from $z=0$ to $z=2\pi$ is

$$\int_0^{2\pi} \|\mathbf{r}'(t)\|\,dt = \int_0^{2\pi} \sqrt{(-\sin z)^2 + \cos^2 z + 1^2}\,dz = \int_0^{2\pi} \sqrt{2}\,dz = 2\sqrt{2}\,\pi.$$

Concepts

1. (a) True. By Definition 14.3, the result of such an integral is

$$\int_a^b f(x(t), y(t), z(t)) \, \|\mathbf{r}'(t)\| \, dt,$$

   which is an ordinary single-variable integral, producing a scalar.

   (b) False. By Definition 14.4, the result of such an integral is

$$\int_C \left(F_1(x, y, z)x'(t) + F_2(x, y, z)y'(t) + F_3(t)z'(t) \right) \, dt,$$

   which is a scalar.

   (c) True. That is part of the definition; see the discussion in the text.

   (d) True. Break the curve up into its smooth pieces, integrate over each piece separately using Definition 14.3 or 14.4 as appropriate, and add the results.

   (e) False. A conservative vector field may be integrated along a curve just like any other vector field. The difference is that a simpler method is available for evaluating such an integral; see Theorem 14.5.

   (f) True. It provides a simpler method of integrating such a vector field along a smooth curve. See Theorem 14.5.

   (g) False. In order for the Fundamental Theorem of Line Integrals to apply, the vector field must be conservative. See Theorem 14.5.

   (h) False. In general this is true only for conservative vector fields, where the Fundamental Theorem of Line Integrals applies. For example, let $\mathbf{F}(x, y) = \langle y, 0 \rangle$, and let $\mathbf{r}(t) = \langle \cos t, \sin t \rangle$ for $t \in [0, 2\pi]$ be the unit circle. Then since $y = \sin t$ along $\mathbf{r}$, we get

$$\int_C \mathbf{F}(x, y) \cdot d\mathbf{r} = \int_0^{2\pi} (\sin t \cdot (-\sin t) + 0 \cdot (\cos t)) \, dt = -\int_0^{2\pi} \sin^2 t \, dt = -\pi.$$

3. Since $\mathbf{r}'(t) = \langle 1, 2t, 4t^3 \rangle$, we have

$$\|\mathbf{r}'(t)\| = \sqrt{1^2 + (2t)^2 + (4t^3)^2} = \sqrt{16t^6 + 4t^2 + 1},$$

   so that by Definition 14.3 the line integral is

$$\int_a^b e^{t \cdot t^2 \cdot t^4} \sqrt{16t^6 + 4t^2 + 1} \, dt = \int_a^b e^{t^7} \sqrt{16t^6 + 4t^2 + 1} \, dt.$$

5. Considering Definition 14.4 and the comment immediately following, the vector field components are $F_1(x, y, z) = x + y$, $F_2(x, y, z) = xy$, and $F_3(x, y, z) = 0$, so the vector field is $\langle x + y, xy, 0 \rangle$. This can also be thought of more simply as a vector field $\langle x + y, xy \rangle$ in $\mathbb{R}^2$.

7. Considering Definition 14.4 and the comment immediately following, the vector field components are $F_1(x, y, z) = xy^2$, $F_2(x, y, z) = xy - z$, and $F_3(x, y, z) = \cos y$, so the vector field is $\langle xy^2, xy - z, \cos y \rangle$.

9. With $\mathbf{F}(x, y) = \langle -y, x \rangle$ and C_1 as in Example 3, we get

$$\int_{C_1} \mathbf{F}(x, y) \cdot d\mathbf{r} = \int_{C_1} (-y \, dx + x \, dy).$$

11. By Definition 14.3, $ds = \|\mathbf{r}'(t)\| \, dt$.

13. On $\mathbf{r}(t)$, we have $x = 2t$ and $y = e^t$, so that $\mathbf{F}(x, y) = \langle x^2 y, x - y \rangle = \langle 4t^2 e^t, 2t - e^t \rangle$. Also, $\mathbf{r}'(t) = \langle x'(t), y'(t) \rangle = \langle 2, e^t \rangle$. Then by Definition 14.4,

$$\int_C \mathbf{F}(x, y) \cdot d\mathbf{r} = \int_a^b \left(F_1(x, y) x'(t) + F_2(x, y) y'(t) \right) dt$$

$$= \int_a^b \left(4t^2 e^t \cdot 2 + (2t - e^t) e^t \right) dt = \int_a^b e^t (8t^2 + 2t - e^t) \, dt.$$

15. This is the usual parametrization, $\mathbf{r}(\theta) = \langle \cos\theta, \sin\theta \rangle$ for $\theta \in [0, 2\pi)$, since increasing angles result in counterclockwise motion. Also, $\mathbf{r}(0) = \langle \cos 0, \sin 0 \rangle = \langle 1, 0 \rangle$. This parametrization is smooth since $\mathbf{r}'(\theta) = \langle -\sin\theta, \cos\theta \rangle$ is continuous everywhere, and $\|\mathbf{r}'(\theta)\| = \sqrt{\sin^2\theta + \cos^2\theta} = 1 \neq 0$ everywhere.

17. The helix $\langle \cos t, t, \sin t \rangle$ has radius 1, and is centered on the y axis, since $x^2 + z^2 = 1$ everywhere. To change its radius to 2, we multiply the trigonometric terms by 2 to get $\mathbf{r}(t) = \langle 2\cos t, t, 2\sin t \rangle$. At $t = 0$, we have $\mathbf{r}(0) = \langle 2, 0, 0 \rangle$. Since one revolution is from $t = 0$ to $t = 2\pi$, two full revolutions is $t \in [0, 4\pi)$. This parametrization is smooth since

$$\mathbf{r}'(t) = \langle -2\sin t, 1, 2\cos t \rangle$$

is continuous everywhere, and $\|\mathbf{r}'(t)\| = \sqrt{4\sin^2 t + 1 + 4\cos^2 t} = \sqrt{5}$ is nonzero everywhere.

19. Apply the same concept as for the average of anything else: the average is the sum, total, or in this case integral, divided by the number of elements or length. In our case, the length is just the length of the curve, so we get

$$f_{\text{avg}}(x, y, z) = \frac{\int_C f(x, y, z) \|\mathbf{r}'(t)\| \, dr}{\int_C \|\mathbf{r}'(t)\| \, dr} = \frac{\int_C f(x, y, z) \, ds}{\int_C ds}.$$

Skills

21. Parametrize the curve by $\mathbf{r}(t) = \langle t, t^2 \rangle$ for $t \in [0, 2]$. Then $\mathbf{r}'(t) = \langle 1, 2t \rangle$ so that $\|\mathbf{r}'(t)\| = \sqrt{4t^2 + 1}$. On $\mathbf{r}(t)$, we have $g(x, y) = x = t$, so the line integral is

$$\int_0^2 t\sqrt{4t^2 + 1} \, dt = \left[\frac{1}{12} (4t^2 + 1)^{3/2} \right]_0^2 = \frac{1}{12} \left(17\sqrt{17} - 1 \right).$$

23. Parametrize C by $\mathbf{r}(t) = \langle \cos t, \sin t \rangle$ for $t \in [0, 2\pi]$. Then $\mathbf{r}'(t) = \langle -\sin t, \cos t \rangle$ and $\|\mathbf{r}'(t)\| = \sqrt{\sin^2 t + \cos^2 t} = 1$. On the unit circle, $f(x, y) = x^2 + y^2 = 1$, so the line integral is

$$\int_0^{2\pi} 1 \cdot 1 \, dt = 2\pi.$$

25. Parametrize C by $\mathbf{r}(t) = t \langle 1, 2, 3 \rangle$ for $0 \leq t \leq 1$, so that $\mathbf{r}'(t) = \langle 1, 2, 3 \rangle$ and thus $\|\mathbf{r}'(t)\| = \sqrt{1 + 4 + 9} = \sqrt{14}$. Also, $f(x, y, z) = e^{x+y+z} = e^{6t}$. Hence the line integral is

$$\int_0^1 e^{6t} \cdot \sqrt{14} \, dt = \left[\frac{\sqrt{14}}{6} e^{6t} \right]_0^1 = \frac{\sqrt{14}}{6} (e^6 - 1).$$

27. With the given parametrization, $\mathbf{r}'(t) = \langle -\sin t, \cos t, 1 \rangle$, so that $\|\mathbf{r}'(t)\| = \sqrt{\sin^2 t + \cos^2 t + 1} = \sqrt{2}$. Also,

$$f(x,y,z) = e^{x^2+y^2+z} = e^{\cos^2 t + \sin^2 t + t} = e^{t+1},$$

so that the line integral is

$$\int_0^\pi \sqrt{2}\, e^{t+1}\, dt = \sqrt{2} \left[e^{t+1} \right]_0^\pi = \sqrt{2}(e^{\pi+1} - e).$$

29. The given curve is $y = x^2 + 1$ for $x \in [5, 10]$; parametrize it with $x = t$, so we get $\mathbf{r}(t) = \langle t, t^2 + 1 \rangle$ for $t \in [5, 10]$. Then $\mathbf{r}'(t) = \langle 1, 2t \rangle$. Further, $\mathbf{F}(x,y) = \mathbf{i} - \mathbf{j} = \langle 1, -1 \rangle$. Then the line integral is

$$\int_C \mathbf{F}(x,y) \cdot d\mathbf{r} = \int_5^{10} (1 \cdot 1 + 2t \cdot (-1))\, dt = \int_5^{10} (1 - 2t)\, dt = \left[t - t^2 \right]_5^{10} = -70.$$

31. The given curve is $y = x^2 + 1$ for x from 10 to 5; parametrize it with $x = -t$, so we get $\mathbf{r}(t) = \langle -t, (-t)^2 + 1 \rangle = \langle -t, t^2 + 1 \rangle$ for $t \in [-10, -5]$. Then $\mathbf{r}'(t) = \langle -1, 2t \rangle$. Further, $\mathbf{F}(x,y) = \mathbf{i} - \mathbf{j} = \langle 1, -1 \rangle$. Then the line integral is

$$\int_C \mathbf{F}(x,y) \cdot d\mathbf{r} = \int_{-10}^{-5} (-1 \cdot 1 + 2t \cdot (-1))\, dt = \int_{-10}^{-5} (-1 - 2t)\, dt = \left[-t - t^2 \right]_{-10}^{-5} = 70.$$

33. This vector field is conservative, since

$$\frac{\partial}{\partial y}(y \sin xy) = \sin xy + xy \cos xy, \qquad \frac{\partial}{\partial x}(x \sin xy) = \sin xy + xy \cos xy.$$

Since the cardioid is a closed curve that is differentiable everywhere, the integral is zero by the Fundamental Theorem of Line Integrals.

35. The given curve is parametrized by $\mathbf{r}(t) = \langle \cos t, \sin t, t \rangle$ for $t \in [2\pi, 3\pi]$. Differentiating gives $\mathbf{r}'(t) = \langle -\sin t, \cos t, 1 \rangle$. Also,

$$\mathbf{F}(x,y,z) = \left\langle -x^2 y, x^3, y^2 \right\rangle = \left\langle -\sin t \cos^2 t, \cos^3 t, \sin^2 t \right\rangle.$$

Then the line integral is

$$\int_{2\pi}^{3\pi} \left((-\sin t \cos^2 t) \cdot (-\sin t) + \cos^3 t \cdot \cos t + \sin^2 t \cdot 1 \right) dt$$

$$= \int_{2\pi}^{3\pi} \left(\cos^2 t \left(\sin^2 t + \cos^2 t \right) + \sin^2 t \right) dt$$

$$= \int_{2\pi}^{3\pi} \left(\cos^2 t + \sin^2 t \right) dt$$

$$= \int_{2\pi}^{3\pi} 1\, dt = \pi.$$

37. Since

$$\frac{\partial}{\partial y} \left(\frac{1}{x} + \ln y \right) = \frac{1}{y}, \qquad \frac{\partial}{\partial x} \left(\frac{1}{y} + \frac{x}{y} \right) = \frac{1}{y},$$

the vector field is conservative. To find a potential function, integrate the first component with respect to x to get $\ln x + x \ln y$. Since the first term of the second component does not depend on x, we must integrate that term with respect to y to get $\ln y$. Adding these results gives a potential function $f(x,y) = \ln x + \ln y + x \ln y$. Then by the Fundamental Theorem of Line Integrals, the value of the integral is

$$f(1, \pi) - f(\pi, e) = \ln 1 + \ln \pi + \ln \pi - \ln \pi - \ln e - \pi \ln e = \ln \pi - \pi - 1.$$

39. Since $\frac{\partial F_1}{\partial z} = -1$ while $\frac{\partial F_3}{\partial x} = 1$, the mixed partials for the first and third components are unequal, so this vector field is not conservative.

41. Differentiating gives

$$\frac{\partial F_1}{\partial y} = z^{xy} \ln z + xyz^{xy} \ln^2 z,$$

$$\frac{\partial F_2}{\partial x} = z^{xy} \ln z + xyz^{xy} \ln^2 z,$$

$$\frac{\partial F_1}{\partial z} = xy^2 z^{xy-1} \ln z + yz^{xy} \cdot \frac{1}{z} = xy^2 z^{xy-1} \ln z + yz^{xy-1},$$

$$\frac{\partial F_3}{\partial x} = yz^{xy-1} + xy^2 z^{xy-1} \ln z,$$

$$\frac{\partial F_2}{\partial z} = x^2 yz^{xy-1} \ln z + xz^{xy} \cdot \frac{1}{z} = x^2 yz^{xy-1} \ln z + xz^{xy-1},$$

$$\frac{\partial F_3}{\partial y} = xz^{xy-1} + x^2 yz^{xy-1} \ln z.$$

Since the mixed partials are equal, the vector field is conservative. To find a potential function start with the third component; integrating with respect to z gives z^{xy}. Since the only term of each of the remaining components depends on z, we are done, and $f(x, y, z) = z^{xy}$ is a potential function. Then by the Fundamental Theorem of Line Integrals, the value of the integral is

$$f(2, 16, 3) - f(0, 0, 1) = 3^{2 \cdot 16} - 1^{0 \cdot 0} = 3^{32} - 1.$$

43. Since $\frac{\partial F_1}{\partial y} = 0$ while $\frac{\partial F_2}{\partial x} = \ln(y+4) \neq 0$, this vector field is not conservative. (None of the other mixed partials are equal either: $\frac{\partial F_1}{\partial z} = \sin z$ while $\frac{\partial F_3}{\partial x} = 0$, and $\frac{\partial F_2}{\partial z} = 0$ while $\frac{\partial F_3}{\partial y} = 1$).

45. This vector field is not conservative, since $\frac{\partial F_1}{\partial y} = -1$ while $\frac{\partial F_2}{\partial x} = 1$. We have

$$\mathbf{r}'(t) = \langle \cos t - t \sin t, \sin t + t \cos t \rangle, \qquad \mathbf{F}(x, y, z) = \langle -y, x \rangle = \langle -t \sin t, t \cos t \rangle.$$

Then the line integral is

$$\int_\pi^{2\pi} ((-t \sin t)(\cos t - t \sin t) + (t \cos t)(\sin t + t \cos t)) \, dt = \int_\pi^{2\pi} \left(t^2 \sin^2 t + t^2 \cos^2 t \right) dt$$

$$= \int_\pi^{2\pi} t^2 \, dt$$

$$= \left[\frac{1}{3} t^3 \right]_\pi^{2\pi} = \frac{7}{3} \pi^3.$$

47. With $\mathbf{r}(t) = \langle t \sin t, t \cos t \rangle$ for $t \in [0, 4\pi]$, we have $\mathbf{r}'(t) = \langle \sin t + t \cos t, \cos t - t \sin t \rangle$, so that

$$\|\mathbf{r}'(t)\| = \sqrt{(\sin t + t \cos t)^2 + (\cos t - t \sin t)^2}$$

$$= \sqrt{\sin^2 t + 2t \sin t \cos t + t^2 \cos^2 t + \cos^2 t - 2t \sin t \cos t + t^2 \sin^2 t}$$

$$= \sqrt{t^2 + 1}.$$

On $\mathbf{r}(t)$, we have

$$f(x, y) = (x^2 + y^2)^{1/2} = \sqrt{t^2 (\sin^2 t + \cos^2 t)} = t,$$

where we are justified in setting the square root to t since the range of t is nonnegative. Then the line integral is

$$\int_0^{4\pi} t\sqrt{t^2+1}\,dt = \left[\frac{1}{3}(t^2+1)^{3/2}\right]_0^{4\pi} = \frac{1}{3}\left((16\pi^2+1)^{3/2}-1\right).$$

49. Solving the equation of the plane for z gives $z = \frac{1}{2}x - \frac{1}{2}y + 5$, so the intersection of the surface and the plane is the curve $\sqrt{x^2+y^2} = \frac{1}{2}x - \frac{1}{2}y + 5$. Square both sides and simplify to get

$$\frac{3}{4}x^2 + \frac{3}{4}y^2 = 5x - 5y - \frac{xy}{2} + 25.$$

A plot of this curve is

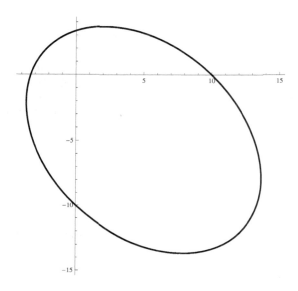

It is a closed curve (in fact, it is an ellipse). Now, the vector field is conservative, since

$$\frac{\partial F_1}{\partial y} = ze^{xyz} + xyz^2e^{xyz}, \qquad \frac{\partial F_2}{\partial x} = ze^{xyz} + xyz^2e^{xyz},$$

$$\frac{\partial F_1}{\partial z} = ye^{xyz} + xy^2ze^{xyz}, \qquad \frac{\partial F_3}{\partial x} = ye^{xyz} + xy^2ze^{xyz},$$

$$\frac{\partial F_2}{\partial z} = xe^{xyz} + x^2yze^{xyz}, \qquad \frac{\partial F_3}{\partial y} = xe^{xyz} + x^2yze^{xyz}.$$

So by the Fundamental Theorem of Line Integrals, the integral of this vector field along a closed curve is zero.

Applications

51. On the unit circle, $\sqrt{x^2+y^2} = 1$, so that the density is $\rho(x,y,z) = \ln\left(e^2\sqrt{x^2+y^2}\right) = \ln e^2 = 2$, so the density is constant. Thus the mass of the wire is twice its length. The circumference of the unit circle is 2π, so the mass is $\frac{2\pi}{4}\cdot 2 = \pi$.

53. We have $\mathbf{r}'(t) = \langle 2, -1, t \rangle$, so that $\|\mathbf{r}'(t)\| = \sqrt{t^2 + 5}$. Then the mass of the wire is

$$\int_C \rho(x, y, z)\, ds = \int_0^1 \sqrt{2\left(\frac{t^2}{2}\right) + 5}\left((2t + 1) + 2(10 - t)\right)\sqrt{t^2 + 5}\, dt = \int_0^1 (t^2 + 5)(21)\, dt$$

$$= 21\left[\frac{1}{3}t^3 + 5t\right]_0^1 = 112.$$

55. We want to find the line integral $\int_C \mathbf{F}(x, y) \cdot d\mathbf{r}$. Parametrize C as $\mathbf{r}(t) = \langle 2\cos t, 2\sin t \rangle$ for $t \in [0, \frac{\pi}{2}]$, so that $\mathbf{r}'(t) = \langle -2\sin t, 2\cos t \rangle$. Then the work is

$$W = \int_0^{\pi/2} \langle 2\sin t, -2\cos t \rangle \cdot \langle -2\sin t, 2\cos t \rangle\, dt = \int_0^{\pi/2} (-4\sin^2 t - 4\cos^2 t)\, dt$$

$$= \int_0^{\pi/2} (-4)\, dt = -2\pi.$$

57. We have $\mathbf{r}'(t) = \langle 1, e^t, -e^{-t} \rangle$, so that the work is

$$W = \int_1^{\ln 2} \left\langle t\ln(e^t), e^{-t}, e^{-2t} \right\rangle \cdot \langle 1, e^t, -e^{-t} \rangle\, dt$$

$$= \int_1^{\ln 2} \left\langle t^2, e^{-t}, e^{-2t} \right\rangle \cdot \langle 1, e^t, e^{-t} \rangle\, dt$$

$$= \int_1^{\ln 2} (t^2 + 1 + e^{-3t})\, dt$$

$$= \left[\frac{1}{3}t^3 + t - \frac{1}{3}e^{-3t}\right]_1^{\ln 2}$$

$$= \frac{1}{3}(\ln 2)^3 + \ln 2 - \frac{1}{3} - 1 - \frac{1}{3}\left(e^{-3\ln 2} - e^{-3}\right)$$

$$= \frac{1}{3}(\ln 2)^3 + \ln 2 - \frac{4}{3} - \frac{1}{3}\left(\frac{1}{8} - e^{-3}\right)$$

$$= \frac{1}{3}(\ln 2)^3 + \ln 2 + \frac{1}{3e^3} - \frac{11}{8}.$$

59. Parametrize the path by $\mathbf{r}(x, y) = \langle \cos t, \sin t \rangle$ for $t \in [0, \frac{\pi}{2}]$. Then $\mathbf{r}'(t) = \langle -\sin t, \cos t \rangle$, so that the average value is

$$f_{avg}(C) = \frac{\int_0^{\pi/2} \sin^{-1}(\sin t)\cos^{-1}(\cos t) \cdot \|\mathbf{r}'(t)\|\, dt}{\int_0^{\pi/2} \|\mathbf{r}'(t)\|\, dt} = \frac{\int_0^{\pi/2} t^2\, dt}{\int_0^{\pi/2} 1\, dt} = \frac{2}{\pi}\left[\frac{1}{3}t^3\right]_0^{\pi/2}$$

$$= \frac{2}{\pi} \cdot \frac{\pi^3}{24} = \frac{\pi^2}{12}.$$

61. Since $\mathbf{n} = \langle x, y \rangle$, the integral is

$$\int_C \mathbf{F}(x, y) \cdot \mathbf{n}\, ds = \int_C \left\langle xe^{x^2 + y^2}, ye^{x^2 + y^2} \right\rangle \cdot \langle x, y \rangle\, ds = \int_C (x^2 + y^2)e^{x^2 + y^2}\, ds.$$

To perform this integration, use polar coordinates; then $x = \cos\theta$ and $y = \sin\theta$, so that $x^2 + y^2 = 1$. The factor of r in the volume element is 1 in this integration, since we are on a circle of radius 1. So we get for the integral

$$\int_C \mathbf{F}(x, y) \cdot \mathbf{n}\, ds = \int_0^{2\pi} e\, d\theta = 2\pi e.$$

63. (a) This is not a conservative vector field. Since its first coordinate is zero, and this would be the x derivative of a potential function, the potential function would have to be a function purely of y. But looking at the second coordinate, it clearly is not. Thus this is not a conservative field.

   (b) Using the given parametrization for her path, the current velocity along her path is found by substituting t (the parametrization for the x coordinate) for x in the current velocity vector, so that $V = \langle 0, 0.9t(1.2 - t) \rangle$. We have $\mathbf{r}'(t) = \langle 1, 0.24 - 0.4t \rangle$. Since her path starts at $t = 0$ and ends at $t = 1.2$, where her velocity again becomes zero, so the line integral is

$$\int_0^{1.2} \langle 0, 0.9t(1.2 - t) \rangle \cdot \langle 1, 0.24 - 0.4t \rangle \, dt = \int_0^{1.2} (1.08t - 0.9t^2)(0.24 - 0.4t) \, dt$$

$$= \int_0^{1.2} (0.36t^3 - 0.648t^2 + 0.2592t) \, dt$$

$$= \left[0.09t^4 - 0.216t^3 + 0.1296t^2 \right]_0^{1.2}$$

$$= 0.$$

   (c) Since the line integral is zero, Annie does not expend any work in making the crossing.

Proofs

65. Suppose $\mathbf{F}$ is conservative with potential function f, and let C be a smooth curve parametrized as $\mathbf{r}(t) = \langle x(t), y(t), z(t) \rangle$ for $t \in [a, b]$. Then by the Fundamental Theorem of Line Integrals,

$$\int_C \mathbf{F}(x, y, z) \cdot d\mathbf{r} = f(x(b), y(b), z(b)) - f(x(a), y(a), z(a)).$$

But the right-hand side depends only on the endpoints $(x(b), y(b), z(b))$ and $(x(a), y(a), z(a))$ of the path C, not on the path itself.

Thinking Forward

Integration across a surface

▷ For example, suppose the surface is defined by $z = f(x, y)$ (this is a special case, but easier to understand than the general case, which is discussed in the next section). Then look at the projection of the surface in the xy plane and break it up into a number of subrectangles. For each subrectangle, approximate the area of the surface over that subrectangle by the area of the quadrilateral whose vertices are the points of the surface lying over the vertices of the subrectangle. Adding all of these up approximates the area of the surface; letting the number of subrectangles go to infinity gives us a Riemann sum whose value is the surface area.

▷ The integral of the function on the surface involves computing a Riemann sum similar to that above, but instead of adding up the areas of each subrectangle, we add up the areas multiplied by the value of the given function f at some arbitrary point in each subrectangle. This is analogous to the processes we used in the one and two-dimensional cases. Dividing by the area of the surface gives the average value of the function over the surface.

▷ The process of integration along a curve involved choosing a vector associated with the curve, $\mathbf{r}'(t)$, and integrating its dot product with the vector field along the range of parametrization of the curve. A natural vector to associate with any point on its surface is its unit normal, $\mathbf{n}$. So we might expect $\int_S \mathbf{F} \cdot \mathbf{n} \, ds$ to be a useful concept; it turns out to be the *flux* of the vector field through the surface.

14.3   Surfaces and Surface Integrals

Thinking Back

Area

The curve $x = \sqrt{y}$ is the curve $y = x^2$, so these two curves are the same and the area bounded by them is zero.

Double integrals

The x axis, which is $y = 0$, intersects $y = 4 - x^2$ at $x = \pm 2$. So the integral is

$$
\begin{aligned}
\int_{-2}^{2} \int_{0}^{4-x^2} (x^2 - y^2) \, dy \, dx &= \int_{-2}^{2} \left[x^2 y - \frac{1}{3} y^3 \right]_{y=0}^{y=4-x^2} dx \\
&= \int_{-2}^{2} \left(x^2 (4 - x^2) - \frac{1}{3}(4 - x^2)^3 \right) dx \\
&= \int_{-2}^{2} \left(-x^4 + 4x^2 + \frac{1}{3} x^6 - 4x^4 + 16x^2 - \frac{64}{3} \right) dx \\
&= \int_{-2}^{2} \left(\frac{1}{3} x^6 - 5x^4 + 20x^2 - \frac{64}{3} \right) dx \\
&= \left[\frac{1}{21} x^7 - x^5 + \frac{20}{3} x^3 - \frac{64}{3} x \right]_{-2}^{2} \\
&= -\frac{640}{21}.
\end{aligned}
$$

Average value

In the first quadrant the lines $y = x$ and $y = x^2$ meet at $(0,0)$ and $(1,1)$. Thus

$$
\iint_{\Omega} g(x,y) \, dA = \int_{0}^{1} \int_{x^2}^{x} xy \, dy \, dx = \int_{0}^{1} \left[\frac{1}{2} xy^2 \right]_{y=x^2}^{y=x} dx = \frac{1}{2} \int_{0}^{1} (x^3 - x^5) \, dx = \frac{1}{2} \left[\frac{1}{4} x^4 - \frac{1}{6} x^6 \right]_{0}^{1} = \frac{1}{24}
$$

$$
\iint_{\Omega} dA = \int_{0}^{1} \int_{x^2}^{x} dy \, dx = \int_{0}^{1} [y]_{y=x^2}^{y=x} dx = \int_{0}^{1} (x - x^2) \, dx = \left[\frac{1}{2} x^2 - \frac{1}{3} x^3 \right]_{0}^{1} = \frac{1}{6}.
$$

Thus the average value is

$$
\frac{\iint_{\Omega} g(x,y) \, dA}{\iint_{\Omega} dA} = \frac{1/24}{1/6} = \frac{1}{4}.
$$

Concepts

1. (a) False. Just as in the case of integration over curves, the result is a scalar. See Definition 14.8 and the discussion.

    (b) True. By Definition 14.7, this integral is defined to be a double integral of the function over a particular domain defined by the parametrization of the surface, so it is a scalar.

    (c) True. By Definition 14.6(b), since the xy plane is defined by $z = 0$, we have

$$
dS = \sqrt{ \left(\frac{\partial z}{\partial x} \right)^2 + \left(\frac{\partial z}{\partial y} \right)^2 + 1 } \, dA = 1 \, dA = dA.
$$

(d) True. The integral is given, by Definition 14.7, by an integral involving only the function itself and the parametrization of the surface, not the normal to the surface.

(e) False. This formula has nothing to do with flow. It is the integral of the function over the surface.

(f) True. See Definition 14.8.

(g) False. Changing the direction of the normal vector multiplies the result by -1; see Definition 14.8 and the discussion following.

(h) True. Since $dS = \|\mathbf{r}_u \times \mathbf{r}_v\| \, dA$ and $\mathbf{n} = \pm \frac{\mathbf{r}_u \times \mathbf{r}_v}{\|\mathbf{r}_u \times \mathbf{r}_v\|}$, the two copies of $\|\mathbf{r}_u \times \mathbf{r}_v\|$ will cancel.

3. The extra term $\|\mathbf{r}_u \times \mathbf{r}_v\|$ in a surface integral of a multivariable function is there to account for the distortion of the surface area compared to the surface area in the parameter plane (the uv plane) — essentially, this is a kind of Jacobian that modifies the area element. When integrating a vector field, however, we are trying to understand the action of $\mathbf{F}(x, y, z)$ through the surface, and motion through the surface is defined by motion in a direction normal to the surface.

5. The plane $ax + by + cz = k$ when $c \neq 0$ may be written $z = \frac{k - ax - by}{c}$, so that by Definition 14.6(b),

$$dS = \sqrt{\left(\frac{\partial z}{\partial x}\right)^2 + \left(\frac{\partial z}{\partial y}\right)^2 + 1}\, dA = \sqrt{\frac{a^2}{c^2} + \frac{b^2}{c^2} + 1}\, dA = \frac{\sqrt{a^2 + b^2 + c^2}}{|c|}\, dA.$$

7. Since the upper half of the unit sphere is the surface given by $x^2 + y^2 + z^2 = 1$ for $z \geq 0$, we can parametrize the upper hemisphere as $\mathbf{r}(x, y) = \left\langle x, y, \sqrt{1 - x^2 - y^2}\right\rangle$ for (x, y) in the unit disk $\{(x, y) \mid x^2 + y^2 \leq 1\}$. Note that this parametrization is smooth since

$$\mathbf{r}_x \times \mathbf{r}_y = \left\langle 1, 0, -\frac{x}{\sqrt{1 - x^2 - y^2}}\right\rangle \times \left\langle 0, 1, -\frac{y}{\sqrt{1 - x^2 - y^2}}\right\rangle$$

$$= \left\langle \frac{x}{\sqrt{1 - x^2 - y^2}}, \frac{y}{\sqrt{1 - x^2 - y^2}}, 1\right\rangle,$$

which is never the zero vector since the third coordinate is 1.

9. Use spherical coordinates for the parametrization. Then $\rho = k$ since the sphere has radius k. Write u for θ and v for ϕ; then the sphere is parametrized by

$$\mathbf{r}(u, v) = \langle k \cos u \sin v, k \sin u \sin v, k \cos v\rangle, \quad 0 \leq u \leq 2\pi, \; 0 \leq v \leq \pi.$$

11. Parametrize the curve as $\mathbf{r}(x, y) = \langle x, y, f(x, y)\rangle$. Then

$$\mathbf{n} = \pm \frac{\mathbf{r}_x \times \mathbf{r}_y}{\|\mathbf{r}_x \times \mathbf{r}_y\|}.$$

Now,

$$\mathbf{r}_x = \langle 1, 0, f_x\rangle, \qquad \mathbf{r}_y = \langle 0, 1, f_y\rangle,$$

so that

$$\mathbf{r}_x \times \mathbf{r}_y = \langle -f_x, -f_y, 1\rangle, \qquad \|\mathbf{r}_x \times \mathbf{r}_y\| = \sqrt{f_x^2 + f_y^2 + 1}.$$

Then since we want the normal to be upwards-pointing, we choose $\langle -f_x, -f_y, 1\rangle$ rather than its negation, $\langle f_x, f_y, -1\rangle$, which is downwards-pointing. Thus the upwards-pointing unit normal is

$$\mathbf{n} = \frac{1}{\sqrt{f_x^2 + f_y^2 + 1}} \langle -f_x, -f_y, 1\rangle.$$

13. We have

$$\mathbf{r}_u \times \mathbf{r}_v = \langle 1, 2u, 0 \rangle \times \langle 0, 0, 1 \rangle = \langle 2u, -1, 0 \rangle,$$

so that $\|\mathbf{r}_u \times \mathbf{r}_v\| = \sqrt{4u^2 + 1}$. Thus

$$\mathbf{n} = \pm \frac{1}{\sqrt{4u^2 + 1}} \langle 2u, -1, 0 \rangle,$$

where the sign may be chosen to produce the desired orientation.

15. Analogously with the definition in the one and two-dimensional cases, the average value should be the integral of the function over the surface, divided by the area of the surface:

$$f_{\text{avg}} = \frac{\int_S f(x, y, z) \, dS}{\int_S 1 \, dS}.$$

17. If the parametrization is defined by a function, say $\mathbf{r}(x, y) = \langle x, y, z(x, y) \rangle$, then the area of the part of the surface lying over the nth subrectangle in some decomposition of the surface is approximated by the area of a parallelogram with sides $\langle 1, 0, z_x(x, y) \rangle$ and $\langle 0, 1, z_y(x, y) \rangle$, which are simply $\mathbf{r}_x(x, y)$ and $\mathbf{r}_y(x, y)$. The area of this parallelogram is the norm of the cross product of these vectors, so it is $\|\mathbf{r}_x(x, y) \times \mathbf{r}_y(x, y)\|$. For a general surface, parametrized by $\mathbf{r}(u, v) = \langle x(u, v), y(u, v), z(u, v) \rangle$, we generalize this formula and say that the volume element is modified by the factor $\|\mathbf{r}_u \times \mathbf{r}_v\|$. This can also be justified geometrically — note that both $\mathbf{r}_u$ and $\mathbf{r}_v$ lie in the plane tangent to the surface at $\mathbf{r}(u, v)$, so again this norm is the area of a parallelogram that approximates the area of the surface near the given point.

19. Parametrize the surface by $\mathbf{r}(y, z) = \langle f(y, z), y, z \rangle$. Then

$$\mathbf{r}_y \times \mathbf{r}_z = \langle f_y, 1, 0 \rangle \times \langle f_z, 0, 1 \rangle = \langle 1, -f_y, -f_z \rangle,$$
$$\|\mathbf{r}_y \times \mathbf{r}_z\| = \sqrt{f_y^2 + f_z^2 + 1}.$$

Thus a normal vector is

$$\mathbf{n} = \pm \frac{1}{\sqrt{f_y^2 + f_z^2 + 1}} \langle 1, -f_y, -f_z \rangle.$$

Skills

21. With $z = y + \frac{\pi}{2}$ and $D = [0, 4] \times [3, 6]$, Definition 14.6 gives for the area

$$\int_S 1 \, dS = \iint_D \sqrt{\left(\frac{\partial z}{\partial x}\right)^2 + \left(\frac{\partial z}{\partial y}\right)^2 + 1} \, dA = \int_0^4 \int_3^6 \sqrt{2} \, dy \, dx = 12\sqrt{2}.$$

23. We have

$$\int_S 1 \, dS = \iint_D \sqrt{\left(\frac{\partial z}{\partial x}\right)^2 + \left(\frac{\partial z}{\partial y}\right)^2 + 1} \, dA = \iint_D \sqrt{4x^2 + 4y^2 + 1} \, dA.$$

Using cylindrical coordinates, we have $4x^2 + 4y^2 + 1 = 4r^2 + 1$, and the region D is given by

$\theta \in [0, 2\pi]$ and $r \in \left[\frac{\sqrt{3}}{2}, \sqrt{2}\right]$. Then $dA = r\,dr\,d\theta$, so the area is

$$\iint_D \sqrt{4x^2 + 4y^2 + 1}\,dA = \int_0^{2\pi} \int_{\sqrt{3}/2}^{\sqrt{2}} r\sqrt{4r^2 + 1}\,dr\,d\theta$$

$$= \int_0^{2\pi} \left[\frac{1}{12}(4r^2 + 1)^{3/2}\right]_{r=\sqrt{3}/2}^{r=\sqrt{2}} d\theta$$

$$= \int_0^{2\pi} \left(\frac{1}{12}(27 - 8)\right) d\theta$$

$$= \frac{19\pi}{6}.$$

25. We have

$$\|\mathbf{r}_u \times \mathbf{r}_v\| = \|\langle 3, 1, -1\rangle \times \langle -1, 1, 1\rangle\| = \|\langle 2, -2, 4\rangle\| = \sqrt{4 + 4 + 16} = 2\sqrt{6}.$$

Then the area is

$$\int_S 1\,dS = \iint_D \|\mathbf{r}_u \times \mathbf{r}_v\|\,dA = \int_0^1 \int_0^1 2\sqrt{6}\,dv\,du = 2\sqrt{6}.$$

27. Parametrize the surface by $\mathbf{r}(x, z) = \langle x, 2, z\rangle$ for $x^2 + z^2 \leq 1$. Then

$$\|\mathbf{r}_x \times \mathbf{r}_z\| = \|\langle 1, 0, 0\rangle \times \langle 0, 0, 1\rangle\| = \|\langle 0, -1, 0\rangle\| = 1.$$

Thus the integral is

$$\int_S f(x, z)\,dS = \iint_D f(x, z)\,dA = \iint_D e^{-(x^2 + z^2)}\,dA.$$

Use cylindrical coordinates in the y direction, so that $r^2 = x^2 + z^2$ and $y = y$. Then the region of integration is $\theta \in [0, 2\pi]$ and $r \in [0, 1]$, so the integral is

$$\int_0^{2\pi} \int_0^1 re^{-r^2}\,dr\,d\theta = \int_0^{2\pi} \left[-\frac{1}{2}e^{-r^2}\right]_0^1 d\theta = \int_0^{2\pi} \frac{1}{2}\left(1 - e^{-1}\right) d\theta = \pi\left(1 - e^{-1}\right).$$

29. On the cone, we have $f(x, y, z) = xyz^2 = xy\left(\frac{\sqrt{x^2 + y^2}}{\sqrt{3}}\right)^2 = \frac{1}{3}xy(x^2 + y^2)$. The cone, which may be written $3z^2 = x^2 + y^2$, intersects the sphere of radius 4, which is $x^2 + y^2 + z^2 = 16$, where $3z^2 + z^2 = 16$, so $z = \pm 2$. Since the cone has $z \geq 0$, we choose $z = 2$ as the only solution. When $z = 2$, the intersection is $2\sqrt{3} = \sqrt{x^2 + y^2}$, or the circle $x^2 + y^2 = 12$. Finally, with $z = \frac{1}{\sqrt{3}}(x^2 + y^2)^{1/2}$, we get $\frac{\partial z}{\partial x} = \frac{x}{\sqrt{3}(x^2 + y^2)^{1/2}}$ and $\frac{\partial z}{\partial y} = \frac{y}{\sqrt{3}(x^2 + y^2)^{1/2}}$. Then the integral is

$$\iint_S f(x, y, z)\,dS = \iint_D f(x, y, z)\sqrt{\left(\frac{\partial z}{\partial x}\right)^2 + \left(\frac{\partial z}{\partial y}\right)^2 + 1}\,dA$$

$$= \iint_D f(x, y, z)\sqrt{\frac{x^2}{3(x^2 + y^2)} + \frac{y^2}{3(x^2 + y^2)} + 1}\,dA$$

$$= \iint_D \frac{1}{3}xy(x^2 + y^2)\sqrt{\frac{4x^2 + 4y^2}{3x^2 + 3y^2}}\,dA$$

$$= \frac{2}{3\sqrt{3}} \iint_D xy(x^2 + y^2)\,dA$$

Now convert to cylindrical coordinates; the integral becomes

$$
\begin{aligned}
\iint_S f(x,y,z)\,dS &= \frac{2}{3\sqrt{3}} \iint_D xy(x^2+y^2)\,dA \\
&= \frac{2}{3\sqrt{3}} \int_0^{\sqrt{12}} \int_0^{2\pi} r^4 \sin\theta\cos\theta \cdot r\,d\theta\,dr \\
&= \frac{2}{3\sqrt{3}} \int_0^{\sqrt{12}} \int_0^{2\pi} r^5 \sin\theta\cos\theta\,d\theta\,dr \\
&= \frac{2}{3\sqrt{3}} \int_0^{\sqrt{12}} \left[\frac{1}{2} r^5 \sin^2\theta \right]_{\theta=0}^{\theta=2\pi} dr \\
&= \frac{2}{3\sqrt{3}} \int_0^{\sqrt{12}} 0\,dr = 0.
\end{aligned}
$$

(Note that the surface is symmetric around the z axis, and the function is "odd" in the sense that it is negative for (x,y) in the second and fourth quadrants, and positive in the first and third, and that $f(x,y,z) = -f(-x,y,z)$ and so on, so that everything cancels).

31. On the paraboloid, we have $f(x,y,z) = \frac{y}{x}\sqrt{4z+1} = \frac{y}{x}\sqrt{4(x^2+y^2)+1}$. Then

$$
\begin{aligned}
\int_S f(x,y,z)\,dS &= \iint_D f(x,y,z)\sqrt{\left(\frac{\partial z}{\partial x}\right)^2 + \left(\frac{\partial z}{\partial y}\right)^2 + 1}\,dA \\
&= \int_1^e \int_0^2 \frac{y}{x}\sqrt{4x^2+4y^2+1}\sqrt{4(x^2+y^2)+1}\,dy\,dx \\
&= \int_1^e \int_0^2 \frac{y}{x}(4x^2+4y^2+1)\,dy\,dx \\
&= \int_1^e \int_0^2 \left(4xy + \frac{4y^3}{x} + \frac{y}{x} \right) dy\,dx \\
&= \int_1^e \left[2xy^2 + \frac{y^4}{x} + \frac{y^2}{2x} \right]_{y=0}^{y=2} dx \\
&= \int_1^e \left(8x + \frac{16}{x} + \frac{2}{x} \right) dx \\
&= \left[4x^2 + 18\ln|x| \right]_1^e = 4e^2 + 18 - 4 = 4e^2 + 14.
\end{aligned}
$$

33. The region in the uv plane is the Type I region bounded by $u = 0$ and $u = 1$, and by $v = u^2$ and $v = u$. In uv coordinates, we get

$$
f(x,y,z) = x - z + y^2 = (u+v) - (u-v) + 4(u^2+v^2) = 4u^2 + 4v^2 + 2v.
$$

Finally, for the volume element we have

$$
\begin{aligned}
\|\mathbf{r}_u \times \mathbf{r}_v\| &= \left\| \left\langle 1, 2u(u^2+v^2)^{-1/2}, 1 \right\rangle \times \left\langle 1, 2v(u^2+v^2)^{-1/2}, -1 \right\rangle \right\| \\
&= \left\| \left\langle -2(u+v)(u^2+v^2)^{-1/2}, 2, 2(v-u)(u^2+v^2)^{-1/2} \right\rangle \right\| \\
&= \sqrt{4\frac{(u+v)^2}{u^2+v^2} + 4 + 4\frac{(v-u)^2}{u^2+v^2}} \\
&= 2\sqrt{\frac{2u^2 + 2v^2}{u^2+v^2} + 1} = 2\sqrt{3}.
\end{aligned}
$$

Then the integral is

$$\int_S f(x,y,z)\,dS = 2\sqrt{3}\int_0^1\int_{u^2}^u (4u^2 + 4v^2 + 2v)\,dv\,du$$

$$= 2\sqrt{3}\int_0^1\left[4u^2v + \frac{4}{3}v^3 + v^2\right]_{v=u^2}^{v=u}\,du$$

$$= 2\sqrt{3}\int_0^1\left(4u^3 + \frac{4}{3}u^3 + u^2 - 4u^4 - \frac{4}{3}u^6 - u^4\right)\,du$$

$$= 2\sqrt{3}\int_0^1\left(\frac{16}{3}u^3 + u^2 - 5u^4 - \frac{4}{3}u^6\right)\,du$$

$$= 2\sqrt{3}\left[\frac{4}{3}u^4 + \frac{1}{3}u^3 - u^5 - \frac{4}{21}u^7\right]_0^1$$

$$= 2\sqrt{3}\cdot\frac{10}{21} = \frac{20}{21}\sqrt{3}.$$

35. Since $z = y^3 - y^2$, then from Chapter 12, or simply by using the parametrization

$$\mathbf{r}(x,y) = \left\langle x, y, y^3 - y^2\right\rangle,$$

we see that

$$\left\langle 1, 0, \frac{\partial z}{\partial x}\right\rangle \times \left\langle 0, 1, \frac{\partial z}{\partial y}\right\rangle = \left\langle -\frac{\partial z}{\partial x}, -\frac{\partial z}{\partial y}, 1\right\rangle = \left\langle 0, 2y - 3y^2, 1\right\rangle$$

is normal to the surface. Since it has a positive z component, we simply scale it to make it a unit vector:

$$\mathbf{n} = \frac{1}{\sqrt{1 + (2y - 3y^2)^2}}\left\langle 0, 2y - 3y^2, 1\right\rangle.$$

We also have for the volume element

$$\sqrt{\left(\frac{\partial z}{\partial x}\right)^2 + \left(\frac{\partial z}{\partial y}\right)^2 + 1} = \sqrt{1 + (2y - 3y^2)^2}.$$

Finally, on the surface, $\mathbf{F}(x,y,z) = \langle\cos(xyz), 1, -yz\rangle = \langle\cos(xy(y^3 - y^2)), 1, y^3 - y^4\rangle$. Thus the flux is

$$\int_S \mathbf{F}(x,y,z)\cdot\mathbf{n}\,dS = \int_{-3}^2\int_{-1}^1 \left\langle\cos(xy(y^3 - y^2)), 1, y^3 - y^4\right\rangle\cdot\frac{1}{\sqrt{1 + (2y - 3y^2)^2}}\left\langle 0, 2y - 3y^2, 1\right\rangle$$

$$\cdot\sqrt{1 + (2y - 3y^2)^2}\,dy\,dx$$

$$= \int_{-3}^2\int_{-1}^1 \left(2y - 3y^2 + y^3 - y^4\right)\,dy\,dx$$

$$= \int_{-3}^2\left[y^2 - y^3 + \frac{1}{4}y^4 - \frac{1}{5}y^5\right]_{y=-1}^{y=1}\,dx$$

$$= \int_{-3}^2\left(-\frac{12}{5}\right)\,dx = -12.$$

37. A normal vector to the surface is

$$\left\langle -\frac{\partial z}{\partial x}, -\frac{\partial z}{\partial y}, 1\right\rangle = \left\langle -x(x^2 + y^2)^{-1/2}, -y(x^2 + y^2)^{-1/2}, 1\right\rangle.$$

This points upwards, which is into the cone, so we negate and scale it to make it a unit vector:

$$\mathbf{n} = \frac{1}{\sqrt{x^2(x^2+y^2)^{-1} + y^2(x^2+y^2)^{-1} + 1}} \left\langle x(x^2+y^2)^{-1/2}, y(x^2+y^2)^{-1/2}, -1 \right\rangle$$
$$= \frac{1}{\sqrt{2}} \left\langle x(x^2+y^2)^{-1/2}, y(x^2+y^2)^{-1/2}, -1 \right\rangle.$$

We have for the volume element

$$\sqrt{\left(\frac{\partial z}{\partial x}\right)^2 + \left(\frac{\partial z}{\partial y}\right)^2 + 1} = \sqrt{x^2(x^2+y^2)^{-1} + y^2(x^2+y^2)^{-1} + 1} = \sqrt{2}.$$

On the surface, $\mathbf{F}(x,y,z) = \left\langle -xz, -yz, z^2 \right\rangle = \left\langle -x\sqrt{x^2+y^2}, -y\sqrt{x^2+y^2}, x^2+y^2 \right\rangle$. Thus the flux is

$$\int_S \mathbf{F}(x,y,z) \cdot \mathbf{n}\, dS = \iint_D \left(\left\langle -x\sqrt{x^2+y^2}, -y\sqrt{x^2+y^2}, x^2+y^2 \right\rangle \right.$$
$$\left. \cdot \left\langle x(x^2+y^2)^{-1/2}, y(x^2+y^2)^{-1/2}, -1 \right\rangle \right) dA$$
$$= \iint_D \left(-x^2 - y^2 - x^2 - y^2 \right) dA = -2 \iint_D (x^2+y^2)\, dA.$$

Convert to polar coordinates. Then D is defined by $2 \le r \le 4$ and $\theta \in [0, 2\pi]$, so we get

$$-2 \int_0^{2\pi} \int_2^4 r \cdot r^2\, dr\, d\theta = -2 \int_0^{2\pi} \int_2^4 r^3\, dr\, d\theta$$
$$= -2 \int_0^{2\pi} \left[\frac{1}{4} r^4 \right]_{r=2}^{r=4} d\theta$$
$$= -2 \int_0^{2\pi} 60\, d\theta = -240\pi.$$

39. Solving the equation of the surface for x gives $x = 4y + 5z + 21$, so that a normal is

$$\left\langle 1, -\frac{\partial x}{\partial y}, -\frac{\partial x}{\partial z} \right\rangle = \langle 1, -4, -5 \rangle.$$

Assuming we want an upward-pointing normal, we get

$$\mathbf{n} = \frac{1}{\sqrt{25 + 16 + 1}} \langle -1, 4, 5 \rangle = \frac{1}{\sqrt{42}} \langle -1, 4, 5 \rangle.$$

The volume element is

$$\sqrt{\left(\frac{\partial x}{\partial y}\right)^2 + \left(\frac{\partial x}{\partial z}\right)^2 + 1} = \sqrt{42}.$$

The flux is then

$$
\begin{aligned}
\int_S \mathbf{F}(x,y,z) \cdot \mathbf{n}\, dS &= \int_0^\pi \int_{-\pi}^\pi \langle 0, z\cos(yz), z\sin(yz)\rangle \cdot \langle -1,4,5\rangle \; du\, dz \\
&= \int_0^\pi \int_{-\pi}^\pi (4z\cos(yz) + 5z\sin(yz))\, dy\, dz \\
&= \int_0^\pi [4\sin(yz) - 5\cos(yz)]_{z=-\pi}^{z=\pi}\, dz \\
&= \int_0^\pi (4\sin \pi z - 5\cos \pi z - 4\sin(-\pi z) + 5\cos(-\pi z))\, dz \\
&= \int_0^\pi 8\sin \pi z\, dz \\
&= \left[-\frac{8}{\pi}\cos \pi z \right]_0^\pi = \frac{8 - 8\cos \pi^2}{\pi}.
\end{aligned}
$$

41. Parametrize the unit sphere using spherical coordinates, i.e., $\mathbf{r}(u,v) = \langle \cos u \sin v, \sin u \sin v, \cos v\rangle$ for $u \in [0,2\pi]$ and $v \in [0,\pi]$. Then as usual, the volume element is $\rho^2 \sin v = \sin v$, since $\rho = 1$, and a normal is

$$
\begin{aligned}
\mathbf{r}_u \times \mathbf{r}_v &= \langle -\sin u \sin v, \cos u \sin v, 0\rangle \times \langle \cos u \cos v, \sin u \cos v, -\sin v\rangle \\
&= \left\langle -\cos u \sin^2 v, -\sin u \sin^2 v, -\sin v \cos v \right\rangle.
\end{aligned}
$$

For $u = 0$ and $v = \frac{\pi}{2}$, i.e., the point $(1,0,0)$, this is $\langle -1,0,0\rangle$, so this normal points inwards and we must negate it. Its norm is $\sqrt{\sin^2 v} = \sin v$, which cancels with the volume element. Thus the flux is

$$
\begin{aligned}
\int_S \mathbf{F}(x,y,z) \cdot \mathbf{n}\, dS &= \iint_D \langle 5,13,2\rangle \cdot \left\langle \cos u \sin^2 v, \sin u \sin^2 v, \sin v \cos v \right\rangle dA \\
&= \int_0^\pi \int_0^{2\pi} (5\cos u \sin^2 v + 13\sin u \sin^2 v + 2\sin v \cos v)\, du\, dv \\
&= \int_0^\pi \left[5\sin u \sin^2 v - 13\cos u \sin^2 v + 2u\sin v \cos v \right]_{u=0}^{u=2\pi} dv \\
&= \int_0^\pi 4\pi \sin v \cos v\, dv \\
&= \left[2\pi \sin^2 v \right]_0^\pi = 0.
\end{aligned}
$$

43. The region in the xy plane is a Type I region bounded by $x = 0$ and $x = 3$, and by $y = 0$ and $y = 2x$. The volume element for the surface is

$$
\sqrt{\left(\frac{\partial z}{\partial x}\right)^2 + \left(\frac{\partial z}{\partial y}\right)^2 + 1} = \sqrt{4x^2 + 3 + 1} = 2\sqrt{x^2 + 1}.
$$

Thus the integral, which is the area of this portion of the surface, is

$$
\begin{aligned}
\int_S 1\, dS &= \int_0^3 \int_0^{2x} 2\sqrt{x^2 + 1}\, dy\, dx = \int_0^3 \left[2y\sqrt{x^2+1} \right]_{y=0}^{y=2x} dx \\
&= \int_0^3 4x\sqrt{x^2+1}\, dx = \left[\frac{4}{3}(x^2+1)^{3/2} \right]_0^3 = \frac{4}{3}(10^{3/2} - 1).
\end{aligned}
$$

45. On the unit sphere, $z = \sqrt{1 - x^2 - y^2}$, so that

$$f(x, y, z) = z^3 + z(x^2 + 2^y) = (1 - x^2 - y^2)^{3/2} + (x^2 + 2^y)\sqrt{1 - x^2 - y^2}$$
$$= (1 - y^2 + 2^y)\sqrt{1 - x^2 - y^2}.$$

We have for the volume element

$$\sqrt{\left(\frac{\partial z}{\partial x}\right)^2 + \left(\frac{\partial z}{\partial y}\right)^2 + 1} = \sqrt{\frac{x^2}{1 - x^2 - y^2} + \frac{y^2}{1 - x^2 - y^2} + 1} = \frac{1}{\sqrt{1 - x^2 - y^2}}.$$

Thus the integral is

$$\int_S f(x, y, z)\, dS = \int_0^{1/2} \int_0^{1/3} (1 - y^2 + 2^y)\sqrt{1 - x^2 - y^2} \cdot \frac{1}{\sqrt{1 - x^2 - y^2}}\, dy\, dx$$
$$= \int_0^{1/2} \int_0^{1/3} (1 - y^2 + 2^y)\, dy\, dx$$
$$= \int_0^{1/2} \left[y - \frac{1}{3}y^3 + \frac{1}{\ln 2}2^y \right]_{y=0}^{y=1/3} dx$$
$$= \int_0^{1/2} \left(\frac{1}{3} - \frac{1}{81} + \frac{1}{\ln 2}2^{1/3} - \frac{1}{\ln 2} \right) dx$$
$$= \frac{1}{2}\left(\frac{26}{81} + \frac{1}{\ln 2}(2^{1/3} - 1) \right).$$

47. Parametrize the surface S using cylindrical coordinates: $\mathbf{r}(r, \theta) = \left\langle r\cos\theta, r\sin\theta, -\sqrt{r^2 - 9} \right\rangle$ (we use the negative square root since we are interested in values of z below the xy plane). The plane $z = -4$ corresponds to $x^2 + y^2 - 9 = 16$, which is a circle of radius 5, while $z = 0$ corresponds to $x^2 + y^2 - 9 = 0$, which is a circle of radius 3. Thus the portion of the hyperboloid we are interested in is $r \in [3, 5]$, $\theta \in [0, 2\pi]$. A normal vector is

$$\mathbf{r}_r \times \mathbf{r}_\theta = \left\langle \cos\theta, \sin\theta, -r(r^2 - 9)^{-1/2} \right\rangle \times \left\langle -r\sin\theta, r\cos\theta, 0 \right\rangle$$
$$= \left\langle r^2(r^2 - 9)^{-1/2}\cos\theta, r^2(r^2 - 9)^{-1/2}\sin\theta, r \right\rangle.$$

The point $(3, 0, 0)$ on the hyperboloid corresponds to $r = 3$ and $\theta = 0$; there, the normal is $\langle 0, 0, 3 \rangle$, which points outwards, so this is the desired orientation. As usual, the volume element multiplier and the norm of the normal vector cancel. Finally, with the given parametrization,

$$\mathbf{F}(x, y, z) = \langle 2xz, 2yz, -18 \rangle = \left\langle -2r\cos\theta\sqrt{r^2 - 9}, -2r\sin\theta\sqrt{r^2 - 9}, -18 \right\rangle,$$

so we get for the flux

$$\int_S \mathbf{F}(x,y,z) \cdot \mathbf{n}\, dS = \int_0^{2\pi} \int_3^5 \left\langle -2r\cos\theta\sqrt{r^2-9}, -2r\sin\theta\sqrt{r^2-9}, -18 \right\rangle \cdot$$

$$\left\langle r^2(r^2-9)^{-1/2}\cos\theta, r^2(r^2-9)^{-1/2}\sin\theta, r \right\rangle dr\, d\theta$$

$$= \int_0^{2\pi} \int_3^5 \left(-2r^3(\sin^2\theta + \cos^2\theta) - 18r \right) dr\, d\theta$$

$$= \int_0^{2\pi} \int_3^5 \left(-2r^3 - 18r \right) dr\, d\theta$$

$$= \int_0^{2\pi} \left[-\frac{1}{2}r^4 - 9r^2 \right]_3^5 d\theta$$

$$= \int_0^{2\pi} (-416)\, d\theta = -832\pi.$$

Applications

49. If the density is ρ, then the mass is ρ times the surface area of the lamina. Using cylindrical coordinates, $x = r\cos\theta$ and $y = r\sin\theta$, so the lamina is parametrized by

$$\mathbf{r}(r,\theta) = \left\langle r\cos\theta, r\sin\theta, r^2(\cos^2\theta - \sin^2\theta) \right\rangle = \left\langle r\cos\theta, r\sin\theta, r^2\cos 2\theta \right\rangle.$$

We have (using identities for the sine and cosine of the difference of two angles)

$$\mathbf{r}_r = \left\langle \cos\theta, \sin\theta, 2r\cos 2\theta \right\rangle, \qquad \mathbf{r}_\theta = \left\langle -r\sin\theta, r\cos\theta, -2r^2\sin 2\theta \right\rangle,$$

$$\mathbf{r}_r \times \mathbf{r}_\theta = \left\langle -2r^2\cos\theta, 2r^2\sin\theta, r \right\rangle.$$

Then by Definition 14.6, the mass is

$$SA = \rho \int_S 1\, dS = \int_0^2 \int_0^{2\pi} \|\mathbf{r}_r \times \mathbf{r}_\theta\|\, d\theta\, dr$$

$$= \rho \int_0^2 \int_0^{2\pi} \sqrt{4r^4\cos^2\theta + 4r^4\sin^2\theta + r^2}\, d\theta\, dr$$

$$= \rho \int_0^2 \int_0^{2\pi} \sqrt{4r^4 + r^2}\, d\theta\, dr$$

$$= \rho \int_0^2 \int_0^{2\pi} r\sqrt{4r^2 + 1}\, d\theta\, dr$$

$$= 2\pi\rho \int_0^2 r\sqrt{4r^2 + 1}\, dr$$

$$= 2\pi\rho \cdot \frac{1}{8} \int_{r=0}^{r=2} \sqrt{u}\, du$$

$$= \frac{\pi\rho}{4} \left[\frac{2}{3}u^{3/2} \right]_{r=0}^{r=2}$$

$$= \frac{\pi\rho}{6} \left[(4r^2 + 1)^{3/2} \right]_0^2$$

$$= \frac{\pi\rho}{6} \left(17^{3/2} - 1 \right).$$

51. A normal to $\mathcal{S}$ is

$$\left\langle -\frac{\partial z}{\partial x}, -\frac{\partial z}{\partial y}, 1 \right\rangle = \left\langle -8x, -2y, 1 \right\rangle.$$

Assuming that we want an upward pointing normal, this is the desired orientation. Then the flux is

$$\int_S \mathbf{F}(x, y, z) \cdot \mathbf{n} \, dS = \iint_D \langle y, x, 1 \rangle \cdot \langle -8x, -2y, 1 \rangle \, dA = \iint_D (1 - 10xy) \, dA.$$

Parametrize D using polar coordinates; the region of integration is $\theta \in [0, 2\pi]$ and $r \in [1, 2]$, so the flux is

$$
\begin{aligned}
\iint_D (1 - 10xy) \, dA &= \int_0^{2\pi} \int_1^2 (1 - 10r^2 \sin\theta \cos\theta) r \, dr \, d\theta \\
&= \int_0^{2\pi} \int_1^2 (r - 10r^3 \sin\theta \cos\theta) \, dr \, d\theta \\
&= \int_0^{2\pi} \left[\frac{1}{2} r^2 - \frac{5}{2} r^4 \sin\theta \cos\theta \right]_{r=1}^{r=2} d\theta \\
&= \int_0^{2\pi} \left(\frac{3}{2} - \frac{75}{2} \sin\theta \cos\theta \right) d\theta \\
&= \left[\frac{3}{2}\theta - \frac{75}{4} \sin^2\theta \right]_0^{2\pi} = 3\pi.
\end{aligned}
$$

53. The flux across the surface of the cube is the sum of the fluxes across each of the eight faces.

   - On the bottom face $z = 0$, the outward-pointing normal is $\mathbf{n} = -\mathbf{k} = \langle 0, 0, -1 \rangle$, and the field becomes $\mathbf{E} = \langle 2y, 2xy, 0 \rangle$. Then $\mathbf{E} \cdot \mathbf{n} = 0$, so the flux is zero across this face.

   - On the top face $z = 1$, the outward-pointing normal is $\mathbf{n} = \mathbf{k} = \langle 0, 0, 1 \rangle$, and the field becomes $\mathbf{E} = \langle 2y, 2xy, y \rangle$. Then $\mathbf{E} \cdot \mathbf{n} = y$, so the flux is

   $$\int_0^1 \int_0^1 y \, dy \, dx = \int_0^1 \left[\frac{1}{2} y^2 \right]_{y=0}^{y=1} dx = \int_0^1 \frac{1}{2} \, dx = \frac{1}{2}.$$

   - On the right face $x = 1$, the outward-pointing normal is $\mathbf{n} = \mathbf{i} = \langle 1, 0, 0 \rangle$, and the field becomes $\mathbf{E} = \langle 2y, 2y, yz \rangle$. Then $\mathbf{E} \cdot \mathbf{n} = 2y$, so the flux is

   $$\int_0^1 \int_0^1 2y \, dy \, dz = \int_0^1 \left[y^2 \right]_{y=0}^{y=1} dz = \int_0^1 1 \, dz = 1.$$

   - On the left face $x = 0$, the outward-pointing normal is $\mathbf{n} = -\mathbf{i} = \langle -1, 0, 0 \rangle$, and the field becomes $\mathbf{E} = \langle 2y, 0, yz \rangle$. Then $\mathbf{E} \cdot \mathbf{n} = -2y$, so the flux is

   $$\int_0^1 \int_0^1 (-2y) \, dy \, dz = -\int_0^1 \left[y^2 \right]_{y=0}^{y=1} dz = -\int_0^1 1 \, dz = -1.$$

   - On the front face $y = 0$, the outward-pointing normal is $\mathbf{n} = -\mathbf{j} = \langle 0, -1, 0 \rangle$, and the field becomes $\mathbf{E} = \langle 0, 0, 0 \rangle$, so the flux is 0.

   - On the back face $y = 1$, the outward-pointing normal is $\mathbf{n} = \mathbf{j} = \langle 0, 1, 0 \rangle$, and the field becomes $\mathbf{E} = \langle 2, 2x, z \rangle$. Then $\mathbf{E} \cdot \mathbf{n} = 2x$, so the flux is

   $$\int_0^1 \int_0^1 2x \, dx \, dz = \int_0^1 \left[x^2 \right]_{x=0}^{x=1} dz = \int_0^1 1 \, dz = 1.$$

Adding these up gives a total flux of $\frac{3}{2}$.

55. With $z = 0.24\sqrt{x^2 + y^2}$ and $D = [0, 0.25] \times [0, 0.25]$, Definition 14.6 gives for the area

$$\int_S 1 \, dS = \iint_D \sqrt{\left(\frac{\partial z}{\partial x}\right)^2 + \left(\frac{\partial z}{\partial y}\right)^2 + 1} \, dA$$

$$= \int_0^{0.25} \int_0^{0.25} \sqrt{\left(0.24x(x^2 + y^2)^{-1/2}\right)^2 + \left(0.24y(x^2 + y^2)^{-1/2}\right)^2 + 1} \, dy \, dx$$

$$= \int_0^{0.25} \int_0^{0.25} \sqrt{\frac{0.0576x^2 + 0.0576y^2}{x^2 + y^2} + 1} \, dy \, dx$$

$$= \int_0^{0.25} \int_0^{0.25} \sqrt{1.0576} \, dy \, dx = 0.25^2 \sqrt{1.0576} \approx 0.0643 \text{ square miles.}$$

Proofs

57. Suppose S is parametrized by $\mathbf{r}(u, v)$. Then the volume element is $\|\mathbf{r}_u \times \mathbf{r}_v\|$, and the two unit normals are

$$\mathbf{n} = \frac{\mathbf{r}_u \times \mathbf{r}_v}{\|\mathbf{r}_u \times \mathbf{r}_v\|}, \text{ and } -\mathbf{n} = -\frac{\mathbf{r}_u \times \mathbf{r}_v}{\|\mathbf{r}_u \times \mathbf{r}_v\|}$$

Then

$$\int_S \mathbf{F}(x, y, z) \cdot (-\mathbf{n}) \, dS = \iint_D \mathbf{F}(x, y, z) \cdot \left(-\frac{\mathbf{r}_u \times \mathbf{r}_v}{\|\mathbf{r}_u \times \mathbf{r}_v\|}\right) \|\mathbf{r}_u \times \mathbf{r}_v\| \, dA$$

$$= \iint_D \mathbf{F}(x, y, z) \cdot (-\mathbf{r}_u \times \mathbf{r}_v) \, dA$$

$$= -\iint_D \mathbf{F}(x, y, z) \cdot (\mathbf{r}_u \times \mathbf{r}_v) \, dA$$

$$= -\iint_D \mathbf{F}(x, y, z) \cdot \left(\frac{\mathbf{r}_u \times \mathbf{r}_v}{\|\mathbf{r}_u \times \mathbf{r}_v\|}\right) \|\mathbf{r}_u \times \mathbf{r}_v\| \, dA$$

$$= -\int_S \mathbf{F}(x, y, z) \cdot \mathbf{n} \, dS.$$

59. For the paraboloid,

$$dS = \sqrt{\left(\frac{\partial z}{\partial x}\right)^2 + \left(\frac{\partial z}{\partial y}\right)^2 + 1} \, dA = \sqrt{(2x)^2 + (2y)^2 + 1} \, dA = \sqrt{4x^2 + 4y^2 + 1} \, dA,$$

and for the saddle,

$$dS = \sqrt{\left(\frac{\partial z}{\partial x}\right)^2 + \left(\frac{\partial z}{\partial y}\right)^2 + 1} \, dA = \sqrt{(2x)^2 + (-2y)^2 + 1} \, dA = \sqrt{4x^2 + 4y^2 + 1} \, dA.$$

Then the area of the paraboloid and of the saddle are each equal to

$$\iint_R \sqrt{4x^2 + 4y^2 + 1} \, dA.$$

Thinking Forward

Integrating vector fields over three-dimensional regions

The integral will represent the amount that the fluid or gas is expanding (if the integral is positive) or compressing (if it is negative) in W.

Comparing double integrals and surface integrals

When integrating multivariable functions over surfaces, the idea is very similar to double integrals in the plane. In both cases, we are evaluating a function over a two-dimensional region. The only difference is that in the surface integral case, the region might not be flat. That lack of flatness affects the area of the region, and thus the integral of the function, and the lack of flatness is reflected in the volume element that is introduced when converting dS to dA.

Computations in $\mathbb{R}^4$

In $\mathbb{R}^4$, one can still think of a normal vector, just as one can in $\mathbb{R}^2$ and $\mathbb{R}^3$. Of course, it is a vector with four components, but it still provides a normal direction to the solid W. An integral to determine the volume of W looks much like the similar thing in lower dimensions: $\int_W 1 \, dW$. This is evaluated by parametrizing W using three variables and converting the integral over W to an integral in $\mathbb{R}^3$, such as $\int_W 1 \, dW = \int_R v \, dA$, where v is the volume element. If $w = f(x, y, z)$, it is reasonable to expect that

$$dW = \sqrt{\left(\frac{\partial w}{\partial x}\right)^2 + \left(\frac{\partial w}{\partial y}\right)^2 + \left(\frac{\partial w}{\partial z}\right)^2 + 1} \, dA.$$

Finally, integrating a function $f(x, y, z)$ over W looks the same as well

$$\int_W f(x, y, z) \, dW = \int_R f(x, y, z) v \, dA.$$

14.4 Green's Theorem

Thinking Back

Remembering the Fundamental Theorem of Calculus

▷ The boundary of the region is the set of points such that any open ball around one of the points contains points both in the interval and not in the interval. This set is clearly $\{a, b\}$, since any point in (a, b) has a small open ball around it wholly contained in (a, b), and any point outside of $[a, b]$ has a small open ball around it wholly outside of $[a, b]$. So the boundary is $\{a, b\}$, and the interior (which is the set of points such that there is an open ball around each wholly contained in $[a, b]$) is (a, b).

▷ By the Fundamental Theorem of Calculus, $g(b) - g(a) = \int_a^b g'(t) \, dt$.

▷ If F is an antiderivative of f, then $\int_a^b f(x) \, dx = F(b) - F(a)$.

Intractable Integrals

For example, $\int\int e^{x^2 + y^2} \, dy \, dx$.

Concepts

1. (a) False. The del operator ∇ can be applied in different ways to get different results, but in all cases it acts on vector fields. $\nabla \cdot \mathbf{F}$ is a multivariable function, while $\nabla \times \mathbf{F}$ is a multivariable vector-valued function (i.e., a vector field).

    (b) False. The curl, $\nabla \times \mathbf{F}$, measures the rotation of $\mathbf{F}$.

    (c) True. The divergence of a vector field is the sum of various partial derivatives of the components of the field (see Definition 14.10), so it is a scalar function.

(d) True. The curl of a vector field is a vector field whose components are differences of partial derivatives of the components of the field (see Definition 14.11), so it is a vector-valued function.

(e) True. This is the content of Theorem 14.12(b); see Exercise 56 as well.

(f) True. The Fundamental Theorem of Calculus does, as it relates $\int_a^b f(x)\,dx$ to $F(b) - F(a)$ where F is an antiderivative of f. Green's Theorem does since it relates the line integral of a vector field on a closed curve to the surface integral of a function on the region bounded by the curve. (See Theorem 14.13).

(g) False. The divergence measures compression or expansion; the curl measures rotation.

(h) False. Since one side of Green's Theorem is a line integral, reversing the orientation of that curve will reverse the sign of the integral. So Green's Theorem does depend on the choice of direction of parametrization of the boundary curve.

3. The terms in the integrand are partial derivatives of the component functions of the vector field, so the integrand of the double integral is a kind of derivative of the vector field.

5. From Section 2, $\int_C \mathbf{F} \cdot d\mathbf{r}$ may be used to compute the work required to move along the curve C in the presence of a force field given by $\mathbf{F}$. From Green's Theorem, $\int_C \mathbf{F}(x,y) \cdot d\mathbf{r}$ where C is a closed curve may be used to compute the surface integral of a particular function, $\frac{\partial F_2}{\partial x} - \frac{\partial F_1}{\partial y}$, on the region bounded by the curve.

7. The third example in Exercise 6 shows that one application of $\iint_{\mathcal{R}} G(x,y)\,dA$ is to find the value of a line integral, using Green's Theorem.

9. A vector field that is compressing has negative divergence, so in this case, div $\mathbf{F}$ at the point in question will be negative. See the discussion preceding Definition 14.10.

11. A vector field that is expanding has positive divergence, so in this case, div $\mathbf{F}$ at the point in question will be positive. See the discussion preceding Definition 14.10.

13. If C is the unit circle, then R, the region bounded by the unit circle, is the unit disc $\{(x,y) \mid x^2 + y^2 \le 1\}$. Then Green's Theorem says that

$$\int_C \mathbf{F}(x,y) \cdot d\mathbf{r} = \iint_R \left(\frac{\partial F_2}{\partial x} - \frac{\partial F_1}{\partial y} \right) dA = \iint_R \left(-\sin x \sin y - x^2 e^y \right) dA.$$

15. With $\mathbf{F}(x,y)$ as in Exercise 13, we have by Definition 14.11

$$\operatorname{curl} \mathbf{F} = \left(\frac{\partial F_2}{\partial x} - \frac{\partial F_1}{\partial y} \right) \mathbf{k} = \left(-\sin x \sin y - x^2 e^y \right) \mathbf{k}.$$

Skills

17. By Definition 14.10,

$$\operatorname{div} \mathbf{F}(x,y) = \nabla \cdot \mathbf{F}(x,y) = \frac{\partial F_1}{\partial x} + \frac{\partial F_2}{\partial y} = 2yx + x \sin y = 2xy + x \sin y.$$

19. By Definition 14.10,

$$\operatorname{div} \mathbf{G}(x,y,z) = \nabla \cdot \mathbf{G}(x,y,z) = \frac{\partial G_1}{\partial x} + \frac{\partial G_2}{\partial y} + \frac{\partial G_3}{\partial z} = 0 - 0 + 0 = 0.$$

21. By Definition 14.10,

$$\text{div } \mathbf{F}(x,y,z) = \nabla \cdot \mathbf{F}(x,y,z) = \frac{\partial F_1}{\partial x} + \frac{\partial F_2}{\partial y} + \frac{\partial F_3}{\partial z} = \frac{y}{\sqrt{1-x^2y^2}} + \frac{1}{y+z} - \frac{5}{(2x+3y+5z+1)^2}.$$

23. By Definition 14.11,

$$\text{curl } \mathbf{F}(x,y,z) = \nabla \times \mathbf{F}(x,y,z) = \left\langle \frac{\partial F_3}{\partial y} - \frac{\partial F_2}{\partial z}, \frac{\partial F_1}{\partial z} - \frac{\partial F_3}{\partial x}, \frac{\partial F_2}{\partial x} - \frac{\partial F_1}{\partial y} \right\rangle$$

$$= \left\langle -\frac{3}{(2x+3y+5z+1)^2} - \frac{1}{y+z}, 0 + \frac{2}{(2x+3y+5z+1)^2}, 0 - \frac{x}{\sqrt{1-x^2y^2}} \right\rangle$$

$$= \left\langle -\frac{3}{(2x+3y+5z+1)^2} - \frac{1}{y+z}, \frac{2}{(2x+3y+5z+1)^2}, -\frac{x}{\sqrt{1-x^2y^2}} \right\rangle.$$

25. By Definition 14.11,

$$\text{curl } \mathbf{F}(x,y,z) = \nabla \times \mathbf{F}(x,y,z) = \left\langle \frac{\partial F_3}{\partial y} - \frac{\partial F_2}{\partial z}, \frac{\partial F_1}{\partial z} - \frac{\partial F_3}{\partial x}, \frac{\partial F_2}{\partial x} - \frac{\partial F_1}{\partial y} \right\rangle$$

$$= \langle xze^{xy} - xye^{xz}, xye^{yz} - yze^{xy}, yze^{xz} - xze^{yz} \rangle.$$

27. By Definition 14.11,

$$\text{curl } \mathbf{F}(x,y) = \nabla \times \langle F_1(x,y), F_2(x,y), 0 \rangle = \left(\frac{\partial F_2}{\partial x} - \frac{\partial F_1}{\partial y} \right) \mathbf{k} = (\cos(x-y) + \sin(x+y)) \mathbf{k}.$$

29. The unit circle bounds (for example) the unit disk D. Since $\frac{\partial F_2}{\partial x} - \frac{\partial F_1}{\partial y} = 4y^2 + 4x^2$, we have by Green's Theorem

$$\int_C \mathbf{F} \cdot d\mathbf{r} = \iint_D (4x^2 + 4y^2) \, dA.$$

Convert to polar coordinates, giving

$$\iint_D (4x^2 + 4y^2) = \int_0^{2\pi} \int_0^1 4r^3 \, dr \, d\theta = \int_0^{2\pi} \left[r^4 \right]_{r=0}^{r=1} d\theta = \int_0^{2\pi} 1 \, d\theta = 2\pi.$$

31. The region of which C is the boundary is the Type II region D bounded by $y = -2$ and $y = 2$, and by $x = y^2$ and $x = 4$. Since

$$\frac{\partial F_2}{\partial x} - \frac{\partial F_1}{\partial y} = 1 - 2 = -1,$$

we have by Green's Theorem

$$\int_C \mathbf{F} \cdot d\mathbf{r} = \iint_D (-1) \, dA = \int_{-2}^2 \int_{y^2}^4 (-1) \, dx \, dy = \int_{-2}^2 [-x]_{x=y^2}^{x=4} \, dy$$

$$= \int_{-2}^2 (y^2 - 4) \, dy = \left[\frac{1}{3} y^3 - 4y \right]_{-2}^2 = -\frac{32}{3}.$$

33. The region is the triangle with vertices $[-1, 0]$, $[-1, -1]$, and $[0, 0]$, so it is a Type I region bounded by $x = -1$ and $x = 0$, and by $y = x$ and $y = 0$. Since

$$\frac{\partial F_2}{\partial x} - \frac{\partial F_1}{\partial y} = 1 - 2xye^{x^2+y^2},$$

we have by Green's Theorem

$$
\begin{aligned}
\int_C \mathbf{F} \cdot d\mathbf{r} &= \iint_D \left(1 - 2xye^{x^2+y^2}\right) dA = \int_{-1}^0 \int_x^0 \left(1 - 2xye^{x^2+y^2}\right) dy\, dx \\
&= \int_{-1}^0 \left[y - xe^{x^2+y^2}\right]_{y=x}^{y=0} dx \\
&= \int_{-1}^0 \left(-xe^{x^2} - x + xe^{2x^2}\right) dx \\
&= \left[-\frac{1}{2}e^{x^2} - \frac{1}{2}x^2 + \frac{1}{4}e^{2x^2}\right]_{-1}^0 \\
&= -\frac{1}{2} - 0 + \frac{1}{4} + \frac{1}{2}e + \frac{1}{2} - \frac{1}{4}e^2 \\
&= \frac{1}{4}(1 + 2e - e^2).
\end{aligned}
$$

35. Note that $\mathbf{F}$ is not conservative, since $\frac{\partial F_2}{\partial x} = -y \neq \frac{\partial F_1}{\partial y} = 2y$. So we must actually perform an integration. With $\mathbf{r}(t) = \langle \cos t, \sin t\rangle$ for $t \in [0, 2\pi]$, we have $\mathbf{r}'(t) = \langle -\sin t, \cos t\rangle$, and

$$
\mathbf{F}(x,y) = \left\langle y^2, -xy \right\rangle = \left\langle \sin^2 t, -\sin t \cos t\right\rangle.
$$

Then

$$
\begin{aligned}
\int_C \mathbf{F}(x,y) \cdot d\mathbf{r} &= \int_0^{2\pi} \left\langle \sin^2 t, -\sin t \cos t\right\rangle \cdot \langle -\sin t, \cos t\rangle\, dt \\
&= \int_0^{2\pi} \left(-\sin^3 t - \sin t \cos^2 t\right) dt \\
&= \int_0^{2\pi} \left(-\sin t(\sin^2 t + \cos^2 t)\right) dt \\
&= \int_0^{2\pi} (-\sin t)\, dt \\
&= [\cos t]_0^{2\pi} = 0.
\end{aligned}
$$

37. Note that $\mathbf{F}$ is not conservative, since $\frac{\partial F_2}{\partial x} = 2 \neq \frac{\partial F_1}{\partial y} = 1$. The vertices of the given triangle are $(0,0)$, $(1,0)$, and $(0,1)$. Parametrize C by splitting it up into three separate lines, one between each pair of vertices:

- $\mathbf{r}_1(t) = \langle t, 0\rangle$ for $t \in [0,1]$. This traverses the segment from $(0,0)$ to $(1,0)$. We have $\mathbf{r}_1'(t) = \langle 1, 0\rangle$, and

$$
\mathbf{F}(x,y) = \langle 2^x + y, 2x - y\rangle = \left\langle 2^t, 2t\right\rangle,
$$

  so that

$$
\int_{C_1} \mathbf{F}(x,y) \cdot d\mathbf{r}_1 = \int_0^1 \langle 2^t, 2t\rangle \cdot \langle 1, 0\rangle\, dt = \int_0^1 2^t\, dt = \left[\frac{2^t}{\ln 2}\right]_0^1 = \frac{1}{\ln 2}.
$$

- $\mathbf{r}_2(t) = \langle 1-t, t\rangle$ for $t \in [0,1]$, This traverses the segment from $(1,0)$ to $(0,1)$. We have $\mathbf{r}_2'(t) = \langle -1, 1\rangle$, and

$$
\mathbf{F}(x,y) = \langle 2^x + y, 2x - y\rangle = \left\langle 2^{1-t} + t, 2(1-t) - t\right\rangle = \left\langle 2^{1-t} + t, 2 - 3t\right\rangle.
$$

Then

$$\int_{C_2} \mathbf{F}(x,y) \cdot d\mathbf{r}_2 = \int_0^1 \left\langle 2^{1-t} + t, 2 - 3t \right\rangle \cdot \langle -1, 1 \rangle \, dt = \int_0^1 \left(2 - 4t - 2^{1-t} \right) dt$$

$$= \left[2t - 2t^2 + \frac{2^{1-t}}{\ln 2} \right]_0^1 = -\frac{1}{\ln 2}.$$

- $\mathbf{r}_3(t) = \langle 0, 1 - t \rangle$ for $t \in [0,1]$. This traverses the segment from $(0,1)$ to $(0,0)$. We have $\mathbf{r}_3'(t) = \langle 0, -1 \rangle$, and

$$\mathbf{F}(x,y) = \langle 2^x + y, 2x - y \rangle = \langle 1 + 1 - t, t - 1 \rangle = \langle 2 - t, t - 1 \rangle.$$

Then

$$\int_{C_3} \mathbf{F}(x,y) \cdot d\mathbf{r}_3 = \int_0^1 \langle 2 - t, t - 1 \rangle \cdot \langle 0, -1 \rangle \, dt = \int_0^1 (1 - t) \, dt = \left[t - \frac{1}{2}t^2 \right]_0^1 = \frac{1}{2}.$$

Adding these three up gives

$$\int_C \mathbf{F}(x,y) \cdot d\mathbf{r} = \int_{C_1} \mathbf{F}(x,y) \cdot d\mathbf{r} + \int_{C_2} \mathbf{F}(x,y) \cdot d\mathbf{r} + \int_{C_3} \mathbf{F}(x,y) \cdot d\mathbf{r} = \frac{1}{\ln 2} - \frac{1}{\ln 2} + \frac{1}{2} = \frac{1}{2}.$$

39. Using polar coordinates, $\mathcal{R}$ is parametrized by $\theta \in [0, \pi]$ and $r \in [0,2]$. Then the integral is

$$\iint_{\mathcal{R}} (3y - 3x) \, dA = \int_0^\pi \int_0^2 (3r\sin\theta - 3r\cos\theta) r \, dr \, d\theta$$

$$= \int_0^\pi \int_0^2 3r^2 (\sin\theta - \cos\theta) \, d\theta$$

$$= \int_0^\pi \left[r^3 (\sin\theta - \cos\theta) \right]_{r=0}^{r=2} d\theta$$

$$= 8 \int_0^\pi (\sin\theta - \cos\theta) \, d\theta$$

$$= 8 \left[-\cos\theta - \sin\theta \right]_0^\pi = 16.$$

41. Integrating directly gives

$$\iint_{\mathcal{R}} (e^x + e^y) \, dA = \int_1^{\ln 2} \int_0^2 (e^x + e^y) \, dy \, dx$$

$$= \int_1^{\ln 2} [ye^x + e^y]_{y=0}^{y=2} \, dx$$

$$= \int_1^{\ln 2} \left(2e^x + e^2 - 1 \right) dx$$

$$= \left[2e^x + (e^2 - 1)x \right]_1^{\ln 2}$$

$$= 4 + (e^2 - 1)\ln 2 - 2e - (e^2 - 1) = 4 - 2e + (e^2 - 1)(\ln 2 - 1).$$

43. The region itself is clearly a more convenient region for integration than is its boundary, so we evaluate the integral using Green's Theorem. We have

$$\frac{\partial F_2}{\partial x} - \frac{\partial F_1}{\partial y} = y^e - 2^x.$$

Now, C is traversed clockwise, so that introduces a minus sign into the integral using Green's Theorem:

$$\int_C \mathbf{F} \cdot d\mathbf{r} = -\iint_D (y^e - 2^x)\, dA = -\int_0^1 \int_0^1 (y^e - 2^x)\, dy\, dx$$

$$= -\int_0^1 \left[\frac{1}{e+1} y^{e+1} - y2^x \right]_{y=0}^{y=1} dx$$

$$= -\int_0^1 \left(\frac{1}{e+1} - 2^x \right) dx$$

$$= -\left[\frac{1}{e+1} x - \frac{1}{\ln 2} 2^x \right]_0^1 = \frac{1}{\ln 2} - \frac{1}{e+1}.$$

45. Integrate directly using polar coordinates, integrating first with respect to θ:

$$\iint_{\mathcal{R}} \left(2xe^{x^2+y^2} + 2ye^{-(x^2+y^2)} \right) dA = \int_1^2 \int_0^{2\pi} r \left(2r\cos\theta e^{r^2} + 2r\sin\theta e^{-r^2} \right) d\theta\, dr$$

$$= \int_1^2 \int_0^{2\pi} \left(2r^2 e^{r^2} \cos\theta + 2r^2 e^{-r^2} \sin\theta \right) d\theta\, dr$$

$$= \int_1^2 \left[2r^2 e^{r^2} \sin\theta - 2r^2 e^{-r^2} \cos\theta \right]_{\theta=0}^{\theta=2\pi} dr$$

$$= \int_1^2 0\, dr = 0.$$

Applications

47. The work done is the integral of $\mathbf{F}$ along the unit circle, which is $\int_C \mathbf{F}(x,y) \cdot d\mathbf{r}$. By Green's Theorem, since all the functions involved are continuous and differentiable on the unit circle and its interior, this integral is equal to

$$\iint_R \left(\frac{\partial F_2}{\partial x} - \frac{\partial F_1}{\partial y} \right) dA = \iint_R (-8xy + 8xy)\, dA = \iint_R 0\, dA = 0,$$

where R is the unit disk. So no work is done.

49. The work done is the integral of $\mathbf{F}$ along the unit circle, which is $\int_C \mathbf{F}(x,y) \cdot d\mathbf{r}$. By Green's Theorem, since all the functions involved are continuous and differentiable on the unit circle and its interior, this integral is equal to

$$\iint_R \left(\frac{\partial F_2}{\partial x} - \frac{\partial F_1}{\partial y} \right) dA.$$

But on the unit circle, we have $\sqrt{x^2 + y^2} = 1$, so that $\mathbf{F} = ex\mathbf{i} + ey\mathbf{j}$, so that $\frac{\partial F_1}{\partial y} = \frac{\partial F_2}{\partial x} = 0$ and thus the integral is zero. No work is done.

51. (a) Since $\frac{\partial F_2}{\partial x} = -1.6x$ and $\frac{\partial F_1}{\partial y} = 0$, the integral is (using polar coordinates)

$$\int_0^1 \int_0^{2\pi} (-1.6r\cos\theta) r\, d\theta\, dr = \int_0^1 \left[-1.6r^2 \sin\theta \right]_0^{2\pi} dr = 0.$$

(b) The outward-pointing unit normal to the unit circle is $\langle x, y \rangle$, so that

$$\mathbf{F} \cdot \mathbf{n} = \left\langle 0, 1.152 - 0.8x^2 \right\rangle \cdot \langle x, y \rangle = (1.152 - 0.8x^2)y.$$

To compute the integral, we convert to polar coordinates. Since we are integrating around the unit circle, the volume element is $r = 1$, and $x = r \cos\theta = \cos\theta$, $y = r \sin\theta = \sin\theta$, so the integral is

$$\int_{\partial R} \mathbf{F} \cdot \mathbf{n}\, ds = \int_0^{2\pi} (1.152 - 0.8 \cos^2\theta) \sin\theta\, d\theta$$

$$= \int_0^{2\pi} (1.152 \sin\theta - 0.8 \cos^2\theta \sin\theta)\, d\theta$$

$$= \left[-1.152 \cos\theta + \frac{0.8}{3} \cos^3\theta \right]_0^{2\pi} = 0.$$

(c) Since the flux across the boundary is zero from part (b), this tells us that as much water flows in as flows out.

53. (a) Since speed increases as we move left among the lanes, we know that $\frac{\partial v_2}{\partial x} < 0$. Since $v_1 = 0$, we know that $\frac{\partial v_1}{\partial y} = 0$. Thus $\frac{\partial v_2}{\partial x} - \frac{\partial v_1}{\partial y} < 0$; that is, the curl is always negative and traffic tends to have a rotation to it. If the curl is low, then $\frac{\partial v_2}{\partial x}$ is close to zero, so that all lanes are running at roughly the same speed, while if the curl is large, then there is a large variation in lane speed. Presumably low curl is safer, since then all cars on the road are traveling at more or less the same speed.

(b) On the right hand side we have

$$\int_0^{0.0113} \int_0^{0.5} \left(\frac{\partial}{\partial x}(75 - \alpha x) - \frac{\partial}{\partial y}(0) \right) dy\, dx = \int_0^{0.0113} \int_0^{0.5} (-\alpha)\, dy\, dx = -0.00565\alpha.$$

On the left-hand side, we divide the boundary into four segments, one for each edge of the rectangle.

- $\mathbf{r}_1(t) = \langle t, 0 \rangle$ for $t \in [0, 0.0113]$ traverses the edge from $(0,0)$ to $(0.0113, 0)$. Then $\mathbf{r}_1'(t) = \langle 1, 0 \rangle$, and
$$\mathbf{v}(x, y) = \langle 0, 75 - \alpha x \rangle = \langle 0, 75 - \alpha t \rangle.$$

Thus
$$\int_{C_1} \mathbf{v}(x, y) \cdot d\mathbf{r}_1 = \int_0^{0.0113} 0\, dt = 0.$$

- $\mathbf{r}_2(t) = \langle 0.0113, t \rangle$ for $t \in [0, 0.5]$ traverses the edge from $(0.0113, 0)$ to $(0.0113, 0.5)$. Then $\mathbf{r}_2'(t) = \langle 0, 1 \rangle$, and
$$\mathbf{v}(x, y) = \langle 0, 75 - \alpha x \rangle = \langle 0, 75 - 0.0113\alpha \rangle.$$

Thus
$$\int_{C_2} \mathbf{v}(x, y) \cdot d\mathbf{r}_2 = \int_0^{0.5} (75 - 0.0113\alpha)\, dt = 37.5 - 0.00565\alpha.$$

- $\mathbf{r}_3(t) = \langle 0.0113 - t, 0.5 \rangle$ for $t \in [0, 0.0113]$ traverses the edge from $(0.0113, 0.5)$ to $(0, 0.5)$. Then $\mathbf{r}_3'(t) = \langle -1, 0 \rangle$, and
$$\mathbf{v}(x, y) = \langle 0, 75 - \alpha x \rangle = \langle 0, 75 - \alpha(0.0113 - t) \rangle.$$

Thus
$$\int_{C_3} \mathbf{v}(x, y) \cdot d\mathbf{r}_3 = \int_0^{0.0113} 0\, dt = 0.$$

- $\mathbf{r}_4(t) = \langle 0, 0.5 - t \rangle$ for $t \in [0, 0.5]$ traverses the edge from $(0, 0.5)$ to $(0, 0)$. Then $\mathbf{r}_4'(t) = \langle 0, -1 \rangle$, and

$$\mathbf{v}(x, y) = \langle 0, 75 - \alpha x \rangle = \langle 0, 75 \rangle.$$

Thus

$$\int_{C_4} \mathbf{v}(x, y) \cdot d\mathbf{r}_4 = \int_0^{0.5} (-75) \, dt = -37.5.$$

Adding these four partial results gives for the total integral -0.00565α, which equals the value from integrating the right-hand side.

(c) These integrals describe the total curl, or rotation, over this stretch of road. This may have implications for traffic safety; see part (a). The integral on the right is clearly easier to compute, depending on the specific formula for $\mathbf{v}_2$, since it is a simple integral over a rectangular region.

Proofs

55. In either case, we are assuming, per the hypotheses of Theorem 14.12, that $\mathbf{F}$ is a vector field having continuous second order partial derivatives. Note that this implies that the value of a second partial derivative does not depend on the order of differentiation, so for example $\frac{\partial^2 F_2}{\partial y \, \partial x} = \frac{\partial^2 F_2}{\partial x \, \partial y}$.

Suppose first that $\mathbf{F}$ is a vector field in $\mathbb{R}^2$. Then

$$\operatorname{div} \operatorname{curl} \mathbf{F} = \operatorname{div} \left(0\mathbf{i} + 0\mathbf{j} + \left(\frac{\partial F_2}{\partial x} - \frac{\partial F_1}{\partial y} \right) \mathbf{k} \right)$$
$$= \frac{\partial}{\partial x}(0) + \frac{\partial}{\partial y}(0) + \frac{\partial}{\partial z} \left(\frac{\partial F_2}{\partial x} - \frac{\partial F_1}{\partial y} \right)$$
$$= 0,$$

where the last equality holds since neither F_1 nor F_2 depends on z.

If $\mathbf{F}(x, y, z)$ is a vector field in $\mathbb{R}^3$ having continuous second order partial derivatives, then

$$\operatorname{div} \operatorname{curl} \mathbf{F} = \operatorname{div} \left(\left(\frac{\partial F_3}{\partial y} - \frac{\partial F_2}{\partial z} \right) \mathbf{i} + \left(\frac{\partial F_1}{\partial z} - \frac{\partial F_3}{\partial x} \right) \mathbf{j} + \left(\frac{\partial F_2}{\partial x} - \frac{\partial F_1}{\partial y} \right) \mathbf{k} \right)$$
$$= \frac{\partial}{\partial x} \left(\frac{\partial F_3}{\partial y} - \frac{\partial F_2}{\partial z} \right) + \frac{\partial}{\partial y} \left(\frac{\partial F_1}{\partial z} - \frac{\partial F_3}{\partial x} \right) + \frac{\partial}{\partial z} \left(\frac{\partial F_2}{\partial x} - \frac{\partial F_1}{\partial y} \right)$$
$$= \frac{\partial^2 F_3}{\partial x \, \partial y} - \frac{\partial^2 F_2}{\partial x \, \partial z} + \frac{\partial^2 F_1}{\partial y \, \partial z} - \frac{\partial^2 F_3}{\partial y \, \partial x} + \frac{\partial^2 F_2}{\partial z \, \partial x} - \frac{\partial^2 F_1}{\partial z \, \partial y}$$
$$= \frac{\partial^2 F_3}{\partial x \, \partial y} - \frac{\partial^2 F_3}{\partial y \, \partial x} - \frac{\partial^2 F_2}{\partial x \, \partial z} + \frac{\partial^2 F_2}{\partial z \, \partial x} + \frac{\partial^2 F_1}{\partial y \, \partial z} - \frac{\partial^2 F_1}{\partial z \, \partial y}$$
$$= 0.$$

57. If R consists of multiple connected pieces, we can repeat the following argument for each of them. So we may assume that R is connected, and the hypotheses of the problem say that R is also simply connected. So by Theorem 14.5, or Exercise 64 in Section 14.2, if C is the boundary of R, then $\int_C \mathbf{F}(x, y) \cdot d\mathbf{r} = 0$ since C is a closed curve. But then by Green's Theorem, since C is smooth or piecewise smooth,

$$0 = \int_C \mathbf{F}(x, y) \cdot d\mathbf{r} = \iint_R \left(\frac{\partial F_2}{\partial x} - \frac{\partial F_1}{\partial y} \right) dA.$$

Thinking Forward

Generalizing Green's Theorem

Green's Theorem says that integrating a vector field in $\mathbb{R}^2$ around a closed curve is the same as integrating the curl of the vector field over the region bounded by the closed curve, with appropriate conditions on the region and the curve. The language of this statement generalizes to $\mathbb{R}^3$ in the following way: integrating a vector field in $\mathbb{R}^3$ around a closed curve is the same as integrating the curl of the vector field over any surface bounded by the closed curve. (Note that there are many surfaces in $\mathbb{R}^3$ bounded by a given closed curve; for example, the unit circle $r = 1$ bounds both the unit disk and the upper half of the unit sphere). This is Stokes' Theorem; see Section 14.5.

Another generalization of Green's Theorem

The basic idea of Green's Theorem is that integrating a vector field on a boundary is the same as integrating something else on the interior. If the boundary is a closed surface, then the integral we are familiar with along a surface is the flux integral, $\iint_S \mathbf{F}(x, y, z) \cdot \mathbf{n} \, dS$. In two dimensions, the divergence form of Green's Theorem says that

$$\int_C \mathbf{F}(x, y) \cdot \mathbf{n} \, ds = \iint_R \operatorname{div} \mathbf{F}(x, y) \, dA,$$

so we might expect that formula to generalize to three dimensions as

$$\iint_S \mathbf{F}(x, y, z) \cdot \mathbf{n} \, dS = \iiint_W \operatorname{div} \mathbf{F}(x, y, z) \, dV.$$

This is the Divergence Theorem; see Section 14.6.

14.5   Stokes' Theorem

Thinking Back

Antecedents

▷ Let P and Q be the common initial and terminal points of C_1 and C_2. Since $\mathbf{F}$ is conservative, let f be a potential function. Then the Fundamental Theorem of Line Integrals says that each of these integrals is equal to $f(Q) - f(P)$, so they are equal.

▷ In Stokes' Theorem, we can think of the surface S as the interval, the boundary curve C as the "endpoints" of the surface (just as the endpoints of an interval are its boundary), and $\mathbf{F}(x, y, z)$ as the antiderivative of curl $\mathbf{F}(x, y, z)$. Thus the integrand of the surface integral in Stokes' theorem corresponds to the integrand in the Fundamental Theorem of Calculus.

Concepts

1. (a) False. It asserts that the flux of *the curl of* a vector field through a smooth surface with smooth boundary is equal to the line integral of the field around the boundary of the surface.

   (b) True. See the discussion following Theorem 14.15, or the first Thinking Forward exercise for Section 14.4.

   (c) False. By the statement of Theorem 14.15, it applies to any (integrable) vector field.

   (d) False. In some cases, it may be easier to simply evaluate the surface integral. Also, Stokes' Theorem may be used in reverse, to turn a difficult integral around a closed curve into a simpler integral over a surface.

(e) True. The Fundamental Theorem of Line Integrals gives a way to evaluate a line integral for a conservative vector field using the value of the potential function at the endpoints. Stokes' Theorem gives a way to evaluate a surface integral of a curl using the vector field on the boundary.

(f) True. Suppose $\mathbf{F}(x, y, z)$ is conservative with potential function $f(x, y, z)$, and let C be a simple, smooth or piecewise smooth, closed curve. Assume C bounds some surface S. Then

$$\int_C \mathbf{F}(x, y, z) \cdot d\mathbf{r} = \iint_S \operatorname{curl} \mathbf{F} \cdot \mathbf{n}\, dS = \iint_S (\operatorname{curl} \nabla f) \cdot \mathbf{n}\, dS \stackrel{14.12}{=} \iint_S 0 \cdot \mathbf{n}\, dS = 0.$$

(g) True. Since by part (e) it generalizes the Fundamental Theorem of Line Integrals, and that theorem generalizes the Fundamental Theorem of Calculus (it simply restates the Fundamental Theorem of Calculus for arbitrary smooth curves as opposed to intervals in $\mathbb{R}^1$), Stokes' Theorem generalizes the Fundamental Theorem of Calculus.

(h) True. Suppose $\operatorname{curl} \mathbf{F}(x, y, z) \cdot \mathbf{n}$ is a constant k. Then $\iint_S \operatorname{curl} \mathbf{F}(x, y, z) \cdot \mathbf{n}\, dS = k$ times the surface area of S. But by Stokes Theorem, this is also equal to $\int_C \mathbf{F}(x, y, z) \cdot d\mathbf{r}$ where C is the boundary of the surface.

3. Two possible normals are

$$\left\langle -\frac{\partial g}{\partial x}, -\frac{\partial g}{\partial y}, 1 \right\rangle, \qquad \left\langle \frac{\partial g}{\partial x}, \frac{\partial g}{\partial y}, -1 \right\rangle.$$

Other normals are scalar multiples of these.

5. Note that an outward-pointing normal $\mathbf{n}_1$ on the upper half of the unit sphere leads to a counterclockwise parametrization of the unit circle viewed from above, which is the same parametrization as that given for the balloon-shaped surface. Then from two applications of Stokes' Theorem, we have

$$\iint_{\mathcal{S}_1} \operatorname{curl} \mathbf{F} \cdot \mathbf{n}_1\, d\mathcal{S}_1 = \int_C \mathbf{F} \cdot d\mathbf{r} = \iint_{\mathcal{S}_2} \operatorname{curl} \mathbf{F} \cdot \mathbf{n}_2\, d\mathcal{S}_2.$$

Thus the two surface integrals are equal.

7. By Stokes' Theorem,

$$\iint_S 1\, d\mathcal{S} = \iint_S \operatorname{curl} \mathbf{F}(x, y, z) \cdot \mathbf{n}\, d\mathcal{S} = \int_C \mathbf{F}(x, y, z) \cdot d\mathbf{r}.$$

9. If the orientation of S is reversed, the sign of the right-hand side of Stokes' theorem will change, since $d\mathcal{S}$ will be negated. If the parametrization of C is reversed, the sign of the line integral on the left-hand side of Stokes' theorem will change. Thus the orientation of the surface and the direction of parametrization of the boundary must be consistent.

11. Let S be the unit disk, whose boundary is the unit circle, and let $\mathbf{F}(x, y, z) = \langle 1, 0, 0 \rangle$. Then $\operatorname{curl} \mathbf{F}(x, y, z) = \langle 0, 0, 0 \rangle$, so that the right-hand side of Stokes' Theorem will be zero independent of the choice of $\mathbf{k}$ or $-\mathbf{k}$ as unit normal. On the left-hand side, we may parametrize the unit circle by $\mathbf{r}_1(\theta) = \langle \cos \theta, \sin \theta, 0 \rangle$ or $\mathbf{r}_2(\theta) = \langle \cos \theta, -\sin \theta, 0 \rangle$ for $\theta \in [0, 2\pi]$, corresponding to the choices $\mathbf{k}$ and $-\mathbf{k}$ for normals. Then

$$\mathbf{r}_1'(\theta) = \langle -\sin \theta, \cos \theta, 0 \rangle, \qquad \mathbf{r}_2'(\theta) = \langle -\sin \theta, -\cos \theta, 0 \rangle,$$

so that $\mathbf{F}(x, y, z) \cdot d\mathbf{r}_1 = -\sin \theta$ and $\mathbf{F}(x, y, z) \cdot d\mathbf{r}_2 = -\sin \theta$, so that the two line integrals will be equal as well (and both will be zero since $\int_0^{2\pi} (-\sin \theta)\, d\theta = 0$).

13. In Green's Theorem, the region is a flat region lying in the xy plane, so its area element is denoted by dA. In Stokes' Theorem, the surface is an arbitrary surface in $\mathbb{R}^3$, so its area element is denoted by dS.

15. If the boundary is difficult to integrate over, but it bounds a surface that is more easily described, one might choose to integrate over the surface using Stokes' Theorem (for example, if the boundary is a square, so that the line integral would require four separate parametrizations and integrations). Alternatively, if curl $\mathbf{F}$ is significantly simpler than $\mathbf{F}$, then performing the surface integral rather than the line integral is also called for.

17. Recall that the proof of Stokes' Theorem requires the use of Green's Theorem in showing the equality of surface and line integrals on each of the small tangent planes. If the surface S is not piecewise smooth, there is no guarantee that we can choose subrectangles to subdivide S so that each is smooth. Thus Green's Theorem might not apply.

Skills

19. Evaluating the line integral as given requires three integrations, one along each edge. Using Stokes' Theorem we can change this to an integral over the inside of the triangle, $\mathcal{T}$, which can be expressed as a Type I region in the plane $y = 2$, bounded by $x = 0$ and $x = 1$ and by $z = 0$ and $z = 1 - x$. Note that

$$\text{curl } \mathbf{F} = \left\langle \frac{\partial}{\partial y}(x^2 z) - \frac{\partial}{\partial z}(e^x), \frac{\partial}{\partial z}(3yz) - \frac{\partial}{\partial x}(x^2 z), \frac{\partial}{\partial x}(e^x) - \frac{\partial}{\partial y}(3yz) \right\rangle$$
$$= \langle 0, 3y - 2xz, e^x - 3z \rangle.$$

We are on the plane $y = 2$, so we have curl $\mathbf{F} = \langle 0, 6 - 2xz, e^x - 3z \rangle$, and $\mathbf{n} = \mathbf{j}$, so that by Stokes' Theorem

$$\int_C \mathbf{F}(x, y, z) \cdot d\mathbf{r} = \iint_{\mathcal{T}} \text{curl } \mathbf{F} \cdot \mathbf{n} \, d\mathcal{T}$$
$$= \iint_{\mathcal{T}} \langle 0, 6 - 2xz, e^x - 3z \rangle \cdot \mathbf{j} \, d\mathcal{T}$$
$$= \int_0^1 \int_0^{1-x} (6 - 2xz) \, dz \, dx$$
$$= \int_0^1 \left[6z - xz^2 \right]_{z=0}^{z=1-x} dx$$
$$= \int_0^1 \left(6 - 6x - x(1 - x)^2 \right) dx$$
$$= \int_0^1 \left(-x^3 + 2x^2 - 7x + 6 \right) dx$$
$$= \left[-\frac{1}{4}x^4 + \frac{2}{3}x^3 - \frac{7}{2}x^2 + 6x \right]_0^1 = \frac{35}{12}.$$

21. Note that $\mathbf{F}$ is conservative, with potential function $f(x, y, z) = e^{xyz}$. Thus curl $\mathbf{F} = \text{curl } \nabla f = \mathbf{0}$, so that
$$\iint_S \text{curl } \mathbf{F} \cdot \mathbf{n} \, dS = \iint_S \mathbf{0} \cdot \mathbf{n} \, dS = 0.$$

23. In the xz plane, the region of integration is a Type I region bounded by $x = 0$ and $x = 1$, and by $z = x^2$ and $z = x$. We have

$$\text{curl } \mathbf{F} = \left\langle \frac{\partial}{\partial y}(3xyz) - \frac{\partial}{\partial z}(-z), \frac{\partial}{\partial z}(7xy) - \frac{\partial}{\partial x}(3xyz), \frac{\partial}{\partial x}(-z) - \frac{\partial}{\partial y}(7xy) \right\rangle$$
$$= \langle 3xz + 1, -3yz, -7x \rangle.$$

On the plane $y = x$, this becomes $\langle 3xz + 1, -3xz, -7x \rangle$. Then using Stokes' Theorem, with $\mathbf{n} = \langle 1, -1, 0 \rangle$, we get

$$\int_C \mathbf{F}(x, y, z) \cdot d\mathbf{r} = \iint_S \text{curl } \mathbf{F}(x, y, z) \cdot \mathbf{n} \, dS$$

$$= \int_0^1 \int_{x^2}^x \langle 3xz + 1, -3xz, -7x \rangle \cdot \langle 1, -1, 0 \rangle \, dz \, dx$$

$$= \int_0^1 \int_{x^2}^x (6xz + 1) \, dz \, dx$$

$$= \int_0^1 \left[3xz^2 + z \right]_{z=x^2}^{z=x} dx$$

$$= \int_0^1 (3x^3 + x - 3x^5 - x^2) \, dx$$

$$= \left[\frac{3}{4}x^4 + \frac{1}{2}x^2 - \frac{1}{2}x^6 - \frac{1}{3}x^3 \right]_0^1 = \frac{5}{12}.$$

25. The cylinder $x^2 + y^2 = \frac{1}{9}$ and the sphere $x^2 + y^2 + z^2 = 1$ intersect when $z^2 = \frac{8}{9}$, so that (since we are considering the cap below the xy plane) $z = -\frac{2\sqrt{2}}{3}$. The intersection is a circle C of radius $\frac{1}{3}$ centered on the z axis, at $z = -\frac{2\sqrt{2}}{3}$. Since the normal points outwards, it points down, so that we want to parametrize the circle clockwise as viewed from above. So choose the parametrization as

$$\mathbf{r}(\theta) = \left\langle \frac{1}{3} \cos \theta, \frac{1}{3} \sin(-\theta), -\frac{2\sqrt{2}}{3} \right\rangle = \left\langle \frac{1}{3} \cos \theta, -\frac{1}{3} \sin \theta, -\frac{2\sqrt{2}}{3} \right\rangle.$$

Then on that circle,

$$\mathbf{F}(x, y, z) = \left\langle -yz^2, xz^2, 3^{-xyz} \right\rangle = \left\langle \frac{8}{27} \sin \theta, \frac{8}{27} \cos \theta, 3^{-2\sqrt{2} \sin \theta \cos \theta / 27} \right\rangle.$$

Finally, $\mathbf{r}'(\theta) = \left\langle -\frac{1}{3} \sin \theta, -\frac{1}{3} \cos \theta, 0 \right\rangle$, so that by Stokes' Theorem,

$$\iint_S \text{curl } \mathbf{F}(x, y, z) \cdot \mathbf{n} \, dS = \int_C \mathbf{F}(x, y, z) \cdot d\mathbf{r}$$

$$= \int_0^{2\pi} \left\langle \frac{8}{27} \sin \theta, \frac{8}{27} \cos \theta, 3^{-2\sqrt{2} \sin \theta \cos \theta / 27} \right\rangle \cdot \left\langle -\frac{1}{3} \sin \theta, -\frac{1}{3} \cos \theta, 0 \right\rangle d\theta$$

$$= \int_0^{2\pi} \left(-\frac{8}{81} \sin^2 \theta - \frac{8}{81} \cos^2 \theta \right) d\theta$$

$$= -\int_0^{2\pi} \frac{8}{81} \, d\theta$$

$$= -\frac{16}{81} \pi.$$

27. Use Stokes' Theorem. We have

$$\text{curl } \mathbf{F} = \left\langle \frac{\partial}{\partial y}(x + 4y + 2z) - \frac{\partial}{\partial z}(5x + y - z), \frac{\partial}{\partial z}(2x - 3y + 4z) - \frac{\partial}{\partial x}(x + 4y + 2z), \right.$$

$$\left. \frac{\partial}{\partial x}(5x + y - z) - \frac{\partial}{\partial y}(2x - 3y + 4z) \right\rangle$$

$$= \langle 5, 3, 8 \rangle.$$

In the xy plane, the region is a Type I region bounded by $x = -2$ and $x = 2$, and by $y = x^2$ and $y = 4$. Thus the integral is

$$\int_C \mathbf{F}(x,y,z) \cdot d\mathbf{r} = \iint_{\mathcal{S}} \operatorname{curl} \mathbf{F}(x,y,z) \cdot \mathbf{n}\, d\mathcal{S}$$

$$= \int_{-2}^{2} \int_{x^2}^{4} \langle 5,3,8 \rangle \cdot \langle 1,-1,1 \rangle \; dy\, dx$$

$$= \int_{-2}^{2} \int_{x^2}^{4} 10\, dy\, dx$$

$$= \int_{-2}^{2} [10y]_{y=x^2}^{y=4} \; dx$$

$$= \int_{-2}^{2} (40 - 10x^2)\, dx$$

$$= \left[40x - \frac{10}{3}x^3 \right]_{-2}^{2} = \frac{320}{3}.$$

29. The surface is the half of the sphere of radius 2 centered at the origin where $y \geq 0$, so $y = 2$ corresponds to the top of the sphere (and the level curves for $y > 2$ are empty). The intersection of the sphere with $y = \sqrt{3}$ is a circle C of radius 1 (since $\sqrt{3} = \sqrt{4 - x^2 - z^2}$) in the plane $y = \sqrt{3}$, centered on the y axis. Parametrize C by $\mathbf{r}(t) = \left\langle \cos t, \sqrt{3}, \sin t \right\rangle$, so that $\mathbf{r}'(t) = \langle -\sin t, 0, \cos t \rangle$, and on C,

$$\mathbf{F}(x,y,z) = \left\langle -4z - xz^2, \sin(xyz), 4x + x^2z \right\rangle$$

$$= \left\langle -4\sin t - \cos t \sin^2 t, \sin(\sqrt{3}\,\sin t \cos t), 4\cos t + \cos^2 t \sin t \right\rangle.$$

Then by Stokes' Theorem,

$$\iint_{\mathcal{S}} \operatorname{curl} \mathbf{F}(x,y,z) \cdot \mathbf{n}\, d\mathcal{S} = \int_C \mathbf{F}(x,y,z) \cdot d\mathbf{r}$$

$$= \int_0^{2\pi} \left\langle -4\sin t - \cos t \sin^2 t, \sin(\sqrt{3}\,\sin t \cos t), 4\cos t + \cos^2 t \sin t \right\rangle$$

$$\cdot \langle -\sin t, 0, \cos t \rangle \; dt$$

$$= \int_0^{2\pi} \left(4\sin^2 t + \cos t \sin^3 t + 4\cos^2 t + \sin t \cos^3 t \right) dt$$

$$= \int_0^{2\pi} \left(4(\sin^2 t + \cos^2 t) + \sin t \cos t (\sin^2 t + \cos^2 t) \right) dt$$

$$= \int_0^{2\pi} (4 + \sin t \cos t)\, dt$$

$$= \left[4t + \frac{1}{2}\sin^2 t \right]_0^{2\pi} = 8\pi.$$

31. Since

$$\operatorname{curl} \mathbf{F} = \left\langle \frac{\partial}{\partial y}(2x + y) - \frac{\partial}{\partial z}(4x + \ln(y^2 + 1) - z), \; \frac{\partial}{\partial z}(3x + 3) - \frac{\partial}{\partial x}(2x + y), \right.$$

$$\left. \frac{\partial}{\partial x}(4x + \ln(y^2 + 1) - z) - \frac{\partial}{\partial y}(3x + 3) \right\rangle$$

$$= \langle 2, -2, 4 \rangle$$

is quite simple, we use Stokes' Theorem. The curve of intersection is the circle of radius 3 in the plane $z = e^{-9}$ centered on the z axis. Since we are free to choose any surface bounded by that curve, choose the disc of radius 3 in the same plane also centered on the z axis. That disc is parametrized by $\mathbf{r}(t) = \langle r\cos\theta, r\sin\theta, e^{-9} \rangle$. The normal to the disc is $\mathbf{k}$, so that the integral is

$$
\begin{aligned}
\int_C \mathbf{F} \cdot d\mathbf{r} &= \iint_S \operatorname{curl} \mathbf{F} \cdot \mathbf{n}\, dS \\
&= \int_0^{2\pi} \int_0^3 r\,\langle 2, -2, 4\rangle \cdot \mathbf{k}\, dr\, d\theta \\
&= \int_0^{2\pi} \int_0^3 4r\, dr\, d\theta \\
&= \int_0^{2\pi} \left[2r^2 \right]_0^3 d\theta \\
&= \int_0^{2\pi} 18\, d\theta = 36\pi.
\end{aligned}
$$

33. The vector field $\mathbf{F}$ is not defined on the z axis, where $x = y = 0$. But any surface S of which C is a boundary must include at least one point on the z axis. Thus Stokes' Theorem will not apply in this case.

35. The cone $z = \sqrt{x^2 + y^2}$ is not smooth, since it has a sharp point at $(0, 0, 0)$. More formally, since the normal to the surface is

$$
\left\langle -\frac{\partial z}{\partial x}, -\frac{\partial z}{\partial y}, 1 \right\rangle = \left\langle -\frac{x}{\sqrt{x^2 + y^2}}, -\frac{y}{\sqrt{x^2 + y^2}}, 1 \right\rangle,
$$

we see that the normal is not well-defined at the origin, so that the surface is not smooth.

Applications

37. By Stokes' Theorem, which applies since all the functions involved are continuous and differentiable everywhere,

$$
\int_C \mathbf{F}(x, y, z) \cdot d\mathbf{r} = \iint_S \operatorname{curl} \mathbf{F}(x, y, z) \cdot \mathbf{n}\, dS,
$$

where S is the interior of the curve C. However, since all derivatives of the component functions of $\mathbf{F}$ are zero, $\operatorname{curl} \mathbf{F} = 0$, so that the circulation is zero.

39. Since

$$
\begin{aligned}
\operatorname{curl} &\langle -\sin x,\ 3y^3, 4z + 12 \rangle \\
&= \left\langle \frac{\partial}{\partial y}(4z + 12) - \frac{\partial}{\partial z}(3y^3),\ \frac{\partial}{\partial z}(-\sin x) - \frac{\partial}{\partial x}(4z + 12),\ \frac{\partial}{\partial x}(3y^3) - \frac{\partial}{\partial y}(-\sin x) \right\rangle \\
&= \langle 0, 0, 0 \rangle,
\end{aligned}
$$

$\mathbf{F}$ is irrotational everywhere.

41. Since

$$
\begin{aligned}
\operatorname{curl} \langle 2xyz, x^2z, x^2y \rangle &= \left\langle \frac{\partial}{\partial y}(x^2y) - \frac{\partial}{\partial z}(x^2z),\ \frac{\partial}{\partial z}(2xyz) - \frac{\partial}{\partial x}(x^2y),\ \frac{\partial}{\partial x}(x^2z) - \frac{\partial}{\partial y}(2xyz) \right\rangle \\
&= \left\langle x^2 - x^2, 2xy - 2xy, 2xz - 2xz \right\rangle = \langle 0, 0, 0 \rangle,
\end{aligned}
$$

$\mathbf{F}$ is irrotational everywhere.

43. We have curl $\mathbf{F} = \langle -x, 0, z \rangle$, so that curl $\mathbf{F} \cdot \mathbf{n} = \langle -x, 0, z \rangle \cdot \langle 12, 2, -1 \rangle = -12x - z$. By Stokes' Theorem, since the functions involved are continuous and differentiable everywhere, the work done is

$$\int_C \mathbf{F}(x, y, z) \cdot d\mathbf{r} = \iint_S \text{curl } \mathbf{F}(x, y, z) \cdot \mathbf{n}\, dS = \iint_S (-12x - z)\, dS,$$

where S is the interior of C; thus S is the disk bounded by C. Parametrize using cylindrical coordinates, but with $y = r \cos \theta$, $z = r \sin \theta$, and x. Then S is given by $x = \frac{15 - 2r \cos \theta + r \sin \theta}{12}$ for $r \in [0, 2]$, $\theta \in [0, 2\pi]$, so that

$$\iint_S (-12x - z)\, dS = \int_0^2 \int_0^{2\pi} (-(15 - 2r \cos \theta + r \sin \theta) - r \sin \theta)\, r\, d\theta\, dr$$

$$= \int_0^2 \int_0^{2\pi} (-15r + 2r^2 \cos \theta - 2r^2 \sin \theta)\, d\theta\, dr$$

$$= \int_0^2 \left[-15r\theta + 2r^2 \sin \theta + 2r^2 \cos \theta \right]_0^{2\pi}\, dr$$

$$= \int_0^2 (-30\pi r)\, dr$$

$$= \left[-15\pi r^2 \right]_0^2 = -60\pi.$$

45. (a) To compute $\nabla \times \mathbf{F}$, we regard $\mathbf{F}$ as the vector $\langle 0, 1.152 - 0.8x^2, 0 \rangle$ in $\mathbb{R}^3$. Then the curl is $\nabla \times \mathbf{F} = \langle 0, 0, -1.6x \rangle$. The normal vector is $\mathbf{n} = \langle 0, 0, 1 \rangle$, so that (using polar coordinates)

$$\iint_R \nabla \times \mathbf{F} \cdot \mathbf{n}\, dA = \iint_R -1.6x\, dA = -1.6 \int_0^1 \int_0^{2\pi} r \cos \theta \cdot r\, d\theta\, dr$$

$$= -1.6 \int_0^1 \left[r^2 \sin \theta \right]_0^{2\pi}\, dr = 0.$$

(b) The outward-pointing unit normal to the unit circle is $\langle x, y \rangle$, so that

$$\mathbf{F} \cdot \mathbf{n} = \left\langle 0, 1.152 - 0.8x^2 \right\rangle \cdot \langle x, y \rangle = (1.152 - 0.8x^2)y.$$

To compute the integral, we convert to polar coordinates. Since we are integrating around the unit circle, the volume element is $r = 1$, and $x = r \cos \theta = \cos \theta$, $y = r \sin \theta = \sin \theta$, so the integral is

$$\int_{\partial R} \mathbf{F} \cdot \mathbf{n}\, ds = \int_0^{2\pi} (1.152 - 0.8 \cos^2 \theta) \sin \theta\, d\theta$$

$$= \int_0^{2\pi} (1.152 \sin \theta - 0.8 \cos^2 \theta \sin \theta)\, d\theta$$

$$= \left[-1.152 \cos \theta + \frac{0.8}{3} \cos^3 \theta \right]_0^{2\pi} = 0.$$

(c) Since the flux across the boundary is zero from part (b), this tells us that as much water flows in as flows out.

Proofs

47. Since C is in the plane $z = r$, any parametrization of C looks like $\mathbf{r}(t) = \langle f(t), g(t), r \rangle$, so that $\mathbf{r}'(t) = \langle f'(t), g'(t), 0 \rangle$. Then

$$\int_C \mathbf{F} \cdot d\mathbf{r} = \int_C \langle F_1(x, y, z), F_2(x, y, z), F_3(x, y, z) \rangle \cdot \langle f'(t), g'(t), 0 \rangle\, dt$$

$$= \int_C (F_1(x, y, z)f'(t) + F_2(x, y, z)g'(t))\, dt.$$

This is independent of F_3.

49. Let the region in the given plane that is enclosed by C be S. A normal to S is the vector $\langle a, b, c \rangle$, which is a constant vector. Since $\mathbf{F}(x, y, z)$ consists of linear functions, its curl will also be constant since (for example) $\frac{\partial F_1}{\partial y} = \frac{\partial(\alpha x + \beta y + \gamma z)}{\partial y} = \beta$. Thus

$$\int_C \mathbf{F} \cdot d\mathbf{r} = \iint_S \operatorname{curl} \mathbf{F} \cdot \mathbf{n} \, dS = \iint_S \langle \text{constant} \rangle \, dS = \langle \text{constant} \rangle \iint_S dS.$$

But that integral gives the area of S, so that the line integral is a constant times the area of S.

51. Since the boundary of each surface is the curve C, and since the given orientations of the surfaces are compatible with the counterclockwise parametrization of C, two applications of Stokes' Theorem gives

$$\iint_{S_1} \operatorname{curl} \mathbf{F}(x, y, z) \cdot \mathbf{n} \, dS = \int_C \mathbf{F}(x, y, z) \cdot d\mathbf{r} = \iint_{S_2} \operatorname{curl} \mathbf{F}(x, y, z) \cdot \mathbf{n} \, dS.$$

Thinking Forward

Generalizing to Higher Dimensions

▷ For $\mathbb{R}^4$, the theorem would look much the same except for notation: Suppose $\mathcal{S}$ is a smooth or piecewise smooth oriented surface with a smooth or piecewise-smooth boundary curve C. Suppose that $\mathcal{S}$ has an oriented unit normal vector $\mathbf{n}$ and that C has a parametrization that traverses C in the counterclockwise direction with respect to $\mathbf{n}$ (*Note:* This concept would likely have to be thought through. In three dimensions, its meaning is clear: look "down" on C from the direction of the normal. It's not completely obvious that this makes sense in higher dimensions). If $\mathbf{F}(x, y, z, w) = \langle F_1(x, y, z, w), F_2(x, y, z, w), F_3(x, y, z, w), F_4(x, y, z, w) \rangle$ is a vector field on an open region containing $\mathcal{S}$, then

$$\int_C \mathbf{F}(x, y, z, w) \, d\mathbf{r} = \iint_{\mathcal{S}} \operatorname{curl} \mathbf{F}(x, y, z, w) \cdot \mathbf{n} \, d\mathcal{S}.$$

▷ This generalization too is pretty similar to the stated version of Stokes' Theorem. The boundary of a solid is a surface, so: Suppose W is a smooth or piecewise smooth oriented solid with a smooth or piecewise-smooth boundary surface S. Suppose that W has an oriented unit normal vector $\mathbf{n}$ and that S has a parametrization $S = \mathbf{r}(u, v)$ such that its normal vector $\mathbf{r}_u \times \mathbf{r}_v$ is consistently oriented with the normal $\mathbf{n}$. If $\mathbf{F}(x, y, z, w) = \langle F_1(x, y, z, w), F_2(x, y, z, w), F_3(x, y, z, w), F_4(x, y, z, w) \rangle$ is a vector field on an open region containing W, then

$$\iint_S \mathbf{F}(x, y, z, w) \, dS = \iiint_W \operatorname{curl} \mathbf{F}(x, y, z, w) \cdot \mathbf{n} \, dW.$$

14.6   The Divergence Theorem

Thinking Back

A Long Look Back

▷ Many answers are possible. All of our definitions of differentiation and integration are given as limiting processes: derivatives are limits of difference quotients, while integrals are limits of one form or another of Riemann sums.

▷ Many answers are possible. These definitions are all variant forms of the basic subdivide, approximate, add strategy for integrals in earlier sections of this book.

▷ Many answers are possible. For example, how can integrals be defined over objects of dimension higher than two and how does the Fundamental Theorem extend to those cases?

▷ Many answers are possible. For example, all of these theorems, starting with the Fundamental Theorem, are similar in that they relate integration on a region to integration along a boundary. Is there a framework in which all of these theorems can be expressed in a single form, applying to regions of any dimension?

▷ Many answers are possible. See the previous part.

Concepts

1. (a) True. All of these theorems, starting with the Fundamental Theorem, are similar in that they relate integration on a region to integration along a boundary.

   (b) True. The Divergence Theorem relies on Green's Theorem, which imposes that condition, on each slice of the region W parallel to the xy plane.

   (c) False. That is certainly one application of the theorem, but it can also be used in reverse, to evaluate a difficult triple integral as a surface integral.

   (d) True. Quite often, regions in $\mathbb{R}^3$ have simpler descriptions than their boundaries.

   (e) False. It is a generalization, ultimately, of the Fundamental Theorem of Calculus.

   (f) False. Stokes' Theorem applies to surfaces in $\mathbb{R}^3$, while the Divergence Theorem applies to solids in $\mathbb{R}^3$. If it can be said to be a consequence of any theorem in this chapter, it would be Green's Theorem, since that is used in the proof; however, the proof is not obvious even assuming Green's Theorem.

   (g) True. The flux through the two closed surfaces is equal to a triple integral over the solid bounded by those surfaces, by the Divergence Theorem.

   (h) True. The Divergence Theorem relies on Green's Theorem, which is used (in the simple case in the text) on each slice of the region W parallel to the xy plane.

3. No. The Divergence Theorem requires that the region W be bounded by a smooth or piecewise-smooth surface.

5. The paraboloid can be parametrized by $\langle x, y, x^2 + y^2 \rangle$, which is clearly a smooth parametrization since all the functions involved are polynomials, and since $\mathbf{r}_x \times \mathbf{r}_y = \langle -2x, -2y, 1 \rangle \neq \mathbf{0}$ anywhere. However, the rectangular solid cannot be smoothly parametrized along its edges or at its vertices since these are sharp corners and thus no parametrization will have a well-defined normal vector there. It is, however, piecewise smooth, since each of its six sides is a plane and thus can be smoothly parametrized.

7. (a) A plot of the vector field is:

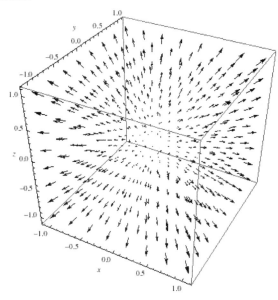

Since the vectors all point outwards, we would expect that (assuming an outward-pointing normal) the integral, which is the flux of **F** through the sphere, would be positive.

(b) A plot of the vector field is

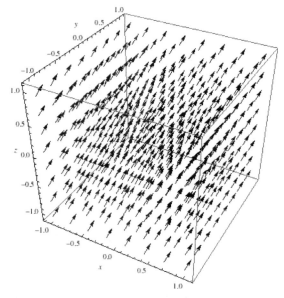

Since the vectors are all the same, and point in the same direction, any positive flux at one point on the sphere will be exactly balanced by a corresponding negative flux on the antipodal point. So we would expect this flux integral to be zero.

(c) For the first vector field,

$$\text{div}\, \langle x, y, z \rangle = \frac{\partial x}{\partial x} + \frac{\partial y}{\partial y} + \frac{\partial z}{\partial z} = 3,$$

so that by the Divergence Theorem,

$$\iint_{\mathcal{S}} \mathbf{F} \cdot \mathbf{n}\, d\mathcal{S} = \iiint_{W} \text{div}\, \mathbf{F}\, dV = 3 \iiint_{W} dV,$$

which is three times the volume of the sphere, so it is positive (and equal to 4π).

For the second vector field, since all three components are constants, div $\mathbf{F} = 0$, so that by the Divergence Theorem,

$$\iint_{\mathcal{S}} \mathbf{F} \cdot \mathbf{n} \, d\mathcal{S} = \iiint_{W} \operatorname{div} \mathbf{F} \, dV = \iiint_{W} 0 \, dV = 0.$$

9. Since $\nabla f = \left\langle \frac{\partial f}{\partial x}, \frac{\partial f}{\partial y}, \frac{\partial f}{\partial z} \right\rangle$, we have

$$\operatorname{div} \nabla f = \frac{\partial}{\partial x} \frac{\partial f}{\partial x} + \frac{\partial}{\partial y} \frac{\partial f}{\partial y} + \frac{\partial}{\partial z} \frac{\partial f}{\partial z} = f_{xx} + f_{yy} + f_{zz}.$$

11. For example, $\mathbf{F}(x, y, z) = \langle 1, 1, 1 \rangle$ is conservative, since if $f(x, y, z) = x + y + z$, then $\nabla f = \mathbf{F}$. However,

$$\operatorname{div} \mathbf{F} = \frac{\partial}{\partial x}(1) + \frac{\partial}{\partial y}(1) + \frac{\partial}{\partial z}(1) = 0.$$

13. For example, $\mathbf{F}(x, y, z) = \langle x, x + y, z \rangle$ is not conservative since $\frac{\partial F_1}{\partial y} \neq \frac{\partial F_2}{\partial x}$. However,

$$\operatorname{div} \mathbf{F} = \frac{\partial}{\partial x}(x) + \frac{\partial}{\partial y}(x + y) + \frac{\partial}{\partial z}(z) = 3.$$

15. It cannot, since $\mathcal{S}$ is neither smooth nor piecewise smooth — there is no parametrization of the cone, and no way to break it up into multiple regions, that addresses the sharp point at $(0, 0, 0)$.

17. No, it cannot, since $\mathbf{F}$ is not defined at $(0, 0, 0)$, and $(0, 0, 0)$ is in the region W enclosed by $\mathcal{S}$. Thus the hypotheses of the Divergence Theorem are not satisfied.

19. Yes, it can. The surface given bounds a solid, and the surface is piecewise smooth (it consists of two planes and a piece of the hyperboloid, each of which is smooth). Also, the vector field $\mathbf{F}$ is defined and continuous (and differentiable) everywhere.

Skills

21. We have

$$\operatorname{div} \mathbf{F} = \frac{\partial x}{\partial x} + \frac{\partial y}{\partial y} + \frac{\partial z}{\partial z} = 3.$$

23. We have

$$\operatorname{div} \mathbf{F} = \frac{\partial}{\partial x}(xe^{xyz}) + \frac{\partial}{\partial y}(ye^{xyz}) + \frac{\partial}{\partial z}(ze^{xyz})$$
$$= e^{xyz} + xyze^{xyz} + e^{xyz} + xyze^{xyz} + xyz + xyze^{xyz}$$
$$= 3e^{xyz}(xyz + 1).$$

25. Since $\mathcal{S}$ is piecewise smooth and $\mathbf{F}$ is defined everywhere, the Divergence Theorem applies. We have

$$\operatorname{div} \mathbf{F} = \frac{\partial}{\partial x}(4x^3yz) + \frac{\partial}{\partial y}(6x^2y^2z) + \frac{\partial}{\partial z}(6x^2yz^2) = 12x^2yz + 12x^2yz + 12x^2yz = 36x^2yz.$$

Then

$$\iint_{\mathcal{S}} \mathbf{F}(x,y,z) \cdot \mathbf{n} \, d\mathcal{S} = \iiint_{W} \operatorname{div} \mathbf{F} \, dV$$

$$= \int_{0}^{\pi} \int_{0}^{\pi} \int_{0}^{\pi} 36x^2 yz \, dz \, dy \, dx$$

$$= \int_{0}^{\pi} \int_{0}^{\pi} \left[18x^2 yz^2\right]_{z=0}^{z=\pi} dy \, dx$$

$$= \int_{0}^{\pi} \int_{0}^{\pi} 18\pi^2 x^2 y \, dy \, dx$$

$$= \int_{0}^{\pi} \left[9\pi^2 x^2 y^2\right]_{y=0}^{y=\pi} dx$$

$$= \int_{0}^{\pi} 9\pi^4 x^2 \, dx$$

$$= \left[3\pi^4 x^3\right]_{0}^{\pi} = 3\pi^7.$$

27. Since $\mathcal{S}$ is piecewise smooth and $\mathbf{F}$ is defined everywhere, the Divergence Theorem applies. We have

$$\operatorname{div} \mathbf{F} = \frac{\partial}{\partial x}(e^z x \sin y) + \frac{\partial}{\partial y}(e^z \cos y) + \frac{\partial}{\partial z}(e^x \tan^{-1} y) = e^z \sin y - e^z \sin y + 0 = 0.$$

Thus the integral is

$$\iint_{\mathcal{S}} \mathbf{F}(x,y,z) \cdot \mathbf{n} \, d\mathcal{S} = \iiint_{W} \operatorname{div} \mathbf{F} \, dV = \iiint_{W} 0 \, dV = 0.$$

29. Since $\mathcal{S}$ is piecewise smooth and $\mathbf{F}$ is defined everywhere, the Divergence Theorem applies. We have

$$\operatorname{div} \mathbf{F} = \frac{\partial}{\partial x}(15xz^2) + \frac{\partial}{\partial y}(15yx^2) + \frac{\partial}{\partial z}(15zy^2) = 15(x^2 + y^2 + z^2).$$

Parametrize the given hemisphere using spherical coordinates, so that $\phi \in \left[\frac{\pi}{2}, \pi\right]$, and $\operatorname{div} \mathbf{F} = 15\rho^2$. Then the integral is

$$\iint_{\mathcal{S}} \mathbf{F}(x,y,z) \cdot \mathbf{n} \, d\mathcal{S} = \iiint_{W} \operatorname{div} \mathbf{F} \, dV$$

$$= \int_{0}^{2\pi} \int_{\pi/2}^{\pi} \int_{0}^{1} 15\rho^2 \cdot \rho^2 \sin\phi \, d\rho \, d\phi \, d\theta$$

$$= 15 \int_{0}^{2\pi} \int_{\pi/2}^{\pi} \int_{0}^{1} \rho^4 \sin\phi \, d\rho \, d\phi \, d\theta$$

$$= 15 \int_{0}^{2\pi} \int_{\pi/2}^{\pi} \left[\frac{1}{5}\rho^5 \sin\phi\right]_{\rho=0}^{\rho=1} d\phi \, d\theta$$

$$= 3 \int_{0}^{2\pi} \int_{\pi/2}^{\pi} \sin\phi \, d\phi \, d\theta$$

$$= 3 \int_{0}^{2\pi} \left[-\cos\phi\right]_{\phi=\pi/2}^{\phi=\pi} d\theta$$

$$= 6 \int_{0}^{2\pi} d\theta = 12\pi.$$

31. Since $\mathcal{S}$ is piecewise smooth and $\mathbf{F}$ is defined everywhere, the Divergence Theorem applies. We have

$$\operatorname{div} \mathbf{F} = \frac{\partial}{\partial x}(xz) + \frac{\partial}{\partial y}(yz) + \frac{\partial}{\partial z}(xyz) = xy + 2z.$$

Parametrize the solid bounded by the cylinder using cylindrical coordinates; then it is defined by $\theta \in [0, 2\pi]$, $z \in [-2, 2]$, and $r \in [0, 3]$. Also, div $\mathbf{F} = xy + 2z = r^2 \sin\theta \cos\theta + 2z$. Then the integral is

$$\iint_{\mathcal{S}} \mathbf{F}(x, y, z) \cdot \mathbf{n} \, d\mathcal{S} = \iiint_W \operatorname{div} \mathbf{F} \, dV$$

$$= \int_0^{2\pi} \int_0^3 \int_{-2}^2 (r^2 \sin\theta \cos\theta + 2z) \cdot r \, dz \, dr \, d\theta$$

$$= \int_0^{2\pi} \int_0^3 \int_{-2}^2 (r^3 \sin\theta \cos\theta + 2rz) \, dz \, dr \, d\theta$$

$$= \int_0^{2\pi} \int_0^3 \left[r^3 z \sin\theta \cos\theta + 2rz^2 \right]_{z=-2}^{z=2} dr \, d\theta$$

$$= \int_0^{2\pi} \int_0^3 4r^3 \sin\theta \cos\theta \, dr \, d\theta$$

$$= \int_0^{2\pi} \left[r^4 \sin\theta \cos\theta \right]_{r=0}^{r=3} d\theta$$

$$= \int_0^{2\pi} (81 \sin\theta \cos\theta) \, d\theta$$

$$= \left[\frac{81}{2} \sin^2\theta \right]_0^{2\pi} = 0.$$

33. Since $\mathcal{S}$ is piecewise smooth and $\mathbf{F}$ is defined everywhere, the Divergence Theorem applies. We have

$$\operatorname{div} \mathbf{F} = \frac{\partial}{\partial x}(\sin y \cos z) + \frac{\partial}{\partial y}(yz^2) + \frac{\partial}{\partial z}(zx^2) = x^2 + z^2.$$

In cylindrical coordinates (modified to be centered on the y axis), then, div $\mathbf{F} = r^2$, and the region is defined by $\theta \in [0, 2\pi]$, $y \in [1, 4]$, and (since $r = \sqrt{x^2 + z^2} = \sqrt{y}$) $r \in [0, \sqrt{y}]$. Then the integral becomes

$$\iint_{\mathcal{S}} \mathbf{F}(x, y, z) \cdot \mathbf{n} \, d\mathcal{S} = \iiint_W \operatorname{div} \mathbf{F} \, dV$$

$$= \int_0^{2\pi} \int_1^4 \int_0^{\sqrt{y}} r^2 \cdot r \, dr \, dy \, d\theta$$

$$= \int_0^{2\pi} \int_1^4 \left[\frac{1}{4} r^4 \right]_{r=0}^{r=\sqrt{y}} dy \, d\theta$$

$$= \int_0^{2\pi} \int_1^4 \frac{1}{4} y^2 \, dy \, d\theta$$

$$= \int_0^{2\pi} \left[\frac{1}{12} y^3 \right]_{y=1}^{y=4} dy \, d\theta$$

$$= \int_0^{2\pi} \frac{21}{4} \, d\theta = \frac{21}{2} \pi.$$

35. Since $\mathcal{S}$ is piecewise smooth and $\mathbf{F}$ is defined everywhere, the Divergence Theorem applies. We have

$$\operatorname{div} \mathbf{F} = \frac{\partial}{\partial x}(x^2 y^2) + \frac{\partial}{\partial y}(2xy) + \frac{\partial}{\partial z}\left(\frac{2}{3} xz^3 \right) = 2xy^2 + 2x + 2xz^2 = 2x(1 + y^2 + z^2).$$

In cylindrical coordinates (modified to be centered on the x axis), then, div $\mathbf{F} = 2x(1 + r^2)$. The region is defined by $\theta \in [0, 2\pi]$, $x \in [1, 4]$, and $r \in [0, \sqrt{x^2 - 1}]$. Then the integral becomes

$$
\iint_{\mathcal{S}} \mathbf{F}(x, y, z) \cdot \mathbf{n} \, d\mathcal{S} = \iiint_{W} \operatorname{div} \mathbf{F} \, dV
$$

$$
= \int_{0}^{2\pi} \int_{1}^{4} \int_{0}^{\sqrt{x^2 - 1}} 2x(1 + r^2) \cdot r \, dr \, dx \, d\theta
$$

$$
= \int_{0}^{2\pi} \int_{1}^{4} \int_{0}^{\sqrt{x^2 - 1}} 2x(r + r^3) \, dr \, dx \, d\theta
$$

$$
= \int_{0}^{2\pi} \int_{1}^{4} \left[2x \left(\frac{1}{2} r^2 + \frac{1}{4} r^4 \right) \right]_{r=0}^{r=\sqrt{x^2 - 1}} dx \, d\theta
$$

$$
= \frac{1}{2} \int_{0}^{2\pi} \int_{1}^{4} \left(2x(x^2 - 1) + x(x^2 - 1)^2 \right) dx \, d\theta
$$

$$
= \frac{1}{2} \int_{0}^{2\pi} \int_{1}^{4} \left(x^5 - x \right) dx \, d\theta
$$

$$
= \frac{1}{2} \int_{0}^{2\pi} \left[\frac{1}{6} x^6 - \frac{1}{2} x^2 \right]_{x=1}^{x=4} d\theta
$$

$$
= \frac{1}{2} \int_{0}^{2\pi} 675 \, d\theta = 675\pi.
$$

37. Use the Divergence Theorem to convert this to an integral over the interior of the sphere. We have

$$
\operatorname{div} \mathbf{F} = \frac{\partial}{\partial x} (2xe) + \frac{\partial}{\partial y} (zx - ye) + \frac{\partial}{\partial z} \left(\ln(x^2 + y^2 + 2) \right) = 2e - e + 0 = e.
$$

Then

$$
\int_{\mathcal{S}} \mathbf{F}(x, y, z) \cdot \mathbf{n} \, d\mathcal{S} = \iiint_{W} \operatorname{div} \mathbf{F} \, dV = e \iiint_{W} 1 \, dV,
$$

which is just e times the volume of the sphere. So the answer is

$$
e \cdot \frac{4}{3} \pi \cdot 3^3 = 36e\pi.
$$

39. Since the surface is clearly piecewise smooth, and $\mathbf{F}$ is defined everywhere, the Divergence Theorem applies. We have

$$
\operatorname{div} \mathbf{F} = \frac{\partial}{\partial x} (8x) + \frac{\partial}{\partial y} (-13y) + \frac{\partial}{\partial z} (13z - 12e^y) = 8.
$$

Now, by the Divergence Theorem,

$$
\iint_{\mathcal{S}} \mathbf{F}(x, y, z) \cdot \mathbf{n} \, d\mathcal{S} = \iiint_{W} \operatorname{div} \mathbf{F} \, dV = 8 \iiint_{W} dV,
$$

so it is 8 times the volume of the region. A plot of the region is

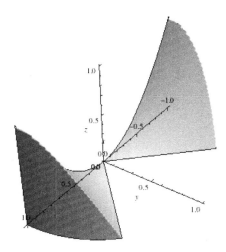

Clearly, from both the graph and the equations defining the region, the volumes of the two halves are equal, so we need only compute the volume of the right half (say), for $x \geq 0$, and multiply by 16. Using cylindrical coordinates, the base of the right half is defined by $0 \leq \theta \leq \frac{\pi}{4}$ and $0 \leq r \leq 1$. Its height is $0 \leq z \leq x^2 - y^2 = r^2(\cos^2\theta - \sin^2\theta) = r^2\cos 2\theta$. Thus the integral is

$$8 \iiint_W dV = 16 \int_0^{\pi/4} \int_0^1 \int_0^{r^2\cos 2\theta} r\, dz\, dr\, d\theta$$

$$= 16 \int_0^{\pi/4} \int_0^1 [rz]_{z=0}^{z=r^2\cos 2\theta} \, dr\, d\theta$$

$$= 16 \int_0^{\pi/4} \int_0^1 r^3 \cos 2\theta\, dr\, d\theta$$

$$= 16 \int_0^{\pi/4} \left[\frac{1}{4}r^4 \cos 2\theta\right]_{r=0}^{r=1} d\theta$$

$$= 4 \int_0^{\pi/4} \cos 2\theta\, d\theta$$

$$= 4 \left[\frac{1}{2}\sin 2\theta\right]_0^{\pi/4} = 2.$$

41. Since the surface is clearly piecewise smooth, and $\mathbf{F}$ is defined everywhere, the Divergence Theorem applies. We have

$$\operatorname{div}\mathbf{F} = \frac{\partial}{\partial x}(xe^y) + \frac{\partial}{\partial y}(\ln(xyz)) + \frac{\partial}{\partial z}(xyz^2) = e^y + \frac{1}{y} + 2xyz.$$

Then we have

$$\iint_{\mathcal{S}} \mathbf{F}(x,y,z) \cdot \mathbf{n}\, d\mathcal{S} = \iiint_W \operatorname{div} \mathbf{F}\, dV$$

$$= \int_1^7 \int_1^5 \int_1^4 \left(e^y + \frac{1}{y} + 2xyz \right) dz\, dy\, dx$$

$$= \int_1^7 \int_1^5 \left[ze^y + \frac{z}{y} + xyz^2 \right]_{z=1}^{z=4} dy\, dx$$

$$= \int_1^7 \int_1^5 \left(3e^y + \frac{3}{y} + 15xy \right) dy\, dx$$

$$= \int_1^7 \left[3e^y + 3\ln|y| + \frac{15}{2}xy^2 \right]_{y=1}^{y=5} dx$$

$$= \int_1^7 \left(3e^5 - 3e + 3\ln 5 + 180x \right) dx$$

$$= \left[\left(3e^5 - 3e + 3\ln 5 \right) x + 90x^2 \right]_1^7$$

$$= 18(e^5 - e + \ln 5) + 90 \cdot 48 = 18(e^5 - e + \ln 5) + 4320.$$

43. Since the surface is clearly piecewise smooth, and $\mathbf{F}$ is defined everywhere, the Divergence Theorem applies. We have

$$\operatorname{div} \mathbf{F} = \frac{\partial}{\partial x}(y\cos z) + \frac{\partial}{\partial y}(3y) + \frac{\partial}{\partial z}(\sin(xy)) = 3.$$

Thus

$$\iint_{\mathcal{S}} \mathbf{F} \cdot \mathbf{n}\, d\mathcal{S} = \iiint_W \operatorname{div} \mathbf{F}\, dV = 3 \iiint_W dV.$$

Thus the flux is three times the volume of the pyramid. But the volume of a pyramid is $\frac{1}{3}bh$ where b is the area of the base and h is the height. Here, b is the area of a 2×2 square, so is 4, and $h = 4$. Hence the integral is

$$3 \cdot \frac{1}{3} \cdot 4 \cdot 4 = 16.$$

Applications

45. (a) Computing,

$$\iint_R \nabla \cdot \mathbf{F}\, dA = \iint_R \left(\frac{\partial}{\partial x}(0) + \frac{\partial}{\partial y}(1.152 - 0.8x^2) \right) dA = \iint_R 0\, dA = 0.$$

(b) The outward-pointing unit normal to the unit circle is $\langle x, y \rangle$, so that

$$\mathbf{F} \cdot \mathbf{n} = \left\langle 0, 1.152 - 0.8x^2 \right\rangle \cdot \langle x, y \rangle = (1.152 - 0.8x^2)y.$$

To compute the integral, we convert to polar coordinates. Since we are integrating around the unit circle, the volume element is $r = 1$, and $x = r\cos\theta = \cos\theta$, $y = r\sin\theta = \sin\theta$, so the integral is

$$\int_{\partial R} \mathbf{F} \cdot \mathbf{n}\, ds = \int_0^{2\pi} (1.152 - 0.8\cos^2\theta)\sin\theta\, d\theta$$

$$= \int_0^{2\pi} (1.152\sin\theta - 0.8\cos^2\theta\sin\theta)\, d\theta$$

$$= \left[-1.152\cos\theta + \frac{0.8}{3}\cos^3\theta \right]_0^{2\pi} = 0.$$

(c) Since the flux across the boundary is zero from part (b), this tells us that as much water flows in as flows out.

47. (a) Along the left and right edges we have $\mathbf{n} = \langle \pm 1, 0 \rangle$, so that $\mathbf{v} \cdot \mathbf{n} = \langle 0, v_2(x,y) \rangle \cdot \langle 1, 0 \rangle = 0$. Thus the line integral is zero along the left and right edges. Parametrize the bottom edge by $\mathbf{r}_1(t) = \langle t, 0 \rangle$ and the top edge by $\mathbf{r}_2(t) = \langle 0.113 - t, 1 \rangle$ (remember we need to parametrize the entire curve counterclockwise), for $t \in [0, 0.113]$. Along the top edge, $\mathbf{n} = \langle 0, 1 \rangle$, while along the bottom edge it is $\langle 0, -1 \rangle$. Then we get (using the substitution $u = 0.113 - t$)

$$\int_C \mathbf{v} \cdot \mathbf{n} \, ds = \int_0^{0.113} \langle 0, v_2(t,0) \rangle \cdot \langle 0, -1 \rangle \, dt + \int_0^{0.113} \langle 0, v_2(0.113 - t, 1) \rangle \cdot \langle 0, 1 \rangle \, dt$$

$$= \int_0^{0.113} (-v_2(t,0)) \, dt + \int_0^{0.113} v_2(0.113 - t, 1) \, dt$$

$$= -\int_0^{0.113} v_2(t,0) \, dt - \int_{0.113}^0 v_2(u,1) \, du$$

$$= -\int_0^{0.113} v_2(t,0) \, dt + \int_0^{0.113} v_2(u,1) \, du$$

$$= -\int_0^{0.113} v_2(t,0) \, dt + \int_0^{0.113} (v_2(u,0) - 5) \, du$$

$$= -\int_0^{0.113} 5 \, du = -0.565.$$

Since the horizontal velocity is zero, no cars are leaving the roadway to the left or right. However, the flux is negative, which means that the number of cars in the roadway is increasing. This means that the cars are slowing down as they move north along the roadway (so that more enter at the bottom than leave at the top).

(b) By the Divergence Theorem,

$$\frac{1}{A} \iint_S \nabla \cdot \mathbf{v} \, dx \, dy = \frac{1}{A} \iint_S \operatorname{div} \mathbf{v} \, dx \, dy = \frac{1}{A} \int_C \mathbf{v} \cdot \mathbf{n} \, dx \, dy = \frac{1}{A} (-0.565) = \frac{-0.565}{0.113} = -5.$$

But the divergence of $\mathbf{v}$ is $\frac{\partial v_2}{\partial y}$, which is the acceleration along the road. Since the average acceleration is negative, this confirms that traffic is slowing down.

Proofs

49. If $f(x,y,z) = x^2 + y^2 + z^2$, then

$$\mathbf{F}(x,y,z) = \nabla f(x,y,z) = \left\langle \frac{\partial f}{\partial x}, \frac{\partial f}{\partial y}, \frac{\partial f}{\partial z} \right\rangle = \langle 2x, 2y, 2z \rangle.$$

Then $\operatorname{div} \mathbf{F} = 6$. Since S is obviously smooth and $\mathbf{F}$ is defined everywhere, we have by the Divergence Theorem

$$\iint_S \mathbf{F}(x,y,z) \cdot \mathbf{n} \, dS = \iiint_W \operatorname{div} \mathbf{F} \, dV = 6 \iiint_W dV.$$

Thus the flux is 6 times the volume of the unit sphere, so it is 8π.

51. For the given vector field,

$$\operatorname{div} \mathbf{F} = \frac{\partial}{\partial x} F_1 + \frac{\partial}{\partial y} F_2 + \frac{\partial}{\partial z} F_3 = a + c + e.$$

By the Divergence Theorem, if $\mathcal{S}$ is a smooth or piecewise smooth closed surface enclosing a solid W, then

$$\iint_{\mathcal{S}} \mathbf{F} \cdot \mathbf{n}\, d\mathcal{S} = \iiint_W \operatorname{div} \mathbf{F}(x,y,z)\, dV = \iiint_W (a+c+e)\, dV = (a+c+e) \iiint_W dV.$$

Since this is $a+c+e$ times the volume of W, it follows that the original integral is positive for every such $\mathcal{S}$ if and only if $a+c+e > 0$.

53. For example, $\mathbf{F}(x,y,z) = \langle y, x, y \rangle$ is not conservative since $\frac{\partial F_2}{\partial z} \neq \frac{\partial F_3}{\partial y}$. However,

$$\operatorname{div} \mathbf{F} = \frac{\partial}{\partial x}(y) + \frac{\partial}{\partial y}(x) + \frac{\partial}{\partial z}(y) = 0,$$

so that with appropriate conditions on $\mathcal{S}$ we can apply the Divergence Theorem to get

$$\iint_{\mathcal{S}} \mathbf{F}(x,y,z) \cdot \mathbf{n}\, d\mathcal{S} = \iiint_W \operatorname{div} \mathbf{F}\, dV = 0.$$

Thinking Forward

Calculus in $\mathbb{R}^4$ and beyond

▷ The Divergence Theorem in $\mathbb{R}^3$ relates the 3 dimensional volume integral of the divergence of a vector field $\mathbf{F}$ over some region W to the 2 dimensional surface integral of $\mathbf{F}$ over $U = \partial W$. So in $\mathbb{R}^n$, we would expect that the Divergence Theorem would say that if W is a sufficiently nice n-dimensional solid in $\mathbb{R}^n$, then

$$\int_W \operatorname{div} \mathbf{F}\, dW_n = \int_{\partial W} \mathbf{F} \cdot \mathbf{n}\, dS_{n-1},$$

where the integral on the right is an integral over an n dimensional object and that on the right is an integral over an $(n-1)$ dimensional object. (For $\mathbb{R}^4$, just substitute 4 for n in the discussion above).

Calculus for functions from $\mathbb{R}^n$ to $\mathbb{R}^m$

Since $f : \mathbb{R}^4 \to \mathbb{R}^5$, the image of f is some subset W of $\mathbb{R}^5$, so that f is a parametrization in $\mathbb{R}^4$ (i.e., using four variables) of W. Then the integral of f is a way of determining the volume of W by integrating in $\mathbb{R}^4$ rather than considering some possibly very complicated subset of $\mathbb{R}^5$.

Suppose that

$$f(x_1, x_2, x_3, x_4) = (f_1(x_1, x_2, x_3, x_4), f_2(x_1, x_2, x_3, x_4), f_3(x_1, x_2, x_3, x_4), f_4(x_1, x_2, x_3, x_4), f_5(x_1, x_2, x_3, x_4)).$$

Considering the partial derivatives, we could define a 5×4 matrix Df whose entry in the ith column and j^{th} row is

$$(Df)_{ji} = \frac{\partial f_i}{\partial x_j}.$$

Since this is a matrix, at any point $P = (x_1, x_2, x_3, x_4)$ in $\mathbb{R}^4$ it is a the same (from linear algebra) as a linear map, and it can be shown that with appropriate definitions, this linear map is the linear map from $\mathbb{R}^4$ to $\mathbb{R}^5$ whose image contains $f(P)$ and that most closely approximates f near P (in the same sense that the ordinary derivative of a function $g : \mathbb{R} \to \mathbb{R}$ is the line that best approximates g at a point). This idea can be generalized for $f : \mathbb{R}^n \to \mathbb{R}^m$.

Analysis

The approach taken to integration in this book, the *Riemann integral*, proceeds by partitioning the domain of a function into small pieces, approximating, and adding (note that even when you reverse the order of integration, you are still partitioning the domain — it is just that you have changed your view of which variable is the independent variable).

Another approach would have been to partition the *range* of the function rather than its domain. So for example, consider the size of the region under the graph of $y = f(x)$ between $y = 1$ and $y = 1.1$. This could consists of lots of separated regions each of which looks close to a rectangle, and the approximation then is the sum of the areas of those rectangles. Since f can be pretty complicated (for example, the function given in the problem statement), this requires a more sophisticated notion of the lengths of those small rectangles. This concept is called the *measure*, and there is a lot of mathematical rigor around defining what it means for a set of points to be measurable, and what its measure is, so that the notion corresponds to our intuition in simple cases. This kind of integral is called a *Lebesgue integral*.

For the function given, it turns out that on $[0, 1]$, say, the irrational numbers have measure 1 (so that the rationals have measure 0). Thus any small interval of height δy around $y = 1$ will have area approximately $\delta y \cdot 1$, and other intervals not including $y = 1$ will have area approximately zero. Hence the Lebesgue integral of this function on $[0, 1]$, is 1. Note that this function is not Riemann integrable. To see this, suppose we have any subdivision of $[0, 1]$ along the x axis, and consider any subrectangle. Then there is at least one rational and one irrational number in that subrectangle, so that the lower bound of $f(x)$ will be zero for every subrectangle and the upper bound will be 1. Thus the lower Riemann sum for any subdivision is 0 while the upper Riemann sum is 1. Since they do not converge to each other, the function is not Riemann integrable.

Chapter Review and Self-Test

1. Integrate the first component with respect to x to get $x^3 y^2$. Since the second component has only one term, and that term depends on x, we are done, and $x^3 y^2$ is a potential function. As a check, note that $\frac{\partial}{\partial y}(x^3 y^2) = 2x^3 y$, which is the second component of **F**.

3. Integrating the first component with respect to x gives xze^{y^2}. Since the only term of each of the other components depends on x, we are done, and $f(x, y, z) = xze^{y^2}$ is a potential function. As a check, note that

$$\frac{\partial}{\partial y}\left(xze^{y^2}\right) = 2xyze^{y^2}, \qquad \frac{\partial}{\partial z}\left(xze^{y^2}\right) = xe^{y^2}.$$

5. Since $\frac{\partial}{\partial y}\left(\frac{2x}{y}\right) = -\frac{2x}{y^2} \neq \frac{\partial}{\partial x}\left(\frac{x^2}{y^2}\right) = \frac{2x}{y^2}$, this is not conservative.

7. We have

$$\frac{\partial F_1}{\partial y} = \frac{\partial}{\partial y}\left(y^3 e^{xy^2}\right) = 3y^2 e^{xy^2} + y^3 \cdot 2xye^{xy^2} = \left(3y^2 + 2xy^4\right)e^{xy^2}$$

$$\frac{\partial F_2}{\partial x} = \frac{\partial}{\partial x}\left((1 + 2xy^2)e^{xy^2}\right) = 2y^2 e^{xy^2} + (1 + 2xy^2)\cdot y^2 e^{xy^2} = \left(3y^2 + 2xy^4\right)e^{xy^2}.$$

These are equal, so **F** is conservative. To find a potential function, integrate F_1 with respect to x to get ye^{xy^2}. Since all terms in F_2 depend on x, we are done, and ye^{xy^2} is a potential function. As a check, note that

$$\frac{\partial}{\partial y}(ye^{xy^2}) = e^{xy^2} + y \cdot 2xye^{xy^2} = (1 + 2xy^2)e^{xy^2} = F_2(x, y).$$

9. Since

$$\frac{\partial G_1}{\partial z} = \cos y \cos z \neq \frac{\partial G_3}{\partial x} = -\cos y \cos z,$$

G is not conservative.

11. With $\mathbf{r}(t) = \langle t, 3t \rangle$ for $0 \leq t \leq 1$, we have $\mathbf{r}'(t) = \langle 1, 3 \rangle$, and

$$\mathbf{F}(x,y) = \left\langle 7y^2, -3xy \right\rangle = \left\langle 7 \cdot (3t)^2, -3t(3t) \right\rangle = \left\langle 63t^2, -9t^2 \right\rangle.$$

Then

$$\int_C \mathbf{F}(x,y) \cdot d\mathbf{r} = \int_0^1 \left\langle 63t^2, -9t^2 \right\rangle \cdot \langle 1, 3 \rangle \, dt = \int_0^1 36t^2 \, dt = \left[12t^3 \right]_0^1 = 12.$$

13. Parametrize C by $\mathbf{r}(t) = \langle \cos t, \sin t \rangle$ for $t \in [0, 2\pi]$. Then $\mathbf{r}'(t) = \langle -\sin t, \cos t \rangle$, and

$$\mathbf{F}(x,y) = \left\langle \frac{x}{x^2+y^2}, -\frac{y}{x^2+y^2} \right\rangle = \left\langle \frac{\cos t}{\cos^2 t + \sin^2 t}, -\frac{\sin t}{\cos^2 t + \sin^2 t} \right\rangle = \langle \cos t, -\sin t \rangle.$$

Then

$$\int_C \mathbf{F}(x,y) \cdot d\mathbf{r} = \int_0^{2\pi} \langle \cos t, -\sin t \rangle \cdot \langle -\sin t, \cos t \rangle \, dt = \int_0^{2\pi} (-2 \sin t \cos t) \, dt = \left[\cos^2 t \right]_0^{2\pi} = 0.$$

15. Parametrize C by $\mathbf{r}(t) = \langle t, t, t \rangle$ for $0 \leq t \leq 1$. Then $\mathbf{r}'(t) = \langle 1, 1, 1 \rangle$, and

$$\mathbf{F}(x,y,z) = \langle e^x, e^y, e^z \rangle = \langle e^t, e^t, e^t \rangle.$$

Then

$$\int_C \mathbf{F}(x,y,z) \cdot d\mathbf{r} = \int_0^1 \langle e^t, e^t, e^t \rangle \cdot \langle 1, 1, 1 \rangle \, dt = \int_0^1 3e^t \, dt = \left[3e^t \right]_0^1 = 3e - 3.$$

17. This plane intersects the xy plane when $z = 0$, so on the line $2x + 3y = 12$. Thus the range of integration in the xy plane is $0 \leq x \leq 6$ and $0 \leq y \leq 4 - \frac{2}{3}x$. The area element is

$$dS = \sqrt{\left(\frac{\partial z}{\partial x}\right)^2 + \left(\frac{\partial z}{\partial y}\right)^2 + 1} \, dA = \sqrt{\left(-\frac{1}{2}\right)^2 + \left(-\frac{3}{4}\right)^2 + 1} \, dA = \frac{\sqrt{29}}{4} \, dA.$$

Thus the area is

$$\int_S dS = \int_0^6 \int_0^{4-(2/3)x} \frac{\sqrt{29}}{4} \, dy \, dx = \frac{\sqrt{29}}{4} \int_0^6 [y]_{y=0}^{y=4-(2/3)x} \, dx$$

$$= \frac{\sqrt{29}}{4} \int_0^6 \left(4 - \frac{2}{3}x\right) dx$$

$$= \frac{\sqrt{29}}{4} \left[4x - \frac{1}{3}x^2 \right]_0^6 = 3\sqrt{29}.$$

19. With $z = \sqrt{16 - x^2 - y^2}$, we get

$$dS = \sqrt{\left(\frac{\partial z}{\partial x}\right)^2 + \left(\frac{\partial z}{\partial y}\right)^2 + 1} \, dA$$

$$= \sqrt{\left(\frac{-x}{\sqrt{16-x^2-y^2}}\right)^2 + \left(\frac{y}{\sqrt{16-x^2-y^2}}\right)^2 + 1} \, dA$$

$$= \sqrt{\frac{16}{16-x^2-y^2}} \, dA = \frac{4}{\sqrt{16-x^2-y^2}} \, dA.$$

Then using polar coordinates, we get

$$\int_S dS = \iint_D \frac{4}{\sqrt{16 - x^2 - y^2}} \, dA$$

$$= \int_0^{2\pi} \int_0^3 \frac{4r}{\sqrt{16 - r^2}} \, dr \, d\theta$$

$$= \int_0^{2\pi} \left[-4\sqrt{16 - r^2} \right]_{r=0}^{r=3} \, d\theta$$

$$= \int_0^{2\pi} \left(16 - 4\sqrt{7} \right) d\theta$$

$$= (32 - 8\sqrt{7})\pi.$$

21. A normal vector to the surface is

$$\left\langle -\frac{\partial z}{\partial x}, -\frac{\partial z}{\partial y}, 1 \right\rangle = \left\langle -x(x^2 + y^2)^{-1/2}, -y(x^2 + y^2)^{-1/2}, 1 \right\rangle.$$

This points upwards, which is into the cone, so we negate and scale it to make it a unit vector:

$$\mathbf{n} = \frac{1}{\sqrt{x^2(x^2 + y^2)^{-1} + y^2(x^2 + y^2)^{-1} + 1}} \left\langle x(x^2 + y^2)^{-1/2}, y(x^2 + y^2)^{-1/2}, -1 \right\rangle$$

$$= \frac{1}{\sqrt{2}} \left\langle x(x^2 + y^2)^{-1/2}, y(x^2 + y^2)^{-1/2}, -1 \right\rangle.$$

We have for the volume element

$$\sqrt{\left(\frac{\partial z}{\partial x} \right)^2 + \left(\frac{\partial z}{\partial y} \right)^2 + 1} = \sqrt{x^2(x^2 + y^2)^{-1} + y^2(x^2 + y^2)^{-1} + 1} = \sqrt{2}.$$

On the surface, $\mathbf{F}(x, y, z) = \langle x, y, z \rangle = \left\langle x, y, \sqrt{x^2 + y^2} \right\rangle$. Thus the flux is

$$\int_S \mathbf{F}(x, y, z) \cdot \mathbf{n} \, dS = \iint_D \left(\left\langle x, y, \sqrt{x^2 + y^2} \right\rangle \cdot \left\langle x(x^2 + y^2)^{-1/2}, y(x^2 + y^2)^{-1/2}, -1 \right\rangle \right) dA$$

$$= \iint_D \left((x^2 + y^2)(x^2 + y^2)^{-1/2} - \sqrt{x^2 + y^2} \right) dA = 0.$$

23. Since $z = 4 - x - y$, we have for the volume element

$$\sqrt{\left(\frac{\partial z}{\partial x} \right)^2 + \left(\frac{\partial z}{\partial y} \right)^2 + 1} = \sqrt{3}.$$

The normal is $\langle 1, 1, 1 \rangle$, which is outward-pointing, and has norm $\sqrt{3}$, which will cancel with the volume element. Now, the plane $x + y + z = 4$ intersects the xy plane along the line $x + y = 4$. Finally, on the plane, we have

$$\mathbf{F}(x, y, z) = \langle xyz, 0, 0 \rangle = \langle xy(4 - x - y), 0, 0 \rangle,$$

so the flux is

$$\int_S \mathbf{F}(x,y,z) \cdot \mathbf{n}\, dS = \iint_D \langle xy(4-x-y), 0, 0 \rangle \cdot \langle 1,1,1 \rangle\, dA$$

$$= \int_0^4 \int_0^{4-x} (4xy - x^2 y - xy^2)\, dy\, dx$$

$$= \int_0^4 \left[2xy^2 - \frac{1}{2}x^2 y^2 - \frac{1}{3}xy^3 \right]_{y=0}^{y=4-x}\, dx$$

$$= \int_0^4 \left(2x(4-x)^2 - \frac{1}{2}x^2(4-x)^2 - \frac{1}{3}x(4-x)^3 \right)\, dx$$

$$= \int_0^4 \left(-\frac{1}{6}x^4 + 2x^3 - 8x^2 + \frac{32}{3}x \right)\, dx$$

$$= \left[-\frac{1}{30}x^5 + \frac{1}{2}x^4 - \frac{8}{3}x^3 + \frac{16}{3}x^2 \right]_0^4$$

$$= \frac{128}{15}.$$

25. We have

$$\operatorname{div} \mathbf{F} = \frac{\partial F_1}{\partial x} + \frac{\partial F_2}{\partial y} = \frac{\partial}{\partial x}(3x+4y) + \frac{\partial}{\partial y}(x-5y) = -2$$

$$\operatorname{curl} \mathbf{F} = \left(\frac{\partial F_2}{\partial x} - \frac{\partial F_1}{\partial y} \right) \mathbf{k} = \left(\frac{\partial}{\partial x}(x-5y) - \frac{\partial}{\partial y}(3x+4y) \right) \mathbf{k} = -3\mathbf{k}.$$

27. We have

$$\operatorname{div} \mathbf{F} = \frac{\partial F_1}{\partial x} + \frac{\partial F_2}{\partial y} + \frac{\partial F_3}{\partial z} = \frac{\partial}{\partial x}(x-y) + \frac{\partial}{\partial y}(y-z) + \frac{\partial}{\partial z}(z-x) = 3$$

$$\operatorname{curl} \mathbf{F} = \left\langle \frac{\partial F_3}{\partial y} - \frac{\partial F_2}{\partial z}, \frac{\partial F_1}{\partial z} - \frac{\partial F_3}{\partial x}, \frac{\partial F_2}{\partial x} - \frac{\partial F_1}{\partial y} \right\rangle$$

$$= \left\langle \frac{\partial}{\partial y}(z-x) - \frac{\partial}{\partial z}(y-z), \frac{\partial}{\partial z}(x-y) - \frac{\partial}{\partial x}(z-x), \frac{\partial}{\partial x}(y-z) - \frac{\partial}{\partial y}(x-y) \right\rangle$$

$$= \langle 1,1,1 \rangle.$$

29. We have

$$\frac{\partial F_2}{\partial x} - \frac{\partial F_1}{\partial y} = \frac{\partial}{\partial x}(2xy) - \frac{\partial}{\partial y}(y^2+1) = 2y - 2y = 0.$$

So by Green's Theorem,

$$\int_C \mathbf{F}(x,y) \cdot d\mathbf{r} = \iint_R \left(\frac{\partial F_2}{\partial x} - \frac{\partial F_1}{\partial y} \right) dA = \iint_R 0\, dA = 0.$$

31. We have

$$\frac{\partial F_2}{\partial x} - \frac{\partial F_1}{\partial y} = \frac{\partial}{\partial x}(xy) - \frac{\partial}{\partial y}(0) = y.$$

Using Green's Theorem gives

$$\int_C \mathbf{F}(x,y) \cdot d\mathbf{r} = \iint_R y\, dA = \int_{-3}^3 \int_{-3}^3 y\, dy\, dx = \int_{-3}^3 \left[\frac{1}{2}y^2 \right]_{y=-3}^{y=3}\, dx = \int_{-3}^3 0\, dx = 0.$$

33. Since evaluating the line integral would require three integrations, use Stokes' Theorem. We have

$$\operatorname{curl} \mathbf{F}(x,y,z) = \left\langle \frac{\partial F_3}{\partial y} - \frac{\partial F_2}{\partial z}, \frac{\partial F_1}{\partial z} - \frac{\partial F_3}{\partial x}, \frac{\partial F_2}{\partial x} - \frac{\partial F_1}{\partial y} \right\rangle = \langle -2z, -2x, -2y \rangle,$$

The three vertices determine the plane $3x + y + 2z = 6$, with upward-pointing normal $\langle 3,1,2 \rangle$. The given triangle sits over the triangle determined by $(2,0)$, $(0,6)$, and $(0,0)$ in the xy plane. Finally, on the given plane,

$$\operatorname{curl} \mathbf{F}(x,y,z) = \langle -2y - 2z, -2x, -2y \rangle = \left\langle -2y - 2\frac{6 - 3x - y}{2}, -2x, -2y \right\rangle = \langle 6x - 12, -2x, -2y \rangle.$$

So by Stokes' Theorem,

$$\begin{aligned}
\int_C \mathbf{F}(x,y,z) \cdot d\mathbf{r} &= \iint_S \langle 6x - 12, -2x, -2y \rangle \cdot \langle 3,1,2 \rangle \, dA \\
&= \int_0^2 \int_0^{6-3x} (18x - 36 - 2x - 4y) \, dy \, dx \\
&= \int_0^2 \int_0^{6-3x} (16x - 4y - 36) \, dy \, dx \\
&= \int_0^2 \left[16xy - 2y^2 - 36y \right]_{y=0}^{y=6-3x} dx \\
&= \int_0^2 \left(16x(6 - 3x) - 2(6 - 3x)^2 - 36(6 - 3x) \right) dx \\
&= \int_0^2 \left(-66x^2 + 276x - 288 \right) dx \\
&= \left[-22x^3 + 138x^2 - 288x \right]_0^2 = -200.
\end{aligned}$$

35. Since

$$\begin{aligned}
\operatorname{curl} \mathbf{F}(x,y,z) &= \left\langle \frac{\partial F_3}{\partial y} - \frac{\partial F_2}{\partial z}, \frac{\partial F_1}{\partial z} - \frac{\partial F_3}{\partial x}, \frac{\partial F_2}{\partial x} - \frac{\partial F_1}{\partial y} \right\rangle \\
&= \left\langle \frac{\partial}{\partial y}(z^2) - \frac{\partial}{\partial z}(5x), \frac{\partial}{\partial z}(5y) - \frac{\partial}{\partial x}(z^2), \frac{\partial}{\partial x}(5x) - \frac{\partial}{\partial y}(5y) \right\rangle \\
&= \langle 0,0,0 \rangle,
\end{aligned}$$

we have $\iint_S \operatorname{curl} \mathbf{F}(x,y,z) \cdot \mathbf{n} \, dS = 0$.

37. We have

$$\operatorname{div} \mathbf{F}(x,y,z) = \frac{\partial}{\partial x}(x^2 y) + \frac{\partial}{\partial y}(y^2 z) + \frac{\partial}{\partial z}(xz^2) = 2xy + 2yz + 2xz.$$

Parametrize the upper half-sphere using spherical coordinates; then

$$\operatorname{div} \mathbf{F}(x,y,z) = 2xy + 2yz + 2xz = 2\rho^2 \sin^2 \phi \sin \theta \cos \theta + 2\rho^2 \sin \phi \cos \phi \sin \theta + 2\rho^2 \sin \phi \cos \phi \cos \theta.$$

With a volume element of $\rho^2 \sin \phi$, we get

$$
\iint_{\mathcal{S}} \mathbf{F}(x, y, z) \cdot \mathbf{n} \, d\mathcal{S} = \iiint_{W} \operatorname{div} \mathbf{F}(x, y, z) \, dV
$$

$$
= 2 \iiint_{W} \rho^2 (\sin^2 \phi \sin \theta \cos \theta + \sin \phi \cos \phi \sin \theta + \sin \phi \cos \phi \cos \theta) \rho^2 \sin \phi \, dV
$$

$$
= 2 \int_0^1 \int_0^{\pi/2} \int_0^{2\pi} \rho^4 (\sin^3 \phi \sin \theta \cos \theta + \sin^2 \phi \cos \phi \sin \theta
$$

$$
+ \sin^2 \phi \cos \phi \cos \theta) \, d\theta \, d\phi \, d\rho
$$

$$
= 2 \int_0^1 \int_0^{\pi/2} \left[\rho^4 \left(\sin^3 \phi \cdot \frac{1}{2} \sin^2 \theta - \sin^2 \phi \cos \phi \cos \theta \right. \right.
$$

$$
\left. \left. + \sin^2 \phi \cos \phi \sin \theta \right) \right]_{\theta=0}^{\theta=2\pi} d\phi \, d\rho
$$

$$
= 2 \int_0^1 \int_0^{\pi/2} 0 \, d\phi \, d\rho = 0.
$$

39. Using cylindrical coordinates centered on the x axis, the solid W bounded by the given surface is described by $0 \le \theta \le 2\pi$, $2 \le x \le 5$, and $0 \le r = \sqrt{y^2 + z^2} \le x$. We have

$$
\operatorname{div} \mathbf{F}(x, y, z) = \frac{\partial}{\partial x}(x^3 z) + \frac{\partial}{\partial y}(xy) + \frac{\partial}{\partial z}(4yz) = 3x^2 z + x + 4y.
$$

In cylindrical coordinates, this is

$$
3x^2 r \sin \theta + x + 4r \cos \theta.
$$

Then by the Divergence Theorem,

$$
\iint_{\mathcal{S}} \mathbf{F}(x, y, z) \cdot \mathbf{n} \, d\mathcal{S} = \iiint_{W} \operatorname{div} \mathbf{F}(x, y, z) \, dV
$$

$$
= \int_2^5 \int_0^x \int_0^{2\pi} (3x^2 r \sin \theta + x + 4r \cos \theta) r \, d\theta \, dr \, dx
$$

$$
= \int_2^5 \int_0^x \int_0^{2\pi} (3x^2 r^2 \sin \theta + rx + 4r^2 \cos \theta) \, d\theta \, dr \, dx
$$

$$
= \int_2^5 \int_0^x \left[-3x^2 r^2 \cos \theta + rx\theta + 4r^2 \sin \theta \right]_{\theta=0}^{\theta=2\pi} dr \, dx
$$

$$
= \int_2^5 \int_0^x 2\pi rx \, dr \, dx
$$

$$
= \int_2^5 \left[\pi r^2 x \right]_{r=0}^{r=x} dx
$$

$$
= \int_2^5 \pi x^3 \, dx
$$

$$
= \left[\frac{1}{4} \pi x^4 \right]_2^5 = \frac{609}{4} \pi.
$$